Advanced Bayesian Methods for Medical Test Accuracy

Chapman & Hall/CRC Biostatistics Series

Chapman & Hall/CRC Biostatistics Series

Adaptive Design Theory and Implementation Using SAS and R
Mark Chang

Advanced Bayesian Methods for Medical Test Accuracy
Lyle D. Broemeling

Advances in Clinical Trial Biostatistics
Nancy L. Geller

Applied Statistical Design for the Researcher
Daryl S. Paulson

Basic Statistics and Pharmaceutical Statistical Applications, Second Edition
James E. De Muth

Bayesian Adaptive Methods for Clinical Trials
Scott M. Berry, Bradley P. Carlin,
J. Jack Lee, and Peter Muller

Bayesian Analysis Made Simple: An Excel GUI for WinBUGS
Phil Woodward

Bayesian Methods for Measures of Agreement
Lyle D. Broemeling

Bayesian Missing Data Problems: EM, Data Augmentation and Noniterative Computation
Ming T. Tan, Guo-Liang Tian,
and Kai Wang Ng

Bayesian Modeling in Bioinformatics
Dipak K. Dey, Samiran Ghosh,
and Bani K. Mallick

Causal Analysis in Biomedicine and Epidemiology: Based on Minimal Sufficient Causation
Mikel Aickin

Clinical Trial Data Analysis using R
Ding-Geng (Din) Chen and Karl E. Peace

Clinical Trial Methodology
Karl E. Peace and Ding-Geng (Din) Chen

Computational Methods in Biomedical Research
Ravindra Khattree and Dayanand N. Naik

Computational Pharmacokinetics
Anders Källén

Controversial Statistical Issues in Clinical Trials
Shein-Chung Chow

Data and Safety Monitoring Committees in Clinical Trials
Jay Herson

Design and Analysis of Animal Studies in Pharmaceutical Development
Shein-Chung Chow and Jen-pei Liu

Design and Analysis of Bioavailability and Bioequivalence Studies, Third Edition
Shein-Chung Chow and Jen-pei Liu

Design and Analysis of Clinical Trials with Time-to-Event Endpoints
Karl E. Peace

Design and Analysis of Non-Inferiority Trials
Mark D. Rothmann, Brian L. Wiens,
and Ivan S. F. Chan

Difference Equations with Public Health Applications
Lemuel A. Moyé and Asha Seth Kapadia

DNA Methylation Microarrays: Experimental Design and Statistical Analysis
Sun-Chong Wang and Arturas Petronis

DNA Microarrays and Related Genomics Techniques: Design, Analysis, and Interpretation of Experiments
David B. Allsion, Grier P. Page,
T. Mark Beasley, and Jode W. Edwards

Dose Finding by the Continual Reassessment Method
Ying Kuen Cheung

Elementary Bayesian Biostatistics
Lemuel A. Moyé

Frailty Models in Survival Analysis
Andreas Wienke

Generalized Linear Models: A Bayesian Perspective
Dipak K. Dey, Sujit K. Ghosh,
and Bani K. Mallick

Handbook of Regression and Modeling: Applications for the Clinical and Pharmaceutical Industries
Daryl S. Paulson

Measures of Interobserver Agreement and Reliability, Second Edition
Mohamed M. Shoukri

Medical Biostatistics, Second Edition
A. Indrayan

Meta-Analysis in Medicine and Health Policy
Dalene Stangl and Donal A. Berry

Monte Carlo Simulation for the Pharmaceutical Industry: Concepts, Algorithms, and Case Studies
Mark Chang

Multiple Testing Problems in Pharmaceutical Statistics
Alex Dmitrienko, Ajit C. Tamhane,
and Frank Bretz

Sample Size Calculations in Clinical Research, Second Edition
Shein-Chung Chow, Jun Shao
and Hansheng Wang

Statistical Design and Analysis of Stability Studies
Shein-Chung Chow

Statistical Evaluation of Diagnostic Performance: Topics in ROC Analysis
Kelly H. Zou, Aiyi Liu, Andriy Bandos,
Lucila Ohno-Machado, and Howard Rockette

Statistical Methods for Clinical Trials
Mark X. Norleans

Statistics in Drug Research: Methodologies and Recent Developments
Shein-Chung Chow and Jun Shao

Statistics in the Pharmaceutical Industry, Third Edition
Ralph Buncher and Jia-Yeong Tsay

Translational Medicine: Strategies and Statistical Methods
Dennis Cosmatos and Shein-Chung Chow

Chapman & Hall/CRC Biostatistics Series

Advanced Bayesian Methods for Medical Test Accuracy

Lyle D. Broemeling

CRC Press
Taylor & Francis Group
Boca Raton   London   New York

CRC Press is an imprint of the
Taylor & Francis Group, an **informa** business
A CHAPMAN & HALL BOOK

CRC Press
Taylor & Francis Group
6000 Broken Sound Parkway NW, Suite 300
Boca Raton, FL 33487-2742

First issued in paperback 2020

Version Date: 20110608

ISBN-13: 978-0-367-57690-5 (pbk)
ISBN-13: 978-1-4398-3878-5 (hbk)

Visit the Taylor & Francis Web site at
http://www.taylorandfrancis.com

and the CRC Press Web site at
http://www.crcpress.com

Contents

Preface   xv

Acknowledgments   xvii

Author   xix

1 Introduction   1
   1.1 Introduction   1
   1.2 Statistical Methods in Medical Test Accuracy   2
   1.3 Datasets for This Book   4
   1.4 Software   4
   1.5 Bayesian Approach   4
        References   6

2 Medical Tests and Preliminary Information   9
   2.1 Introduction   9
   2.2 Medical Imaging Tests   9
        2.2.1 X-ray   9
        2.2.2 Computed tomography   10
        2.2.3 Mammography   11
        2.2.4 Magnetic resonance imaging   11
        2.2.5 Nuclear medicine   11
        2.2.6 Ultrasound   12
        2.2.7 Combined medical images   13
   2.3 Other Medical Tests   13
        2.3.1 Introduction   13
        2.3.2 Sentinel lymph node biopsy for melanoma   13
        2.3.3 Tumor depth to diagnose metastatic melanoma   14
        2.3.4 Interventional radiology: A biopsy for non-small cell
               lung cancer   15
        2.3.5 Coronary artery disease   16
        2.3.6 Type 2 diabetes   16
        2.3.7 Other medical tests   17
               2.3.7.1 Tests for HIV   17
               2.3.7.2 Tests for ovarian cancer   17

	2.3.7.3	Prostate-specific antigen test for prostate cancer	18
	2.3.7.4	Bacterial infection with *Strongyloides*	18
	2.3.7.5	Tuberculosis	18
2.4	Activities Involved in Medical Testing		19
2.5	Accuracy and Agreement		20
2.6	Developmental Trials for Medical Devices		22
2.7	Literature		23
	References		23

3 Preview of the Book — **27**

3.1	Introduction		27
3.2	Preliminary Information		27
	3.2.1	Chapter 1: Introduction	27
	3.2.2	Chapter 2: Medical tests and preliminary information	28
	3.2.3	Chapter 3: Preview of the book	29
3.3	Fundamentals of Test Accuracy		29
	3.3.1	Chapter 4: Fundamentals of medical test accuracy	29
	3.3.2	Chapter 5: Regression and medical test accuracy	30
	3.3.3	Chapter 6: Agreement and test accuracy	31
3.4	Advanced Methods for Test Accuracy		34
	3.4.1	Chapter 7: Estimating test accuracy with an imperfect reference standard	34
	3.4.2	Chapter 8: Verification bias and test accuracy	35
	3.4.3	Chapter 9: Test accuracy and medical practice	37
	3.4.4	Chapter 10: Accuracy of combined tests	39
	3.4.5	Chapter 11: Meta-analysis for test accuracy	42
	References		43

4 Fundamentals of Diagnostic Accuracy — **45**

4.1	Introduction		45
4.2	Study Design		46
	4.2.1	Protocol	47
	4.2.2	Objectives	47
	4.2.3	Background	47
	4.2.4	Patient and reader selection	48
	4.2.5	Study plan	49
	4.2.6	Number of patients	50
	4.2.7	Statistical design and analysis	50
	4.2.8	References	51
4.3	Bayesian Methods for Test Accuracy: Binary and Ordinal Data		51
	4.3.1	Introduction	51
	4.3.2	Classification probabilities	52

      4.3.3   Predictive values 55

      4.3.4   Diagnostic likelihood ratios 56

      4.3.5   Receiver operating characteristic curve 57

  4.4 Bayesian Methods for Test Accuracy: Quantitative

      Variables 61

      4.4.1   Introduction 61

      4.4.2   Shields computed tomography study of

             heart disease 62

      4.4.3   Receiver operating characteristic area 63

      4.4.4   Definition of the receiver operating characteristic

             curve 70

      4.4.5   Choice of optimal threshold value 70

  4.5 Clustered Data: Detection and Localization 72

      4.5.1   Introduction 72

      4.5.2   The Bayesian receiver operating characteristic curve

             for clustered information 73

      4.5.3   Clustered data in mammography 74

  4.6 Comparing Accuracy between Modalities with

      Ordinal Scores 78

      4.6.1   Comparing true and false positive fractions 78

      4.6.2   Comparing the receiver operating characteristic areas

             of two modalities with ordinal scores 82

      4.6.3   Comparing receiver operating characteristic areas

             for continuous scores 87

  4.7 Exercises 92

      References 96

5 Regression and Medical Test Accuracy   **99**

  5.1 Introduction 99

  5.2 Audiology Study 100

      5.2.1   Introduction 100

      5.2.2   Log link function 101

      5.2.3   Logistic link 106

      5.2.4   Diagnostic likelihood ratio 108

  5.3 Receiver Operating Characteristic Area and Patient

      Covariates 110

      5.3.1   Introduction 110

      5.3.2   Ordinal regression methods 111

      5.3.3   Staging metastasis for melanoma: Accuracy of

             four radiologists 117

  5.4 Regression Methods for Continuous Test Scores 124

      5.4.1   Induced receiver operating characteristic curves 127

      5.4.2   Diagnosing prostate cancer with total

             prostate-specific antigen 127

      5.4.3   Stover audiology study 133

5.5   Exercises   136
      References   147

6 Agreement and Test Accuracy   151
6.1   Introduction   151
6.2   Ordinal Scores with a Gold Standard   152
6.3   Continuous Scores with a Gold Standard   155
6.4   Agreement with Ordinal Scores and No Gold Standard . . .   162
      6.4.1   Precursors of Kappa   162
      6.4.2   Chance corrected measures of agreement   165
      6.4.3   Conditional Kappa   166
      6.4.4   Kappa and stratification   169
      6.4.5   Weighted Kappa   171
      6.4.6   Intraclass Kappa   173
6.5   Other Measures of Agreement   178
6.6   Agreement and Test Accuracy   180
6.7   Kappa and Association   183
6.8   Consensus   185
6.9   Agreement with Multiple Raters and Ordinal Scores—
      No Gold Standard   186
      6.9.1   Kappa with many raters   187
      6.9.2   Partial agreement   190
      6.9.3   Stratified Kappa   195
6.10 Conclusions for Agreement and Accuracy   202
6.11 Exercises   203
      References   207

**7 Estimating Test Accuracy with an Imperfect Reference
    Standard   209**
7.1   Introduction   209
7.2   Two Binary Tests   212
7.3   Posterior Distribution for Two Binary Tests   214
7.4   Posterior Distribution without Conditional Independence . .   215
7.5   Posterior Distribution Assuming Conditional Independence .   216
7.6   Example of Accuracy for Diagnosing a Bacterial Infection . .   218
7.7   Accuracies of Two Binary Tests for Several Populations
      with Conditional Independence   226
7.8   Accuracies of Two Binary Tests without Conditional
      Independence: Two Populations   231
7.9   Multiple Tests in a Single Population   233
7.10 Multiple Tests without Conditional Independence   239
7.11 Two Ordinal Tests and the Receiver Operating
      Characteristic Area   244
7.12 Exercises   249
      References   257

8 Verification Bias and Test Accuracy   **259**

8.1 Introduction 259

8.2 Verification Bias and Binary Tests 261

8.3 Two Binary Tests 264

8.4 Ordinal Tests and Verification Bias 268

8.5 Two Ordinal Tests and Verification Bias 273

8.6 Two Ordinal Tests and Covariates 279

8.7 Inverse Probability Weighting 279

8.8 Without the Missing at Random Assumption 283

8.9 One Ordinal Test and the Receiver Operating
Characteristic Area 290

8.10 Comments and Conclusions 292

8.11 Exercises 292

References 298

9 Test Accuracy and Medical Practice   **301**

9.1 Introduction 301

9.2 Choice of Optimal Threshold 302

    9.2.1 Optimal threshold for the prostate-specific antigen
test for prostate cancer 303

    9.2.2 Test accuracy of blood glucose for diabetes 310

9.3 Test Accuracy with Bayesian Decision Curves 312

    9.3.1 Introduction 312

    9.3.2 Decision curves 314

    9.3.3 Bayesian inferences for decision curves 315

    9.3.4 Bayesian decision curves for prostate cancer 317

    9.3.5 Maximum likelihood and Bayes estimators 323

    9.3.6 Another example of decision curves 324

    9.3.7 Comments and conclusions 325

9.4 Test Accuracy and Clinical Trials 326

    9.4.1 Introduction 326

    9.4.2 Clinical trials 327

    9.4.3 Phase I designs 327

    9.4.4 Phase II trials 328

    9.4.5 Phase III trials 330

    9.4.6 Protocol 331

    9.4.7 Guidelines for tumor response 332

    9.4.8 Bayesian sequential stopping rules 333

    9.4.9 Software for clinical trials 337

    9.4.10 CRM simulator for Phase I trials 338

    9.4.11 Multc Lean for Phase II trials 339

    9.4.12 A Phase I trial for renal cell carcinoma 340

    9.4.13 An ideal Phase II trial 341

    9.4.14 A Phase II trial for advanced melanoma 344

9.5 Summary and Conclusions 346

9.6 Exercises 348
     References 350

10 Accuracy of Combined Tests   **353**
10.1 Introduction 353
10.2 Two Binary Tests 355
     10.2.1 Two binary tests for heart disease 358
     10.2.2 Computed tomography and magnetic resonance
            imaging and coronary stenosis 360
10.3 Two Binary Tests and Several Readers 362
10.4 Accuracy of Combined Binary Tests with Verification Bias . 368
10.5 Likelihood Ratio, the Risk Score, the Neyman–Pearson Lemma,
     and the Accuracy of Multiple Ordinal Tests 374
     10.5.1 Magnetic resonance imaging and computed tomography
            determination of lung cancer risk 379
     10.5.2 Body array and endorectal coil magnetic resonance
            imaging for localization of prostate cancer 387
     10.5.3 Accuracy of combined test with a covariate for
            lung cancer study 391
     10.5.4 Accuracy of a combined test with two components
            and verification bias 395
10.6 Accuracy of the Combined Test for Continuous Scores 397
     10.6.1 Two biomarkers for pancreatic cancer 398
     10.6.2 Blood glucose tests and type 2 diabetes 400
10.7 Observations and Conclusions 402
10.8 Exercises 404
     References 411

11 Bayesian Methods for Meta-Analysis   **413**
11.1 Introduction 413
11.2 Summary Receiver Operating Characteristic Curve
     and Bilogistic Regression 415
11.3 Bayesian Analysis for Summary Accuracy 417
     11.3.1 A meta-analysis with one test 417
     11.3.2 Summary accuracy for scintigraphy 420
11.4 Meta-Analysis with Two Tests 422
11.5 Meta-Analysis with Study Covariates and One Test 427
11.6 Meta-Analysis with Covariates for Several Tests 430
11.7 Other Meta-Analyses 435
11.8 Comments and Conclusions 438
11.9 Exercises 439
     References 448

Appendix: Introduction to WinBUGS **451**

A.1 Introduction 451
A.2 Download 452
A.3 Essentials 452
    A.3.1 Main body 452
    A.3.2 List statements 452
A.4 Execution of Analysis 454
    A.4.1 Specification tool 454
    A.4.2 Sample monitor tool 455
    A.4.3 Update tool 455
A.5 Output 456
A.6 Another Example 457
A.7 Summary 459
    References 460

Index **461**

Preface

Bayesian methods are being used more and more in medicine and biology. For example, at the University of Texas MD Anderson Cancer Center and other institutions, Bayesian sequential stopping rules are implemented somewhat routinely in the design of clinical trials. Also, Bayesian techniques are being used more frequently in diagnostic medicine, such as estimating the accuracy of diagnostic tests and for screening large populations for various diseases. Bayesian methods are quite attractive in many areas of medicine because they are based on prior information, which is usually available in the form of related previous studies. An example of this is in the planning of Phase II clinical trials, where a new therapy will be administered to patients who have advanced disease. Such therapies are developed by pharmaceutical companies and their success depends on the success of previous Phase I or other relevant Phase II trials. Bayes theorem allows a logical way to incorporate the previous information with the information that will accrue in the future. Accuracy of a medical test is an essential component of the diagnostic process and is the key issue of this book. Of course, medicine and biology are not the only areas where the concept of test accuracy plays a paramount role. For example, in the area of sports (e.g., cycling or baseball), the accuracy of a test for "doping" is of extreme importance for maintaining the integrity of the sport.

Advanced Bayesian Methods for Medical Test Accuracy is intended as a textbook for graduate students in statistics and as a reference for consulting statisticians. It will be an invaluable resource especially for biostatistics students who will be working in the various areas of diagnostic medicine (e.g., pathology and/or diagnostic imaging). The book is very practical and the student will learn many useful methods for measuring the accuracy of various medical tests. Most of the book is focused on Bayesian inferential procedures, but some is devoted to the design of such studies. A student should have completed a year of introductory probability and mathematical statistics, several introductory methods courses, such as regression and the analysis of variance, and a course that is primarily an introduction to Bayesian inference.

Consulting statisticians working in the areas of medicine and biology will have an invaluable reference with *Advanced Bayesian Methods for Medical Test Accuracy*, which will supplement the books *Statistical Methods for Diagnostic Medicine* by Zhou, Obuchowski, and McClish, and *The Statistical Evaluation of Medical Tests for Classification and Prediction* by Pepe. The two references are not presented from a Bayesian viewpoint; thus, the present volume is unique and will develop methods of test accuracy that should prove to be very useful to the consultant. Another unique feature of the book is that all computing and analysis is based on the WinBUGS package, which will allow the user a platform that efficiently uses prior information. Many of the ideas in the present volume are presented for the first time and go far beyond the two standard references. For the novice, an appendix introduces the fundamentals of programming and executing BUGS, and as a result, the reader will have the tools and experience to successfully analyze studies for medical test accuracy.

A very attractive feature of the book is that the author's blog: http://medtestacc.blogspot.com provides the BUGS code, which can be executed as one progresses through the book and as one does the exercises at the end of each chapter. Note, each chapter includes the code labeled as BUGS CODE 4.1, BUGS CODE 4.2, etc., and this is also included in the author's blog; thus, the student can cycle between the book and the blog, which reinforces the subject in a beneficial manner.

Acknowledgments

The author gratefully acknowledges the many departments and people at the University of Texas MD Anderson Cancer Center in Houston, who assisted him during the writing of this book. Many of the analyses that appear in *Advanced Bayesian Methods for Medical Test Accuracy* are based on studies performed at the Division of Diagnostic Imaging, and in particular, he would like to thank Drs. Gayed, Munden, Kundra, Marom, Ng, Tamm, and Gupta. Special thanks to Ana Broemeling for the editing and organization of the early versions of the manuscript and for her encouragement during the writing of this book.

Author

Lyle D. Broemeling, PhD, is director of Broemeling and Associates Inc., and is a consulting biostatistician. He has been involved with academic health science centers for over twenty years and has taught and been a consultant at the University of Texas Medical Branch in Galveston, the University of Texas MD Anderson Cancer Center, and the University of Texas School of Public Health. During his tenure at the University of Texas, he developed an interest in medical test accuracy, which resulted in various publications in the medical literature as well as in statistical journals. His main interest is in developing Bayesian methods for use in medical and biological problems and in authoring textbooks in statistics. His previous books are *Bayesian Analysis of Linear Models*, *Econometrics and Structural Change*, written with Hiroki Tsurumi, *Bayesian Biostatistics and Diagnostic Medicine*, and *Bayesian Methods for Measures of Agreement*.

Chapter 1

Introduction

1.1 Introduction

This book describes Bayesian statistical methods for the design and analysis of studies involving medical test accuracy. It grew out of the author's experience in consulting with many investigators of the Division of Diagnostic Imaging at the University of Texas MD Anderson Cancer Center (MDACC) in Houston, Texas. In a modern medical center, medical test accuracy is crucial for patient management, from the initial diagnosis to assessing the extent of disease as the patient is being treated.

Why a book on medical test accuracy? The short answer is that your life depends on it! Every visit to the doctor involves the use of some medical test, from measuring blood pressure and temperature to perhaps more expensive follow-up tests. If you go to the doctor complaining of chest pain, the doctor might refer you for an exercise stress test, and if that test is positive, you might undergo a heart catheterization to detect coronary artery disease. Suppose the exercise stress test is mistakenly negative, and the doctor does not order any follow-up procedures, then you go home with a hidden heart disease. You pay the price later when you are belatedly treated for the disease. On the other hand, suppose the exercise stress test is mistakenly positive, and the doctor orders an unnecessary heart catheterization, which gives the correct diagnosis, namely, that you do not have heart disease. Then you have paid for an unnecessary test. Diabetes is another example, where misdiagnosis can lead to expensive treatment. This can happen when the blood glucose test mistakenly indicates that you have type 2 diabetes, when in fact your blood glucose is elevated, but not to the extent that drugs are needed to control the disease. Of course, accuracy depends not only on the medical test, but also on the interpretation of the test results. There are two sources of error, the inherent variability of the medical test and the subjectivity involved in interpreting the test output. Both of the above examples will be dealt with in more detail later in the book, namely, the blood glucose test for type 2 diabetes, and the tests for coronary artery disease. As a patient, you should know the accuracy of the medical tests that will be administered to you when you go to the doctor for routine visits and, more importantly, when you are undergoing treatment for disease. An informative reference for the patient is Johnson, Sandmire, and Klein [1].

1.2   Statistical Methods in Medical Test Accuracy

Biostatistics plays a pivotal role in the assessment of the accuracy of medical tests, as can be discerned by reading papers in mainline journals, such as *Radiology* and *The Journal of Pathology*, and more specialized journals, such as *The Journal of Computed Assisted Tomography*, *The Journal of Magnetic Resonance Imaging*, *The Journal of Nuclear Medicine*, and *The Journal of Infectious Diseases*. As we will see, the usual methods, ranging from the t-test and chi-square test to others such as the analysis of variance and various regression techniques, are standard fare for assessing the accuracy of medical tests.

However, there are also some methods that are somewhat unique to the field, including ways to estimate diagnostic test accuracy and methods to measure the agreement between various tests and/or readers. This topic will be addressed throughout the book. The most basic indicators of test accuracy are the true and false positive fractions for medical tests that have binary scores or score where a cutoff is used to declare disease, in effect providing binary scores. For those patients with the disease, the fraction that test positive is referred to as the true positive fraction. The fraction that test positive, among the non-diseased patients, is called the false positive fraction. From the viewpoint of the patient, the positive predictive value is important, because it is the fraction of patients that have disease, among those that test positive for disease, but also important is the negative predictive value, which is the fraction of subjects who do not have the disease, among those that test negative. In most situations for a medical test, these four values indicating test accuracy do not lead to an unambiguous declaration that the test is a good one. Those factors that affect the various measures of test accuracy will be described in the book.

An overall measure of a medical test with continuous scores is provided by the area under the receiver operating characteristic (ROC) curve, which is defined as follows. The area can vary from 0 to 1 and, in general, the area is defined as follows:

$$\text{ROC area} = P[Y > X], \tag{1.1}$$

where Y is the test score of an individual selected at random from the diseased and X is the score of a subject selected at random from the non diseased. When the area is 1, the test scores discriminate perfectly among the diseased and non-diseased subjects, and if the area is 0.5, the scores are not at all informative for discriminating between the two groups. The above definition is changed to

$$\text{ROC area} = P[Y > X] + (1/2)P[Y = X], \tag{1.2}$$

when the test scores are ordinal.

The types of tests employed in this book are used for diabetes, heart disease, various forms of cancer, and tests for infectious diseases. A large proportion of the tests are imaging tests for cancer, while tests for heart disease are also represented. There are only a few examples of diabetes and infectious diseases. One very important case is for medical imaging tests, which are employed in cancer clinical trials, such as Phase II trials, where the main objective is to determine the response to a new therapy, where response is based on an image measurement. Computed tomography (CT; a form of x-ray) is used to measure the tumor size at baseline before the trial begins and is used at various times throughout the trial. Thus, there are several measurements of tumor size for each patient, where all the measurements are used to classify the patient's response. Thus, it is of paramount importance that the CT measurements of tumor size be accurate, because inaccurate measurements could lead to false declarations about the success or failure of a particular therapy! This will be described in more detail later on when the Erasmus et al. [2] study is explained. Also involved in this type of trial is the error introduced by the several radiologists (readers) who are interpreting the CT tumor measurements.

Along with the basic indicator of test accuracy, various statistical methodologies will be employed. For example, patient covariates will be taken into account by regression techniques. Obviously, the true and false positive fractions are affected by patient covariates, such as age, gender, and medical history, and these regression techniques are described and illustrated in the book. When the scores are ordinal or continuous, the appropriate regression techniques are employed to measure accuracy by the area under the ROC curve. Regression also plays an important role when comparing several readers and when estimating the agreement between readers who are interpreting the test results.

Often, not all patients are subject to the gold standard, the test that is used as a reference to compute the accuracy of medical tests. For example, in an exercise stress test, those that test negative are usually not referred to the gold standard (heart catheterization) compared to those that test positive and are usually given a heart catheterization. This is a special situation, called verification bias, which requires specialized methods to estimate test accuracy involving an application of Bayes theorem. When this is considered, various generalizations will be implemented, including the consideration of several tests and several readers and regression to take into account other patients covariates.

Patient sample size for a clinical trial, based on Bayesian sequential stopping rules, is another application that has proven to be quite beneficial in the development of new medical therapies, where the accuracy of the medical tests is key in the development of "new" therapies. The Bayesian approach will be used throughout this book and is the foundation for estimating medical test accuracy.

1.3   Datasets for This Book

The datasets used in this book come from the following sources: (1) the protocol review process of clinical trials at MDACC, where the author was either a reviewer or a collaborator on the protocol; (2) the author's consultations with the scientific and clinical faculty of the Division of Diagnostic Imaging at MDACC with some 32 datasets; (3) the several datasets accompanying the excellent book by Pepe [3] (these can be downloaded at http:// www.fhcrc.org/labs/pepe/Book) *The Statistical Evaluation of Medical Tests for Classification and Prediction*; (4) the information contained in the examples of the WinBUGS package; and (5) other miscellaneous sources, including the examples and problems in *Statistical Methods in Diagnostic Medicine* by Zhou, McClish, and Obushowski [4].

1.4   Software

WinBUGS will be used for the Bayesian analysis for sampling from the posterior distribution, and the appendix, which introduces the reader to the basic elements of using the software, including many examples. The WinBUGS code is clearly labeled in each chapter, e.g., BUGS CODE 4.1 and BUGS CODE 4.2 in Chapter 4, and the code can be downloaded from the author's blog (http://www.medtestacc.blogspot.com). The reader can easily reproduce the many analyses included in the book, which should greatly facilitate the reader's understanding of the Bayesian approach to estimating medical test accuracy. The blog also contains a detailed example of how to execute the Bayesian analysis using WinBUGS.

Many specialized Bayesian programs for the design and analysis of clinical trials have been developed at the Department of Biostatistics and Applied Mathematics at MDACC, some of which will be used for the design of clinical trials as well as for many other analyses involved in biostatistics. These can be accessed at http://biostatistics.mdanderson.org/SoftwareDownload/.

1.5   Bayesian Approach

Why is the Bayesian approach taken here? The author has been a Bayesian for many years, since 1974 when he took leave to study at University College London. Dennis Lindley persuaded him of the advantages of such an approach and, of course, the main advantage is that it is a practical way to utilize prior information. Prior information, especially in a medical setting, is ubiquitous and should be used to one's advantage as it would be a pity not to use it. It is assumed that the reader is familiar with the Bayesian approach to inference, but a brief introduction will be given here.

Suppose X is a continuous observable random vector and $\theta \in \Omega \subset R^m$ is an unknown parameter vector, and suppose the conditional density of X given θ is denoted by $f(x/\theta)$. If $x = (x_1, x_2, \ldots, x_n)$ represents a random sample of size n from a population with density $f(x/\theta)$, and $\xi(\theta)$ is the prior density of θ, then Bayes theorem is given by

$$\xi(\theta/x) = c \prod_{i=1}^{i=n} f(x_i/\theta)\xi(\theta), \quad x_i \in R \text{ and } \theta \in \Omega,$$

where the proportionality constant is c and the term

$$\prod_{i=1}^{i=n} f(x_i/\theta),$$

is called the likelihood function. The density $\xi(\theta)$ is the prior density of θ and represents the knowledge one possesses about the parameter before one observes X. Such prior information is most likely available to the experimenter from other previous related experiments. Note that θ is considered a random variable and that Bayes theorem transforms one's prior knowledge of θ, represented by its prior density, to the posterior density, and that the transformation is the combining of the prior information about θ with the sample information represented by the likelihood function.

"An essay toward solving a problem in the doctrine of chances" by the Reverend Thomas Bayes appeared and was the beginning of our subject. He considered a binomial experiment with n trials and assumed that the probability θ of success was uniformly distributed (by constructing a billiard table). Bayes presented a way to calculate $P(a \leq \theta \leq b/x = p)$, where x is the number of successes in n independent trials. This was a first in the sense that Bayes was making inferences via $\xi(\theta/x)$, the conditional density of θ given x. Also, by assuming that the parameter was uniformly distributed, he was assuming vague prior information for θ. In what follows, the components of the parameter vector θ will be various measures of medical test accuracy.

It can well be argued that Laplace [5] made many significant contributions to inverse probability (he did not know of Bayes), beginning in 1774 with his own version of Bayes theorem, "Memorie sur la probabilite des causes par la evenemens" and over a period of some 40 years culminating in "Theorie analytique des probabilites." See Stigler [6] and Chapters 9–20 of Hald [7] for the history of Laplace's many contributions to inverse probability.

It was in modern times that Bayesian statistics began its resurgence with Lhoste [8], Jeffreys [9], Savage [10], and Lindley [11]. According to Broemeling and Broemeling [12], Lhoste was the first to justify non-informative priors by invariance principals, a tradition carried on by Jeffreys. Savage's book was a major contribution in that Bayesian inference and decision theory was put on a sound theoretical footing as a consequence of certain axioms of probability and utility, while Lindley's two volumes showed the relevance of Bayesian inference to everyday statistical problems and was quite influential, setting the tone and style for later books such as Box and Tiao [13], Zellner [14], and Broemeling [15]. Box and Tiao and Broemeling were essentially works that presented Bayesian methods for the usual statistical problems of the analysis of variance and regression, while Zellner focused Bayesian methods primarily on certain regression problems in econometrics. During this period, inferential problems were solved analytically or by numerical integration. Models with many parameters (such as hierarchical models with many levels) were difficult to use because at that time numerical integration methods had limited capability in higher dimensions. For a good history of inverse probability, see Chapter 3 of Stigler [6], and the two volumes of Hald [7], which present a comprehensive history and are invaluable as a reference. Dale [16] gives a very complete and interesting account of Bayes' contributions.

The last 20 years are characterized by the rediscovery and development of resampling techniques, where samples are generated from the posterior distribution via Markov Chain Monte Carlo (MCMC) methods, such as Gibbs sampling. Large samples generated from the posterior make it possible to make statistical inferences and to employ multi-level hierarchical models to solve complex, but practical problems, because computing technology is available. See Leonard and Hsu [17], Gelman et al. [18], Congdon [19–21], Carlin, Gelfand, and Smith [22], Gilks, Richardson, and Spiegelhalter [23], who demonstrate the utility of MCMC techniques in Bayesian statistics. Of course, in using WinBUGS, this book employs MCMC techniques to estimate the parameters of the model. The output of the analysis typically includes the posterior mean, standard deviation, median, and the upper and lower 2 1/2 percentiles. Also included is the MCMC error, which is an important component of the analysis, and tells one how close the MCMC estimate is to the "true" posterior characteristic, consequently, the MCMC error can be utilized to adjust the sample size for the simulation.

References

[1] Johnson, D., Sandmire, D., and Klein, D. *Medical Tests that can Save your Life: 21 Tests your Doctor won't Order... Unless You Know to Ask.* Rodale, Emmaus, PA, 2004.

[2] Erasmus, J.J., Gladish, G.W., Broemeling, L., Sabloff, B.S., Truong, M.T., Herbst, R.S., and Munden, R.F. Interobserver variability in measurement of non-small cell carcinoma of the lung lesions: Implications

for assessment of tumor response. *Journal of Clinical Oncology*, 21:2574, 2003.

[3] Pepe, M.S. *The Statistical Evaluation of Medical Tests for Classification and Prediction*. Oxford University Press, Oxford, UK, 2000.

[4] Zhou, H.H., McClish, D.K., and Obuchowski, N.A. *Statistical Methods for Diagnostic Medicine*. John Wiley, New York, 2002.

[5] Laplace, P.S. Memorie des les probabilities. *Memories de l'Academie des Sciences de Paris*, 227, 1778.

[6] Stigler, M. *The History of Statistics. The Measurement of Uncertainty before 1900*. The Belknap Press of Harvard University Press, Cambridge, MA, 1986.

[7] Hald, A. *A History of Mathematical Statistics from 1750 to 1930*. Wiley Interscience, London, 1990.

[8] Lhoste, E. Le calcul des probabilities appliqué a l'artillerie, lois de probabilite a prior. *Revue d'Artillerie*, 91:405–423, 1923.

[9] Jeffreys, H. *An Introduction to Probability*. Clarendon Press, Oxford, 1939.

[10] Savage, L.J. *The Foundation of Statistics*. John Wiley, New York, 1954.

[11] Lindley, D.V. *Introduction to Probability and Statistics from a Bayesian Viewpoint*, volumes I and II. Cambridge University Press, Cambridge, 1965.

[12] Broemeling, L.D. and Broemeling, A.L. Studies in the history of probability and statistics XLVIII: The Bayesian contributions of Ernest Lhoste. *Biometrika*, 90(3):728–731, 2003.

[13] Box, G.E.P. and Tiao, G.C. *Bayesian Inference in Statistical Analysis*. Addison Wesley, Reading, MA, 1973.

[14] Zellner, A. *An Introduction to Bayesian Inference in Econometrics*. John Wiley, New York, 1971.

[15] Broemeling, L.D. *The Bayesian Analysis of Linear Models*. Marcel-Dekker, New York, 1985.

[16] Dale, A. *A History of Inverse Probability from Thomas Bayes to Karl Pearson*. Springer-Verlag, Berlin, 1991.

[17] Leonard, T. and Hsu, J.S.J. *Bayesian Methods. An Analysis for Statisticians and Interdisciplinary Researchers*. Cambridge University Press, Cambridge, 1999.

[18] Gelman, A., Carlin, J.B., Stern, H.S., and Rubin, D.B. *Bayesian Data Analysis*. Chapman & Hall/CRC and Taylor & Francis, Boca Raton, 1997.

[19] Congdon, P. *Bayesian Statistical Modeling*. John Wiley, London, 2001.

[20] Congdon, P. *Applied Bayesian Modeling*. John Wiley, New York, 2003.

[21] Congdon, P. *Bayesian Models for Categorical Data*. John Wiley, New York, 2005.

[22] Carlin, B.P., Gelfand, A.E., and Smith, A.F.M. Hierarchical Bayesian analysis of changepoint problems. *Applied Statistics*, 41:389–405, 1992.

[23] Gilks, W.R., Richardson, S., and Spiegelhalter, D.J. *Markov Chain Monte Carlo in Practice*. Chapman & Hall/CRC, New York, 1996.

Chapter 2

Medical Tests and Preliminary Information

2.1 Introduction

This chapter gives a brief description of medical imaging tests and other tests routinely used at a major health care institution. Diagnostic imaging plays an extremely important role in the overall care of the patient, including diagnosis, staging, and monitoring of the patient during their stay in hospital. Some of the examples in this book are taken from diagnostic imaging studies for cancer, however, there are many other ways to perform diagnoses, and some of these will also be explained. In addition to cancer, medical tests for heart disease and stroke will be described, as well as some medical tests for diagnosing diabetes and infectious diseases. Most of the medical tests described in this chapter will appear in the many examples to follow in later chapters.

2.2 Medical Imaging Tests

The primary tests for diagnostic imaging are x-ray, fluoroscopy, mammography, computed tomography (CT), ultrasonography (US), magnetic resonance imaging (MRI), and nuclear medicine. Each test has advantages and disadvantages with regard to image quality, depending on the particular clinical situation. Broadly speaking, image quality consists of three components. The first is contrast. Contrast is good when important physical differences in anatomy and tissue are displayed with corresponding different shades of gray levels. The ability to display fine detail is another important aspect of image quality and is called resolution. Anything that interferes with image quality is referred to as noise, which is the third component of image quality. Obviously, noise should be minimized in order to improve image quality.

2.2.1 X-ray

Medical images are best thought of as being produced by tracking certain probes as they pass through the body. A stream of x-rays are passed through the patient and captured on film as the stream exits. An x-ray is a stream of photons, which are discrete packets of energy. As they pass through the body, various tissues interact with the photons and these collisions remove and scatter some of the photons. The various tissues reduce the amount of energy in various parts of the stream by different amounts. A shadow is produced that appears on a special photographic plate, producing an image. If the density of the target object is much higher than that of the surrounding environment (such as a bone), an x-ray does a good job of locating it. Some lesions have densities that are quite similar to the surrounding medium and are difficult to detect. Generally speaking, an x-ray has very good resolution and the noise is easy to control, but has low contrast in certain cases. An x-ray is routine in all medical settings, and is the most utilized of all imaging devices.

A close relative of x-rays is fluoroscopy. In this modality, the exiting beam is processed further by projecting it onto an image intensifier, which is a vacuum tube that transforms the x-ray shadow onto an optical image. This mode has about the same image quality as an x-ray, but allows the radiologist to manage images in real time. For example, it allows the operator to visualize the movement of a contrast agent passing certain landmark locations in the gastrointestinal tract or vascular system.

2.2.2   Computed tomography

Another variation of the x-ray is CT, which overcomes some of the limitations of x-rays. The superimposition of shadows of overlapping tissues and other anatomical structures often obscures detail in the image. CT produces images quite differently from x-rays, however it does use x-rays, but the detection and processing of the shadows is quite sophisticated and is the distinctive feature of the modality that vastly improves the image over that of an x-ray. CT has good contrast among soft tissues (e.g., lung and brain tissue) and good resolution. An x-ray takes information from a three-dimensional structure and projects it onto a two-dimensional image, which causes the loss of detail due to overlapping tissues. How does CT overcome this problem? The patient is placed in a circle; inside the circle is an x-ray source and embedded in the circle is an array of detectors that capture the shadow of the x-ray beam. The x-ray source irradiates a thin slice of tissue across the patient and the detector captures the shadow. The x-ray source moves to an adjacent location and the process is repeated, say 700 times. The x-ray source circumscribes the patient through 360 degrees. The source then repeats the above process with another thin slice. For a given slice, there are 700 projections of that slice and these 700 projections are processed via computer and back projection algorithms to produce the two-dimensional representation. The computer works backward from the projections to reconstruct the spatial distribution of the structure of the thin slice. In other words, CT answers the following question: what does the original structure have to look like in order to produce the 700 generated projections?

A good example of CT (using the Imatron C-100 Ultrafast) is screening for coronary heart disease, where the coronary artery calcium (CAC) score indicates the degree of disease severity. See Mielke, Shields, and Broemeling [1, 2], DasGupta et al. [3], and Broemeling and Mielke [4], where the accuracy of CAC to diagnose heart disease is estimated by the area under the receiver operating characteristic (ROC) curve. These examples will be examined again in later chapters from a Bayesian perspective.

2.2.3   Mammography

Mammography is yet another variation of the x-ray. While some small masses can be detected by a physician or by self-examination, mammography has the ability to detect very small lesions. However, the smaller they are, the more difficult they are to detect. The set up for mammography consists of a specialized x-ray tube and generator, a breast compression device, an anti-scatter grid, and film. The procedure must be able to reveal small differences in breast density, possibly indicative of a suspicious mass, and it must also be able to detect small calcifications that may be important to diagnosis. All the attributes of good image quality are required, namely, high contrast, good resolution, and low noise. Later in this book, the role of mammography in screening for breast cancer will be described.

2.2.4   Magnetic resonance imaging

A completely different form of imaging is MRI. A beam of photons is not passed through the body, but instead the body is placed in a large magnet and hydrogen atoms (in the water molecules) line up in the same direction as the magnetic field. When the magnetic field is disrupted by directing radio energy into the field, the magnetic orientation of the hydrogen atoms is disrupted. The radio source is switched off and the magnetic orientation of the hydrogen atoms returns to the original state. The manner (referred to as T1 and T2 relaxation times) and the way in which they return to the original state produces the image. Essentially, what is being measured is the proton density per unit volume of imaged material. The actual image looks like an x-ray, however the principal foundations of MRI are completely different. The same image processing technology used in CT can be used in MRI to process the images. For example, thin slices and backward projection methods are often used to improve MRI image quality. MRI has excellent resolution and contrast among soft tissue, and displays good anatomical detail.

2.2.5   Nuclear medicine

Nuclear medicine is the joining of nuclear physics, nuclear chemistry, and radiation detection. A radioactive chemical substance, called a radiopharmaceutical, is injected, usually by intravenous (IV), where it concentrates in a particular tissue or organ of interest. The substance emits gamma rays, which are detected by gamma cameras. The gamma camera counts the number of gamma particles it captures. There are two principal gamma cameras—positron emission tomography (PET) and single photon emission tomography (SPECT). Nuclear imaging is often used to view physiological processes. For example, FDG-PET is often used to measure glucose metabolism, where the radiopharmaceutical (18) F-florodeoxyglucose is absorbed by every cell in the body. The higher the observed radioactivity as measured by PET, the higher the glucose metabolism. In some cancer studies, the malignant lesion has an increased glucose metabolism compared to the adjacent non-malignant tissue, thus FDG-PET is useful in the diagnosis and staging of disease. Another area where nuclear medicine is useful is in cardiac perfusion studies. For example, radiation therapy of esophageal cancer often induces damage to the heart in the form of ischemia and scarring. The damage can be assessed by a nuclear medicine procedure such as an exercise stress test, where thallium is administered via IV to the patient and concentrates in the heart muscle; the resulting radioactivity is counted by SPECT to produce the image. Among the soft tissues, nuclear medicine procedures have fair to good contrast but poor resolution, and noise can be a problem for image quality. Another very important instance where MRI is used is to diagnose coronary heart disease via the exercise stress test; if the test is positive, the patient can be referred to coronary angiography for a more accurate assessment of the disease.

2.2.6   Ultrasound

US is the last modality to be described. It is based on a physical stream of energy passing through the body to be imaged. The source is a transducer that converts electrical energy into a brief pulse of high-frequency acoustical energy to be transmitted into the patient's tissues. The transducer acts as a transmitter and receiver. The receiver detects echoes of sound deflected from the tissues, where the depth of a particular echo is measured by the round trip time of the transmitted emission. The images are viewed in real time on a monitor and are produced by interrogating patient tissue in the field of view. The real-time images are rapidly produced on the monitor, allowing one to view moving tissue such as respiration and cardiac motion. The US examination consists of applying the US transducer to the patient's skin using a water-soluble gel to make the connection secure for good transmission of the signal. Image quality is adversely affected by bone and gas-filled structures such as the bowel and lung. For example, bone causes almost complete absorption of the signal, producing an acoustic shadow on the image that hides the detail of tissues near the bone, while soft tissue gas-filled objects produce a complete reflection of sound energy that eliminates visualization of deep structures. Despite these drawbacks, the mode has many advantages, one of which is the non-invasive nature of the procedure. US is used to image a multitude of clinical challenges and is very beneficial when solving a particular clinical problem, such as viewing the development of the fetus.

2.2.7 Combined medical images

Various modalities are often combined to improve overall diagnostic accuracy. For example, recently, PET and CT have been combined to diagnose and stage esophageal cancer. When two modalities are combined, one must formulate certain rules to decide when the combined procedure is deemed to produce a "positive" or "negative" determination. In another interesting study, US and CT were combined and their accuracy compared to FDG-PET. The ideas involved in measuring the accuracy of combined modalities will be outlined in the Chapter 10.

It is important to remember that the imaging device does not make the diagnosis, rather the radiologist and others make the diagnosis! The modality is an aid to the radiologist and to others who are responsible for the diagnosis. After the radiologist reads the image, how is this information transformed to a scale where the biostatistician and others are able to use it for their own purposes?

For a non-technical introduction to medical imaging tests, Wolbarst [5] presents a very readable account. In addition, Jawad [6], Chandra [7], Seeram [8], and Markisz and Aguilia [9] are standard references to cardiac ultrasound, nuclear medicine, CT, and MRI, respectively.

2.3 Other Medical Tests

2.3.1 Introduction

In order to establish a definitive diagnosis, there are several phases. For example, a screening mammography might reveal a suspicious lesion, and this will be followed with a biopsy of the suspected lesion. Many caregivers are involved in the diagnostic process, and as has been emphasized, diagnostic imaging plays a major role in that effort. However, they are just one of many groups, including oncologists, surgeons, nurses, pathologists, geneticists, microbiologists, and many more. The pathologist plays a crucial role in performing the histologic tests on cell specimens taken for biopsy, as does the microbiologist and geneticist, who are developing new techniques to measure gene sensitivity from deoxyribo nucleic acid (DNA) specimens, etc. Two examples are described below: (1) metastasis of the primary melanoma lesion to the lymph nodes, and (2) biopsy of lung nodules.

2.3.2 Sentinel lymph node biopsy for melanoma

This technique involves the cooperation of a melanoma oncologist, a surgical team to dissect the lymph nodes, diagnostic radiologists who will perform the nuclear medicine procedure, and pathologists who will do the histology of the lymph node samples. The following description of the technique is based on Pawlik and Gershenwald [10]. The early procedures are described by Morton et al. [11] and consist of injecting a blue dye intradermally around the primary lesion and biopsy site, where the lymphatic system takes up the dye and carries it, via afferent lymphatics, to the draining regional node basins. Surgeons then explore the draining nodal basin; the first draining lymph nodes, the sentinel lymph nodes (SLNs), are identified by their uptake of blue dye, then dissected and sent to pathology for histological examination of malignancy.

These early methods were recently revised to include a nuclear medicine application using a handheld gamma camera. (See Gershenwald et al. [12] for a good explanation of this.) With this technique, intraoperative mapping uses a handheld gamma probe, where 0.5–1.0 mCi of a radiopharmaceutical is injected intradermally around the intact melanoma. The gamma camera monitors the level of radioactivity from the injection sites to the location of the SLNs and is also employed to assist the surgeons with the dissection of the lymph nodes. This probe is used transcutaneously prior to surgery and has an accuracy of 96–99% in correctly identifying the SLNs. Histological examination of the lymph node specimens determine if the lymph node basin has malignant melanoma cells.

2.3.3 Tumor depth to diagnose metastatic melanoma

An SLN biopsy for melanoma metastasis is illustrated with a recent study by Rousseau et al. [13], where the records of 1376 melanoma patients were reviewed. The main objective was to diagnose metastasis to the lymph nodes, where the gold standard is the outcome of the SLN biopsy and the diagnosis is made on the basis of tumor depth of the primary lesion, the Clark level of the primary lesion, the age and gender of the patient, the presence of an ulcerated primary lesion, and the site (axial or extremity) of the primary lesion. The overall incidence of a positive biopsy was 16.9%, the median age was 51 years, and 58% were male. A multivariate analysis with logistic regression showed that tumor thickness and ulceration were highly significant in predicting SLN status. For additional details about this study, refer to Rousseau et al., but for the present the focus will be on tumor thickness for the diagnosis of lymph node metastasis.

How accurate is tumor thickness for the diagnosis of lymph node metastasis? The original measurement of tumor thickness was categorized into four groups: (1) ≤1 mm, (2) 1.01–2.00 mm, (3) 2.01–4.00 mm, and (4) >4.00 mm. If groups 3 and 4 are used to designate a positive (lymph node metastasis) test, and groups 1 and 2 a negative test, the sensitivity and specificity are calculated as $156/234 = 0.666$ and $832/1147 = 0.725$, respectively. There were 156 patients with a tumor thickness >2 mm among 234 patients with a positive SLN biopsy, on the other hand, there were 832 patients with a tumor thickness ≤2 mm among 1147 with a negative SLN biopsy. Also, using the original continuous measurement and a conventional estimation method, the area under the ROC curve is 0.767 with a standard deviation of 0.016. This is the type of problem that will be studied in the following chapters, but from a Bayesian perspective.

2.3.4 Interventional radiology: A biopsy for non-small cell lung cancer

At the MD Anderson Cancer Center (MDACC), the Department of Interventional Radiology is part of the Division of Diagnostic Imaging, and they perform invasive biopsy procedures. For example, they perform biopsies of lung lesions using a CT-guided technique, see Gupta et al. [14]. The Gupta example described below compared two methods of biopsy, short vs. long needle path, for target lesions <2 cm in size. The objective is to retrieve a specimen of the lesion to be examined for malignancy by a cytopathologist.

Many people are involved, including those assisting the interventional radiologist in guiding the needle to the target lesion, which was earlier detected and located by various imaging modalities. Of main concern is the occurrence of a pneumothorax, which can result in a collapsed lung and bleeding, sometimes requiring a chest tube to drain fluid from the chest cavity.

This cohort study included 176 patients, 79 men and 97 women, with an age range from 18 to 84 years. This was not a randomized study, and patient information came from all persons who underwent a CT-guided biopsy for lung nodules during the period from November 1, 2000 to December 31, 2002. There were two groups: Group A with 48 patients, where the needle path was <1 cm in length of aerated lung; and Group B with 128 patients, where the needle path length was >1 cm.

The two groups were similar with regard to age, gender, lesion size, and lesion location, and the major endpoints were diagnostic yield (number of diagnostic samples and test accuracy, measured by sensitivity and specificity) and frequency of pneumothorax. The pathology report served as a gold standard for test accuracy.

The statistical analysis consisted of estimating the test accuracy of the two methods and comparing accuracy via the chi-square test. There was no significant difference between the two groups with regard to sensitivity and specificity, however, there were significant differences between the two with regard to complications from the procedure. For example, the pneumothorax rate of $35/48 = 0.73$ was larger for the short needle path group compared to $38/128 = 0.29$ for the long needle path group.

As a follow up to this, Gupta et al. [15] recently studied 191 lung biopsy patients who experienced a pneumothorax. In that study, the principal aim was to identify those factors that significantly impact the development of a persistent air leak of the lung.

A conventional statistical analysis was performed for these studies, but later in this book, we will revisit them with a Bayesian approach for the analysis.

2.3.5   Coronary artery disease

A common scenario in the diagnosis of coronary artery disease is: following complaints of chest pain, the patient undergoes an exercise stress test and, if necessary, followed by an angiogram, a catheterization of the coronary arteries. There are several experimental studies that involve a CT determination of the CAC in the coronary arteries. One such study involved 1958 men and 1281 women, who were referred to the Shields Coronary Artery Center in Spokane, Washington, from January 1990 to May 1998. Some of the subjects had been diagnosed with coronary artery disease, while others were referred because they were suspected of having the disease. Measurements of CAC were made with the Imatron C-100 Ultrafast CT Scanner. In Chapter 4, the diagnostic accuracy of CAC is examined with a Bayesian technique for this study.

Another way to diagnose coronary artery disease is to measure the degree of stenosis in the arteries by magnetic resonance angiography, where Obuchowski [16] used the results of a study by Masaryk et al. [17] to illustrate a non-parametric way of estimating the area under the ROC curve for clustered data. There were two readers and two measurements per patient, one for the left and one for the right coronary arteries, and the correlation introduced by this clustering effect was taken into account by Obuchowski's analysis.

2.3.6   Type 2 diabetes

There are several tests for type 2 diabetes, including a random plasma glucose test, a fasting blood glucose test, and an oral glucose tolerance test. The first does not require fasting and can be given at any time, even after a meal. If the amount of glucose is >200 mg/dL, the subject is considered to be diabetic.

A better method to test for type 2 diabetes is the fasting blood glucose test, which requires the subject to fast for approximately 8 hours before the test. The test is usually done in the morning before breakfast, where a blood glucose level between 70 and 110 mg/dL is considered normal; however, a level between 111 and 125 mg/dL indicates some problems with glucose metabolism. Levels in excess of 126 mg/dL are usually an indication that the subject has diabetes.

Perhaps if the fasting blood glucose test indicates that the subject has the disease, the doctor will order an oral glucose tolerance test, which requires that the subject fast for 10 hours before the test. At baseline, a blood glucose test is given, then the subject is given a high amount of sugar and the blood glucose level is measured 30 minutes later, 1 hour later, and 2 and 3 hours later, thus, there are four measurements taken after baseline. In a person without diabetes, the glucose level rises immediately after taking the sugar load, but then falls back to "normal" as insulin is produced. On the other hand, in diabetics the glucose levels rise higher than normal after drinking the sugar load. A person is said to have impaired glucose tolerance if the 2-hour level is between 140 and 200 mg/dL and is referred to as prediabetes. A person with a 2-hour level in excess of 200 mg/dL is considered to be diabetic, and one should seek a physician's advice in order to treat the disease.

The fasting blood glucose test and the glucose tolerance test will be considered several times in later chapters as an illustration for estimating the accuracy of medical tests.

2.3.7 Other medical tests

Johnson, Sandmire, and Klein [18] should be read for additional information about the accuracy of medical tests. Of course, there are many other medical tests that can be presented, but for the present, those for human immunodeficiency virus (HIV), prostate, and ovarian cancer will be described.

2.3.7.1 Tests for HIV

There are several tests for HIV, including enzyme linked immunosorbent assay (ELISA) and oral tests.

The ELISA test is the most commonly used test to look for HIV antibodies and if present, a confirmatory test called the Western blot analysis is done. Once an antibody test shows that the subject has been exposed to HIV, a plasma viral load (PVL) test can be performed and will often be ordered to measure the amount of HIV virus in the blood. Three different PVL tests are commonly used: the reverse transcription polymerase chain reaction (RT-PCR) the branched DNA (bDNA), and the nucleic acid sequence-based amplification (NASBA) test. All these tests work well and measure the same thing, the amount of HIV virus in the blood, but they can differ in the recorded amounts, thus, one test should be used throughout the treatment for the disease. It is comforting to know that the risk of a false positive with ELISA is quite low. Note that several tests are administered in order to diagnose the disease and their accuracy plays an important role both for diagnosis and treatment.

If you are at a high risk for HIV and you have a negative ELISA, the test should be repeated every 6 months. False negatives using RT-PCR are also rare because of prior testing using ELISA.

2.3.7.2 Tests for ovarian cancer

The carcinogenic antigen (CA) 125 blood test measures the levels of a protein that is normally confined to the cell wall, but if the wall is inflamed or damaged, the protein may be released into the blood stream. Ovarian cancer cells may produce an excess of these protein molecules, thus a test involving CA 125 can help in the diagnosis and monitoring of the disease. It is important to remember that basing the diagnosis of ovarian cancer only on CA 125 is prone to error because the levels of CA 125 are not present in the early stages of the disease and false positives can occur. Used together with transvaginal ultrasound, CA 125 can be quite effective in detecting the disease. A transvaginal ultrasound involves the use of sound waves to delineate internal structures with a transducer placed in the vagina. An example is given later in the book using CA 125 to detect ovarian cancer.

2.3.7.3   Prostate-specific antigen test for prostate cancer

Prostate specific antigen (PSA), discovered in 1979, is a protein produced by the cells that line the inside of the prostate gland. The cancer causes cell changes to the cellular barriers that normally keep PSA within the ductal system of the gland, and PSA is released into the blood stream in higher than normal quantities. The total PSA test measures the total amount of PSA in the blood, where the results are given in nanograms per millimeter and a level in excess of 4 ng/mL is considered a possible sign of prostate cancer. The total PSA test and the digital rectal examination are considered the first line of defense against the disease and if suspicious findings are found in either examination, follow-up tests are ordered, including the percent-free PSA test and transrectal prostate ultrasound. Medical tests involving PSA and transrectal ultrasound will be presented in later chapters. The percent-free PSA test is mainly used as a follow-up test when the total PSA is found in the gray area, between 4 and 9.9 ng/mL, to help determine who should undergo a biopsy of the prostate. Currently, a biopsy is ordered if the percent-free PSA level is >25%. The PSA test has problems with accuracy, where only 15–25% of men who have elevated levels of total PSA in excess of 4 ng/dL develop prostate cancer. Also, 30% of men who have prostate cancer have normal PSA levels!

2.3.7.4   Bacterial infection with *Strongyloides*

Strongyloides is an infectious organism that affects certain groups and is used as an example in this book where a gold standard is not available. A group of Cambodian refugees immigrating to Canada is tested for the disease with two medical tests, a serology test and a test based on a stool example. This example relies on prior information about the accuracy of the two tests, and Bayesian inference is used to correct the observed accuracy of the two tests.

2.3.7.5   Tuberculosis

Another case of infectious disease used in this book is a study of two tests to diagnose tuberculosis, at two different sites, the first is a southern school district in the 1940s and the second is a tuberculosis sanatorium. The scenario is a case when there is no gold standard, and Bayesian methods are used to correct the accuracy of the two tests, namely, the Mantour and Tine tests, where the Mantour test is based on a sputum sample, and the Tine is a tuberculin skin test.

2.4 Activities Involved in Medical Testing

As stated earlier, medical tests are ubiquitous in the health care system. These activities will be divided generally into two categories: (1) screening for preclinical disease, such as breast cancer, heart disease, or lung cancer; and (2) as part of patient management during the patient's stay in a large, modern health care facility. The emphasis in this book will be on the latter, where the patient has been diagnosed with the help of imaging, and are then followed and monitored during their stay in the hospital. During the patient's stay, the following imaging activities are usually involved: primary diagnosis or confirmation of earlier diagnoses, diagnostic imaging to determine the extent of disease including biopsy procedures, so-called staging studies, and follow-up medical procedures, such as surgery for biopsy or other forms of therapy, and monitoring the progression of the disease during therapy, such as in Phase II clinical trials.

Screening is performed to detect disease in the early phase, before symptoms appear. The main objective of screening is the early detection of disease when treatment is more effective and less expensive. It is assumed that early detection will lead to a more favorable diagnosis, and that early treatment will be more effective than treatment given after symptoms appear. Another important goal of screening is to identify risk factors that would predispose the subject to a higher than average risk of developing disease. Imaging is almost always involved in the diagnosis of disease, but mammography is the only examination in wide use today as a screening tool. There are some other areas where screening is being tested, namely, in lung cancer with multidetector CT, and in the detection of colorectal adenomatous polyps. One of the most important and difficult problems in clinical medicine is making recommendations for imaging studies for disease screening.

Screening should only be performed if the disease is serious and in the preclinical phase, and on a population that is at relatively high risk for developing the disease. Screening would not be effective if the disease can be treated effectively after the appearance of symptoms. If a false positive occurs, the patient is subjected to unnecessary follow-up procedures, such as surgery, additional imaging, and pathological testing for extent of disease.

A medical test like mammography is efficacious only if it is accurate, if it has good diagnostic characteristics like high sensitivity, specificity, and positive predictive value, and if a survival advantage can be demonstrated. How should a study be designed in order to evaluate the effectiveness of an imaging screening procedure? Of course, randomized studies have an advantage and are the basis for a recent paper by Shen et al. [19], who reported on the survival advantage of screening detected cases over control groups. This investigation used data from three randomized studies with a total of 65,170 patients, and it used Cox regression techniques to control for the so-called lead time bias (detection of early stage disease with screening), tumor size, stage of disease, lymph node status, and age. They conclude that mammography screening is indeed effective. For additional information on the advantages of mammography, see Berry et al. [20]. For recent Bayesian contributions to the estimation of sensitivity and lead time in mammography, see Wu, Rosner, and Broemeling [21, 22].

The whole area of diagnostic screening has a voluminous literature. This book will not focus on screening and the reader is referred to Shen et al. [19], who cite the most relevant studies.

2.5   Accuracy and Agreement

How good is a diagnostic procedure? For example, suppose one is using mammography to diagnose breast cancer, then how well does it correctly classify patients who have disease and those who do not have disease? Among those patients who have been classified with disease, what proportion actually have it? And, among those who were designated without disease, how many actually do not have it? To answer these questions, one must have a gold standard by which the true status of disease is determined. Thus, the gold standard will divide the patients into two groups: those with and those without the disease.

Another question is how does the radiologist decide when to classify an image as showing a malignant lesion? Often a confidence level scale is used, where 1 designates definitely no malignancy, 2 probably no malignancy, 3 indeterminate, 4 probably a malignant lesion, and 5 definitely a malignant lesion. Given this diagnostic ordinal scale, how does the reader decide when to designate a patient as diseased? In the case of mammography, a score of 4 or 5 is often used to classify a patient as having the disease, in which case each image can be classified as either: (1) a true positive, (2) a true negative, (3) a false positive, and (4) a false negative. Of course, these four possibilities can only be used if one knows the true status of the disease as given by the gold standard. Given these four outcomes, one may estimate the accuracy of the procedure with the usual measures of sensitivity, specificity, and positive and negative predictive values. For example, the specificity is estimated as the proportion of patients who test negative, among those that do have the disease. There are many statistical methods to estimate test accuracy and these will be explained in detail in Chapter 4. The idea of the area under the ROC will be explained and many examples introduced to demonstrate its use as an overall measure of test accuracy.

Other factors that need to be taken into account are: (1) the design of the study, (2) the gold standard and how it is utilized, and (3) the variability among and between observers and the input of others involved in diagnostic decisions.

With regard to the design, several questions must be asked: How are the patients selected? Is one group of patients selected at random from some population, or are two groups of patients, diseased and non diseased, selected? Or are they selected from patient charts, such as in a retrospective review? Along with this is the nature of the population from which the patients are selected. Is it a screening population, a community clinic, or a group of patients undergoing biopsy? These factors all affect the final determination of the accuracy as well as what biases will be introduced.

The gold standard often depends on surgery for biopsy, the pathology report from the laboratory, and additional imaging procedures. When and how the gold standard is used, frequently depends on the results of the diagnostic test. Often, only those who test positive for disease are subjected to the gold standard, while those that test negative are not. For example, with mammography those that test positive are tested further with biopsy and tests for histology. While among those that test negative, follow up of patient status is the gold standard.

Lastly, with regard to reader variability, it is important to remember that the medical test is an aid for the people who make the diagnosis, and that the diagnosis is made by a group (e.g., cardiologists, oncologists, surgeons, radiologists, and pathologists). All of this introduces variability and error into the final determination of disease status. Is agreement between and among observers (radiologists, pathologists, surgeons, etc.) an important component of diagnostic medicine? Of course it is, for suppose a Phase II clinical trial is being conducted to determine the efficacy of new treatment for advanced prostate cancer with, say, 35 patients. The major endpoint is tumor response to therapy, which is based on the change in tumor size from baseline to some future time point. Often, the percentage change from baseline is used and, furthermore, this determination depends on the readings of the same images by several radiologists. Since they differ in regard to training and experience, their determination of the percentage change varies from reader to reader. How is this taken into account? How is a consensus reached?

Statistical methods that take into account and measure agreement are well developed. For example, with ordinal test scores, agreement between observers is often measured by the Kappa statistic, while if the test score is continuous, regression techniques for calibration (e.g., Bland-Altman) are frequently done to assess accuracy within and between observers. Analysis of variance techniques that account for various sources (patients, readers, modalities, replications, etc.) of variability help in estimating the between and within reader variability, via the intra class correlation coefficient. In Chapters 4 and 5, test accuracy and agreement between observers will be revealed in detail. See Broemeling [23] for a Bayesian approach to the study of agreement.

Kundel and Polansky [24] give a brief introduction to the various issues concerning the measurement of agreement between observers in diagnostic imaging, and Shoukri [25] has an excellent book on the subject.

2.6   Developmental Trials for Medical Devices

When developing a new imaging modality, the test must pass three phases labeled I, II, and III. This is similar to the designation for patient clinical trials, but what is being referred to here is the development of medical devices. The different phases are for different objectives of test accuracy and are as follows.

Phase I trials are exploratory and are usually retrospective with 10–50 patients and 2–3 readers. There are two populations, a homogenous group of diseased subjects who are definitely known to have the disease, and a second group of homogenous people who are definitely known not to have the disease. The key word here is homogenous, where the manifestations of the disease are more or less the same among diseased patients, while among the non diseased, their health status is the same. The accuracy is measured by true positive and false positive rates, as well as the area under the ROC curve. Thus, if the accuracy is not good, the modality needs to be improved. See Bogaert et al. [26] for a good example of Phase I developmental trial involving MRI angiography.

If a device has sufficient accuracy during Phase I, it is studied as a Phase II trial, and is called a challenge trial, with 50–200 cases and 5–10 observers. They are also retrospective, but with a wide spectrum of the disease in the two groups. Thus, if the disease is, say, non-small cell lung cancer, patients with different manifestations (different ages, different stages of disease, and patients who have disease similar to non-small cell lung cancer) of disease are included. Thus, it is more difficult for the device to distinguish between diseased and non-diseased subjects. Among the non diseased, the patients are also heterogeneous. Test accuracy is measured as in a Phase I trial, and the association between accuracy and the pathological, clinical, and co-morbid features of the patient can be investigated with regression modeling. A comparison between digital radiography and conventional chest imaging was performed as a Phase II trial by Theate et al. [27].

Beam, Lyde, and Sullivan [28] investigated the interpretation of screening mammograms as a Phase III trial using 108 readers, 79 images read twice by each reader, and many health care centers. The sensitivity ranged from 0.47 to 1 and specificity from 0.36 to 0.99 across the readers. Phase III trials are prospective and are designed to estimate test performance in a well-defined clinical population and involve at least 10 observers, several hundred cases, and competing modalities. A device should pass all three phases before becoming standard in a general clinical setting.

Note that it is important to know the inter observer variability in these trials, because the accuracy of the modality depends not only on the device, but also the interpretation of the image via the various readers. In Chapter 8, Pepe [29] provides a more detailed description of developmental trials, and

Obuchowski [30] provides sample size tables for the number of observers and the number of patients in trials for device development.

2.7 Literature

As mentioned earlier, biostatistics plays a pivotal role in the imaging literature, as can be discerned by reading papers in the mainline journals, such as *Academic Radiology, The American Journal of Roentgenology,* and *Radiology,* and the more specialized journals, such as *The Journal of Computed Assisted Tomography, The Journal of Magnetic Resonance Imaging, The Journal of Nuclear Medicine,* and *Ultrasound in Medicine.* For non-imaging studies, the journal *Pathology* provides many examples of studies for medical test accuracy.

For some reference books in the area of general diagnostic imaging, the standard one is *Fundamentals of Diagnostic Radiology* (1999 Second Edition), edited by Brant and Helms [31]. Both references are for radiologists and give the fundamentals of imaging principals plus a description of the latest clinical applications. For some good general information for the patient, Johnson, Sandmire, and Klein [18] describe medical tests for a large number of diseases, including those for cancer, stroke, heart disease, diabetes, and infectious diseases.

Two statistical books are relevant: *The Statistical Evaluation of Medical Tests for Classification and Prediction* by Pepe [29], and *Statistical Methods in Diagnostic Medicine* by Zhou, McClish, and Obuchowski [32]. Both are excellent and are intended for biostatisticians.

References

[1] Mielke, C.H., Shields, J.P., and Broemeling, L.D. Coronary artery calcium, coronary artery disease, and diabetes. *Diabetes Research and Clinical Practice,* 53:55, 2001.

[2] Mielke, C.H., Shields, J.P., and Broemeling, L.D. Risk factors and coronary artery disease for asymptomatic women using electron beam computed tomography. *Journal of Cardiovascular Risk,* 8:81, 2001.

[3] DasGupta, N., Xie, P., Cheney, M.O., Broemeling, L., and Mielke, C.H. The Spokane heart study: Weibull regression and coronary artery disease. *Communications in Statistics,* 29:747, 2000.

[4] Broemeling, L.D. and Mielke, C.H. Coronary risk assessment in women. *The Lancet*, 354:426, 1999.

[5] Wolbarst, A.B. *Looking Within: How X-ray, CT, MRI, and Ultrasound and Other Medical Images are Created and How They Help Physicians Save Lives.* University of California Press, Berkeley, 1999.

[6] Jawad, I.J. *A Practical Guide to Echocardiography and Cardiac Doppler Ultrasound* (2nd ed.). Little, Brown & Co., Boston, 1996.

[7] Chandra, R. *Nuclear Medicine Physics: The Basics.* Williams & Wilkins, Baltimore, MD, 1998.

[8] Seeram, E. *Computed Tomography, Clinical Applications, and Quality Control* (2nd ed.). W.B. Saunders Company, Philadelphia, 2001.

[9] Markisz, J.A. and Aguilia, M. *Technical Magnetic Imaging.* Appleton & Lange, Stamford, CT, 1996.

[10] Pawlik, T.M. and Gershenwald, J.E. Sentinel lymph node biopsy for melanoma. *Contemporary Surgery*, 61(4):175, 2005.

[11] Morton, D.L., Wanek, L., Nizze, J.A., Elashoff, R.M., and Wong, J.H. Improved long term survival lymphadenectomy of melanoma metastatic to regional lymph nodes: Analysis of prognostic factors in 1134 patients from the John Wayne Cancer Center Institute. *Annals of Surgery*, 214:491, 1991.

[12] Gershenwald, J.E., Tseng, C.H., Thompson, W., Mansfield, P., Lee, J.E., Bouvet, M., Lee, J.J., and Ross, M.I. Improved sentinel lymph node localization in patients with primary melanoma with the use of radiolabeled colloid. *Surgery*, 124:203, 1998.

[13] Rousseau, D.L., Ross, M.I., Johnson, M.M., Prieto, V.G., Lee, J.E., Mansfield, P.F., and Gershenwald, J.E. Revised American Joint Committee on Cancer staging criteria accurately predict sentinel lymph node positivity in clinically node negative melanoma patients. *Annals of Surgical Oncology*, 10(5):569, 2003.

[14] Gupta, S., Krishnamurth, S., Broemeling, L.D., Morello, F.A., Wallace, M.J., Ahrar, K., Madoff, D.L., Murthy, R., and Hicks, M.E. Small (<2 cm) subpleural pulmonary lesions; short versus long needle path, CT-guided biopsy: Comparison of diagnostic yields and complications. *Radiology*, 234:631, 2005.

[15] Gupta, S., Kobayashi, S., Phongkitkarun, S., Broemeling, L.D., and Kun, S. Effect of trans catheter hepatic arterial embolization on angiogenesis in an animal model. *Investigative Radiology*, 41(6):516, 2006.

[16] Obuchowski, N.A. Non parametric analysis of clustered ROC curve data. *Biometrics*, 53:567, 1997.

[17] Masaryk, A.M., Ross, J.S., DiCello, M.C., Modic, M.T., Paranandi, L., and Masaryk, T.J. Angiography of the carotid bifurcation: Potential and limitations as a screening examination. *Radiology*, 121:337, 1991.

[18] Johnson, D., Sandmire, D., and Klein, D. *Medical Tests that Can Save Your Life: 21 Tests Your Doctor Won't Order... Unless You Know to Ask.* Rodale, New York, 2004.

[19] Shen, Y., Inoue, L.Y.T., Munsell, M.F., Miller, A.B., and Berry, D.A. Role of detection method in predicting breast cancer survival: analysis of randomized screening trials. *Journal of the National Cancer Institute*, 97:1195, 2005.

[20] Berry, D.A., Cronin, K.A., and Plevritis, S.K. Effect of screening and adjuvant therapy on mortality from breast cancer. *The New England Journal of Medicine*, 353(17):1784, 2005.

[21] Wu, D., Rosner, G., and Broemeling, L.D. MLE and Bayesian inferences of age-dependent sensitivity and transition probability in periodic screening. *Biometrics*, 61:1056, 2005.

[22] Wu, D., Rosner, G., and Broemeling, L.D. Bayesian inference for the lead time in periodic cancer screening. *Biometrics*, 63:873, 2005.

[23] Broemeling, L.D. *Bayesian Methods for Measures of Agreement.* Chapman & Hall/CRC, Boca Raton, 2010.

[24] Kundel, H.L. and Polansky, M. Measure of observer agreement. *Radiology*, 228:303, 2003.

[25] Shoukri, M.M. *Measures of Interobserver Agreement.* Chapman & Hall/ CRC, Boca Raton, 2002.

[26] Bogaert, J., Kuzo, R., Dymarkowski, S., Becke, R., Plessens, J., and Rademakers, F.E. Coronary artery imaging with real-time navigator three dimensional turbo field echo MR coronary angiography: Initial experience. *Radiology*, 226:707, 2003.

[27] Thaete, F.L., Fuhrman, C.R., Oliver, J.H., Britton, C.A., Campbell, W.L., Feist, J.H., Staub, W.H., Davis, P.L., and Plunkett, M.B. Digital radiography and conventional imaging of the chest: a comparison of observer performance. *American Journal of Roentgeneology*, 162:575, 1994.

[28] Beam, C.A., Lyde, P.M., and Sullivan, D.C. Variability in the interpretation of screening mammograms by US radiologists. *Archives of Internal Medicine*, 156:209, 1996.

[29] Pepe, M.S. *The Statistical Evaluation of Medical Tests for Classification and Prediction.* Oxford University Press, Oxford, UK, 2003.

[30] Obuchowski, N.A. Sample size tables for receiver operating characteristic studies. *American Journal of Roentgenology*, 175:603, 2000.

[31] Brant, W.E. and Helms, C.A. *Fundamentals of Diagnostic Imaging* (2nd ed.). Lippincott, Williams & Wilkins, New York, 1999.

[32] Zhou, H.H., McClish, D.K., and Obuchowski, N.A. 2002. *Statistical Methods for Diagnostic Medicine.* John Wiley, New York, 2002.

Chapter 3

Preview of the Book

3.1 Introduction

This chapter should give the reader a good idea of what this book is about. In one sentence, this book introduces the reader to the design and analysis of medical test accuracy, with emphasis on a Bayesian analysis. A Bayesian approach is taken where the foundation is based on Bayes theorem, and all inferences are expressed as posterior distributions of the relevant parameters. WinBUGS is the software that will execute Bayesian inferences for medical test accuracy and the associated code is labeled in the book and also appears on the author's blog. In what follows, I will carefully describe the contents of each chapter, so that the reader will know what to expect.

3.2 Preliminary Information

The first three chapters present the preliminary information necessary for the reader to understand the importance of knowing the accuracy of a medical test.

3.2.1 Chapter 1: Introduction

The chapter begins with a short introduction previewing the chapter, followed by a very brief introduction to the indicators of accuracy, including the four basic measures: true positive fraction (TPF), false positive fraction (FPF), positive predictive value, and negative predictive value. Such measures are applicable if the test scores are binary or if the scores have been dichotomized with a cutoff value. The area under the receiver operating characteristic (ROC) curve is described as a measure of overall accuracy for medical tests that have ordinal or continuous scores. The next part of the chapter explains the various datasets that are used for the examples. For example, some of the datasets in the book by Pepe [1] will be used, as will some examples from the book by Zhou, McClish, and Obuchowski [2]. Also included for analysis is information that the author obtained while consulting at the University of Texas MD Anderson Cancer Center (MDACC). The information quite valuable and contains many examples of imaging studies for cancer, including studies involving x-ray, computed tomography (CT), magnetic resonance imaging (MRI), nuclear medicine, and ultrasound. The various forms of cancer include breast, prostate, lung, ovarian, etc., and will give the reader a good idea of the important role played by the accuracy of a particular medical test. Other sources used in the book are papers appearing in the *Journal of Radiology*, with an emphasis on procedures that combine two or more tests.

The software employed in this book is WinBUGS and is most appropriate for our purposes of expressing accuracy inferences via the posterior distribution of the appropriate parameter. Inference is expressed by computing the posterior mean, median, standard deviation, and the lower and upper 2 1/2 percentiles of the posterior distribution. WinBUGS generates samples from the posterior distribution, via Markov Chain Monte Carlo (MCMC), where the simulation sample size can be adjusted by referring to the MCMC error. The reader is expected to have some knowledge of Bayesian inference, but a brief introduction is presented and some history from Bayes to the present day is given.

3.2.2   Chapter 2: Medical tests and preliminary information

Knowing the various medical tests used in health care is essential to understanding the value of medical test accuracy, and this chapter gives brief descriptions of several medical devices. First to be considered are the standard imaging tests found in the diagnostic radiology department of a modern hospital and include descriptions of x-ray, CT, mammography, MRI, nuclear medicine, and ultrasonography (US). Sometimes, more than one test is used to give a better picture of the extent of the disease, e.g., MRI and CT to monitor lung cancer patients. All the tests mentioned are used to diagnose and monitor cancer patients, however, they are also used to diagnose and monitor heart disease and other maladies. Next to be portrayed are some specialized tests for cancer, including nuclear medicine procedures for detecting metastasis of melanoma from the primary tumor to the lymph nodes. Another diagnostic test used for melanoma metastasis is using the depth of the primary tumor.

Switching from cancer to other diseases, the use of CT for screening and monitoring coronary heart disease is characterized. There are many tests for diagnosing heart disease, including the exercise stress test, followed if necessary by coronary angiography, but a promising CT test measures the amount of calcium in the coronary arteries. The advantage of the CT test is that it is safer than the stress test or coronary angiography and is in the experimental stage in order to assess its accuracy. Type 2 diabetes is becoming more of a problem and is diagnosed by the fasting blood glucose test and the blood glucose tolerance test. Both these blood tests are explained in Chapter 2, and will be used in a later chapter as a way to combine two tests to achieve better accuracy. The remaining medical tests to be portrayed are the enzyme linked immunosorbent assay (ELISA) test to detect antibodies for human immuno-deficiency virus (HIV), the biomarker CA 125 test to detect ovarian cancer, and the prostate-specific antigen (PSA) biomarker to diagnose prostate cancer.

The chapter continues by characterizing the interplay between agreement and medical test accuracy. It is important to remember that several people are sometimes involved in interpreting the output of a medical test. With the aid of medical test(s), several health care workers use the medical test output to give a diagnosis or to monitor the progress of the patient under treatment, thus agreement or disagreement between readers is present in the treatment of the patient. Agreement among readers will be explicated in more detail in Chapter 6, but is given a brief introduction in Chapter 2.

Developmental trials for medical devices, including medical tests, are briefly explained at the end of the chapter. In this part of the book, the design aspects of the subject are mentioned for the first time, where a promising medical device is first examined with two different populations, a population with the disease and the other without the disease. Under such conditions, the test, if it has any accuracy, should be able to discriminate between the two populations. If the test passes the Phase I trial, it is subject to a more stringent challenge involving many readers and institutions. This part of the chapter describes in detail Phase I, II, and III studies for medical devices.

3.2.3 Chapter 3: Preview of the book

This chapter gives a preview of the book.

3.3 Fundamentals of Test Accuracy

Chapters 4 through 6 present the basics for understanding the measurement of medical test accuracy, with Chapter 4 describing the four fundamental indicators: the TPF and the FPF, and the positive and negative predictive values. Chapter 5 is largely devoted to regression techniques for incorporating covariate information, while Chapter 6 stresses the study of agreement between several readers who are interpreting the output of medical tests. How does agreement or disagreement between readers affect the accuracy of a medical test?

3.3.1 Chapter 4: Fundamentals of medical test accuracy

The chapter begins with an introduction to the design of a study to measure the accuracy of a medical device by outlining the components that are necessary for implementing the study, where the components of a good design are listed as: objectives, background, patient and reader selection, study design, number of patients, statistical design and analysis, and, lastly, a section for the reference of the study.

Next, a description of the four fundamental indicators of test accuracy for binary test scores is given, where the basic theory is presented, followed by a WinBUGS program that illustrates the estimation of test accuracy. The four indicators are the so-called classification probabilities, namely, the true and false positive fractions. This is followed by the positive and negative predictive values that are of interest to the patient, and the four indicators are estimated by an example using the exercise stress test to diagnose coronary artery disease. The Bayesian analysis is executed with BUGS CODE 4.1 using 45,000 observations, a burn in of 5,000 and a refresh of 100, and the results consist of the posterior characteristics for the four indicators and a graph of their posterior densities. The Bayesian approach is continued by defining the area under the ROC curve and illustrated with an example of mammography, where the test scores are ordinal: 1 indicating positively no evidence of malignancy; 2 indicating there is very little evidence of malignancy; 3 implying an ambiguous situation for scoring the lesion malignant; 4 indicating some evidence of malignancy; and 5 indicating that the lesion is definitely malignant. There are 30 patients with the disease and 30 without, and the analysis is executed with BUGS CODE 4.2. Remember, the code is listed in the book and on the author's blog and is easily accessible to the reader. The ROC area is also illustrated with the Shields Heart Study, which uses CT to measure the extent of coronary artery disease. The Bayesian methods for ordinal scores are developed by the author and appear to be unique.

The chapter continues with an interesting generalization of the ROC area when the scores are ordinal, and portrays the case when the scores are clustered, which is the case for mammography, that is, the image is partitioned into several regions and the radiologist assigns a score from 1 to 5 to each region of the mammogram. In this scenario, one would expect the scores to be correlated and the chapter presents the theory and illustrates the idea with an example taken from Zhou, McClish, and Obuchowski [2: 134] involving mammography, where the Bayesian analysis is executed with BUGS CODE 4.4.

With ordinal scores, the subject is expanded to include a comparison between the accuracies of two medical tests to diagnose the same disease, and is illustrated with CT and MRI to detect lung cancer. The two tests are compared based on their ROC areas, and the Bayesian analysis is executed with BUGS CODE 4.6; it is noted that the design is paired in that both tests are administered to the same patients. The chapter concludes by estimating accuracy with ROC areas for tests with continuous scores and comparing two tests via their ROC areas.

3.3.2   Chapter 5: Regression and medical test accuracy

This chapter deals with patient covariate information that can be accounted for in the estimation of test accuracy. Regression techniques for ordinal and continuous test scores are considered, and the chapter begins with an example from audiology, where the subject's covariate information is accounted for in estimating the true and false positive fractions. An example from an audiology study test is considered, where the accuracy of the test that is supposed to detect impaired hearing is taken from Pepe [1], where the dependent variable is the false positive rate and the covariates are the age of the patient, the version of the test, and the location where the test is given. This example is analyzed using two link functions, the first is a log link and the second is a logistic link, and the analysis is based on BUGS CODE 5.1 and 5.2, the former for the log link and the latter for the logistic link. This example is continued by estimating the positive diagnostic likelihood ratios using the same patient covariates. Next to be considered is using patient covariates to estimate the area under the ROC curve with an ordinal regression model formulated by Congdon [3: 108] and illustrated with an example of a clinical trial that measures tumor response to two therapies. The example is from Holtbrugge and Schumacher [4] and is executed with BUGS CODE 5.4. Another example using ordinal regression is a staging study for metastasis of melanoma and involves four radiologists who all see the same information on the same patients.

When the test scores are continuous and normally distributed, the Bayesian regression approach of O'Malley et al. [5] allows covariate information for estimating the ROC area and is illustrated with an example from Pepe [1], involving screening for prostate cancer, where the test scores are the total PSA values. BUGS CODE 5.6 is executed to produce a posterior analysis where the patient covariate is age, resulting in a posterior mean of 0.80 for the area. The chapter concludes with another example with continuous scores using yet another audiology example. There are 17 exercises that give the student additional valuable information about the Bayesian analysis that estimates test accuracy with the aid of regression models for ordinal and continuous observations. When studying the exercises, remember to download the code and data from the blog: http://medtestacc.blogspot.com.

3.3.3 Chapter 6: Agreement and test accuracy

Several readers are usually involved in interpreting the results of a medical test, and this chapter emphasizes how they affect the overall accuracy of the test. Recall the melanoma example of Chapter 4, where four readers were scoring the degree of metastasis of the disease. Since the readers are viewing the same images, one would expect correlation between the reader scores and their results to be similar, however, some readers may have more experience than others, a factor that introduces additional variability to the determination of test accuracy. The ROC area estimates the accuracy of the test, one for each reader, but which areas do we use? All four are reported, but should one employ some type of summary of the four areas?

The first case to be considered is the melanoma metastasis example of Chapter 5, and the four ROC areas are estimated with a Bayesian approach using BUGS CODE 6.1 with 65,000 observations generated from the posterior distribution. It turns out that the four estimated areas varied from a low of 0.64, estimated by reader 3, to a high of 0.80 for reader 2. On the contrary, a second example involving the blood glucose test for type 2 diabetes with three readers, revealed very little difference in the posterior means of the ROC areas. The latter case involves a continuous score and the O'Malley et al. [5] method of estimating the ROC curve, and is continued by expanding the analysis to include patient age and gender as covariates. A Bayesian analysis based on BUGS CODE 6.2 estimates a summary ROC area with a weighted mean, where the posterior mean area of each reader is weighted by the inverse of the posterior variance. The unweighted mean is also computed as 0.8162(0.0130), which compares to the weighted mean of 0.991(0.0022).

A gold standard is present for the above scenarios, and the chapter continues by considering the case when no gold standard is available, and brings the standard approach to estimating the agreement between the readers. Of course, if the gold standard is present, the readers can be compared on the basis of the ROC areas, but when the gold standard is not available, how should agreement be estimated?

There is a long history of statistical agreement based mostly on the Kappa coefficient, and that approach will be taken for the remainder of Chapter 6. The Kappa coefficient is defined and the Bayesian approach to the index is described and illustrated with an example for nominal scores using an example from Von Eye and Mun [6: 12]. The example consists of a 3×3 table with two psychiatrists, who are assigning scores that express the degree of depression in each of 129 patients, where the three scores are defined as: $1 =$ not depressed, $2 =$ mildly depressed, and $3 =$ clinically depressed. The Bayesian analysis is run with BUGS CODE 6.3, using 25,000 observations for the simulation and is available on the author's blog: http://medtestacc.blogspot.com. Also reported is the density of the posterior distribution of conditional Kappa.

Various generalizations of Kappa are presented in the remainder of Chapter 6, including Kappa and stratification where a hypothetical example portrays the essential components for estimating agreement. Suppose that the agreement between x-ray and CT is estimated, where the study is conducted at three different sites and calls for a total enrollment of 2500 subjects, with 1000 each at two sites and 500 patients at a third site. Our objective is to estimate the overall agreement between the two devices, using a weighted Kappa where the weights are the inverse of the posterior variance of Kappa for a particular site. In this case, there is good agreement at each site, consequently the weighted Kappa is very close to the simple average of the posterior mean of the three Kappas.

Chapter 6 continues with various generalizations of Kappa, including an explanation of the Bayesian analysis for the so-called intraclass Kappa. The situation is similar to that of a one-way layout with c groups and an unequal number of binary observations in the various groups; observations between different groups are assumed to be independent. A crucial assumption is that each binary observation has the same probability of being "1." The Bayesian theory is described to estimate the intraclass Kappa, which is the common correlation between the binary observations in the same group. The intraclass correlation is estimated for an interesting example of three groups, where the "subjects" in a group are rabbit fetuses, and each fetus responds or does not respond to a treatment. BUGS CODE 6.3 is executed in order to estimate intraclass Kappa with a posterior mean of 0.0907(0.1063) and a posterior mean of 0.2262(0.0403) for the common probability of a response. In this case, Kappa estimates the common correlation between the binary responses of the fetuses in the same group, which is similar to the case of the usual one-way random model with normally distributed observations.

Other measures of agreement are introduced, including the G coefficient and the Jacquard index, both of which have the value "1" when there is perfect agreement between two binary scores, but the Kappa coefficient remains the index preferred by researchers in the social and medical sciences.

There is a well-known relationship between the Kappa coefficient and the sensitivity and specificity of two readers assigning scores. Kraemer [7] expressed Kappa in terms of the specificity and sensitivity of the two readers and showed the dependence of Kappa on the disease incidence. Ironically, when the disease incidence is low and the specificity and sensitivity are "high," nevertheless, Kappa can be small. A similar situation occurs in diagnostic testing when the disease rate is small, in that it can be true that the positive predictive value can be small even though the sensitivity and specificity are high.

Chapter 6 continues with a discussion of consensus between readers with an example applicable to Phase II clinical trials. In such studies, two or more radiologists grade the response of each patient, and at the end of the trial must come to a conclusion about the success or failure of the trial.

The idea of agreement is generalized to ways to compute Kappa when there are more than two raters with binary scores, and is demonstrated with an example of four students who assign scores to each image where the analysis is executed with BUGS CODE 6.8 using 25,000 observations for the simulation. When there are more than two raters, one can consider several ways to measure partial agreement. For example, when six raters are assigning binary scores, one can consider the agreement between, say, exactly two of six among them. This is accomplished by defining a Kappa coefficient and illustrating the idea with an example of six pathologists who assign a 0 or 1 if there is a certain lesion present or not in the image. The Bayesian analysis is done with BUGS CODE 6.9 and estimates Kappa as 0.6382(0.0598) with the posterior mean. Various other scenarios for partial agreement are discussed and a relevant Kappa defined and further illustrated with real-life examples. The last generalization for agreement is to define a Kappa when there are many raters and ordinal scores. Twenty exercises reinforce the Bayesian analysis for agreement presented in the chapter and are essential for a complete understanding of the subject.

3.4   Advanced Methods for Test Accuracy

3.4.1   Chapter 7: Estimating test accuracy with an imperfect reference standard

If a gold standard does not exist, the accuracy of a test is usually estimated using an imperfect reference test. There are many cases where no gold standard exists, such as in diagnosing depression, where the condition is assessed with a series of questions to the patient, but such assessments are quite subjective and no one test will give a perfect diagnosis. Pepe [1] presents another situation where no gold standard is available where *Chlamydia* is being diagnosed with polymerase chain reaction (PCR) and enzyme-linked immunosorbent assay (ELISA). The approach taken here is Bayesian and employs augmented data and the assumption of conditional independence for the two tests, the new test T and the imperfect reference test R. By conditional independence, it is meant that given D, where D indicates disease, the two tests are independent. Chapter 7 begins with a hypothetical example of a new test T, and an imperfect reference test R, where the disease status is actually known and the "true" sensitivity and specificity can be determined. Relative to the imperfect reference test R, the sensitivity and specificity of T can also be estimated. The accuracies based on R are, of course, misleading, and this is demonstrated with the example in Table 3.1.

Relative to R, the sensitivity of T is 0.80 and its specificity is 0.60, but relative to the gold standard, the sensitivity is 0.80 and the specificity is 0.70! Using the conditional independence assumption

$$P[R, T \mid D] = P[R \mid D]P[T \mid D],$$

Chapter 6 continues with a description of the Bayesian approach that employs augmented data to estimate the accuracies of the two tests, R and T. Then the approach is illustrated with an example of diagnosing *Strongyloides* among 162 Cambodian refugees who are immigrating to Canada. Two tests are used to diagnose the disease, namely, a serology (blood) test and a test based on the subject's stool. The Bayesian analysis is executed with BUGS CODE 7.1 using 125,000 observations, with a burn in of 5,000 and a refresh of 100. The Bayesian analysis relied on prior information and gave a sensitivity of 0.88 for serology and only 0.30 for the stool examination, but on the other hand, the specificity of the stool examination is 0.76 with 0.69 for serology. The code and data can be downloaded from http://medtestacc.blogspot.com.

TABLE 3.1:   Hypothetical example of imperfect reference R.

	$D = 0$	$D = 1$	$R = 0$	$R = 1$
$T = 0$	70	20	74	16
$T = 1$	30	80	46	64
Total	100	100	120	80

The *Strongyloides* example is reanalyzed, but not assuming conditional independence, and the results are surprising, in that the accuracies are somewhat similar compared to those with the original analysis, implying that the conditional independence assumption is valid.

Chapter 7 continues with another example from Zhou, McClish, and Obuchowski [2] of two tests (the Tine and Mantour tests) without a gold standard, but the two tests are for diagnosing tuberculosis at two different sites, one a southern school district, and the other a tuberculosis sanatorium. This is an interesting example, because one would expect the tuberculosis rates to be very different between the two sites, with a much higher rate for the sanatorium, and one would expect this to also affect the accuracy of the two tests at the two sites. Using the conditional independence assumption at both sites, a Bayesian analysis was conducted with BUGS CODE 7.3 using 130,000 observations for the simulation, and it was found that the two sites did differ substantially with regard to the rate of tuberculosis, but that the accuracy of the two tests at the two sites was quite similar! As above, the example is revisited, but not assuming conditional independence. The analysis is executed with BUGS CODE 7.3, and it is interesting to compare the Bayesian analyses with and without the conditional independence assumption.

Chapter 7 portrays an interesting generalization by considering three binary tests with no gold standard, where the theory assumes conditional independence among the tests. The example is of three radiologists who are conducting a study of x-ray to detect pleural thickening of 1692 male asbestos miners in South Africa. Here, the three tests correspond to three radiologists using the same device (x-ray) to diagnose disease with no gold standard. Executing BUGS CODE 7.5, the disease rate (pleural thickening) is estimated as only 0.019, but the specificities of the three radiologists are essentially the same. The sensitivities do differ somewhat, but the difference is not substantial. The reader should note that the analysis relies heavily on prior information, and the example assesses the effect of changing prior information on the posterior analysis. The assumption of conditional independence is dropped and the pleural thickening example reanalyzed by executing BUGS CODE 7.6.

Chapter 7 is concluded by using the previous concepts of augmented data and conditional independence assumption to study two tests with ordinal scores, where the accuracy of the tests is based on the ROC areas. The Bayesian theory is developed and the ideas illustrated with an example of MRI and CT to stage pancreatic cancer. In order to understand the concepts of estimating accuracy without a gold standard, the student should solve the 23 exercises at the end of the chapter.

3.4.2   Chapter 8: Verification bias and test accuracy

The typical set up for estimating the accuracy of a medical test is to have a gold standard, where the disease status of a subject is determined. Subsequently, among those with the disease, a score that can be attributed to the disease status is taken. As for those without the disease, a medical test is also given, however, there are scenarios where the patient is not subject to the gold standard. Consider for example the case of testing for heart disease with the exercise stress test, where among those that test negative, patients are usually not referred for further testing with coronary angiography, but on the other hand, among those that test positive, usually most are referred for coronary angiography, the gold standard. This is, in fact, the normal scenario when screening for disease. The case where the patient is always given the gold standard is unusual and occurs only when experimental conditions call for it. It is sometimes considered unethical to subject a person to the gold standard when they test negative.

Chapter 8 deals with the problem of estimating accuracy when verification bias is present, which occurs when only some of the subjects are verified for disease; the experimental layout is as shown in Table 3.2.

From Table 3.2, $V = 1$ indicates a patient verified for disease, otherwise $V = 0$, and the number of patients who are not verified for disease is u_1 when the test is positive, and u_0 when the test is negative. If the accuracy of the one binary test, e.g., the TPF, is estimated from among only those patients who have been verified, the estimates are biased in the sense that they will not be unbiased estimates of the "true" accuracy. The Bayesian approach to correcting for verification bias is based on missing at random (MAR) and is outlined and illustrated with an example of hepatic scintigraphy to detect liver disease. The example clearly shows the correction for verification bias, and that the accuracy based only on the verified patients is misleading. The analysis is based on BUGS CODE 8.1, which is executed with 45,000 observations generated for the simulation. A plot of the posterior density of the TPF is also provided.

Next on the agenda of Chapter 8 is estimating the accuracy of two binary tests when verification bias is present. The theory is described, again with the MAR assumption, and illustrated with a screening test for Alzheimer's disease conducted by two observers, that is, two readers are diagnosing the disease with binary scores (either the subject has the disease or the subject does not have the disease). The Bayesian analysis is based on BUGS CODE 8.2, which is executed with 30,000 observations generated for the simulation and the posterior mean for the TPF for the first observer is 0.9936(0.007). The analysis for the second observer is left as an exercise.

TABLE 3.2:   One binary test.

	Y	1 (Positive)	0 (Negative)
$V = 1$			
	$D = 1$	s_1	s_0
	$D = 0$	r_1	r_0
$V = 0$		u_1	u_0
Total		m_1	m_0

The Bayesian approach to estimating the accuracy of an ordinal test with verification bias is portrayed with a mammography example involving 1500 subjects where each patient is scored from 1 to 5, and about 11% of the patients are not verified for disease. Again, the MAR assumption is imposed for the Bayesian analysis and accomplished with BUGS CODE 8.3, which is executed with 55,000 observations for the simulation, and the ROC area for mammography is estimated as 0.7762(0.0126) with a 95% credible interval (0.7509,0.8005). Accuracy of mammography is expanded to two sites, where verification bias is present and the MAR assumption imposed. This type of generalization can be extended to several tests and several sites or observers. For example, the Bayesian approach for two ordinal tests is considered next, where the theory is developed and characterized with a dermatologist and surgeon staging melanoma. The code for BUGS CODE 8.4 appears in the book and on the author's blog and is easily executed, resulting in a ROC area of 0.78 compared to 0.63 for the dermatologist. Covariates are included in the example of staging melanoma and provide a nice generalization of the preceding theory. A nice alternative to the above way to correct for verification bias is to use inverse weighting as described by Pepe [1], where it is demonstrated that it is equivalent to that based on the MAR assumption.

Chapter 8 concludes with an approach that does not impose the MAR assumption, but computing the posterior density can be a problem. However, a posterior analysis for one binary test with verification bias is performed. The code makes it possible to test for the MAR assumption. Lastly, the reader is encouraged to solve the 18 exercises, some of which enhance one's knowledge of the subject.

3.4.3   Chapter 9: Test accuracy and medical practice

A different direction is taken for Chapter 9 in that the emphasis will be on the role that the accuracy of medical tests plays in practice, where the first topic describes how to choose the cutoff or threshold score to declare a patient positive, or one who has the disease.

One issue facing the practitioner is the choice of a threshold or cutoff value in order to declare that the patient has disease. For example, when undergoing a fasting blood glucose test, how high does the value of blood glucose have to be in order for the doctor to treat the condition as type 2 diabetes? Another example is testing for coronary artery disease, when the patient complains of chest pain and seeks help from a physician. The doctor might send the subject for an exercise stress test, which involves injecting the patient with a radioactive nucleotide that is designed to target the heart and emit radiation that is detected by a gamma camera (positron emission tomography [PET] or single photon emission computed tomography [SPECT]). At what point is the patient said to have heart disease? This is a case where the choice of a threshold is crucial. Of course, in this situation, the patient might be referred for further testing involving the gold standard, namely, a heart catheterization to examine the coronary arteries. Note, in practice the physician must choose a cutoff value in order to declare the patient positive for disease.

The idea of choosing an optimal threshold value is demonstrated with the PSA test for prostate cancer, where 12,000 men aged 50–65 were randomized into placebo and treatment groups. The PSA values are based on a study by Etzioni et al. [8] and reported by Pepe [1: 10], and the accuracy is based on the ROC area, which was determined by a Bayesian approach described in Chapter 4. In Chapter 9, there are two criteria for choosing the threshold values, namely: (1) the threshold is chosen to correspond to the point on the ROC curve that is closest to the point (0,1), and (2) the threshold is chosen based on cost considerations. For the PSA example and using the first criterion, the closest point to (0,1) has a false positive fraction of 0.20 and a corresponding true positive fraction of 0.707, which corresponds to a log PSA value of 1.96.

For the second approach, the threshold is chosen to minimize a cost function:

$$C = \text{TPF}p(C_{tp} - C_{fn}) + \text{FPF}(1-p)(C_{fp} - C_{tn}) + C_0 + pC_{fn} + (1-p)C_{tn},$$

where p is the disease incidence, FPF and TPF are the false positive and true positive fractions, and c_{tp}, c_{fn}, c_{fp}, and c_{tn} are the costs of a true positive, a false negative, a false positive, and a true negative, respectively. The Bayesian approach to estimating the optimal threshold was derived by Somoza and Mossman [9] and is implemented for the PSA example with BUGS CODE 9.1, resulting in an FPF of 0.16 and a TPF of 0.709. Note, this is very close to the point (0.20,0.70) on the ROC curve selected by the first criterion. The second criterion, based on cost considerations, is much more difficult to apply because one must know the various costs. A second example using the biomarker CB-KK to predict head trauma complications illustrates the idea of choosing the optimal threshold value.

Bayesian decision curves, an idea introduced by Vickers and Elkins [10] to evaluate the clinical benefit of a medical test, are the second topic of Chapter 9. They considered PSA to diagnose prostate cancer, where three scenarios are possible: (1) the decision to biopsy a patient depends on the value of PSA, (2) all patients have a biopsy, and (3) no patient has a biopsy.

For each scenario, a decision curve is determined, which is a plot of the clinical benefit for a range of threshold probabilities, that is, for each threshold probability, a clinical benefit is computed. For example, for scenario (1), the clinical benefit is defined as

$$p_{11} - p_{10}(p_t/(1 - p_t)),$$

where p_{11} is the probability of a true positive, p_{10} is the probability of a false positive, and p_t is the threshold probability. A person opts for a biopsy if their probability of disease exceeds the threshold probability. A threshold probability is chosen from some range, and logistic regression is performed to estimate the probability of disease. For a threshold value, if the probability of disease is greater than the threshold value, the patient is declared positive for disease. After all of this, a 2×2 table is constructed, from which values for p_{11} and p_{10} are determined, thus determining the decision curve above, and a similar procedure is available for the second scenario. Decision curves are demonstrated for the Etzioni et al. [8] prostate cancer example and implemented with BUGS CODE 9.2, using 45,000 observations for the simulation, and the decision curve for scenario (1) dominates that for scenario (2). The Bayesian approach is shown to be comparable to maximum likelihood and a second example with the head trauma data illustrates the concept of decision curves.

Test accuracy in clinical trials is the third topic of Chapter 9. Once the patient is diagnosed with disease and therapy is initiated, the accuracy of various tests to monitor disease progress is crucial. The example considered in this chapter is the case of a Phase II clinical trial for cancer, where the progress of a patient is monitored by measuring the size of the tumor. At baseline, before treatment begins, various imaging devices are used to measure the size of the tumor, such as CT or MRI or both. At various times during treatment the size is measured, and at the termination of the trial, a final determination of the size is made. Now, a team of radiologists must decide for each patient the progress of the disease as measured by tumor size and other characteristics. In many cases, the effect of treatment is categorized as: (a) a complete response, (b) a partial response, (c) no change, or (d) disease progression. Each category is defined in terms of the change in the size of the tumor from the initiation of treatment to the termination of treatment. Such assessments are made for each patient and the total accumulated evidence in turn determines the overall success or failure of the trial. Of course, at issue is the accuracy of the imaging device (say CT) and the agreement between the team of radiologists responsible for declaring the success or failure of the trial. The chapter concludes with an example of a Phase II trial, which is implemented with some special software written at MDACC.

3.4.4   Chapter 10: Accuracy of combined tests

Chapter 10 introduces the reader to the methodology of measuring the accuracy of several medical tests that are administered to the patient. Our main focus is on measuring the accuracy of a combination of two or more tests. For example, to diagnose type 2 diabetes, the patient is given a fasting blood glucose test, which is followed by an oral glucose tolerance test. What is the accuracy (TPF and FPF) of this combination of two tests? Or, in order to diagnose coronary artery disease, the subject's history of chest pain is followed by an exercise stress test. Yet another example is for the diagnosis of prostate cancer, where a digital rectal examination is followed by measuring PSA. The reader is referred to Johnson, Sandmire, and Klein [11] for a description of additional examples of multiple tests to diagnose disease, including those for heart disease, diabetes, lung cancer, breast cancer, etc.

Chapter 10 begins with a discussion of measuring the accuracy of two binary tests and the basic question is how is the accuracy defined when two tests are combined to diagnose disease. There have been several ways to define accuracy where the main concern is when to declare a patient as positive, based on the results of two tests. Pepe [1: 268] presents two approaches: (1) believe the positive rule (BP), and (2) believe the negative rule (BN), where for the former, a patient is declared "positive" if at least one of the tests is positive, and for the latter rule, a patient is declared positive if both tests are "positive." Chapter 10 continues by presenting the Bayesian theory for implementing the BP and BN rules, which lays the foundation for BUGS CODE 10.1, which is executed in order to analyze an example of two tests for diagnosing heart disease, namely, the exercise stress test and the patient's history of chest pain. The analysis is executed with 55,000 observations, with a burn in of 5,000 and a refresh of 100, and the posterior means of the FPF for the BN and BP rules are 0.1568 and 0.6592, respectively, whereas, the TPF for the BN and BP rules are 0.7662 and 0.9747, respectively. Thus, the FPF is greater for the BP rule compared to the BN rule, and the TPF is greater for the BP rule. Note that the BP and BN rules can put one in the predicament of choosing a way to measure test accuracy. An example that combines CT and MRI to detect coronary stenosis further demonstrates the BN and BP rules for measuring the accuracy of a combined test.

Chapter 10 continues with a generalization of the accuracy for the combination of two binary tests to include several readers, and the theory is illustrated with a continuation of the previous example of CT and MRI for coronary stenosis, but now two readers are interpreting the test scores. BUGS CODE 10.2 is executed and the TPF and FPF for the BN and BP rules for the two readers are computed. With more readers, the problems of summarizing the accuracy become more complicated. How does one combine the test accuracies for the two readers based on the BN and BP rules?

An interesting situation is encountered when two binary tests where verification bias is present are considered. The problem of verification bias is studied in Chapter 8, but in Chapter 10 it is being examined in the context of combining two binary tests. Recall that verification bias is present when some of the patients being tested are not subject to the gold standard and recall the example of screening for Alzheimer's disease. What is the test accuracy of the combined test for the two tests (readers) when verification bias is present? The analysis is executed with BUGS CODE 10.3 with 45,000 observations and determines the posterior mean of the TPF for the BN and BP rules as 0.75 and 0.99, respectively. Note for this case, the combined test is the combined reader's interpretations. On the other hand, the posterior mean of the FPF for the BN and BP rules are 0.13 and 0.36, respectively. Which rule do you prefer for reporting the accuracy of the two readers combined?

A change of emphasis from binary to ordinal and continuous test scores brings us to some "new" ideas for measuring the accuracy by combining two tests. For ordinal and continuous scores, the area under the ROC curve measures the intrinsic accuracy of a medical test, but how should the area be computed when two tests are combined? The ROC curve for the risk score is the foundation for measuring the accuracy of the combined test, but in turn, the risk score is a monotone increasing function of the likelihood ratio, which is the optimal way to measure the accuracy of the combined test.

The optimality of the risk function is a consequence of the Neyman-Pearson lemma, which is a familiar result from classical statistics for testing hypotheses.

The chapter continues with the likelihood ratio, which is defined, and the optimality of the ROC curve of the likelihood ratio is demonstrated by referring to the Neyman-Pearson lemma. Then, the risk function will be defined and shown to be a monotone increasing function of the likelihood ratio, thus, the ROC curve of the risk function is the same as the ROC curve of the likelihood ratio. Pepe's [1: 269–274] development of the subject is followed closely but is given a Bayesian emphasis, and the end result will be that the optimal way to measure the accuracy of the combined test is to estimate the area under the ROC curve of the risk function. Determining the risk function is equivalent to performing a logistic regression using the test scores of the two tests as predictors, then the ROC curve of the predicted probabilities (from the logistic regression) is computed, from which the area is then estimated. Such an area is the accuracy of the combined test, and the methodology is illustrated with various examples using ordinal test scores. Note that the risk score is defined as

$$RS(Y) = P[D = 1 \mid Y]$$

where $Y = (Y_1, Y_2, \ldots, Y_p)$ is the vector of scores of p ordinal tests.

It seems that a logistic regression can determine the risk score for each patient, based on the combined scores of Y.

The first example is from an imaging trial using MRI and CT to detect lung cancer, where one radiologist uses a five-point confidence score, and the ROC curve of the risk function of the combined test is computed and compared to the ROC curve of the individual tests. There are 261 patients with the disease and 674 without, and the analysis is executed with BUGS CODE 10.5 using 55,000 observations for the simulation. The analysis consists of a logistic regression with the scores of the two tests as independent variables and the dependent variable is the risk score (the probability that a subject has the disease). After the risk scores are computed, the ROC area is estimated with a posterior mean of 0.72(0.019) compared to ROC areas of 0.6836 for CT and 0.6886 for MRI, thus, the use of the risk score increases the accuracy compared to the component tests. Of course, it needs to be emphasized that the logistic regression should be a good fit to the data, otherwise the ROC area of the risk scores can be misleading. Note that use of the risk function is a statistical procedure that requires close collaboration between the clinician and biostatistician. In essence, the two scores of the component tests are combined to give one number, namely, the estimated probability of disease!

The next example using the risk score for accuracy of the combined test is the use of two versions of MRI for the localization of prostate cancer. Localization means finding where the lesion is within the prostate gland. The gland is partitioned into 14 segments, and a score from 1 to 5 is assigned to each by both devices. After surgery, histopathology of the prostate gland serves as a gold standard. This study is taken from Coakley et al. [12], where 46 patients are enrolled, giving a total of 644 segments, and body-array MRI and endorectal MRI image the same segments of the prostate. A Bayesian logistic regression is performed with the two tests scores for the two images, thus a risk score is assigned to each patient. BUGS CODE 10.4 performs the logistic regression, providing the input for BUGS CODE 10.5, which estimates the ROC area of the combined test. With 45,000 observations generated for the simulation, the posterior mean of the ROC area for the body-array and endorectal MRI are 0.55(0.0257) and 0.7519(0.0216), respectively, which compares to 0.80(0.0223) for the posterior mean of the combined test. Wow, what an improvement over the accuracies of component tests!!

Chapter 10 concludes with an example that generalizes the analysis done for two versions (body-array and endorectal) of MRI by including age as a patient covariate; consequently, the ROC area of the risk score improves from 0.75 to 0.889. Lastly, the accuracy of a combined test with two ordinal components is considered and demonstrated with a staging example of melanoma done by a dermatologist and a surgeon, when verification bias is present. I believe that this contribution has not appeared before. Note that the risk score is also used for continuous tests and the last section of the chapter computes the ROC area of the risk score for two biomarkers for pancreatic cancer. The 17 exercises expand on the subject matter explained in the chapter and the student should attempt to solve all of them. Also, remember that the code and data appear in the book and can be copied from the author's blog http://medtestacc.blogspot.com.

3.4.5   Chapter 11: Meta-analysis for test accuracy

The last chapter of the book presents the fundamentals of the Bayesian approach to executing a meta-analysis for medical test accuracy. Most of the chapters deal with estimating the summary ROC curve, which summarizes the ROC curve for a series of studies that assume a common ROC curve, but where the studies may differ in the reported TPF and FPF, due to different threshold values. The so-called SROC curve is based on a regression of B values on S values, where the B and S values are linear transforms of the logits of the TPF and FPF. The results of the regression allow one to estimate the accuracy of the summary receiver operating characteristic (SROC) curve using the Q parameter, which varies between 0 and 1 in much the same way as the area under the ROC curve. Meta-analysis is introduced with one binary test, then two binary tests are considered for the Bayesian analysis, and finally one and two binary tests with study covariates are analyzed. Test scores are either ordinal or continuous, where a threshold value declares a patient as either positive or negative for the disease, and, in this way, binary tests are induced by the threshold value, which in turn, produces true and false positive rates. Many examples illustrate the Bayesian approach to meta-analysis and include studies for coronary artery disease, inflammatory bowel disease, osteomyelitis, breast cancer, and recurrent colorectal cancer. A non-Bayesian approach to meta-analysis for test accuracy is portrayed by Zhou, McClish, and Obuchowski [2] and this presentation is similar in many respects, except that the analysis is via Bayesian inference.

References

[1] Pepe, M.S. *The Statistical Evaluation of Medical Tests for Classification and Prediction.* Oxford University Press, Oxford, UK, 2003.

[2] Zhou, H.H., McClish, D.K., and Obuchowski, N.A. *Statistical Methods for Diagnostic Medicine.* John Wiley, New York, 2002.

[3] Congdon, P. *Applied Bayesian Modeling.* John Wiley, Chichester, 2004.

[4] Holtbrugge, W. and Schumacher, M.A. A comparison of regression models for the analysis of categorical data. *Applied Statistics*, 40:249, 1991.

[5] O'Malley, J.A., Zou, J.H., Fielding, J.R., and Tampany, C.M.C. Bayesian regression methodology for estimating a receiver operating characteristic curve with two radiologic applications: Prostate biopsy and spiral CT of urethral stones. *Academic Radiology*, 8:713, 2001.

[6] Von Eye, A. and Mun, E.Y. *Analyzing Rater Agreement, Manifest Variable Methods.* Lawrence Erlbaum, Mahwah, NJ, 2005.

[7] Kraemer, H.C. Ramifications of a population model for Kappa as a coefficient for reliability. *Psychometrika*, 44:461, 1979.

[8] Etzioni, R., Pepe, M.S., Longton, G., Hu, C., and Goodman, G. Incorporating the time dimension in receiver operating characteristic curves: A case study of prostate cancer. *Medical Decision Making*, 19:242, 1991.

[9] Somoza, E. and Mossman, D. Biological markers and psychiatric diagnosis: Risk benefit balancing using ROC analysis. *Biological Psychology*, 29:811, 1991.

[10] Vickers, A.J. and Elkins, B. Decision curve analysis: A novel method for evaluating prediction modes. *Medical Decision Making*, 26:565, 2006.

[11] Johnson, D., Sandmire, D., and Klein, D. *Medical Tests That Can Save Your Life: 21 Tests Your Doctor Won't Order... Unless You Know to Ask.* Rodale, Emmaus, PA, 2004.

[12] Coakley, F.V., Teh, H.S., Quayyum, A., Swanson, M.G., Lu, Y., Roach, M., Pickett, B., Shinohara, K., Vigneron, D.B., and Kurhanewicz, J. Endorectal MR imaging and MR spectroscopic imaging for locally recurrent prostate cancer after external beam radiation therapy: Preliminary experience. *Radiology*, 233:441, 2004.

Chapter 4

Fundamentals of Diagnostic Accuracy

4.1 Introduction

This chapter describes the methodology for making inferences with respect to the basic measures of test accuracy and begins with a section on the design of such studies. First, the elements of a good design for the accuracy of a medical test will be explained in the context of a protocol submission of a trial to assess the accuracy of a diagnostic test in a clinical situation. This will include describing in some detail the components of a clinical protocol, such as the objectives of the study, the background, patient and reader selection, the study plan, the number of patients, the statistical design and analysis, and finally the role that references (prior experimental studies) play in the protocol.

After describing the components of designing a diagnostic study, this chapter introduces Bayesian methods for the analysis of diagnostic test accuracy, including the estimation of sensitivity, specificity, the positive predictive value (PPV) and the negative predictive value (NPV), the positive diagnostic likelihood ratio (PDLR) and the negative diagnostic likelihood ratio (NDLR), and receiver operating characteristic (ROC) curves. The basic measures are the true positive fraction (TPF) and the false positive fraction (FPF), where such measures are appropriate for binary test scores, as are the PPV and NPV, where the latter are of paramount importance to the patient. A Bayesian analysis determines the posterior distribution of the accuracy parameter and its characteristics, such as the posterior mean, median, standard deviation, credible intervals, and associated plots of the density.

The analysis of test accuracy data is introduced first with binary and ordinal diagnostic test data, and then the Bayesian analysis is repeated with quantitative scores. When the test scores are ordinal or quantitative, the area under the ROC curve is the accepted way to measure the accuracy of the test. Mammography is a good example, where the radiologist assigns a score from 1 to 5, which represents the likelihood of a lesion appearing in the image, and then the ROC area is defined as the probability that a diseased patient has a score greater than that of a non-diseased patient, and the area is between 0 and 1. An area of 1 implies that the test is discriminating perfectly between diseased and non-diseased patients. Remember that the above discussion of test accuracy always assumes a reference standard or gold standard measure whether the disease is present or not.

An interesting generalization of ordinal test scores is when the test scores are clustered, such as in the study of coronary artery disease (CAD), where the coronary arteries are grouped into segments, thus the left coronary artery is broken into several segments, as is the left anterior descending artery. To each segment is attached an ordinal score, assigned by the reader, which indicates the degree of stenosis of that segment. When the information is clustered, the correlation is present between adjacent segments, which must be modeled for the analysis. This specialized topic includes localization and detection of disease by diagnostic tests, where the image is partitioned into regions of interest (ROI). This is interesting statistically because of the correlation between regions of the same image. The analysis of correlated data in such a scenario has been approached by Obuchowski, Lieber, and Powell [1], and based on their ideas, a Bayesian technique to estimate the ROC area is developed.

For continuous or quantitative scores, the area under the ROC curve measures the accuracy of the test and many interesting examples are presented, including a study of CAD using computed tomography (CT) to measure the extent of plaque build up in the coronary arteries. How accurate is CT in measuring the extent of CAD? This will be answered by examining the CT score for some 4000 patients! Also of interest in connection with the fundamentals of test accuracy is the comparison of accuracies between two different medical tests that are measuring the same thing on the same patients. For example, several tests measure the blood glucose values of subjects screened for type 2 diabetes, and comparing the accuracy between them is of interest to researchers. Another example of comparing the accuracy of medical tests is given by some recent experimental studies comparing CT screening with x-ray for lung cancer.

Some topics of a more specialized nature are also discussed in this chapter, such as choosing the optimal point on the ROC curve (equivalently, choosing the optimal cut point for a positive binary test) based on cost considerations, and comparing two tests with quantitative scores that are correlated.

A new Bayesian method for measuring the accuracy of ordinal tests is developed and extended to clustered data and illustrated with a mammography example, where the breast is divided into ROIs and a score assigned to each region. This chapter lays the foundation for later chapters and should be mastered by the student before proceeding to Chapter 5.

4.2   Study Design

The role that accuracy plays in a typical clinical study is introduced in this section, where the elements of good study design of trials of medical devices are explained in the context of the submission of a protocol at the MD Anderson Cancer Center (MDACC). All protocols are first reviewed by the department and the protocols are essentially of two types: (a) those that originate locally at the institution and (b) those submitted by pharmaceutical or medical device companies. For the latter, the protocol is critiqued and reviewed by a statistician in the department. For those studies originating within the institution, a biostatistician would assist the investigator with the design of the study, but the protocol would be reviewed by a different person and presented to the department for approval. The protocol is reviewed by the department and, if necessary, revised according to the suggestions recommended by departmental consensus. The principal investigator then revises the protocol, often with the assistance of the statistician.

4.2.1 Protocol

There are many types of protocols submitted, thus only those dealing mainly with the accuracy of diagnostic tests are considered. Of course, medical tests are usually a part of all clinical trials, and these will be described in later chapters. Briefly, the protocol consists of the following components: (1) objectives, (2) background, (3) patient and reader selection, (4) study plan, (5) number of patients, (6) statistical design and analysis, and (7) references.

4.2.2 Objectives

The study's primary and secondary aims are given in the first section of the protocol. The study design is illustrated by a protocol with two nuclear medicine procedures: one using an iodine radionuclide with single photon emission tomography (SPECT; I-123 MIBG SPECT) and the other with thallium (Ti-201 SPECT) that will be used to measure the amount of damage (e.g., scarring of the cardiac wall and nerve damage) to the heart caused by radiotherapy to the chest. The main objective is to determine the association between the delivered dose to the target lesion and the nerve damage caused by radiotherapy to the chest. It is an important study because little work has been done in this area. Since the study involves two medical tests (two SPECT images), their accuracy plays a crucial role in achieving the objectives of the study.

4.2.3 Background

The relevant recent literature on previous studies should be cited in the background section of the protocol. This is a very important component because it gives the rationale for doing the study and it often provides information that is essential for sample size estimation. The background information is often a source of preliminary information, which will be employed as prior information for the Bayesian analysis. In the nuclear medicine example, there is a lot of information on cardiac morbidity and mortality due to radiotherapy, but very little on diagnostic imaging procedures that assess the amount of innervation damage. There are only two references citing studies using I-123 MIBG SPECT to assess nerve damage to the heart. Of course, the background should also contain information about the accuracy (TPF, FPF, and ROC areas) of the two SPECT tests involved in the study!

4.2.4   Patient and reader selection

The patient and reader selection component provides the inclusion (who can be admitted) and the ineligibility (who cannot be admitted) criteria. Generally speaking, those to be included are diseased but not too diseased to be admitted, while those that are too sick will be excluded. In diagnostic studies when several readers are involved in interpreting the diagnostic information, the relationship between how the patients are selected and how the readers are selected must be described. For example, in a traditional selection with two imaging modalities, the same readers will be used to interpret both images and the same patients will be imaged by the two modalities. There are many variations to this scenario, including unpaired patient unpaired readers, where there are two sets of different patients, one for image A and the other for image B, and there are two distinct sets of readers, one for image A and one for image B. Also, there are paired patient and unpaired reader selection plans, etc. For additional selection plans, see Chapter 3 of Zhou, McClish, and Obuchowski [2].

If the readers are to interpret two images, is the order randomized to eliminate order bias, and how is a final determination of image interpretation to be handled? How the patients are selected will also affect the sample size estimation, and it could affect any future analysis. For example, the analysis for comparing image accuracy in a paired patient design would be different from that for an unpaired patient selection. Is this a randomized trial, where one set of patients is selected at random from a diseased population and the other set from a non-diseased population?

Patient and reader selection designs often depend on the type of trial. When developing a new imaging modality, the test should pass three phases: I, II, and III. The different phases are for different objectives of test accuracy and are as follows. The relation (i.e., paired or unpaired) between patients and diagnostic modalities and the relation between readers and modalities should be described in the protocol. Phase I, II, and III trials for imaging devices were described in Chapter 3, and one is referred to Bogaert et al. [3] for an example of a Phase I developmental trial involving magnetic resonance imaging (MRI) angiography, Theate et al. [4] for an example of a Phase II trial, and finally to Beam, Lyde, and Sullivan [5], who investigated the interpretation of screening mammograms as a Phase III trial.

Note that it is important to know the inter observer variability in these trials, because the accuracy of the modality depends not only on the device, but also on the interpretation of the image via the various readers. In Chapter 8, Pepe [6] gives more detail on the description of developmental trials, and Zhou,

McClish, and Obuchowski [2] provide the analysis for studies with multiple readers and multiple modalities for trials of device development.

For the nuclear medicine trial, which is used to motivate the steps involved in the design of a protocol, the patients are paired in that all are imaged by both procedures, however the two modalities will not be compared because they are measuring different things. The one with the iodine radionuclide is measuring nerve damage to the heart, while the thallium stress test is measuring cardiac perfusion variables, like wall scarring and left ventricular ejection fraction, which are other indicators of cardiac damage.

4.2.5   Study plan

For this section of the protocol, details of how the diagnostic tests are to be implemented are spelled out.

Returning to the trial being designed, the study plan is as follows. The sympathetic nervous system of the heart will be imaged using I-123 MIBG, while at the same time performing an exercise stress test (EST) using Thallium-201 (Tl-201). The patients will be imaged prior to initiation of radiation therapy (RT) and at 6–12 months after completion of RT. Stress myocardial perfusion imaging is a standard of care test of baseline evaluation of myocardial perfusion and possible radiation-induced CAD after RT for tumors close to the heart. Currently, stress myocardial perfusion is performed using the dual isotope method where the patient is injected with Tl-201 for the resting part of the study and immediately after rest, the patient is injected with technetium-99m (Tc-99m) tetrofosmin at peak stress and imaging is repeated for the stress part of the study.

Next, the plan to image the patients is described. This would include the details of administering the first radiopharmaceutical I-123 MIBG, including the dose injected by IV and the details of how the resulting radioactivity is to be imaged by the gamma camera, in this case SPECT. This would be followed by a similar description of administering the thallium EST for cardiac perfusion. The patient is imaged with both nuclear medicine procedures before and after radiotherapy. There are two types of cardiac damage variables, those for nerve damage and those measuring scarring of the heart wall and left ventricular ejection fraction, a measure of cardiac output. If radiotherapy is damaging the heart, one would expect to observe it by comparing the post therapy measurements of heart damage to the corresponding pre therapy values.

Lastly, the image processing details are given. For the cardiac damage study, standard filtered back projection techniques to obtain SPECT images will be employed for both imaging modalities. The image processing is an important part of the accuracy of the nuclear medicine procedures. In order to obtain wall motion images and ejection fraction values, gated motion images are required, thus cardiac motion will not affect the image quality. In order to detect nerve damage, the uptake of norepinephrine can be estimated with the

I-123 MIBG procedure. This illustrates the ability of a nuclear procedure to measure metabolic processes.

4.2.6 Number of patients

The total number of patients and the monthly accrual rate is described. For multi-institutional trials, the rates for each institution are provided. The total sample size is justified in the power analysis of the statistics section. A maximum of 40 patients accrued at 2–3 per month should be sufficient for the cardiac damage protocol.

4.2.7 Statistical design and analysis

The statistical section should provide a detailed power analysis outlining the justification for the sample size. The power analysis should show how the results of previous related studies are used to predict the results of the planned study. It should also provide a brief description of the design of the study, including how the readers and patients interface (i.e., paired with the modalities) with the diagnostic tests. The phase (I, II, or III) of the study should be identified, as well as an outline of how the study results will be analyzed.

For the planning of the Phase I nuclear medicine protocol, the power analysis is given as follows. The sample size will be based on the expected association between nerve damage measured by uptake of norepinephrine (as determined by I-123 MIBG) and the dose of radiotherapy administered to the target lesion measured in Gray (Gy) units. Note the role that the accuracy of the I-123 MIBG image plays in the determination of the sample size!

If radiotherapy is damaging cardiac innervation, one would expect the mean uptake ratio to be 2.5 before radiotherapy with a range from 1.5 to 3.5, while after therapy, one would expect the average uptake ratio of norepinephrine to be 1.5 with a range from 0.5 to 2.5. Assuming a correlation of 0.5 between pre and post therapy for the uptake values of norepinephrine, the standard deviation of the difference is 0.5.

The independent variable for the association is the radiotherapy delivered dose, which will have a range of 40–60 Gy, with an average dose of 50 Gy and a standard deviation of 5 Gy. The dose is expected to have an effect on the cardiac nerve damage as follows. When the delivered dose is 40 Gy, it is reasonable to expect an average uptake in the difference to be 0, while if the delivered dose is 60 Gy, it is reasonable to expect the difference in the post minus pre uptake values to average 1. Assuming a linear regression between the difference in the uptake values as the dependent variable and the administered dose as the independent variable, the regression line will be approximately

$$Y = 0.05X - 2,$$

where X is the delivered dose in grays, and Y is the difference in the post minus pre RT uptake values. The null hypothesis is that the slope of the regression is zero vs. the alternative that it is positive. Assuming under the alternative that the slope is 0.05, the power of the test with $\alpha = 0.05$ is 0.68, 0.86, and 0.94 corresponding to sample sizes 20, 30, and 40, respectively.

It appears reasonable that 30 patients will show a strong association between damage to the nerves of the heart and the delivered dose to the target lesion.

The power analysis describes what to expect with regard to the nerve damage to the heart in terms of the uptake ratios of norepinephrine, measured before and after radiotherapy. The hypothetical association between the nerve damage and the dose delivered to the target lesion is given by the above regression equation. The power was computed with a standard software package, and gives 30 patients as a reasonable number to detect the desired association. This is somewhat hypothetical in a sense, but is based on previous studies of heart damage caused by radiotherapy to lesions close to the heart. The power analysis could just as well be done from a Bayesian perspective; see Broemeling [7] for the Bayesian analysis of a linear regression model.

Note that the power analysis is based on just two of the many endpoints that could have been used. There are many ways to measure cardiac nerve damage and many ways to measure other damage to the heart, such as left ventricle ejection fraction and scarring to the heart wall. The power analysis should be brief, but at the same time informative, so that other statisticians can review the work.

4.2.8 References

This as a very important part of the protocol, because the study is only fit to be run if previous studies show a need. Also, for the statistician, the results from previous studies are invaluable for the power analysis and for a history of the accuracy of the medical tests involved in the trial. It is important to remember that previous studies addressing the accuracy of the two SPECT studies may not be directly applicable to the trial at hand, therefore the actual accuracy of the images may not be known to some extent at least.

4.3 Bayesian Methods for Test Accuracy: Binary and Ordinal Data

4.3.1 Introduction

This section will introduce Bayesian techniques to estimate and test hypotheses about the basic measures of test accuracy. The measures of test accuracy are: (a) classification probabilities, (b) predictive measures, and

(c) diagnostic likelihood ratios (DLRs). The classification probabilities are the FPF and TPF, while there are two predictive values, the PPV and the NPV. Lastly, there are two DLRs—PDLR and NDLR. These measures will be defined in the next section in the context of a cohort study. Thus, there is a random sample of size n selected from the target population and a gold standard, thus each patient is classified into the four categories in Table 4.1. The n_{ij} are the number of subjects with test score $i = 0$ or 1 and disease status $j = 0$ or 1, while θ_{ij} is the corresponding probability.

4.3.2   Classification probabilities

The basic measures of test accuracy are the TPF (sensitivity) and the FPF $(1 - \text{specificity})$, where

$$\text{TPF}(\theta) = \theta_{11}/(\theta_{11} + \theta_{01}) = P(X = 1 \mid D = 1), \qquad (4.1)$$

and

$$\text{FPF}(\theta) = \theta_{10}/(\theta_{00} + \theta_{10}) = P(X = 1 \mid D = 0). \qquad (4.2)$$

It is important to know that the TPF and FPF are unknown parameters and are functions of θ. The Bayesian analysis determines the posterior distribution of these quantities, from which the parameters are estimated and certain tests of hypotheses performed. Assume that the prior information is based on a previous study, with the results given in Table 4.2, where m subjects have been classified in the same way as those in Table 4.1.

The density based on prior information is

$$\xi(\theta) \propto \theta_{00}^{m_{00}} \theta_{01}^{m_{01}} \theta_{10}^{m_{10}} \theta_{11}^{m_{11}}, \qquad (4.3)$$

TABLE 4.1:   Classification table.

	Disease	
Test	$D = 0$	$D = 1$
$X = 0$	(n_{00}, θ_{00})	(n_{01}, θ_{01})
$X = 1$	(n_{10}, θ_{10})	(n_{11}, θ_{11})

TABLE 4.2:   Classification table of prior information.

	Disease	
Test	$D = 0$	$D = 1$
$X = 0$	(m_{00}, θ_{00})	(m_{01}, θ_{01})
$X = 1$	(m_{10}, θ_{10})	(m_{11}, θ_{11})

thus, the likelihood function for $\theta = (\theta_{00}, \theta_{01}, \theta_{10}, \theta_{11})$ is

$$L(\theta/n) \propto \theta_{00}^{n_{00}} \theta_{01}^{n_{01}} \theta_{10}^{n_{10}} \theta_{11}^{n_{11}}, \tag{4.4}$$

and the posterior distribution is Dirichlet,

$$\theta/(n, m) \sim \text{Dir}(n_{00} + m_{00} + 1, n_{01} + m_{01} + 1, n_{10} + m_{10} + 1, n_{11} + m_{11} + 1).$$

Note, if there is no prior information, m_{ij} is zero, and one in effect is assuming a uniform prior distribution for θ.

Markov Chain Monte Carlo (MCMC) sampling from the Dirichlet distribution, using WinBUGS, will determine the posterior distribution of these classification probabilities. As an example, consider the example examined by Pepe [6] and based on the study by Wiener et al. [8]. This is a cohort study of 1465 subjects, where each is classified as to disease status (CAD via an angiogram) and a diagnostic test (the EST), which is a nuclear medicine procedure; data can be found at http://labs.fhcrc.org/pepe/book.

The analysis is based on the following code.

BUGS CODE 4.1

```
# Measures of accuracy
# Binary Scores
Model;
{
# Dirichlet distribution for cell probabilities
g00~dgamma(a00,2)
g01~dgamma(a01,2)
g10~dgamma(a10,2)
g11~dgamma(a11,2)
h<-g00+g01+g10+g11
# the theta have a Dirichlet distribution
theta00<-g00/h
theta01<-g01/h
theta10<-g10/h
theta11<-g11/h
# the basic test accuracies are below
tpf<-theta11/(theta11+theta01)
se<-tpf
sp<-1-fpf
fpf<-theta10/(theta10+theta00)
tnf<-theta00/(theta00+theta10)
fnf<-theta01/(theta01+theta11)
ppv<-theta11/(theta10+theta11)
npv<-theta00/(theta00+theta01)
pdlr<-tpf/fpf
```

TABLE 4.3:   Exercise stress
test and heart disease.

EST	CAD $D = 0$	$D = 1$
$X = 0$	327	208
$X = 1$	115	818

```
ndlr<-fnf/tnf
}
# Exercise Stress Test Pepe [6]
# Uniform Prior (add one to each cell of the table frequencies!)
list(a00=328,a01=209,a10=116,a11=819)
# chest pain history
# Uniform prior
list(a00=198,a01=55,a10=246,a11=970)
# initial values
list(g00=1,g01=1,g10=1,g11=1)
```

The notes of interest for the code are headed by #. There are three list statements: the first gives the information necessary to generate a Dirichlet distribution for the four cell probabilities. The entries are the cell frequencies of Table 4.3 plus 1! In this way, a uniform prior is assumed. The third list statement gives the initial values for the MCMC procedure, and the second list statement will be used later.

A uniform prior is assumed, resulting in a posterior distribution that is Dirichlet (328, 209, 116, 819). The analysis is executed with 55,000 observations generated from the joint posterior distribution of the cell probabilities, using 5,000 as a burn in and 100 as a refresh, resulting in Table 4.4 for the posterior analysis for the accuracy of the EST.

The sensitivity or TPF is estimated as 0.7967 with an associated posterior standard deviation of 0.0125 and (0.7716,0.8208) as a 95% credible interval. Note that the analysis also includes $5.84*10^{-5}$ as the MCMC error for estimating the TPF, which implies that the estimate of 0.7967 is within $5.84*10^{-5}$ units of the "true" posterior TPF. The WinBUGS output also includes plots of the marginal posterior distribution of the parameters and Figure 4.1 portrays that for the sensitivity of the EST. With an estimated

TABLE 4.4:   Posterior analysis for exercise stress test.

Parameter	Mean	sd	Error	Lower 2 1/2	Median	Upper 2 1/2
TPF	0.7967	0.0125	$5.84*10^{-5}$	0.7716	0.7968	0.8208
FPF	0.2612	0.0208	$9.22*10^{-5}$	0.2215	0.2608	0.3033

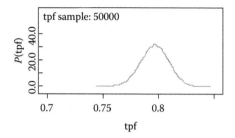

FIGURE 4.1: Posterior density of the true positive fraction exercise stress test.

false positive fraction (FPF) of 0.2612 and an estimated true positive fraction (TPF) of 0.79, the EST has good to fair accuracy, but other measures of accuracy should be considered.

4.3.3 Predictive values

The second set of measures for test accuracy is the PPV and the NPV, defined as follows:

$$\text{PPV}(\theta) = \theta_{11}/(\theta_{01} + \theta_{11}) = P(D = 1 \mid X = 1), \tag{4.5}$$

and

$$\text{NPV}(\theta) = \theta_{00}/(\theta_{00} + \theta_{01}) = P(D = 0 \mid X = 0). \tag{4.6}$$

Since these two quantities depend on disease incidence, it is important that the patients are selected at random from the target population, so that when estimating the predictive values, the disease incidence is estimated without bias. Returning to the example, the posterior distributions of the predictive values are provided in Table 4.5. They answer the question of primary interest to the patient: do I have disease? This, to some extent, is answered by the following posterior analysis.

The distribution of the PPV appears to be symmetric with a mean of 0.8759, which implies that chances of heart disease among those patients that test positive is 0.87, which gives me some confidence in the EST to detect disease. On the other hand, for those that test negative, the chances of not having coronary heart disease is only 0.61. My confidence is somewhat lowered in the ability of the test to discriminate between diseased and non-diseased

TABLE 4.5: Distribution of predictive values.

Parameter	Mean	sd	Error	Lower 2 1/2	Median	Upper 2 1/2
PPV	0.8759	0.0108	<0.0001	0.8538	0.8762	0.8961
NPV	0.6109	0.0211	<0.0001	0.5693	0.611	0.6517

patients. If the test is negative, I am not sure if I have the disease or not! Note that a perfect test occurs when $PPV = NPV = 1$. I did not give the exact figure for the MCMC error, only to note that it is quite small for these two measures of test accuracy. In executing the analysis, one should vary the MCMC sample size to see its effect on the posterior distribution and the error of estimation.

4.3.4   Diagnostic likelihood ratios

The DLRs are a third group of test accuracy measures and are

$$
\begin{aligned}
\mathrm{PDLR}(\theta) &= P(X = 1 \mid D = 1)/P(X = 1 \mid D = 0) \\
&= [\theta_{11}/(\theta_{11} + \theta_{01})]/[\theta_{10}/(\theta_{10} + \theta_{00})] \\
&= \mathrm{TPF}(\theta)/\mathrm{FPF}(\theta) \tag{4.7}
\end{aligned}
$$

and

$$
\begin{aligned}
\mathrm{NDLR}(\theta) &= P(X = 0 \mid D = 1)/P(X = 0 \mid D = 0) \\
&= [\theta_{01}/(\theta_{11} + \theta_{01})]/[\theta_{00}/(\theta_{10} + \theta_{00})] \\
&= \mathrm{FNF}(\theta)/\mathrm{TNF}(\theta). \tag{4.8}
\end{aligned}
$$

With regard to the PDLR, the more accurate the diagnostic test becomes, the numerator (TPF) tends to become larger and the denominator (FPF) tends to become smaller. But for the NDLR, the opposite is true, the numerator (FNF) tends to become smaller and the denominator (TNF) tends to become larger. The range of both is $[0, \infty)$.

For the CASS dataset, the characteristics of the posterior distribution for the likelihood ratios are given in Table 4.6. Note that the estimated error for estimating the PDLR is 0.0011, which implies that the estimate of 3.07 is within 0.0011 of the "true" posterior PDLR and that the test is positive about three times more often among the diseased, compared to those without CAD. On the other hand, among those that have the disease, the test is negative much less often compared to those without the disease. Both measures indicate an accurate test. The larger the PDLR and the smaller the NDLR, the more accurate the test.

In summary, three types of measures of accuracy have been computed for the EST. For the sensitivity and specificity, I am somewhat confident that the test is informative, but with regard to the predictive values, the NPV did not give me high confidence in the test to measure accuracy. For additional

TABLE 4.6:   Distribution of diagnostic likelihood ratios.

Parameter	Mean	sd	Error	Lower 2 1/2	Median	Upper 2 1/2
PDLR	3.07	0.2526	0.0011	2.616	3.055	3.609
NDLR	0.2755	0.0187	<0.0001	0.2399	0.275	0.3135

TABLE 4.7:   Mammogram results.

Status	Normal (1)	Benign (2)	Probably benign (3)	Suspicious (4)	Malignant (5)	Total
Cancer	1	0	6	11	12	30
No cancer	9	2	11	8	0	30

Source: From Zhou, H.H., McClish, D.K., and Obuchowski, N.A. *Statistical Methods for Diagnostic Medicine*. 2002. Copyright Wiley-VCH Verlag GmbH & Co. KGaA. Reproduced with permission.

information about these basic measures of accuracy, Pepe [6: 20] provides a summary.

4.3.5   Receiver operating characteristic curve

Consider the results of mammograms given to 60 women, of which 30 had the disease. This is presented in Zhou, McClish, and Obuchowski [2: 21] (Table 4.7).

The radiologist assigns a score from 1 to 5 to each mammogram, where 1 indicates a normal lesion, 2 a benign lesion, 3 a lesion that is probably benign, 4 indicates suspicious, and 5 malignant. How would one estimate the accuracy of mammography from this information? When the test results are binary, the observed TPF and FPF are calculated, but here there are five possible results for each image. The scores could be converted to binary by designating 4 as the threshold, then scores 1–3 are negative and 4–5 are positive test results. Then estimate the TPF as tpf $= 23/30$ and the specificity $(1 - \text{FPF})$ as $(1 - \text{fpf}) = 21/30$. Another approach would be to use each test result as a threshold and calculate the tpf and fpf, which are depicted in Table 4.8.

Of the 30 diseased, 30 had a score of at least 1, while 23 had a score of at least 4. On the other hand, of the 30 without cancer, 30 had a score of at least 1, and 8 had a score of at least 4, etc. Figure 4.2 is a plot of the observed true and false positive values of Table 4.8. What does this graph tell us about the accuracy of mammography?

The area under the ROC gives the intrinsic accuracy of a diagnostic test and can be interpreted in several ways (see Zhou, McClish, and Obuchowski

TABLE 4.8:   tpf vs. fpf for mammography.

Status	Normal (1)	Benign (2)	Probably benign (3)	Suspicious (4)	Malignant (5)
tpf	30/30 = 1.00	30/30 = 1.00	29/30 = 0.966	23/30 = 0.766	12/30 = 0.400
fpf	30/30 = 1.00	21/30 = 0.700	19/30 = 0.633	8/30 = 0.266	0/30 = 0.000

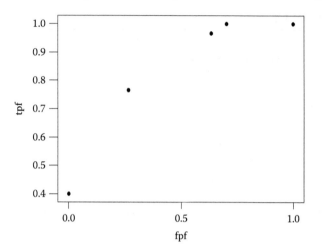

FIGURE 4.2: Empirical ROC graph for mammography.

[2: 28]): as the average sensitivity for all values of specificity, or the average specificity for all values of sensitivity, or as the probability that the diagnostic score of a diseased patient is more of an indication of disease than the score of a patient without the disease or condition. The problem is in determining the area under the curve (AUC). For the Figure 4.2, there are five points corresponding to the five threshold values. If the diagnostic score can be considered continuous (e.g., the coronary artery calcium (CAC) score), then the curve through the points becomes more discernible and the area easier to determine.

In the case of discrete data, the AUC as determined by a linear interpolation of the points on the graph (including (0,0) and (1,1)) have the following interpretation:

$$\text{AUC} = P(Y > X) + (1/2)P(Y = X). \tag{4.9}$$

See Pepe [6: 92], where it is assumed that one patient is selected at random from the population of diseased patients, with a diagnostic score of Y, while another patient, with a score of X, is selected from the population of non-diseased patients. Note that the AUC depends on the parameters of the model. Let us return to the mammography example and estimate the AUC via a Bayesian method.

For the mammography example, the area is defined as

$$\text{AUC}(\theta, \phi) = P(Y > X/\theta, \phi) + (1/2)P(Y = X/\theta, \phi), \tag{4.10}$$

where $Y(=1, 2, 3, 4, 5)$ is the diagnostic score for a person with breast cancer and $X(=1, 2, 3, 4, 5)$ for a person without. It can be shown

$$\text{AUC}(\theta, \phi) \sum_{i=2}^{i=5} \sum_{j=1}^{j=i-1} \theta_i \phi_j + (1/2) \sum_{i=1}^{i=5} \theta_i \phi_i. \tag{4.11}$$

It is assumed that Y and X are independent, given the parameters, and that $P(Y = i) = \theta_i$ and $P(X = j) = \phi_j, i, j = 1, 2, 3, 4, 5$. AUC is a parameter that depends on θ and ϕ. Their posterior distributions are $\theta/\text{data} \sim \text{Dir}(2, 1, 7, 12, 13)$ and independent of $\phi/\text{data} \sim \text{Dir}(10, 3, 12, 9, 1)$, assuming a uniform prior for the parameters (see Table 5.7).

Samples from the posterior distribution of the AUC are generated by sampling from the posterior distributions of θ and ϕ. This is accomplished with WinBUGS, where 55,000 observations are generated from the posterior distribution of all the parameters, with a burn in of 5,000 and a refresh of 100. The code for the operation, BUGS CODE 4.2, follows, and the notes indicated by # identify the important parts of the program. For example, the statements that follow the note "# generate Dirichlet distribution," generate the posterior distribution of the cell probabilities of Table 4.7. The first list statement is the information used to generate the gamma variables that generate the Dirichlet distribution of the cell probabilities. A one is added to each cell frequency of Table 4.7, which induces a uniform prior distribution for the cell probabilities.

BUGS CODE 4.2

```
# Area under the curve
# Ordinal values
# Five values
Model;
{
# generate Dirichlet distribution
g11~dgamma(a11,2)
g12~dgamma(a12,2)
g13~dgamma(a13,2)
g14~dgamma(a14,2)
g15~dgamma(a15,2)
g01~dgamma(a01,2)
g02~dgamma(a02,2)
g03~dgamma(a03,2)
g04~dgamma(a04,2)
g05~dgamma(a05,2)
g1<-g11+g12+g13+g14+g15
g0<-g01+g02+g03+g04+g05
# posterior distribution of probabilities for response of diseased patients
theta1<-g11/g1
theta2<-g12/g1
theta3<-g13/g1
theta4<-g14/g1
theta5<-g15/g1
# posterior distribution for probabilities of response of non-diseased patients
ph1<-g01/g0
ph2<-g02/g0
```

```
ph3<-g03/g0
ph4<-g04/g0
ph5<-g05/g0
# auc is area under ROC curve
#A1 is the P[Y>X]
#A2 is the P[Y=X]
auc<-A1+A2/2
A1<-theta2*ph1+theta3*(ph1+ph2)+theta4*(ph1+ph2+ph3)+
theta5*(ph1+ph2+ph3+ph4)
A2<-theta1*ph1+theta2*ph2+theta3*ph3+theta4*ph4
+theta5*ph5
}
# Mammography Example Zhou et al. [2]
# Uniform Prior
# see Table 4.7
list(a11=2,a12=1,a13=7,a14=12,a15=13,a01=10,a02=3,a03=12,
a04=9,a05=1)
# Gallium citrate Example Zhou et al. [2: 159]
# Uniform Prior
list(a11=13,a12=7,a13=4,a14=2,a15=19,a01=12,a02=3,a03=4,a04=2,a05=4)
# initial values
list(g11=1,g12=1,g13=1,g14=1,g15=1,g01=1,g02=1,g03=1,g04=1,g05=1)
```

The posterior analysis is given by Table 4.9.

Notice that mammography gives fair to good accuracy based on the ROC area, which is estimated as 0.7811(0.0514) with the posterior mean and by (0.6702, 0.8709) using a 95% credible interval. The MCMC error for the parameter based on 50,000 observations is <0.001, but the reader should vary the simulation sample size to see its effect on the MCMC error and posterior mean. The parameter A1 is $P[Y > X]$ and is estimated as 0.688(0.06350) and the probability of a tie, $P[Y = X]$, given by A2, is estimated as 0.1861(0.0307). The distribution of the area appears almost symmetric as evident by a plot of the corresponding density (Figure 4.3). The estimated area of 0.7811 is similar to that computed by Zhou, McClish, and Obuchowski [2: 30].

TABLE 4.9: Posterior distribution of area under ROC curve mammography example.

Parameter	Mean	sd	Error	Lower 2 1/2	Median	Upper 2 1/2
auc	0.7811	0.0514	<0.0001	0.6702	0.7848	0.8709
A1	0.688	0.0635	<0.0001	0.5564	0.6909	0.8036
A2	0.1861	0.0307	<0.0001	0.128	0.1854	0.2484

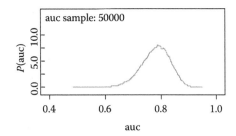

FIGURE 4.3: Posterior density of ROC area for mammography.

Lastly, the mammography example is concluded with a test for the usefulness of the procedure. Obviously, a perfect test has an ROC area of 1, and a useless test has an ROC area of 0.5. Thus, consider a Bayesian test of H: AUC <0.5 vs. the alternative A: AUC ≥ 0.5. How is this performed with WinBUGS? Based on BUGS CODE 5.2, the statement T<-step(auc-.5) will provide a test of the null hypothesis. The mean of T is the probability of the alternative hypothesis, and one can verify that

$$P[\text{AUC}(\theta, \phi) \geq 0.5 \mid \text{data}] = 0.99999(0.0044), \quad (4.12)$$

therefore, the null hypothesis is rejected and one may conclude that mammography is a useful procedure. Also note that the MCMC error for estimating the $P[\text{AUC} > 0.5/\text{data}]$ is <0.0001 with 50,000 observations.

4.4 Bayesian Methods for Test Accuracy: Quantitative Variables

4.4.1 Introduction

The methods introduced previously for discrete diagnostic tests apply to quantitative variables as well. The basic measures of test accuracy, including classification probabilities, predictive measures, and DLRs, all apply to continuous variables like blood glucose levels to diagnose diabetes, the levels of glucose metabolism in nuclear medicine procedures, and the prostate-specific antigen (PSA) levels to help diagnose prostate cancer. Other quantitative variables to be considered in this book are coronary artery calcium (CAC) levels in coronary heart disease, and standardized uptake levels to assess metastasis to the spinal column.

In clinical practice, quantitative variables are often dichotomized. For example, CAC levels in excess of 400, PSA levels in excess of 4 ng/mL, and blood glucose levels in excess of 126 mg/dL are standard threshold values. Of course, with a threshold value and a gold standard, the diagnostic accuracy can be estimated with the Bayesian methods previously introduced. In this section, methods for choosing a threshold value are explained in the context of a cost benefit analysis.

The primary focus on test accuracy will be the area under the ROC curve. Its mathematical properties will be outlined, and Bayesian methods of estimating the area explained.

4.4.2   Shields computed tomography study of heart disease

The Shields Heart Study was conducted at Washington State University and the CT imaging was implemented at the Shields Coronary Artery Center in Spokane, Washington. Over a period of 10 years there were about 4400 patient visits, where the majority of patients were referred by cardiologists and some were self referred. Thus, the relevant population was community based and a comprehensive history was taken of each patient's symptoms. These patients had confirmed CAD or were at high risk for the disease.

At the time, the use of CAC to assist in the diagnosis and patient management was not an accepted standard procedure; however, since then it is gradually being accepted. There are several experimental studies that involve a CT determination of the CAC in the coronary arteries. Measurements of CAC were made with the Imatron C-100 Ultrafast CT Scanner. A description of the Spokane study is given in Mielke, Shields, and Broemeling [9]. The CAC score is a positive score and is the sum of several CAC scores corresponding to the various coronary arteries, giving a measure of the amount of plaque burden.

Rumberger et al. [10] developed a risk index for CAD by categorizing the CAC scores as follows: A value of 1 is a CAC score of zero and indicates very low risk; a value of 2 is assigned for CAC scores between 1 and 10 and represents low risk of disease; CAC scores between 11 and 100 indicate a moderate risk and are assigned a value of 3; a value of 4 is assigned to scores between 101 and 400 for high risk; a very high risk has a value of 5 for CAC scores >400.

With the occurrence of infarction as a gold standard, the 130 patients who had an infarct and the 4263 who did not, were assigned the risk scores recorded in Table 4.10. Assuming a uniform prior distribution for the parameters, the posterior distribution of $\theta = (\theta_1, \theta_2, \theta_3, \theta_4, \theta_5)$ is Dirichlet (13, 7, 28, 41, 46), where θ_1 is the probability that a diseased patient has a low risk of disease, etc., and in a similar fashion, the posterior distribution of $\phi = (\phi_1, \phi_2, \phi_3, \phi_4, \phi_5)$ is Dirichlet (1819, 528, 815, 649, 455), where ϕ_5 is the probability that a

TABLE 4.10:   Spokane heart study.

Risk	Very low	Low	Moderate	High	Very high	Total
No infarction	1818	527	814	648	454	4263
Infarction	12	6	27	40	45	130

patient without an infarct has a very high risk of disease. If high risk is the threshold, it can be shown that TPF(θ) has a beta distribution with mean 0.626, median 0.639, and standard deviation 0.150. On the other hand, FPF(ϕ) has a beta posterior distribution with mean 0.272, median 0.251, and standard deviation 0.133.

It is well known that partitioning a continuous variable into a small number of categories results in a loss of information, therefore, methods developed to estimate the ROC area for quantitative variables are more appropriate and will be introduced in the following section. The Spokane Heart Study will be reanalyzed with these techniques.

4.4.3 Receiver operating characteristic area

The area under the ROC curve gives an intrinsic value to the accuracy of a diagnostic test and has a long history beginning in signal detection theory. See Egan [11] for the early use of the ROC curve in signal detection theory. Also, the books by Pepe [6]; Zhou, McClish, and Obuchowski [2] and Hans et al. [12] provide the history as well as the latest statistical methods (non Bayesian) for using ROC curves in diagnostic medicine. The ROC area is generally accepted as the way to measure diagnostic accuracy in radiology.

Let X be a quantitative variable and r a threshold value, and consider the test positive when $X \geq r$, otherwise negative, then the ROC curve is the set of all points:

$$\text{ROC}(\cdot) = \{[\text{FPF}(r), \ \text{TPF}(r)], r \text{ any real number}\}$$
$$= \{[t, \ \text{ROC}(t)], t \in (0, 1)\}, \tag{4.13}$$

where $t = \text{FPF}(r)$, that is, r is the threshold corresponding to t. As r becomes large, $\text{FPF}(r)$ and $\text{TPF}(r)$ tend to zero, while if r becomes small, $\text{FPF}(r)$ and $\text{TPF}(r)$ tend to 1, thus the ROC curve passes through (0,0) and (1,1). If the AUC is 1, the test is discriminating perfectly between the diseased and non-diseased groups, while if the area is 0.5, the test cannot discriminate between the two groups.

Chapter 4 of Pepe [6] presents several useful properties of the ROC curve, namely: (1) the invariance of the ROC curve under monotone increasing transformations of X, (2) interpreting the ROC area for continuous variables as $\text{AUC} = P(X > Y)$, and (3) a formula for the AUC area when X is normally distributed.

The Bayesian approach to estimating the ROC area is based on

$$\text{AUC} = \Phi\left[a/\sqrt{1 + b^2}\right]. \tag{4.14}$$

where X is normally distributed,

$$a = (\mu_D - \mu_{\bar{D}})/\sigma_D, \tag{4.15}$$

and

$$b = \sigma_D/\sigma_{\bar{D}}. \tag{4.16}$$

The mean and standard deviation of X for the diseased population are μ_D and σ_D, respectively, while $\mu_{\bar{D}}$ and $\sigma_{\bar{D}}$ are the mean and standard deviation of X for the non diseased, and Φ is the cumulative distribution function of the standard normal distribution. Equation 4.15 is the binormal assumption and is cited by many authors, including Pepe [6], who presents a good discussion of its use. Note that the ROC area AUC depends on the unknown parameters of the model.

Bayesian methods for estimating the ROC area will be illustrated by referring to a hypothetical example of diabetes, which involves 41 subjects with diabetes and 19 without, where those with diabetes have a mean blood glucose value of 123.34 mg/dL and those without have a mean value of 107.54 mg/dL. The corresponding standard deviations are 6.76 mg/dL for those with diabetes and 9.09 mg/dL for those without the disease; the actual values from the study are given in Table 4.11.

The Bayesian analysis for the diabetes type 2 data will be done in two ways: first with the basic equation for the area under the ROC curve assuming normality (Equation 4.14), and second with a program by O'Mally et al. [13]. The latter is more general and will also be used in Chapter 5 when covariates are included in the model. The code based on the basic formula is designated by BUGS CODE 4.3a, while the O'Malley method is coded with statements given by BUGS CODE 4.3b.

Consider first the basic formula for the ROC area, which is given by Equation 4.15 and can be expressed as

$$\text{AUC} = \phi\big[(\mu_2 - \mu_1)/\sqrt{1/\tau_1 + 1/\tau_2}\big] \tag{4.17}$$

where ϕ is the distribution function of the standard normal distribution, μ_1 is the mean of the population non diseased, μ_2 is the mean of the diseased population, τ_1 is the precision of the non diseased, and τ_2 is the precision of the diseased population. Given an improper prior distribution for all parameters and assuming that the two populations are independent, it can be shown that:

The conditional posterior distribution of μ_i given τ_i is normal with mean $\bar{x}_i$ and precision $n_i\tau_i$, and the marginal posterior distribution of τ_i is gamma with parameters $a_i = n_i/2$ and $b_i = (n_i - 1)s_i^2/2$, that is to say,

$$\mu_i/\tau_i, \text{data} \sim \text{normal}(\bar{x}_i, n_i\tau_i), \tag{4.18}$$

and

$$\tau_i/\text{data} \sim \text{gamma}(a_i, b_i). \tag{4.19}$$

TABLE 4.11: Data for 78 patients: blood glucose values.

Diabetic patients (59)	Non-diabetic patients (19)
123	109
129	106
115	100
131	88
119	106
111	108
129	110
127	111
118	112
111	94
131	122
118	110
126	113
130	106
122	114
112	101
122	99
128	128
123	106
119	
119	
132	
118	
126	
136	
118	
122	
119	
117	
129	
120	
125	
115	
131	
123	
130	
113	
128	
119	
118	
124	
127	
139	
120	
122	

(continued)

TABLE 4.11 (continued): Data for 78 patients: blood glucose values.

Diabetic patients (59)	Non-diabetic patients (19)
120	
114	
114	
122	
127	
123	
118	
131	
130	
139	
125	
135	
121	
124	

Based on BUGS CODE 4.3a, 65,000 observations are generated from the posterior distribution of all the parameters, with a burn in of 10,000 and a refresh of 100, and the posterior analysis, based on the basic formula (Equation 4.15), is reported in Table 4.12.

The MCMC errors appear to allow one to have confidence in the parameters of the model, including the main parameter, the area under the ROC curve. It is also of interest to know the sample means and standard deviations of the diseased and non-diseased populations. For the diseased population, those with diabetes, the sample mean and sample standard deviation are 123.58 and 6.9 mg/dL, respectively, while those for the patients without disease are 107.54 and 9.09 mg/dL. There is a "large" difference in the sample means, the standard deviations are not too large, and there is some overlap between the two populations, but the estimated area of 0.91 implies some confidence in using the blood glucose test to differentiate between diabetic and

TABLE 4.12: Posterior analysis for diabetes study—the basic Formula 4.15.

Parameter	Mean	sd	Error	Lower 2 1/2	Median	Upper 2 1/2
Area	0.9124	0.0401	<0.0001	0.8156	0.9196	0.9696
μ_1	107.5	2.142	0.0098	103.3	107.5	111.8
μ_2	123.6	0.9087	0.00489	121.8	123.6	125.4
τ_1	0.2421	0.0784	<0.0001	0.1132	0.2333	0.4174
τ_2	1.259	0.2302	0.00112	0.8474	1.245	1.749
a	2.315	0.3964	0.00196	0.7705	1.09	3.135
b	0.7784	0.1462	<0.0001	0.5142	0.7705	1.09

non-diabetic patients. Note that the parameters a and b are quantities used in the calculation of the ROC area (see Equations 4.16 and 4.17).

BUGS CODE 4.3a

```
model
{
# area under ROC curve
auc <-phi((mu2-mu1)/sqrt(1/tau1 + 1/tau2))
# This is for the non diseased patients
mu1~dnorm(113.14, prec1)
# this is for the diseased
mu2~dnorm (122.57, prec2)
prec1<-n1*tau1
prec2<-n2*tau2
# a1=n1/2
# b1=(n1-1) (sample variance 1)/2
tau1~dgamma(a1,b1)
# a2=n2/2
# b2= (n2-1)* (sample variance 2)/2
tau2 ~dgamma(a2,b2)
# binormal parameters
a<-(mu2-mu1)*sqrt(tau2)
b<-sqrt(tau1/tau2)
# c is the proportion of auc values >.80
c<-step(auc-.80)
}
# this is the data for diabetes study
List(n1=19, n2=59, a1=9.5, b1= 745.128, a2=30, b2=1427.74)
# these are the initial values
list(mu1=0, mu2=0, tau1=1, tau2=1)
```

BUGS CODE 4.3b is a slight modification of the O'Malley et al. code [13]. The program is based on the binormal assumption, where the diagnostic score has a normal distribution for both populations. The model is a linear regression model with two regression coefficients, an intercept and one for the group effect (diseased and non diseased). If a covariable is included, add a third regression coefficient beta [3]. This will be done in Chapter 5 on regression techniques. The first level parameters are the mu's and precisions of the blood glucose values, where each patient has a mean expressed as a linear regression on the group effect. The regression coefficients are second level parameters and are given uninformative normal distributions, while the two precision parameters are given non-informative gamma distributions. One would expect the same estimate of the ROC area using the two approaches, namely, the basic formula and the regression approach.

BUGS CODE 4.3b

```
model;
# Calculates posterior distribution of model parameters and the area under
    curve.
y=test
# Based on O'Mally et al. [13] regression method
{
# likelihood function
        for(i in 1:N) {

                y[i]~dnorm(mu[i],precy[d[i]+1]);
#               yt[i] <-log(y[i]); # logarithmic transformation
                mu[i] <-beta[1] + beta[2]*d[i];


                }


# prior distributions - non-informative prior; similarly for informative priors
        for(i in 1:P) {

                beta[i] ~ dnorm(0, 0.000001);


                }


        for (i in 1:K) {

                precy[i]~dgamma(0.001, 0.001);
                vary[i] <-1.0/precy[i];


                }


# calculates area under the curve
        la1 <-beta[2]/sqrt(vary[1]); # ROC curve parameters
        la2 <-vary[2]/vary[1];
        auc <-phi(la1/sqrt(1+la2));


}
# Diabetes data
list(K=2, P=2, N=78, y=c(123,129,115,131,119,111,129,127,118,111,
131,118,126,130,122,112,122,128,123,119,132,118,126,136,118,122,
119,117,129,120,125,115,131,123,130,113,128,138,119,118,124,127,
139,120,122,120,114,114,122,127,123,118,131,130,139,125,135,121,124,
109,106,100,88,106,108,110,111,112,94,122,110,113,106,114,101,99,128,106),
d=c(1,1,1,1,1,1,1,1,1,1,1,1,1,1,1,1,1,1,1,1,1,1,1,1,1,1,1,1,1,1,1,1,1,1,1,1,1,1,1,
1,1,0,0,0,0,0,0,0,0,0,0,0,0,0,0,0,0,0,0,0,0,0,0,0))
# Ordinal values from mammography
```

```
list(K=2,P=2,N=70,
  y=c(1,1,
    2,
    3,3,3,3,3,3,3,
    4,4,4,4,4,4,4,4,4,4,4,4,
    5,5,5,5,5,5,5,5,5,5,5,5,5,
    1,1,1,1,1,1,1,1,1,1,
    2,2,2,
    3,3,3,3,3,3,3,3,3,3,3,3,3,
    4,4,4,4,4,4,4,4,4,
    5),
  d=c(1,1,1,1,1,1,1,1,1,1,1,1,1,1,1,1,1,1,1,1,
    1,1,1,1,1,1,1,1,1,1,1,1,1,1,1,
    0,0,0,0,0,0,0,0,0,0,0,0,0,0,0,0,0,0,0,0,0,
    0,0,0,0,0,0,0,0,0,0,0,0,0,0,0,0))
```

list(beta=c(0,0),precy=c(1,1))

There are two list statements, where the first consists of three vectors for the data. The first list statement provides the blood glucose values and the vector d is a group identification vector. The third list statement specifies the initial values for the MCMC algorithm. The vector beta lists the initial values for the regression coefficients, and the vector precy, the initial values for the two precision parameters.

There are 79 patients, of which 19 do not have diabetes. The primary parameters are the AUC and the regression coefficients. Based on BUGS CODE 4.3b, 75,000 observations are generated from the posterior distribution for the area and regression parameters, with a burn in of 10,000 and a refresh of 100. Table 4.13 presents the posterior analysis.

The estimated area is 0.9082, which is almost identical to that given by the basic formula and reported in Table 4.12, and a plot of the posterior density is shown in Figure 4.4. The second regression coefficient is estimated as 16.13, which implies that the group effect is strong on the blood glucose values,

TABLE 4.13: Posterior distribution for the ROC area—diabetes. Regression approach.

Parameter	Mean	sd	Error	Lower 2 1/2	Median	Upper 2 1/2
beta[1]	107.5	2.246	0.0401	103	107.5	111.9
beta[2]	16.13	2.438	0.2438	11.35	16.11	21.02
precy[1]	0.0120	0.004	<0.0001	0.00544	0.0116	0.0211
precy[2]	0.0205	0.0038	<0.0001	0.0137	0.0203	0.0286
auc	0.9082	0.0424	<0.0001	0.8062	0.9155	0.9689

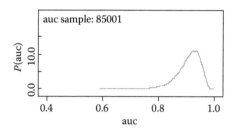

FIGURE 4.4: AUC area head trauma study.

which is one reason why the ROC area is so high. Also, note the variation in the MCMC errors of estimation.

4.4.4   Definition of the receiver operating characteristic curve

The ROC curve is defined by Equation 4.18, and under the assumption of binormality, has the representation:

$$[t, 1 - \Phi(bZ_t - a)], \quad t \in (0,1), \tag{4.20}$$

where t is the FPF, Z_t is the upper t percentage point of the standard normal distribution, a and b are involved in the definition of the ROC area, and Φ is the cumulative distribution function of the standard normal. Note that a and b are unknown parameters and have a posterior distribution. Therefore, the TPR corresponding to the FPR $= t$, is $1 - \Phi(bZ_t - a)$. When plotting the ROC curve, what should be used for the values of a and b in Equation 4.21? If the posterior means of a and b are used (see Table 4.12) in Equation 4.20, the graph for the ROC curve is given in Figure 4.5.

For the points plotted above, the coordinates are (0.025,0), (0.05,0), (0.1,0), (0.3,0.09086), (0.5,0.7928), (0.90,1), (0.95,1), and (0.975,1). Is the area under this curve the same as the estimated value of 0.83 determined by the above analyses? See Table 4.13.

4.4.5   Choice of optimal threshold value

Cost considerations are often used to select a threshold value for a diagnostic test. For example, Chapter 2 of Zhou, McClish, and Obuchowski [2] bases the choice of an optimal cutoff value on minimizing the total cost:

$$C = \text{TPF}p(C_{tp} - C_{fn}) + \text{FPF}(1-p)(C_{fp} - C_{tn}) + C_0 + pC_{fn} + (1-p)C_{tn}, \tag{4.21}$$

where p is the disease incidence, c_0 is the cost of performing the test, and c_{tp}, c_{fn}, c_{fp}, and c_{tn} are the costs of a true positive, false negative, false positive,

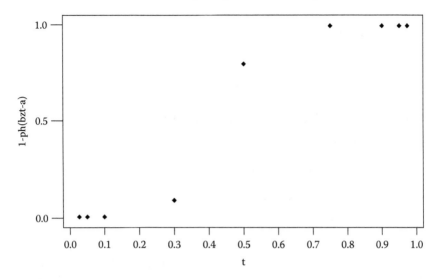

FIGURE 4.5: ROC curve for head trauma data. True positive vs. false positive fraction.

and true negative, respectively. When this expression is differentiated with respect to the FPF, the slope to the curve at the optimal point is

$$\kappa = (1-p)R/p, \tag{4.22}$$

where

$$R = (C_{tn} - C_{fp})/(C_{tp} - C_{fn}). \tag{4.23}$$

Assuming binormality, Somoza and Mossman [14] have shown that the optimal point is [FPF, TPF], where

$$\text{FPF}(a, b) = \Phi\left\{\left[ab - \sqrt{a^2 + 2(1 - b^2)\ln(\kappa/b)}\right]\big/(1 - b^2)\right\},$$

and

$$\text{TPF}(a, b) = \Phi\left\{\left[a - b\sqrt{a^2 + (1 - b^2)\ln(\kappa/b)}\right]\big/(1 - b^2)\right\}. \tag{4.24}$$

Treating κ as a constant, the coordinates of the optimal point are functions of the parameters a and b (see Equations 4.14 and 4.15), and have posterior distributions. Assuming binormality, Zhou, McClish, and Obuchowski [2: 152] compute the coordinates (Equation 4.24) of the optimal point for values of κ

based on $R = 0.5$, 1, 1.5 and $p = 0.2$, 0.5, 0.67. The statements corresponding to Equation 4.24 appear in BUG CODE 4.3b.

4.5   Clustered Data: Detection and Localization

4.5.1   Introduction

In assessing the area under the ROC curve for the accuracy of a diagnostic test, it is imperative to detect and locate multiple abnormalities per image. This approach takes that into account by adopting a statistical model that allows for correlation between the reader scores of several regions of interest (ROIs).

The ROI method of partitioning the image is taken. The readers give a score to each ROI in the image and the statistical model takes into account the correlation between the scores of the ROIs of an image in estimating test accuracy. The test accuracy is given by $P[Y > Z] + (1/2)P[Y = Z]$, where Y is a discrete diagnostic measurement of an affected ROI, and Z is the diagnostic measurement of an unaffected ROI. This way of measuring test accuracy is equivalent to the area under the ROC curve. The parameters are the parameters of a multinomial distribution, then, based on the multinomial distribution, a Bayesian method of inference is adopted for estimating the test accuracy.

Using a multinomial model for the test results, a Bayesian method based on the predictive distribution of future diagnostic scores is employed to find the test accuracy. By resampling from the posterior distribution of the model parameters, samples from the posterior distribution of test accuracy are also generated. Using these samples, the posterior mean, standard deviation, and credible intervals are calculated in order to estimate the area under the ROC curve. A Bayesian way to estimate test accuracy is easy to perform with standard software packages and has the advantage of employing the efficient inclusion of information from prior related imaging studies.

Obuchowski et al. [15] demonstrate how the ROI method is used to estimate the area under the ROC curve. They conclude that the ROI method appropriately captures the detection and localization of multiple abnormalities and is better suited than the free-response ROC curve method. In the ROI approach, the image is partitioned into clinically relevant, mutually exclusive regions. For example, in mammography there are five ROIs: upper outer, upper inner, lower outer, lower inner, and retroareolar. The reader assigns a score to each ROI, ranging from 1 to 5 as to the confidence of the presence of an abnormality; thus, the reader's ability to find abnormalities and to locate them is easily determined. Obuchowski et al. continue by presenting a way of taking into account the correlation between the scores of the several ROIs of the same image. I will not go into the details of the Obuchowski study, but will adopt their ROI approach as the preferred method of assessing test accuracy when there are many ROIs per image.

4.5.2   The Bayesian receiver operating characteristic curve for clustered information

The proposed method is based on Broemeling [16] and the ROI (not on a per patient basis) method and the Bayesian way to make statistical inferences. Suppose that an ROI is selected at random from a group of m affected (based on the gold standard) ROIs. Let Y be the ordinal diagnostic measurement observed on that ROI, and let Z be the measurement of an ROI selected at random from the set of n unaffected ROIs. The accuracy of the test is given by the area under the ROC curve and is estimated by

$$P[Y > Z] + P[Y = Z]/2, \tag{4.25}$$

thus providing the investigator with the overall accuracy of the diagnostic test.

Suppose Y and Z have possible values 1, 2, 3, 4,..., r, where larger values are more of an indication that the ROI is affected, then the study results can be represented by the following likelihood function for θ and ϕ.

$$L(\theta, \phi/y) \propto \prod_{i=1}^{i=r} \theta^i \phi^i, \tag{4.26}$$

where $\theta = (\theta_1, \theta_2, \ldots, \theta_r)$ and $\phi = (\phi_1, \phi_2, \ldots, \phi_r)$. The diagnostic measurement of an affected ROI is such that $Y = i$ with probability θ_i.

Similarly for an unaffected ROI, $Z = i$ with probability ϕ_i, where Y_i is the frequency of $Y = i$ and Z_i is the frequency that $Z = i (i = 1, 2, \ldots, r)$.

Note that this likelihood function is based on the multinomial distribution. We see that

$$\sum_{i=1}^{i=r}(\theta_i + \phi_i) = 1, \quad \sum_{i=1}^{i=r} Y_i = m, \quad \text{and} \quad \sum_{i=1}^{i=r} Z_i = n.$$

For a given study, the values of m, n, Y_i, and Z_i are known, but θ_i and ϕ_i are not and must be estimated from the data. To do this using the Bayesian approach, a prior density for the parameters must be specified. Suppose

$$g(\theta, \phi) \propto \prod_{i=1}^{i=r} \theta_i^{\alpha_i - 1} \phi_i^{\beta_i - 1}, \tag{4.27}$$

is the prior density, then the posterior density of the model parameters is

$$g(\theta, \phi/y, z) \propto \prod_{i=1}^{i=r} \theta_i^{Y_i + \alpha_i - 1} \phi_i^{Z_i + \beta_i - 1}. \tag{4.28}$$

The posterior density is that of a Dirichlet distribution, thus θ_i and ϕ_i are correlated. Because of the constraint, the correlation between the probabilities of the scores of the affected and unaffected ROIs has been taken into account, an essential requirement for the ROI method of detection and localization. The posterior distribution of $\theta = (\theta_1, \theta_2, \ldots, \theta_r)$ and $\phi = (\phi_1, \phi_2, \ldots, \phi_r)$

is Dirichlet with parameter $(Y_1 + \alpha_1, Y_2 + \alpha_2, \ldots, Y_r + \alpha_r, Z_1 + \beta_1, Z_2 + \beta_2, \ldots, Z_r + \beta_r)$.

It should be stressed that the prior distribution must be chosen with care. There are essentially two cases to consider: (a) where prior information from previous related experiments is available, and (b) little prior information is available. We will discuss this further when examples are to be illustrated.

If one lets

$$\theta_i^* = \theta_i \bigg/ \sum_{i=1}^{i=r} \theta_i, \tag{4.29}$$

then θ_i^* is the probability that $Y = i$ when sampling only from the affected ROIs. Suppose an ROI is selected at random from the population of unaffected ROIs, then ϕ_i^* is the probability that $Z = i$, where

$$\phi_i^* = \phi_i \bigg/ \sum_{i=1}^{i=r} \phi_i. \tag{4.30}$$

How does the probability $\Pr[Y > Z] + (1/2)\Pr[Y = Z]$ depend on the model parameters? The following gives the number of ways that $Y \geq Z$ and the corresponding probabilities.

1. $Y = 1$ and $Z = 1$ with probability $\theta_1^* \phi_1^*$ or

2. $Y = 2$ and $Z = 1$ or 2 with probability $\theta_2^*(\phi_1^* + \phi_2^*)$ or

3. $Y = r$ and $Z = 1$ or 2 or, $\ldots$ or r with probability θ_r^*.

In general, the area under the ROC curve is defined as

$$A(\theta, \phi) = \sum_{i=2}^{i=r} \theta_i^* \sum_{j=1}^{j=i-1} \phi_j^* + (1/2) \sum_{j=1}^{i=r} \theta_i^* \phi_i^*. \tag{4.31}$$

Suppose a "large" number M, say 50,000, samples are generated from the posterior Dirichlet distribution of θ and ϕ, then this provides M samples generated from the posterior distribution of θ^* and ϕ^*, via Equations 4.29 through 4.31, and also provides M samples from the posterior distributions of $A(\theta, \phi)$ via Equation 4.32. Based on these samples, the posterior mean, median, standard deviation, and 95% credible interval (or other posterior characteristics) are easily computed.

How large is M? One way to choose M is to choose it large enough so that the MCMC error is less than some specified value. One may vary the MCMC sample size to determine the effect on the MCMC error, and in this way choose an appropriate sample for estimating the parameters of interest.

4.5.3   Clustered data in mammography

A good example of clustered data can be found in Zhou, McClish, and Obuchowski [2: 134], which is based on mammography. In our case, a similar example illustrates the Bayesian method of estimating the ROC area. The example is based on the author's experience with a lung cancer trial, where the

CT image is partitioned into five ROIs: the upper left (UL), upper right (UR), lower left (LL), lower right (LR), and center region (CR). One radiologist assigns a score to each ROI, ranging from 1 to 4, which indicates the degree of malignancy. There are 55 patients and for each a score from 1 to 4 is assigned to each of the five ROIs. As determined by a gold standard, there are 85 abnormal (malignant) ROIs and 190 normal (non malignant) ROIs (Table 4.14).

TABLE 4.14: Diagnostic scores for five regions of the CT image.

	Disease status					Diagnostic scores				
ID	**UL**	**UR**	**LL**	**LR**	**CR**	**UL**	**UR**	**LL**	**LR**	**CR**
1	0	0	0	0	0	1	1	1	1	1
2	0	0	0	0	0	1	1	1	1	1
3	0	0	0	0	0	1	1	1	1	1
4	0	0	0	0	0	1	1	1	1	1
5	0	0	0	0	0	2	1	1	1	1
6	1	1	0	0	0	2	3	1	1	1
7	1	1	0	0	0	3	3	1	1	1
8	1	1	0	0	0	4	1	1	1	1
9	1	1	0	0	0	2	2	1	1	1
10	1	1	0	0	0	1	3	1	1	1
11	0	0	0	1	1	1	2	1	2	2
12	0	0	0	1	1	1	1	2	3	3
13	0	0	0	1	1	1	1	1	4	2
14	0	0	0	1	1	1	1	1	3	3
15	0	0	0	1	1	1	2	1	2	4
16	0	0	1	1	0	1	1	3	2	1
17	0	0	1	1	0	2	1	2	3	1
18	0	0	1	1	0	1	2	2	2	1
19	0	0	1	1	0	1	1	1	4	1
20	0	0	1	1	0	1	1	4	3	1
21	1	0	0	0	1	3	1	1	1	2
22	1	0	0	0	1	2	1	1	2	3
23	1	0	0	0	1	4	1	1	1	4
24	1	0	0	0	1	1	2	1	1	3
25	1	0	0	0	1	3	1	1	1	2
26	0	1	0	0	1	2	3	1	1	3
27	0	1	0	0	1	1	3	1	1	3
28	0	1	0	0	1	1	2	2	1	2
29	0	1	0	0	1	1	4	1	1	4
30	0	1	0	0	1	1	2	1	2	3
31	0	0	1	0	1	1	1	2	2	3
32	0	0	1	0	1	1	1	2	2	3
33	0	0	1	0	1	2	1	4	1	4
34	0	0	1	0	1	2	1	4	2	3
35	0	0	1	0	1	1	1	3	1	2

(continued)

TABLE 4.14 (continued):　Diagnostic scores for five regions of the CT image.

	Disease status					Diagnostic scores				
ID	UL	UR	LL	LR	CR	UL	UR	LL	LR	CR
36	0	1	0	1	0	1	3	1	3	1
37	0	1	0	1	0	1	4	1	4	2
38	0	1	0	1	0	1	3	2	4	1
39	0	1	0	1	0	1	2	1	2	1
40	1	0	0	0	0	4	2	1	1	1
41	1	0	0	0	0	3	1	1	2	2
42	1	0	0	0	0	3	1	1	1	1
43	1	0	0	0	0	2	1	1	2	1
44	1	0	0	0	0	4	1	1	1	1
45	0	0	1	0	0	1	1	4	1	2
46	0	0	1	0	0	1	1	3	1	1
47	0	0	1	0	0	2	2	3	1	1
48	0	0	1	0	0	1	3	2	2	1
49	0	0	1	0	0	1	1	4	1	1
50	0	0	1	0	0	1	1	3	1	1
51	0	0	0	1	0	1	1	3	1	1
52	0	0	0	1	0	1	3	1	4	1
53	0	0	0	1	0	2	1	1	3	2
54	0	0	0	1	0	1	1	2	2	1
55	0	0	0	1	0	1	1	1	2	1

Assuming a uniform prior for the parameters, the joint posterior distribution of $\theta = (\theta_1, \theta_2, \ldots, \theta_4)$ and $\phi = (\phi_1, \phi_2, \ldots, \phi_4)$ is Dirichlet with parameter (5, 27, 35, 22, 161, 29, 3, 1), that is to say, among the abnormal ROIs, there are four with a score of 1, 26 with a score of 2, 34 with a score of 3, and 21 with a score of 4. Among the normal ROIs, there are 160 with a score of 1, 28 with a score of 2, two with a score 3, and none with a score of 4. Increasing values of the diagnostic score indicate a larger chance of malignancy. The AUC is

$$A(\theta, \phi) = \sum_{i=2}^{i=4} \theta_i^* \sum_{j=1}^{j=i-1} \phi_j^* + (1/2) \sum_{i=1}^{i=4} \theta_i^* \phi_i^*. \tag{4.32}$$

Based on BUGS CODE 4.4, 60,000 values were generated from the Dirichlet posterior distribution of θ and ϕ, with a burn in of 5,000 and a refresh of 100.

BUGS CODE 4.4

```
Model;
# Clustered Data ROC Area
{
# Joint Distribution of all cell parameters
# abnormal ROI
  g11~dgamma(a11,2)
```

```
  g12~dgamma(a12,2)
  g13~dgamma(a13,2)
  g14~dgamma(a14,2)
# normal ROI

  g01~dgamma(a01,2)
  g02~dgamma(a02,2)
  g03~dgamma(a03,2)
  g04~dgamma(a04,2)
sg<-g11+g12+g13+g14+g01+g02+g03+g04
the1<-g11/sg
the2<-g12/sg
the3<-g13/sg
the4<-g14/sg
p1<- g01/sg
p2<- g02/sg
p3<- g03/sg
p4<- g04/sg
# sum of the the's
sthe<-the1+the2+the3+the4
# sum of the p's
sp<-p1+p2+p3+p4
# truncated distribution of thetas
     theta1<-the1/sthe
     theta2<-the2/sthe
     theta3<-the3/sthe
     theta4<-the4/sthe

# truncated distributions of the phi's
     ph1<-p1/sp
     ph2<-p2/sp
     ph3<-p3/sp
     ph4<-p4/sp

     # area based on truncated

     Area<-A1+A2/2

     A1<-theta2*ph1+theta3*(ph1+ph2)+theta4*(ph1+ph2+ph3)
     A2<-theta1*ph1+theta2*ph2+theta3*ph3+theta4*ph4

}
# a1j's are cell counts for abnormal ROI
# a0j's are cell counts of normal ROI
# The list below is for the clustered data of the lung cancer study.
list(a11=5,a12=27,a13=34,a14=22,a01=161,a02=29, a03=3,a04=1)
```

TABLE 4.15:   Posterior analysis for clustered lung cancer study.

Parameter	Mean	sd	Error	Lower 2 1/2	Median	Upper 2 1/2
A1	0.8828	0.0262	<0.001	0.8251	0.8852	0.9272
A2	0.0993	0.0211	<0.001	0.0632	0.0975	0.1457
Area	0.9325	0.0159	<0.001	0.8973	0.934	0.9593
ϕ_1	0.8299	0.0269	<0.0001	0.7741	0.831	0.8791
ϕ_2	0.1495	0.0256	<0.0001	0.1031	0.1482	0.203
ϕ_3	0.0154	0.0087	<0.0001	0.0032	0.0138	0.0368
ϕ_4	0.0051	0.0051	<0.0001	0.00013	0.0036	0.0189
θ_1	0.0561	0.0243	<0.001	0.0185	0.05282	0.1125
θ_2	0.3035	0.0484	<0.001	0.2129	0.302	0.4021
θ_3	0.3932	0.0514	<0.001	0.2958	0.3922	0.4956
θ_4	0.2471	0.0452	<0.001	0.1642	0.245	0.3413

```
# initial values
list(g11=1,g12=1,g13=1,g14=1, g01=1,g02=1,g03=1,g04=1)
```

The posterior analysis for the clustered lung cancer study is presented in Table 4.15.

It is interesting to note that the estimated area of 0.9325(0.0159) indicates very good discrimination between the normal and abnormal ROIs. Recall that a score of 1 indicates no evidence of a malignant lesion and that the posterior analysis implies that the probability of a 1 for an abnormal ROI is 0.8255 compared to a posterior mean of 0.0506 for the probability of a 1 for a normal (non-diseased) ROI. The MCMC errors are quite small, giving one confidence that the simulation is providing accurate answers for the posterior characteristics.

The MCMC errors appear to be reasonable. Recall that the MCMC error is an error in the simulation that generates observations from the posterior distribution. It estimates the accuracy of the closeness of the simulated value to the "true" posterior characteristic. For example, in estimating the ROC area, the MCMC error indicates that the posterior mean is within three decimal values of the "true" posterior mean ROC area.

4.6   Comparing Accuracy between Modalities with Ordinal Scores

4.6.1   Comparing true and false positive fractions

In order to compare modalities, the CASS dataset is again used, where the EST and a history of chest pain (CPH) are used to diagnose CAD. The

TABLE 4.16a: CASS study
for diseased subjects.

	CPH		
EST	**0**	**1**	**Total**
0	25	183	208
1	29	786	815
Total	54	969	1023

TABLE 4.16b: CASS study
for non-diseased subjects.

	CPH		
EST	**0**	**1**	**Total**
0	151	176	327
1	46	69	115
Total	197	245	442

example was used to illustrate the Bayesian estimation of the basic measures
of test accuracy, including the classification probabilities, DLRs, and the predictive probabilities. There are 1,465 subjects, all of which had an EST and
a record of their chest pain. This paired study is given in Tables 4.16a and b
(see Pepe [6: 47]). The important question is which modality, EST or CPH,
is most accurate and by how much?

The Bayesian analysis will consist of finding the posterior distribution of
the sensitivity and specificity of the two modalities and comparing them on the
basis of the ratios of the two basic measures. Let θ_{ij} be the probability that
a diseased subject has an EST score of i and a record of chest pain j, where
$i, j = 0, 1$, where 0 indicates a negative outcome and 1 a positive outcome. In
a similar manner, let ϕ_{ij} be the corresponding probability for a non-diseased
subject.

Assuming a uniform prior distribution for $\theta = (\theta_{00}, \theta_{01}, \theta_{10}, \theta_{11})$ and $\phi = (\phi_{00}, \phi_{01}, \phi_{10}, \phi_{11})$, their joint posterior distribution is Dirichlet with parameter (26, 184, 30, 787; 152, 177, 47, 70). Note that there is a joint posterior
distribution of eight parameters. The truncated distribution of the θ's is the
distribution of

$$\theta_{ij}^* = \theta_{ij} \bigg/ \sum_{i=0}^{i=1} \sum_{j=0}^{j=1} \theta_{ij}, \tag{4.33}$$

and the truncated distribution of the ϕ's is the distribution of

$$\phi_{ij}^* = \phi_{ij} \bigg/ \sum_{i=0}^{i=1} \sum_{j=0}^{j=1} \phi_{ij}. \tag{4.34}$$

The sensitivity (TPF) for the EST and CPH is

$$\text{tpfest} = \theta_{1.}^{*} \tag{4.35}$$

and

$$\text{tpfcph} = \theta_{.1}^{*}, \tag{4.36}$$

respectively, where the dot notation indicates summation of θ_{ij}^{*} over the missing subscript. In a similar way, the FPF for the EST and CPH modalities is

$$\text{fpfest} = \phi_{1.}^{*} \tag{4.37}$$

and

$$\text{fpfcph} = \phi_{.1}^{*} \tag{4.38}$$

respectively.

The posterior distribution for the eight multinomial parameters, $\theta = (\theta_{00}, \theta_{01}, \theta_{10}, \theta_{11})$ and $\phi = (\phi_{00}, \phi_{01}, \phi_{10}, \phi_{11})$, are determined with WinBUGS using 70,000 observations, with a burn in of 10,000 and a refresh of 100, and BUGS CODE 4.5 provides the statements for executing the Bayesian analysis.

BUGS CODE 4.5

```
Model;
# Comparing modalities 2 by 2 table
{
# Dirichlet distribution generated
  g00~dgamma(a00,2)
  g01~dgamma(a01,2)
  g10~dgamma(a10,2)
  g11~dgamma(a11,2)
sumall<-g00+g01+g10+g11+
h00+h01+h10+h11
  h00~dgamma(b00,2)
  h01~dgamma(b01,2)
  h10~dgamma(b10,2)
  h11~dgamma(b11,2)
# cell probabilities for diseased
theta00<-g00/sumall
theta01<-g01/sumall
theta10<-g10/sumall
theta11<-g11/sumall
```

```
# cell probabilities for non diseased
ph00<-h00/sumall
ph01<-h01/sumall
ph10<-h10/sumall
ph11<-h11/sumall
# truncated distributions
stheta<-theta00+theta01+theta10+theta11
sph<-ph00+ph01+ph10+ph11
# truncated thetas
th00<-theta00/stheta
th01<-theta01/stheta
th10<-theta10/stheta
th11<-theta11/stheta
# truncated phi's
p00<-ph00/sph
p01<-ph01/sph
p10<-ph10/sph
p11<-ph11/sph
# true positives
# a designates row
# b designates column
tpfa<-th10+th11
tpfb<-th01+th11
# false positive
fpfa<-p10+p11
fpfb<-p01+p11
# ratio of tpf a to b
rtpf<-tpfa/tpfb
#ratio of fpf a to b
rfpf<-fpfa/fpfb
}
# CASS data set
# see Pepe [6: 8] for a description of CASS study
# Comparing est and cph
# a is est
# b is cph
list(a00=26,a01=184,a10=30,a11=787,
    b00=152,b01=177,b10=47,b11=70)
list(g00=1,g01=1,g10=1,g11=1,
    h00=1,h01=1,h10=1,h11=1)
```

The tpf's of the est and the cph can be contrasted with the ratio

$$\text{rtpf(est/cph)} = \text{tpfest/tpfcph},\qquad(4.39)$$

TABLE 4.17:   Posterior analysis for comparing EST and CPH.

Parameter	Mean	sd	Error	Lower 2 1/2	Median	Upper 2 1/2
fpfest	0.2621	0.0207	<0.0001	0.2227	0.2619	0.3036
fpfcph	0.5538	0.0234	<0.0001	0.5075	0.554	0.5944
rfpf(est/cph)	0.4741	0.0416	<0.0001	0.3963	0.473	0.5586
rtpf(est/cph)	0.8413	0.0138	<0.0001	0.814	0.8414	0.8682
tpfest	0.7954	0.0125	<0.0001	0.7704	0.7956	0.8196
tpfcph	0.9455	0.0071	<0.0001	0.9307	0.9457	0.9585

and, in a similar way, the fpf's of the est to cph by

$$rfpf(est/cph) = fpfest/fpfcph. \tag{4.40}$$

See the appropriate statements in BUGS CODE 4.5.

On the basis of the TPF, it appears that the CPH is more accurate, but on the basis of the FPF, it appears that the EST is more accurate. In comparing the sensitivities of the two modalities (Table 4.17), one could use the ratio of the TPFs, given by rtpf(est/cph) with the corresponding density portrayed by Figure 4.6.

4.6.2   Comparing the receiver operating characteristic areas of two modalities with ordinal scores

Suppose one wants to compare the ROC areas of two modalities with ordinal scores. An example involving two imaging modalities, MRI and CT, is given in Tables 4.18a and b.

The scores indicate the confidence that the radiologist has for the presence of a lung cancer lesion, where 1 indicates that there is no evidence of a malignant lesion; 2 designates that there is very little evidence of a lesion; 3 classifies the lesion as benign, but where a follow-up image is indicated; 4 designates the lesion is possibly malignant; and 5 denotes the presence of a definitely

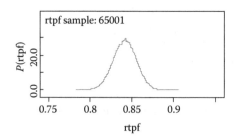

FIGURE 4.6: Posterior density of rtpf(est/cph).

TABLE 4.18a: MRI and CT scores for detecting lung cancer—diseased patients.

Modality	Scores					Total
	1	2	3	4	5	
MRI	0	3	12	10	25	50
CT	1	4	8	11	26	50

TABLE 4.18b: MRI and CT scores for detecting lung cancer—non-diseased patients.

Modality	Scores					Total
	1	2	3	4	5	
MRI	30	18	20	8	2	78
CT	31	22	15	5	5	78

malignant lesion of the lung. Suppose that the scores for non-diseased patients are given in Table 4.18b.

Tables 4.18a and 4.18b are hypothetical example and assumes one radiologist is reading both images per patient, an MRI image and a CT image. One would expect good agreement between the two modalities, but are the two modalities accurate and is one more accurate than the other?

For patients with lung cancer, suppose θ_{ij} is the probability that the radiologist assigns a score of $j(=1,2,3,4,5)$ using modality i, where $i = 1$ indicates an MRI image, and $i = 2$ designates a CT image. In a similar fashion, suppose ϕ_{ij} is the probability for non-diseased patients that the radiologist assign a score of j using modality i.

In order to define an ROC area, truncated probabilities for diseased and non-diseased patients for each modality will be used. For example, for the MRI images of the diseased patients, the cell truncated probabilities are

$$\theta_{1j}^* = \theta_{1j}/\theta_1, \tag{4.41}$$

$$\text{for } j = 1, 2, 3, 4, 5, \tag{4.42}$$

while for the CT images of the diseased patients, the truncated cell probabilities are

$$\theta_{2j}^* = \theta_{2j}/\theta_2, \quad j = 1, 2, 3, 4, 5. \tag{4.43}$$

In a similar way, the truncated cell probabilities for the non-diseased patients are given by

$$\phi_{1j}^* = \phi_{1j}/\phi_1, \quad j = 1, 2, 3, 4, 5, \tag{4.44}$$

for the MRI images, and by

$$\phi_{2j}^* = \phi_{2j}/\phi_2, \quad j = 1, 2, 3, 4, 5, \tag{4.45}$$

for the CT images, where the dot notation means summation over the indicated subscript.

The ROC area for the MRI is defined as

$$\text{Area mri} = \text{Area1 mri} + (1/2)\text{Area2 mri},$$

where

$$\begin{aligned}
\text{Area1 mri} &= \theta_{12}^*\phi_{11}^* + \theta_{13}^* * (\phi_{11}^* + \phi_{12}^*) + \theta_{14}^*(\phi_{11}^* + \phi_{12}^* + \phi_{13}^*) \\
&\quad + \theta_{15}^*(\phi_{11}^* + \phi_{12}^* + \phi_{13}^* + \phi_{14}^*),
\end{aligned} \tag{4.46}$$

and

$$\text{Area2 mri} = \sum_{i=1}^{i=5} \theta_{1i}^* \phi_{1i}^*. \tag{4.47}$$

Note that the area for the CT is defined as

$$\text{Area ct} = \text{Area1 ct} + (1/2)\text{Area2 ct},$$

where

$$\begin{aligned}
\text{Area1 ct} &= \theta_{22}^*\phi_{21}^* + \theta_{23}^* * (\phi_{21}^* + \phi_{22}^*) + \theta_{24}^*(\phi_{21}^* + \phi_{22}^* + \phi_{23}^*) \\
&\quad + \theta_{25}^*(\phi_{21}^* + \phi_{22}^* + \phi_{23}^* + \phi_{24}^*),
\end{aligned} \tag{4.48}$$

and

$$\text{Area2 ct} = \sum_{i=1}^{i=5} \theta_{2i}^* \phi_{2i}^*. \tag{4.49}$$

The above equations have their counterpoints in BUGS CODE 4.6.

BUGS CODE 4.6

```
Model;
# Comparing two modalities
# 2 ROC areas
{
# Dirichlet distribution
# modality a diseased
g11~dgamma(a11,2)
g12~dgamma(a12,2)
g13~dgamma(a13,2)
g14~dgamma(a14,2)
g15~dgamma(a15,2)
# modaltiy b diseased
g21~dgamma(a21,2)
```

```
 g22~dgamma(a22,2)
 g23~dgamma(a23,2)
 g24~dgamma(a24,2)
 g25~dgamma(a25,2)

sg<-g11+g12+g13+g14+g15+
    g21+g22+g23+g24+g25
# modality a non diseased
 h11~dgamma(b11,2)
 h12~dgamma(b12,2)
 h13~dgamma(b13,2)
 h14~dgamma(b14,2)
 h15~dgamma(b15,2)
# modality b non diseased
 h21~dgamma(b21,2)
 h22~dgamma(b22,2)
 h23~dgamma(b23,2)
 h24~dgamma(b24,2)
 h25~dgamma(b25,2)

sh<-h11+h12+h13+h14+h15+
    h21+h22+h23+h24+h25
# the thetas and phis have a Dirichlet distribution
theta11<-g11/sg
theta12<-g12/sg
theta13<-g13/sg
theta14<-g14/sg
theta15<-g15/sg
theta21<-g21/sg
theta22<-g22/sg
theta23<-g23/sg
thcta24<-g24/sg
theta25<-g25/sg
ph11<-h11/sh
ph12<-h12/sh
ph13<-h13/sh
ph14<-h14/sh
ph15<-h15/sh
ph21<-h21/sh
ph22<-h22/sh
ph23<-h23/sh
ph24<-h24/sh
ph25<-h25/sh
# truncate modality a (diseased)
sumad<-theta11+theta12+theta13+theta14+theta15
```

```
th11<-theta11/sumad
th12<-theta12/sumad
th13<-theta13/sumad
th14<-theta14/sumad
th15<-theta15/sumad
# truncate modality b (diseased)
sumbd<-theta21+theta22+theta23+theta24+theta25
th21<-theta21/sumbd
th22<-theta22/sumbd
th23<-theta23/sumbd
th24<-theta24/sumbd
th25<-theta25/sumbd
# truncate modality a (non diseased)
sumand<-ph11+ph12+ph13+ph14+ph15
p11<-ph11/sumand
p12<-ph12/sumand
p13<-ph13/sumand
p14<-ph14/sumand
p15<-ph15/sumand
# truncate modality b (non diseased)
sumbnd<-ph21+ph22+ph23+ph24+ph25
p21<-ph21/sumbnd
p22<-ph22/sumbnd
p23<-ph23/sumbnd
p24<-ph24/sumbnd
p25<-ph25/sumbnd
# area for a
areaa<-areaa1+areaa2/2
areaa1<-th12*p11+th13*(p11+p12)+th14*(p11+p12+p13)+
th15*(p11+p12+p13+p14)
areaa2<-th11*p11+th12*p12+th13*p13+th14*p14+
th15*p15
# area for b
areab<-areab1+areab2/2
areab1<-th22*p21+th23*(p21+p22)+th24*(p21+p22+p23)+
th25*(p21+p22+p23+p24)
areab2<-th21*p21+th22*p22+th23*p23+th24*p24+
th25*p25
# compares areas a and b
diffarea<-areaa-areab
}
# Compares mri and ct for lung cancer lesions
# hypothetical example
# mri is modlaity a
# ct is modality b
```

list(a11=1,a12=4,a13=13,a14=11,a15=26,
  a21=2,a22=5,a23=19,a24=12,a25=27,
  b11=31,b12=19,b13=21,b14=9,b15=3,
  b21=32,b22=23,b23=16,b24=6,b25=6)

The above analysis is executed with 105,000 observations generated from the posterior distribution of the parameters, with a burn in of 5,000 and a refresh of 100. Note that the MCMC simulation error in all cases is accurate to at least three decimal places. The parameters in Table 4.19 are named the same way as in BUGS CODE 4.6, and the code follows the development of Equations 4.42 through 4.49. Both modalities appear to be quite accurate with ROC areas of 0.857 for the MRI and 0.833 for the CT, and they appear to be equivalent areas. A 95% credible interval for the difference in the areas is $(-0.0627, 0.1108)$, implying very little difference (Table 4.19).

4.6.3 Comparing receiver operating characteristic areas for continuous scores

The ROC area of two continuous scores will be compared using O'Malley et al.'s [13] Bayesian approach to computing the area, assuming the scores are normally distributed for the diseased and non-diseased populations. It is a regression approach that regressed the diagnostic score on the disease indicator variable d, namely,

$$Y[i] \sim \text{dnormal}(m(u)), \ \text{precision}[d[i]+1), \tag{4.50}$$

where $i = 1, 2, \ldots, N$,

$$m(u) = \text{beta}[1] + \text{beta}[2]^*d, \tag{4.51}$$

where d is the indicator variable,

$$\text{beta}[i] \sim \text{dnorm}(0, 0.0001)$$
$$\text{precision}[i] \sim \text{dgamma}(0.0001, 0.0001),$$

and $i = 1, 2$.

TABLE 4.19: Bayesian analysis for comparing MRI and CT.

Parameter	Mean	sd	Error	Lower 2 1/2	Median	Upper 2 1/2
area(mri)	0.8573	0.0304	<0.00001	0.7916	0.8595	0.9108
areaa1	0.7963	0.0388	<0.0001	0.7141	0.7984	0.8663
areaa2	0.122	0.0186	<0.00001	0.0871	0.1215	0.1603
areab(ct)	0.8333	0.0320	<0.00001	0.765	0.8353	0.8901
areab1	0.7669	0.0399	<0.0001	0.6833	0.7689	0.8394
areab2	0.1328	0.0175	<0.00001	0.0992	0.1325	0.168
Diff(mri-ct)	0.02404	0.0441	<0.0001	−0.0627	0.0240	0.1108

The AUC is computed as

$$auc<-phi(la1/sqrt(1+la2)), \qquad (4.52)$$

where phi is the distribution function of the standard normal,

$$la1<-beta[2]/sqrt(var[1]),$$
$$la2<-var[2]/var[1].$$

Also

$$var[i] = 1/precison[i],$$

and $i = 1, 2$.

When considering a paired study, where two diagnostic scores, y and z, are expected to be correlated, the approach here is to revise the O'Malley et al. [13] approach by changing Equation 4.52 to

$$m(u) = beta[1] + beta[2]^*d + beta[3]^*z[t]. \qquad (4.53)$$

That is, by regressing one score on the other and adding it as a covariable in the linear regression Equation 4.52, then computing the ROC area, assuming normality. To compute the ROC area for z, the roles of y and z are reversed in Equation 4.53. In this way, correlation between the two diagnostic scores can be taken into account in estimating the ROC area.

The following code will be used to compute the ROC area for two diagnostic scores y and z.

BUGS CODE 4.7

```
model;
# Compares two correlated ROC areas
# Compares two biomarkers
{
    for(i in 1:N) {
    # yt is log of y
      yt[i]< -log(y[i])

         yt[i]~dnorm(mu[i],precy[d[i]+1]);
    # the regression of log y on logz
         mu[i] <-beta[1] + beta[2]*d[i]+beta[3]*zt[i]
    # zt is log of z
         zt[i]< -log(z[i])

         zt[i]~dnorm(vu[i],precz[d[i]+1])
    # the regression of logz on logy
         vu[i]< -delta[1]+delta[2]*d[i]+delta[3]*yt[i]

    }
```

prior distributions - non-informative prior;

```
for (i in 1:3) { beta[i] ~ dnorm(0,0.0001)
                 delta[i]~dnorm(0,0.0001)
}

    for (i in 1:K) {

        precy[i]~dgamma(0.000001,0.000001);
        vary[i]<-1.0/precy[i];
        precz[i]~dgamma(0.000001,0.000001);
        varz[i]<-1.0/precz[i];
            }
```

calculates area under the curve
```
    la1y<-beta[2]/sqrt(vary[1]);
    la2y<-vary[2]/vary[1];
    aucy<-phi(la1y/sqrt(1+la2y));
    la1z<-delta[2]/sqrt(varz[1]);
    la2z<-varz[2]/varz[1];
    aucz<-phi(la1z/sqrt(1+la2z));
    diff<-aucy-aucz
}
```
Wieand et al. [17]
two biomarkers for pancreatic cancer
see Zhou et al. [2: 260]
y is CA19-9 and z is CA125
```
list(K=2, N=141,
y=c(28.00,15.50,8.20,3.40,17.30,15.20,32.90,11.10,87.50,16.20,107.90,5.70,
25.60,31.20,21.60,55.60,8.80,6.50,22.10,14.40,44.20,3.70,7.80,8.90,18.00,6.50,
4.90,10.40,5.00,5.30,6.50,6.90,8.20,21.80,6.60,7.60,15.40,59.20,5.10,10.00,5.30,
32.60,4.60,6,90,4.00,3.65,7.80,32.50,11.50,4.00,10.20,2.40,719.00,2106.67,
24000.00,1715.00,3.60,521.50,1600.00,454.00,109.70,23.70,464.00,9810.00,
255.00,58.70,225.00,90.10,50.00,5.60,4070.00,592.00,28.60,6160.00,1090.00,
10.40,27.30,162.00,3560.00,14.70,83.30,336.00,55.70,1520.00,3.90,5.80,8.45,
361.00,369.00,8230.00,39.30,43.50,361.00,12.80,18.00,9590.00,555.00,60.20,
21.80,900.00,6.60,239.00,3100.00,3275.00,682.00,85.40,10290.00,770.00,247.60,
12320.00,113.10,1079.00,45.60,1630.00,79.40,508.00,3190.00,542.00,1021.00,
235.00,251.00,3160.00,479.00,222.00,15.70,2540.00,11630.00,1810.00,6.90,4.10,
15.60,9820.00,1490.00,15.70,45.80,7.80,12.80,100.53,227.00,70.90,2500.00),
z=c(13.30,11.10,16.70,12.60,7.40,5.50,32.10,27.20,6.60,9.80,10.50,7.80,9.10,
12.30,12.00,42.10,5.90,9.20,7.30,6.80,10.70,15.70,8.00,6.80,47.35,17.90,96.20,
108.90,16.60,9.50,179.00,12.10,35.60,15.00,12.60,5.90,10.10,8.50,11.40,54.65,
9.70,11.20,35.70,22.50,21.20,5.60,9.40,12.00,9.80,17.20,10.60,79.10,31.40,15.00,
77.80,25.70,11.70,8.25,14.95,8.70,14.10,123.90,12.10,99.10,18.60,10.50,6.60,
```

74.00,43.90,45.70,13.00,7.30,8.60,17.20,15.40,14.30,93.10,66.30,26.70,32.40,
9.90,30.30,11.20,202.00,35.70,9.20,103.60,21.40,8.10,29.90,17.50,30.80,57.30,
6.50,33.80,53.60,17.20,94.20,33.50,3.70,11.70,19.90,38.70,27.30,20.10,86.10,
844.00,36.90,6.90,27.70,9.90,38.60,142.60,12.50,11.60,21.20,13.20,19.20,
1024.00,14.10,34.80,35.30,35.00,15.50,12.10,31.60,184.80,24.80,10.40,34.50,
19.40,22.20,53.90,15.40,17.30,36.80,49.80,26.57,9.70,19.20,14.20),
d=
c(0,0,0,0,0,0,0,0,0,0,0,0,0,0,0,0,0,0,0,0,0,0,0,0,0,0,0,0,0,0,0,0,0,0,0,0,0,0,0,0,0,0,0,0,0,0,
0,0,0,0,0,0,0,0,0,1,1,1,1,1,1,1,1,1,1,1,1,1,1,1,1,1,1,1,1,1,1,1,1,1,1,1,1,1,1,1,1,1,1,1,1,1,1,1,
1,1,1,1,1,1,1,1,1,1,1,1,1,1,1,1,1,1,1,1,1,1,1,1,1,1,1,1,1,1,1,1,1,1,1,1,1,1,1,1,1,1,1,1,1,1,1,1,
1,1,1,1,1,1,1,1,1,1,1,1))
list(beta = c(0,0),delta=c(0,0), precy = c(1,1),precz=c(1,1))

Two continuous scores are compared using two biomarkers in a study
by Wieand et al. [17] and analyzed by Pepe [6] and Zhou, McClish, and
Obuchowski [2: 259].Which one of the biomarkers is more accurate in detecting pancreatic cancer? There are 51 control patients—those without pancreatic cancer—and 90 patients with cancer. These values are listed in the
first list statement of BUGS CODE 4.7, and are transformed to logs in the
analysis.

The first biomarker, CA19-9, is designated as y, while z identifies the
second biomarker, CA125. The original values were transformed by logs to
achieve normality. See the P-P plot Figure 4.7. The original values of CA19-9
and CA125 are highly skewed to the right and the P-P plot shows a large
deviation from normality.

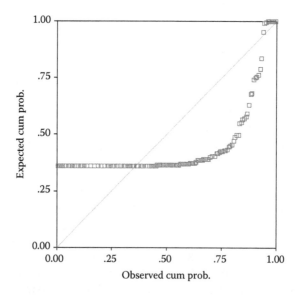

FIGURE 4.7: P-P normal plot of CA19-9.

TABLE 4.20: Posterior analysis for pancreatic cancer.

Parameter	Mean	sd	Error	Lower 2 1/2	Median	Upper 2 1/2
auc CA19-9	0.877	0.0281	<0.0001	0.8157	0.8792	0.9257
auc CA125	0.6443	0.0581	<0.0001	0.526	0.6459	0.7533
beta[1]	2.515	0.387	0.0061	1.752	2.515	3.272
beta[2]	2.95	0.292	0.0015	2.379	2.952	3.525
beta[3]	−0.01574	0.1374	0.0021	−0.285	−0.0155	0.2555
delta[1]	2.568	0.1544	0.0013	2.262	2.568	2.869
delta[2]	0.4768	0.1999	0.0016	0.085	0.4767	0.8699
delta[3]	0.0397	0.0429	<0.0001	−0.0456	0.0396	0.1232
diff	0.2332	0.06447	<0.0001	0.11	0.2321	0.3622

Note: Regression takes into account other biomarker.

The Bayesian analysis consists of generating 45,000 observations from the posterior distribution, with a burn in of 5,000 and a refresh of 100. A uniform prior was assumed and the list statement contains the original values for the two biomarkers. Consider the posterior analysis as portrayed in Table 4.20.

It is seen that for the regression of log CA19-9 on log CA124, the latter is a poor predictor of the former. Note that the 95% credible interval for beta[3] contains zero. On inspection, the 95% credible interval for delta[3] contains zero, which implies that log CA19-9 is a poor predictor of log CA125. Also, the ROC area for CA19-9 is much larger than that for CA125, where the former has an area of 0.877 compared to 0.6443 for CA125. Therefore, the analysis is repeated by not using the biomarkers as covariates in the regression analysis. When the analysis is repeated using the disease indicator as the only predictor, the Bayesian analysis changes (see Table 4.21).

Again, 45,000 observations were generated from the posterior distribution, with a burn in of 5,000 and a refresh of 100. The area for CA19-9 is the same as before with a value of 0.8733, indicating that CA125 was not helpful in predicting CA19-9. On the other hand, the ROC area for CA125 is now

TABLE 4.21: Posterior analysis for pancreatic cancer.

Parameter	Mean	sd	Error	Lower 2 1/2	Median	Upper 2 1/2
auc CA19-9	0.8773	0.0275	<0.0001	0.8177	0.8795	0.9251
auc CA125	0.6786	0.0438	<0.0001	0.5892	0.6799	0.7613
beta[1]	2.472	0.1237	<0.0001	2.228	2.472	2.719
beta[2]	2.944	0.2797	0.00168	2.394	2.943	3.492
delta[1]	2.665	0.1113	0.0010	2.444	2.666	2.883
delta[2]	0.5946	0.1537	0.0014	0.2992	0.594	0.8971
diff	0.1987	0.0518	<0.0001	0.0971	0.1987	0.3018

Note: Other biomarker not taken into account.

estimated as 0.6786, a larger value than in the previous analysis. This is to be expected, because the previous analysis implied that each biomarker was not a good predictor of the other. It should be noted that the Pearson correlation between the two biomarkers is only 0.25! With 40,000 observations, the MCMC errors for the parameters appear to be small enough to trust the analysis, and I would go with the two estimated ROC areas as 0.8733 and 0.6786 for CA19-9 and CA125, respectively. It appears that CA19-9 has much more accuracy in detecting pancreatic cancer than CA125.

4.7 Exercises

1. The following information is taken from Pepe [6: 9], who cites Smith, Bullock, and Catalona [18] who conducted a prostate cancer screening study using PSA and digital rectal examination (DRE) as diagnostic markers, with the results shown in Table 4.22.

   Using BUGS CODE 4.1 with 55,000 observations generated from the joint posterior distribution, with a burn in of 5,000, and a refresh of 500, find the posterior mean, median, standard deviation, MCMC error, and the lower and upper 2 1/2 percentiles of

   (a) The TPF and FPF
   (b) The PPV and NPV
   (c) The PDLR and NDLR
   (d) The MCMC error for estimating the TPF

   Based on the above, is PSA a good test for screening for prostate cancer? Explain your answer in detail.

2. Using BUGS CODE 4.1 and Table 4.3, verify Table 4.4. Generate 55,000 observations from the posterior distribution, with a burn in of 5,000 and a refresh of 100.

3. Using BUGS CODE 4.1 and Table 4.3, verify Tables 4.5 and 4.6. Use 55,000 observations generated from the posterior distribution, with a burn in of 5,000 and a refresh of 100. In addition, plot the posterior density of the PDLR.

TABLE 4.22: Screening for prostate cancer with PSA.

	Prostate cancer	
PSA	**0 (negative)**	**1 (positive)**
0 (negative	1002	145
1 (positive)	899	481

4. Using Table 4.10, find the ROC area for the Spokane Heart Study. Use a uniform prior and generate enough observations from the posterior distribution so that the MCMC error for estimating the ROC area is <0.0001.

   (a) Conduct the usual posterior analysis by calculating the posterior mean, median, standard deviation, MCMC error, and the lower and upper 2 1/2 percentiles.

   (b) Plot the posterior density of the ROC area.

   (c) Construct a boxplot for the ROC area.

5. Based on Table 4.7 and BUGS CODE 4.2, verify Table 4.9. Perform a Bayesian analysis to estimate the ROC area for mammography using 55,000 observations, with a burn in of 5,000 and a refresh of 100. The list statement of the BUGS CODE 4.2 provides the data assuming a uniform prior.

   (a) What is the posterior mean of the ROC area?

   (b) What is the posterior standard deviation of the ROC area?

   (c) What is the MCMC error for estimating the ROC area?

   (d) Is the MCMC error small enough? Explain your answer.

   (e) Plot the posterior density of ROC area.

6. Zhou, McClish, and Obuchowski [2: 159] report a study of McNeil and Hanley [19] who used gallium citrate images to locate the focal source of sepsis using a rectilinear scanner with the following results. The gold standard identified 40 patients with disease and 20 without, and it is believed that increasing scores indicate more severe sepsis (Table 4.23).

   Use 75,000 observations, with a burn in of 10,000 and a refresh of 100 with BUGS CODE 4.2, and estimate the ROC area. Note that a list statement of BUGS CODE 4.2 provides the information for the gallium citrate study, assuming a uniform prior.

   (a) What is the MCMC error for estimating the ROC area?

   (b) What is a 95% credible interval for the ROC area?

TABLE 4.23: Data for sepsis using a rectilinear scanner.

	Score					
	1	2	3	4	5	Total
Disease	12	6	3	1	18	40
No disease	11	2	3	1	3	20

Source: From Zhou, H.H., McClish, D.K., and Obuchowski, N.A. *Statistical Methods for Diagnostic Medicine*. 2002. Copyright Wiley-VCH Verlag GmbH & Co. KGaA. Reproduced with permission.

(c) Is the test for sepsis accurate test? Explain your answer!

(d) Test the null hypothesis that the ROC area is <0.65 vs. the alternative the ROC area if greater than 0.65. You will have to insert a statement involving the step command in the code.

7. Verify Equation 4.12.

8. Based on Table 4.11, verify Table 4.12. Use BUGS CODE 4.3a and generate 65,000 observations from the posterior density, with a burn in of 10,000 and a refresh of 100. Note that a list statement contains the input needed for the head trauma data corresponding to Table 4.11, assuming a uniform prior distribution.

    Suppose you want to decrease the MCMC error for estimating the ROC area (using CK-BB as the diagnostic marker) by 10%, how many additional observations (beyond 60,000) need to be generated from the posterior distribution?

9. Based on Table 4.11 and BUGS CODE 4.3b, verify Table 4.13, Hans et al. [12]. This is the head trauma example using CK-BB as a biomarker and BUGS CODE 4.3b is based on a regression model of O'Malley et al. [13]. The main objective is to estimate the ROC area with a Bayesian analysis. Generate 75,000 observations from the posterior distribution, with a burn in of 10,000 and a refresh of 100. Find the posterior mean, median, MCMC error, and 95% credible interval for

    (a) The ROC area

    (b) The regression coefficients b[1] and b[2]

    (c) The two precision parameters precy[1] and precy[2]

    What is the MCMC error for estimating the ROC area?

10. Perform a Bayesian analysis for the ROC area using the ordinal values of the mammography study. The data are given as a list statement of BUGS CODE 4.3b, assuming a uniform prior. Use 55,000 observations generated from the posterior distribution, with a burn in of 5,000 and a refresh of 100.

    Find the posterior mean, median, MCMC error, and 95% credible interval for

    (a) The ROC area

    (b) The regression coefficients b[1] and b[2]

    What is the MCMC error for estimating the ROC area?

11. Derive the formula:

$$\mathrm{AUC} = \Phi\left[a/\sqrt{1+b^2}\right]$$

where X is normally distributed, $a = (\mu_D - \mu_{\bar{D}})/\sigma_D$, and $b = \sigma_D/\sigma_{\bar{D}}$.

TABLE 4.24a: PSA and DRE for diseased patients.

	DRE	
PSA	**0 (negative)**	**1 (positive)**
0 (negative)	0	1002
1 (positive)	755	141

TABLE 4.24b: PSA and DRE for non-diseased patients.

	DRE	
PSA	**0 (negative)**	**1 (positive)**
0 (negative)	0	1002
1 (positive)	755	141

The mean and standard deviation of X for the diseased population are μ_D and σ_D, respectively, while $\mu_{\bar{D}}$ and $\sigma_{\bar{D}}$ are the mean and standard deviation of X for the non diseased. Φ is the cumulative distribution function of the standard normal distribution.

12. Verify Figure 4.5.

13. Verify Equation 4.32 for the ROC area of the clustered data.

14. Verify Table 4.15, the posterior analysis for the clustered lung cancer study of Table 4.14. Note, this information is contained in a list statement of BUGS CODE 4.4. Use 60,000 observations, a burn in of 5,000 and a refresh of 100.

15. The results in Tables 4.24a and b are found in a study by Smith, Bullock, and Catalona [17] and cited by Pepe [6: 9]. It is a prostate screening study using PSA and DRE (Table 4.24a,b). Perform a Bayesian analysis with a uniform prior using the information from Tables 4.24a and b and BUGS CODE 4.5 with 65,000 observations generated from the posterior distribution. Use a burn in of 5,000 and a refresh of 100 and compute the posterior mean, median, standard deviation, MCMC error, and 95% credible interval for:

(a) The TPF for PSA and DRE

(b) The FPF for PSA and DRE

(c) The PPV and NPV for PSA and DRE

(d) The PDLR and NDLR for PSA and DRE

Based on the above, which one is more accurate, PSA or DRE? Carefully explain your answer! What is the MCMC error for estimating the TPF of DRE? Is the error small enough?

16. Based on Tables 4.16a and b and BUGS CODE 4.5, verify the posterior analysis described in Table 4.17. Generate 70,000 observations, with a burn in of 10,000 and a refresh of 100 and use the list statement in BUGS CODE 4.5. Use a uniform prior for the parameters.

17. Verify Table 4.19, using the information in Table 4.18 and BUGS CODE 4.6. A list statement in the code gives the data (assuming a uniform prior) for the comparison of the MRI and the CT for diagnosing lung cancer. Use 105,000 observations generated from the posterior distribution, with a burn in of 5,000 and a refresh of 100.

18. Verify Table 4.21 using BUGS CODE 4.3b with 45,000 observations generated from the posterior distribution, with a burn in of 5,000 and a refresh of 100. Note that the data for this problem are contained in the first list statement of BUGS CODE 4.7. Use a uniform prior.

References

[1] Obuchowski, N.A., Lieber, M.L., and Powell, K.A. Data analysis for detection and localization of multiple abnormalities with application to mammography. *Academic Radiology*, 7:516, 2000.

[2] Zhou, H.H., McClish, D.K., and Obuchowski, N.A. *Statistical Methods for Diagnostic Medicine*. John Wiley, New York, 2002.

[3] Bogaert, J., Kuzo, R., Dymarkowski, S., Becker, R., Piessens, J., and Rademakers, F.E. Coronary artery imaging with real-time navigator three dimensional turbo field echo MR coronary angiography: Initial experience. *Radiology*, 226:707, 2003.

[4] Thaete, F.L., Fuhrman, C.R., Oliver, J.H., Britton, C.A., Campbell, W.L., Feist, J.H., Staub, W.H., Davis, P.L., and Plunkett, M.B. Digital radiography and conventional imaging of the chest: A comparison of observer performance. *American Journal of Roentgenology*, 162:575, 1994.

[5] Beam, C.A., Lyde, P.M., and Sullivan, D.C. Variability in the interpretation of screening mammograms by US radiologists. *Archives of Internal Medicine*, 156:209, 1996.

[6] Pepe, M.S. *The Statistical Evaluation of Medical Tests for Classification and Prediction.* Oxford University Press, Oxford, UK, 2003.

[7] Broemeling, L.D. *The Bayesian Analysis of Linear Models.* Marcel-Dekker, New York, 1985.

[8] Wiener, D.A., Ryan, T.J., McCabe, C.H., Kennedy, J.W., Schloss, M., Tristani, F., Chaitman, B.R., and Fisher, L.D. Correlations among history of angina, ST-segmented response and prevalence of coronary artery disease. *New England Journal of Medicine*, 301:230, 1979.

[9] Mielke, H.C., Shields, P.J., and Broemeling, L.D. Coronary artery calcium scores for men and women of a large asymptomatic population. *Cardiovascular Disease Prevention*, 2:194, 1999.

[10] Rumberger, J.A., Brundage, B.H., Rader, D.J., and Kondos, G. Electron beam computed tomographic coronary calcium scanning: A review and guidelines for use in asymptomatic patients. *Mayo Clinic Proceedings*, 74:243, 1999.

[11] Egan, J.P. *Signal Detection Theory and ROC Analysis.* Academic Press, New York, 1975.

[12] Hans, P., Albert, A., Born, D., and Chapelle, J.P. Derivation of a biochemical prognostic index in severe head injury. *Intensive Care Medicine*, 11:186, 1985.

[13] O'Malley, J.A., Zou, K.H., Fielding, J.R., and Tampany, C.M.C. Bayesian regression methodology for estimating a receiver operating characteristic curve with two radiologic applications: Prostate biopsy and spiral CT of ureteral stones. *Academic Radiology*, 8:713, 2001.

[14] Somoza, E. and Mossman, D. Biological markers and psychiatric diagnosis: Risk-benefit balancing using ROC analysis. *Biological Psychiatry*, 29:811, 1991.

[15] Obuchowski, N.A. Non parametric analysis of clustered ROC curve data. *Biometrics*, 53:567, 1997.

[16] Broemeling, L.D. Detection and localization in test accuracy: A Bayesian perspective. *Communications in Statistics: Theory and Methods*, 36(8): 1555–1564, 2007.

[17] Wieand, S., Gail, M.H., James, B.R., and James, K.J. A family of nonparametric statistics for comparing diagnostic markers with paired or unpaired data. *Biometrika*, 76:585, 1989.

[18] Smith, D.S., Bullock, A.D., and Catalona, W.J. Racial differences in operating characteristics of prostate cancer screening tests. *Journal of Urology*, 158:1861, 1997.

[19] McNeil, B.J. and Henley, J.A. Statistical approaches to the analysis of receiver operating characteristic (ROC) curves. *Medical Decision Making* 4:190, 1984.

Chapter 5

Regression and Medical Test Accuracy

5.1 Introduction

It is well known that test accuracy depends on many factors, including differences in readers and differences in various patient characteristics. For example, the age of the patient, their gender, the stage of the disease, and the therapy received, all have a bearing on the measured test accuracy. This chapter describes Bayesian regression procedures for estimating the effect of patient and reader covariates on test accuracy, as measured by classification probabilities, predictive probabilities, and diagnostic likelihood ratios. In the case of quantitative diagnostic scores, regression techniques will be used to allow for these patient and reader characteristics when estimating the receiver operating characteristic (ROC) area. For additional information on this, refer to Chapters 3 and 6 of Pepe [1] and Chapter 8 of Zhou et al. [2].

In what follows, Bayesian regression techniques for binary test scores will be illustrated with an audiology example taken from Leisenring et al. [3] and also analyzed by Pepe. In this example, the probability of a false positive on the hearing test of a patient's ear is regressed on patient covariates, including age, severity of disease, and location of the hearing test. Two modeling approaches are taken: (1) using the log linear function illustrated by Pepe, and (2) using a logistic link function. In addition, the effect of patient covariates on other measures of test accuracy, including true positive fraction (TPF) and the positive diagnostic likelihood ratio (PDLR), are examined with regression techniques using log and logit link functions.

When ordinal scores are appropriate, the effect of the covariates are implemented with ordinal regression, based on the cumulative odds model, where the ordinal scores are modeled as a partition of so-called latent variables. Armstrong and Sloan [4] were one of the first to employ latent variables for ordinal regression, while McCullagh [5] gives a good general account of using regression when the dependent variable is ordinal and the covariates are either continuous or discrete type independent variables. Several examples will illustrate the ordinal regression approach to calculating the area under the ROC curve. For example, the first study in this category is taken from Zhou et al., which is based on an earlier study of Rifkin et al. [6], where the effect of several readers on staging prostate cancer is considered. The covariates in this case are the four readers who use ultrasound to stage cancer, and it is of interest to measure the degree of agreement between the readers. A similar example involving staging melanoma is also considered as an illustration estimating the ROC area via Bayesian methods. A third example using ordinal scores that measure the response to two treatments for lung cancer, illustrates the use of ordinal regression to estimate the area under the ROC curve. This example is based on the Gregurich [7] dissertation and shows how the ROC is a way to compare two groups, in this case the two treatments. The main covariate is gender and the Bayesian analysis is based on the WinBUGS code written by Congdon [8].

The Bayesian approach using continuous scores is based on the assumption of normality for the diagnostic scores and the assumption of linearity between the dependent and independent variables, including the covariates. O'Malley et al. [9] present the Bayesian approach to estimating the ROC area when covariates are present and his WinBUGS code will be employed for the analysis of several examples, including another audiology study, which is based on Stover et al. [10]. Another example involving continuous scores is a prostate cancer study of Etzioni et al. [11], which uses prostate-specific antigen (PSA) and age to diagnose prostate cancer. The final example involving continuous scores is a hypothetical example of diagnosing type II diabetes with a fasting blood glucose test. In the latter two studies, age is the principal covariate of interest.

It should be emphasized that the ROC area of a given medical test is an estimate only and corresponds to a specific population appropriate to that study. In a given population, the ROC area changes with various subsets of patients, hence the effect of covariates via regression techniques is an important methodology when assessing the accuracy of a medical test. Covariates can have a dramatic effect on the ROC area and should always be taken into account when possible.

5.2   Audiology Study

5.2.1   Introduction

The dataset for this study can be downloaded at http://www.fhcrc.org/labs/pepe/book and is analyzed extensively in Pepe [1]; earlier analyses appear in Leisenring et al. [3,12,13]. The dataset comprises 3152 cases, where the experimental unit is an ear. There were three modalities, diagnostic tests a, b, and c, and each hearing test took place in either a room or booth. The test result was binary with 1690 tests being designated as positive (hearing impaired) and 1460 being designated as not hearing impaired. According to the gold standard, 1256 ears were indeed impaired, while 1896 were given a non-diseased status. Other patient covariates were age and disease severity.

Among the tests, 1053 were given test b, 1039 test a, and 1060 test c. Among the 1039 ears given test a, 515 took place in a room and 524 in a booth. Also, 633 were declared normal according to the gold standard, while the remaining 406 were designated hearing-impaired ears. Among the 633 who were declared normal ears, the number of false positives that occurred was 253, with 380 true negative. A subset of this dataset is examined below, namely, those ears where there was information on the false positive occurrences, that is, only abnormal (hearing impaired) ears were examined, of which there were 1276. Note that, among these, 633 were given test a, 643 test b, and 651 tests were administered in a room, while the remaining tests took place in a booth. The false positive rates were 37% and 40% for tests a and b, respectively.

5.2.2 Log link function

The conventional analysis for the audiology study appears in Pepe [1: 54], where the true positive occurrence is modeled with a generalized linear model using a log link function with age, location (room or booth), and severity of disease as patient covariates. There are 253 false positives among the 633 normal ears given test a.

The log link model is

$$\phi = \exp(\beta_1 + \beta_2 x_1 + \beta_3 x_2 + \beta_4 x_3 + \beta_5 x_2 x_3), \tag{5.1}$$

where ϕ is the probability of a false positive, x_1 is the age of the patient, x_2 indicates the location (where $x_2 = 1$ for a booth and $x_2 = 0$ for a room) and x_3 indicates either test a or b (where $x_3 = 0$ for test b and $x_3 = 1$ for test a). The covariate severity is not included in the analysis. The program statements appear below.

BUGS CODE 5.1

```
# x1 is age
# x2 is location
# x3 is test a or b
# r ls the false positive occurrence
model
      {
for(i in 1 : N) {
r[i] ~ dbern(p[i])
p[i] <- exp(beta[1]+beta[2]*x1[i]+beta[3]*x2[i]+beta[4]*x3[i]+beta[5]*x2[i]*x3[i])}

phat <- mean(p[])
  for (i in 1:5){

beta[i] ~ dnorm(0.0,0.001)}

 A<-exp(beta[3])
 B<-exp(beta[4])
```

```
C<-exp(beta[5])
    }
Data
```

list(r =

c(1,0,0,0,1,1,0,0,1,1,1,0,0,0,0,0,1,0,1,0,1,1,0,1,0,1,1,0,0,0,0,1,0,0,0,0,0,1,1,0,1,0,
0,0,0,0,1,1,0,1,0,1,0,0,0,1,0,0,0,0,1,1,0,0,0,1,0,0,1,0,1,1,0,0,1,0,1,0,0,1,0,0,1,1,1,
0,1,1,0,0,0,0,0,0,1,1,0,0,0,0,1,0,0,0,1,0,1,0,0,1,0,1,0,1,0,0,0,0,0,1,1,0,0,0,0,0,0,0,
1,1,0,0,0,1,0,0,0,1,1,1,0,0,0,0,1,0,0,0,1,0,0,1,1,1,0,0,1,0,1,1,1,0,1,0,1,0,1,0,0,1,0,
0,0,1,0,0,1,1,0,1,1,1,1,1,1,0,0,0,0,1,0,0,0,0,1,0,0,1,0,0,0,1,0,1,1,1,0,0,1,1,0,0,1,1,
1,1,0,0,0,0,0,1,0,1,0,0,0,0,1,0,0,1,0,0,1,0,0,0,1,0,1,0,0,0,1,0,0,0,1,0,0,1,0,0,1,0,0,1,
0,0,0,0,0,1,0,0,1,0,0,1,1,1,1,0,0,0,0,0,0,1,0,1,0,1,1,1,1,0,0,1,0,0,0,0,1,1,0,0,1,1,
0,1,1,0,1,0,0,0,0,0,0,0,0,1,1,1,0,0,1,0,0,1,0,1,0,0,0,0,0,0,0,0,0,0,1,0,0,0,0,1,0,1,0,
0,1,0,0,0,0,0,1,0,1,0,0,1,0,1,0,0,0,0,0,0,1,0,0,1,0,1,0,0,0,0,0,0,1,1,0,0,0,1,0,1,0,1,
0,0,0,0,1,1,0,0,0,0,1,1,0,0,1,1,0,1,1,1,0,0,0,1,1,0,1,0,1,0,0,0,0,0,0,1,1,0,0,1,0,0,0,
0,0,0,0,1,1,1,0,0,0,1,0,1,0,0,0,0,1,0,0,0,0,0,1,0,0,0,1,1,1,1,1,0,0,0,1,0,1,0,0,0,0,0,
1,0,0,1,0,0,0,0,1,0,1,1,1,0,1,1,0,0,1,0,0,0,0,1,1,1,1,1,0,0,0,0,1,0,0,0,1,1,0,0,0,0,1,
0,0,0,0,1,1,0,1,1,1,0,0,0,1,1,0,0,0,0,1,0,0,0,0,1,0,0,0,1,1,0,0,0,0,0,0,0,0,0,1,0,1,0,0,
1,1,1,0,1,1,1,0,0,1,0,0,0,0,0,0,1,1,0,1,0,0,0,1,1,0,0,0,0,0,1,1,0,0,0,0,0,0,0,1,1,1,
1,1,1,1,1,1,1,0,0,0,1,1,0,0,0,0,0,0,1,1,0,1,1,1,0,0,0,0,0,0,0,0,0,0,1,0,0,0,1,0,0,0,1,0,
0,1,1,0,0,0,0,0,0,0,0,1,0,0,1,1,1,0,0,0,1,0,1,0,0,0,0,1,0,0,1,1,0,0,0,0,0,1,1,0,1,0,0,
1,1,0,0,0,0,1,0,1,0,0,0,0,1,0,1,0,0,0,1,1,0,0,0,1,1,1,1,1,0,0,0,0,1,0,0,0,1,0,0,1,0,1,
0,1,1,1,0,0,0,0,1,1,1,0,0,1,0,0,0,1,1,1,0,1,0,0,1,0,1,1,0,0,0,0,0,1,0,0,0,0,1,1,0,1,0,
0,0,1,0,0,1,1,0,0,0,0,1,1,0,0,0,1,0,1,1,1,1,1,1,1,0,0,0,0,1,0,0,0,0,0,1,0,0,0,0,1,0,0,
0,0,0,1,0,1,1,1,1,0,0,0,0,0,1,1,0,0,1,0,0,0,1,1,1,0,1,1,1,1,1,1,0,1,0,0,0,1,1,1,1,0,
0,1,1,1,0,1,0,0,0,0,1,0,0,1,0,1,0,0,0,1,0,1,0,1,0,0,1,0,0,1,0,0,1,1,1,0,0,1,1,0,0,0,1,
0,1,1,1,0,0,0,0,0,0,1,1,1,0,1,0,0,0,1,0,0,1,0,1,0,1,0,0,1,0,0,1,0,0,1,0,1,0,0,0,0,1,1,0,0,
0,1,1,0,0,0,0,0,0,0,0,0,0,1,0,0,0,0,0,0,0,0,0,1,0,1,1,0,0,0,1,0,0,0,1,0,1,0,0,0,0,0,0,
0,1,0,1,0,0,1,1,0,0,1,0,0,0,1,0,0,0,1,1,0,0,0,0,0,0,0,0,0,1,0,1,1,0,0,1,1,1,0,1,1,0,0,
0,0,0,1,1,1,1,1,1,0,0,1,0,1,1,0,0,0,0,1,1,0,0,0,1,0,0,0,0,1,1,0,0,0,0,1,0,0,0,1,0,0,1,
1,1,1,1,0,0,1,0,0,0,0,1,0,0,0,1,1,0,1,0,0,1,0,1,1,1,1,1,1,0,1,0,1,0,1,0,1,0,0,0,1,0,1,
0,0,0,1,0,0,1,1,0,0,0,0,0,1,0,0,0,0,0,1,1,1,1,1,1,1,1,0,0,1,1,1,0,1,0,1,0,1,0,1,0,0,1,1,0,0,
0,0,1,0,0,1,0,1,0,0,0,1,1,0,0,0,0,0,1,0,0,0,0,0,0,1,0,1,1,0,0,0,1,1,0,0,0,1,0,0,1,1,0,0,
1,1,0,1,0,0,1,0,0,1,1,1,0,0,1,0,0,0,0,1,1,1,0,1,1,1,1,1,0,1,1,1,1,0,1,0,0,1,0,0,0,0,0,0,
1,0,0,0,0,1,1,0,0,1,0,0,0,0,1,1,1,1,1,0,0,1,0,1,1,0,0,1,1,0),

N = 1276,

x3=

c(0,0,0,0,0,0,0,0,0,0,0,0,0,0,0,0,0,0,0,0,0,0,0,0,0,0,0,0,0,0,0,0,0,0,0,0,0,0,0,0,0,0,0,
0,0,0,0,0,0,0,0,0,0,0,0,0,0,0,0,0,0,0,0,0,0,0,0,0,0,0,0,0,0,0,0,0,0,0,0,0,0,0,0,0,0,0,0,0,
0,0,0,0,0,0,0,0,0,0,0,0,0,0,0,0,0,0,0,0,0,0,0,0,0,0,0,0,0,0,0,0,0,0,0,0,0,0,0,0,0,0,0,0,0,
0,0,0,0,0,0,0,0,0,0,0,0,0,0,0,0,0,0,0,0,0,0,0,0,0,0,0,0,0,0,0,0,0,0,0,0,0,0,0,0,0,0,0,0,0,

0,0,0,0,0,0,0,0,0,0,0,0,0,0,0,0,0,0,0,0,0,0,0,0,0,0,0,0,0,0,0,0,0,0,0,0,0,0,0,0,
0,0,0,0,0,0,0,0,0,0,0,0,0,0,0,0,0,0,0,0,0,0,0,0,0,0,0,0,0,0,0,0,0,0,0,0,0,0,0,0,
0,0,0,0,0,0,0,0,0,0,0,0,0,0,0,0,0,0,0,0,0,0,0,0,0,0,0,0,0,0,0,0,0,0,0,0,0,0,0,0,
0,0,0,0,0,0,0,0,0,0,0,0,0,0,0,0,0,0,0,0,0,0,0,0,0,0,0,0,0,0,0,0,0,0,0,0,0,0,0,0,
0,0,0,0,0,0,0,0,0,0,0,0,0,0,0,0,0,0,0,0,0,0,0,0,0,0,0,0,0,0,0,0,0,0,0,0,0,0,0,0,
0,0,0,0,0,0,0,0,0,0,0,0,0,0,0,0,0,0,0,0,0,0,0,0,0,0,0,0,0,0,0,0,0,0,0,0,0,0,0,0,
0,0,0,0,0,0,0,0,0,0,0,0,0,0,0,0,0,0,0,0,0,0,0,0,0,0,0,0,0,0,0,0,0,0,0,0,0,0,0,0,
0,0,0,0,0,0,0,0,0,0,0,0,0,0,0,0,0,0,0,0,0,0,0,0,0,0,0,0,0,0,0,0,0,0,0,0,0,0,0,0,
0,0,0,0,0,0,0,0,0,0,0,0,0,0,0,0,0,0,0,0,0,0,0,0,0,0,0,0,0,0,0,0,0,0,0,0,0,0,0,0,
0,0,0,0,0,0,0,0,0,0,0,0,0,0,0,0,0,0,0,0,0,0,0,0,0,0,0,0,0,0,0,0,0,0,0,0,0,0,0,0,
0,0,0,0,0,0,0,0,0,0,0,0,0,0,0,0,0,0,0,0,0,0,0,0,0,0,0,0,0,0,0,0,0,0,0,0,0,0,0,0,
0,0,0,0,0,0,0,0,0,0,0,0,0,0,0,0,0,0,0,0,0,0,0,0,0,0,0,0,0,0,0,0,0,0,0,0,0,0,0,1,
1,1,1,1,1,1,1,1,1,1,1,1,1,1,1,1,1,1,1,1,1,1,1,1,1,1,1,1,1,1,1,1,1,1,1,1,1,1,1,1,
1,1,1,1,1,1,1,1,1,1,1,1,1,1,1,1,1,1,1,1,1,1,1,1,1,1,1,1,1,1,1,1,1,1,1,1,1,1,1,1,
1,1,1,1,1,1,1,1,1,1,1,1,1,1,1,1,1,1,1,1,1,1,1,1,1,1,1,1,1,1,1,1,1,1,1,1,1,1,1,1,
1,1,1,1,1,1,1,1,1,1,1,1,1,1,1,1,1,1,1,1,1,1,1,1,1,1,1,1,1,1,1,1,1,1,1,1,1,1,1,1,
1,1,1,1,1,1,1,1,1,1,1,1,1,1,1,1,1,1,1,1,1,1,1,1,1,1,1,1,1,1,1,1,1,1,1,1,1,1,1,1,
1,1,1,1,1,1,1,1,1,1,1,1,1,1,1,1,1,1,1,1,1,1,1,1,1,1,1,1,1,1,1,1,1,1,1,1,1,1,1,1,
1,1,1,1,1,1,1,1,1,1,1,1,1,1,1,1,1,1,1,1,1,1,1,1,1,1,1,1,1,1,1,1,1,1,1,1,1,1,1,1,
1,1,1,1,1,1,1,1,1,1,1,1,1,1,1,1,1,1,1,1,1,1,1,1,1,1,1,1,1,1,1,1,1,1,1,1,1,1,1,1,
1,1,1,1,1,1,1,1,1,1,1,1,1,1,1,1,1,1,1,1,1,1,1,1,1,1,1,1,1,1,1,1,1,1,1,1,1,1,1,1,
1,1,1,1,1,1,1,1,1,1,1,1,1,1,1,1,1,1,1,1,1,1,1,1,1,1,1,1,1,1,1,1,1,1,1,1,1,1,1,1,
1,1,1,1,1,1,1,1,1,1,1,1,1,1,1,1,1,1,1,1,1,1,1,1,1,1,1,1,1,1,1,1,1,1,1,1,1,1,1,1,
1,1,1,1,1,1,1,1,1,1,1,1,1,1,1,1,1,1,1,1,1,1,1,1,1,1,1,1,1,1,1,1,1,1,1,1,1,1,1,1,
1,1,1,1,1,1,1,1,1,1,1,1,1,1,1,1,1,1,1,1,1,1,1,1,1,1,1,1,1,1,1,1,1,1,1,1,1,1,1,1,
1,1,1,1,1,1,1,1,1,1,1,1,1,1,1,1,1,1,1,1,1,1,1,1,1,1,1,1,1,1,1),

x1 =

c(35,33,33,38,40,40,38,38,31,44,44,37,38,38,35,31,40,32,32,32,31,34,36,32,41,
46,30,33,33,40,43,32,32,38,38,35,37,34,34,32,40,38,38,35,30,24,38,38,29,34,27,
30,35,35,35,30,40,40,37,35,39,39,32,32,32,39,35,34,33,33,37,34,25,37,32,31,39,
39,33,37,37,34,43,39,39,34,41,46,38,35,32,36,36,44,42,33,37,35,41,41,41,41,30,
36,40,33,40,40,39,39,35,41,33,33,33,33,38,36,36,29,30,34,37,38,38,34,32,32,40,
40,44,44,39,43,43,36,26,41,41,34,34,20,30,34,34,36,34,34,33,33,26,26,33,37,37,
33,33,34,35,35,43,43,44,31,38,36,34,37,37,37,37,36,29,29,32,41,38,33,33,36,37,
37,35,31,26,25,35,35,25,37,48,33,37,41,27,38,38,39,40,40,32,36,36,39,39,35,35,
39,33,37,45,45,33,33,39,30,30,31,31,35,35,30,32,32,40,40,36,33,42,42,27,46,46,
35,31,35,31,32,32,39,34,39,38,36,32,30,30,34,34,37,37,38,29,35,30,30,30,34,37,
37,36,37,38,40,40,29,44,44,27,31,31,40,36,36,34,34,42,34,34,38,37,40,37,34,37,
23,24,38,38,39,28,32,40,38,33,33,36,34,39,34,45,43,43,31,34,34,38,37,37,36,33,
35,28,44,44,28,28,42,34,34,34,34,44,40,33,30,37,44,44,34,34,47,26,38,38,29,29,
43,43,40,42,36,40,40,35,37,30,27,41,41,29,29,29,35,39,39,31,42,28,36,36,38,37,
31,31,34,29,27,39,30,40,40,34,36,32,40,30,43,40,36,36,40,40,36,44,34,34,32,40,
40,31,31,31,34,34,32,46,38,32,32,41,41,40,36,36,30,30,32,38,37,37,41,34,34,32,

34,34,47,38,31,38,34,34,36,34,34,40,37,30,39,36,36,40,40,37,37,32,38,30,28,30,
30,47,40,34,34,32,38,30,39,29,54,54,28,36,35,35,37,35,35,43,31,37,35,26,29,42,
35,34,45,38,38,32,37,41,41,33,36,47,37,38,27,30,

26,26,38,29,38,45,30,38,33,32,40,39,40,42,34,34,42,40,32,32,33,33,40,46,
40,39,38,38,33,33,37,37,34,30,32,32,31,38,38,35,44,44,45,45,31,31,31,32,32,49,
38,39,39,39,23,31,31,30,27,40,40,45,35,33,38,38,32,31,35,35,32,27,32,21,21,34,
34,42,34,38,46,37,39,39,37,36,31,38,38,43,43,31,32,37,32,32,41,41,35,35,31,27,
36,34,28,30,30,37,34,47,47,27,32,36,36,36,36,40,40,34,36,36,40,39,39,41,38,40,
34,37,38,30,30,30,26,26,36,32,42,47,47,27,30,42,27,39,39,38,38,41,37,31,31,42,
39,39,37,37,34,34,33,33,40,40,38,31,44,37,38,38,35,31,32,32,36,34,34,34,36,32,
41,41,46,46,34,33,33,40,35,43,32,38,38,32,35,35,37,34,40,38,38,31,30,24,38,38,
36,29,38,27,30,35,35,35,30,40,40,37,35,39,38,32,32,39,40,34,34,40,33,33,37,34,
34,35,25,37,37,32,31,31,39,39,33,37,37,34,35,43,43,39,41,38,32,36,36,44,37,35,
41,41,41,36,32,33,33,40,40,39,39,35,40,41,31,33,33,33,32,38,36,36,29,30,34,37,
38,34,32,32,40,44,44,39,43,36,36,34,41,34,26,20,20,30,34,36,34,34,30,33,26,26,
33,31,37,37,33,34,35,34,35,35,43,43,44,32,37,38,36,36,34,34,37,37,37,36,29,29,
41,40,38,33,37,37,35,35,26,26,35,37,37,39,33,37,27,38,38,30,39,40,40,32,36,36,
39,35,42,35,39,37,29,45,45,33,33,30,30,31,35,30,32,32,40,40,42,42,27,46,35,38,
32,31,35,32,32,39,38,38,36,32,30,30,30,34,34,37,33,34,29,35,30,30,34,37,37,37,
40,29,44,44,27,27,31,40,40,36,34,34,42,34,34,37,23,23,39,38,38,39,39,28,32,40,
38,33,33,36,34,34,45,49,43,43,34,34,38,38,37,37,33,28,44,44,35,28,42,34,34,35,
34,34,34,40,33,37,37,44,44,34,47,38,38,29,29,43,36,40,40,30,38,30,41,41,29,29,
35,39,31,31,28,33,39,36,36,38,22,36,38,38,31,34,29,27,39,30,30,40,40,34,34,36,
32,40,40,40,36,40,40,34,34,40,31,31,34,32,32,32,30,38,41,41,40,36,30,37,37,41,
34,32,34,34,47,38,38,31,38,38,45,34,38,34,40,37,31,30,30,39,39,36,36,40,37,38,
30,30,47,40,40,32,34,28,32,38,38,30,35,36,39,29,54,54,36,35,35,33,37,35,35,30,
33,43,30,37,35,26,35,34,34,38,38,32,37,29,41,33,33,37,38,35,30,30,26,26,38,38,
38,29,45,30,35,33,40,40,40,34,42,40,32,33,33,40,46,46,39,39,38,38,33,33,37,37,
34,34,30,32,32,38,38,35,35,44,44,31,45,45,31,31,31,32,38,38,39,40,39,23,45,31,
31,30,30,41,43,40,40,45,35,35,33,38,38,32,35,33,32,31,35,35,40,27,27,32,21,36,
34,42,34,38,46,37,38,39,37,36,30,38,38,43,43,38,32,32,41,41,35,35,26,40,36,36,
28,28,30,30,37,37,34,34,47,47,27,36,36,36,36,40,40,34,36,40,40,36,39,39,41,38,
36,40,34,34,37,38,30,30,30,30,26,36,32,32,42,42,47,27,30,30,36,42,42,27,27,39,
39,38,41,41,37,31,31,33,37,39,39,37,37,34,34),

x2 =

c(0,1,1,0,0,0,0,0,0,1,1,1,0,0,1,0,1,0,1,1,0,0,1,0,1,0,0,1,1,1,1,1,1,1,1,0,1,0,0,0,1,1,
1,1,0,0,0,0,0,0,1,0,0,1,1,0,0,0,0,0,0,0,0,0,0,1,0,0,0,1,1,0,1,1,0,0,1,0,0,1,1,1,1,
1,1,0,1,0,1,1,1,1,1,1,0,0,0,0,0,0,1,0,1,0,0,1,1,1,0,1,1,1,1,0,0,0,0,1,1,1,1,1,0,0,0,
0,0,0,0,1,1,0,0,0,0,0,1,1,0,0,1,0,1,1,1,1,1,1,1,1,1,1,1,1,1,1,1,0,0,0,1,0,1,0,0,0,0,0,0,
0,0,0,1,1,1,0,0,0,1,1,0,0,0,1,0,0,0,0,0,1,1,1,1,1,1,1,0,1,1,0,1,1,0,0,0,1,0,0,1,0,0,1,1,
0,0,0,0,0,1,1,1,1,1,1,1,1,1,0,0,1,0,0,1,1,0,1,0,1,1,0,1,0,0,0,0,0,0,0,0,1,1,0,0,0,1,1,1,
1,0,0,1,0,0,1,1,0,0,0,0,1,1,0,1,1,0,0,1,1,1,0,0,1,1,1,1,1,0,1,1,0,1,0,0,1,1,1,1,0,0,0,0,
1,0,0,0,0,0,0,0,0,0,1,0,1,1,1,0,0,1,1,0,0,0,1,1,0,0,0,1,1,1,1,0,0,1,1,0,0,1,1,0,0,0,0,

0,1,1,1,1,0,1,0,0,0,1,0,0,0,0,0,0,1,1,0,0,1,0,0,1,0,1,0,0,0,1,0,1,0,0,1,1,1,0,0,1,0,1,
1,1,0,0,1,0,0,1,1,1,0,1,1,1,1,1,0,0,0,1,1,0,0,0,0,0,1,1,1,1,1,0,0,0,0,1,1,0,1,1,1,1,0,
0,0,0,0,0,1,1,1,1,0,1,0,0,0,1,0,0,0,1,0,1,1,0,0,0,1,0,0,1,1,0,0,1,0,0,0,1,0,1,1,0,1,1,
0,1,1,1,0,1,0,1,1,1,1,0,0,0,0,1,0,1,0,0,0,0,1,1,0,0,0,0,1,1,1,0,0,1,0,0,0,0,0,0,0,0,0,
1,1,0,0,0,0,0,1,1,1,0,0,0,0,1,1,1,0,1,1,0,1,1,1,1,1,0,1,1,1,0,1,0,1,1,1,1,0,1,1,1,1,1,1,
0,0,1,1,0,1,0,1,1,1,1,0,1,1,0,0,0,1,0,0,0,0,0,1,1,1,1,1,1,1,1,1,1,0,1,1,1,1,0,1,1,1,0,0,0,
0,1,1,1,0,1,1,1,0,0,0,1,1,1,1,1,1,1,1,0,0,1,1,1,1,0,0,1,1,0,0,0,0,1,1,0,0,0,1,1,0,0,0,1,
1,0,0,0,0,1,1,0,0,1,0,0,1,0,1,0,1,1,0,1,1,0,0,0,1,1,1,1,1,1,1,1,0,0,0,1,0,1,1,1,1,0,0,
0,0,1,0,1,0,1,0,1,1,1,0,0,0,0,0,1,0,0,0,0,1,1,0,0,0,0,1,1,1,1,0,0,1,1,1,0,0,1,0,0,1,0,
1,1,1,1,1,1,1,1,1,0,0,0,0,0,1,1,1,1,0,0,1,1,1,0,0,1,1,1,1,0,0,0,0,0,1,1,1,1,0,0,0,0,0,
0,0,1,0,0,0,0,1,0,1,1,0,0,0,1,1,0,1,1,1,1,1,1,1,1,1,1,0,0,1,1,0,0,1,0,1,1,0,0,0,0,0,0,
0,0,1,1,0,1,0,1,1,0,0,0,0,0,0,0,1,1,1,1,1,1,0,1,1,0,1,1,0,0,0,1,0,1,1,0,0,1,1,0,0,0,
1,1,1,1,1,1,0,0,1,0,1,0,0,1,0,0,0,1,0,0,0,0,1,0,0,0,0,1,1,1,0,0,1,1,1,0,0,0,1,0,0,0,0,
0,1,0,0,1,0,0,1,1,1,0,1,1,1,1,1,0,0,1,0,0,1,1,1,0,0,1,0,0,0,0,0,0,0,0,0,1,1,1,1,0,0,
1,1,1,1,0,0,0,1,0,0,0,1,1,1,0,1,1,0,0,1,0,0,0,1,0,1,0,1,0,0,1,0,0,0,0,1,0,0,0,1,1,1,1,
1,0,0,0,1,0,1,1,0,0,0,0,1,0,1,1,1,1,0,0,1,1,0,1,0,1,1,1,1,0,1,1,1,0,0,1,0,0,0,1,1,1,1,
0,0,0,0,0,0,1,1,1,1,1,1,1,0,0,0,0,0,0,1,1,0,0,0,1,1,0,0,0,0,1,1,0,1,0,1,1,0,0,1,0,0,
0,1,1,1,1,0,0,0,0,0,1,1,1,1,1,0,1,1,1,0,1,1,1,0,1,1,0,0,0,0,0,0,0,1,0,0,0,1,1,0,0,1,
1,0,0,1,0,0,0,0,0,0,0,0,0,1,1,0,0,0,0,1,1,1,1,0,0,0,0,1,1,1,1,0,0,1,0,1,1,0,0,
0,0,1,1,0,1,1,0,1,1,1,1,0,0,0,1,1,0,1,1,1,1,1,0,1,1,0,1,0,0,1,1,1,0,1,1,0,0,1,0,0,0,0,
1,1,1,0,1,1,1,1,1,1,0,0,1,1,1,1,0,1,1,0,0,0,0,1,1,0,0,1,1,1,1,0,1,0,0,0,1,1,1,1,1,1,1,
1,0,0,0,0,1,1,1,1,1,0,0,0,0,1,1,0,0,0,0,0,1,1,1,0,0,0,0,1,1,0,0,0))

Inits
    list(beta=c(0,0,0,0,0))

The r vector is the occurrence of false positives (1 indicates a false positive and 0 a true negative). The vector xl contains the age of a patient, and the vector of locations (room or booth) is denoted by x2. The vector of false positive probabilities is given by p[], and the beta coefficients are the regression parameters on the log scale. These parameters are given a vague normal prior with mean 0.0 and precision 0.0001, whereas the precision parameters are given non-informative gamma priors, with hyper parameters 0.0001 and 0.0001. The main parameters are A, B, and C, where A estimates the ratio of the false positive rate of a booth to that of a room. The number of samples generated is 65,000, with a burn in of 5,000 and a refresh of 100. The Bayesian analysis is given in Table 5.1 and shows that the Markov Chain Monte Carlo (MCMC) errors produce estimates that are accurate to two decimal places for estimating the "true" posterior characteristics.

Age does not appear to have much of an effect on the probability of a false positive, however as for location, the ratio of the false positive fraction (FPF) of a booth to the FPF of a room is estimated by parameter A, which has a posterior mean of 0.9615 and a 95% credible interval of (0.7805, 1.179). This estimate is adjusted for age, the type of test, and the interaction between location and type of test. It also appears that the effect of the interaction between the location (room or booth) of the test and the type of test (a or b) given by

TABLE 5.1:　Posterior distribution of audiology study—full model.

Parameter	Mean	sd	MCMC error	2.5%	Median	97.5%
A	0.9615	0.1021	0.0035	0.7805	0.9568	1.179
B	0.9841	0.1003	0.0030	0.8053	1.188	1.191
C	1.25	0.1784	0.0060	0.9351	1.239	1.628
beta[1]	−1.312	0.2667	0.0075	−1.657	−1.13	−0.6181
beta[2]	0.00402	0.0070	<0.0001	−0.0096	0.0040	0.0178
beta[3]	−0.0449	0.1059	0.0037	−0.2479	−0.0441	0.165
beta[4]	−0.0212	0.102	0.0030	−0.1265	−0.02	0.1751
beta[5]	0.2134	0.1421	0.0047	−0.0671	0.2143	0.4875

beta[5] and the parameter C is not important, indeed the 95% credible intervals for the primary parameters A, B, and C all include unity. Perhaps the analysis should be done with the interaction term not included? See Exercise 2 at the end of the chapter.

5.2.3　Logistic link

Another plausible approach to assessing the effect of patient covariates on the false positive rate of the audiology example is to use a logistic link function:

$$\log(\phi/(1-\phi)) = \beta_1 + \beta_2 x_1 + \beta_3 x_2 + \beta_4 x_3 + \beta_5 x_2 x_3, \qquad (5.2)$$

where ϕ is the probability of a false positive, x_1 is the vector of ages, x_2 denotes the location of the test, and x_3 denotes the type of test. The regression coefficients are on the logit scale and β_5 measures the effect of the interaction between location and type of test. BUGS CODE 5.2 provides the Bayesian analysis for estimating the plausibility of the logistic model for the audiology study.

BUGS CODE 5.2

```
# x1 is age
# x2 is location
# x3 is test a vs b
# r is the false positive rate
model;
    {
for( i in 1 : N ) {
r[i] ~ dbern(p[i])
# the logistic link
logit( p[i]) <-(beta[1] + beta[2]*x1[i]+beta[3]*x2[i]+beta[4]*x3[i]+beta[5]*x2[i]*x3[i]
    }
```

```
phat <- mean(p[])
for (i in 1:5 ){

beta[i] ~ dnorm(0.0,0.0001)}
A<-exp(beta[2])
B<-exp(beta[3])
C<-exp(beta[4])
D<-exp(beta[5])
    }
```

The first list statement in BUGS CODE 5.1 provides the information on false positive occurrences, location, age, and type of test. The r vector is as before, the vector of false positive occurrences among ears given the hearing test a or b, x2 is the vector of locations (room or booth), and x3 (=1 for test a and 0 for test b) is the indicator vector for test mode. The logistic link is quite natural as a model for binary information, such as the occurrence of a false positive, and has the advantage over the log link, in that with the latter the false positive rate might exceed the value 1. On the other hand, the interpretation of the model parameters is somewhat more complex with the logistic link, that is, with the logistic link, one employs odds ratios, while with the log link, the interpretation of the model coefficient is as a ratio of probabilities.

We now return to a different aspect of the audiology example, which was described by Pepe [1: 57]. Here, the false positive occurrence was regressed on the test modality (a vs. b), the location (room vs. booth), and the location by test interaction. The latter interaction estimates the odds ratio of a false positive for test a vs. b for a booth compared to the odds ratio of a false positive for test a vs. test b for a room. This interaction effect is estimated by the $\exp(beta[5]) = D$, in the above listing of program statements. In this program, one must import the three vectors r, x2, and x3 into the worksheet above, from the internet address given before, and place them in the list statement of BUGS CODE 5.1. Note that the regression coefficients are given vague prior normal distributions that estimate the overall false positive rate among those ears receiving tests a and b hearing modalities. The logistic Bayesian analysis gave the posterior results depicted in Table 5.2.

Regarding D as the main parameter of interest, the odds ratio (of test a vs. test b) for a booth vs. the odds ratio (of test a vs. test b) for a room is estimated as 1.441, and the corresponding credible interval is (0.8927, 2.207), and, of course, this is an adjusted estimate. Thus, the odds ratio (of test a vs. test b) for a booth is 44% larger, compared to the odds ratio (of test a vs. test b) for a room. Figure 5.1 provides a graph of the posterior density of D. The simulation used 65,000 observations from the posterior distribution of D, with a burn in of 5,000 and a refresh of 100. As with the log regression model shown in Equation 5.1, the logistic model estimates that the effects of the various factors (age, location, type of test, and interaction) on the logit of the false positive rate are minimal. For example, the effect of age is quite small, with a 95%

TABLE 5.2:   Posterior distribution of false positive rate.

Parameter	Mean	sd	Error	2.5%	Median	97.5%
A	1.007	0.0119	<0.00001	0.9839	1.007	1.031
B	0.9511	0.157	<0.0001	0.6789	0.9328	1.296
C	0.9846	0.1606	<0.0001	0.7075	0.9721	1.336
D	1.441	0.3379	0.0016	0.8927	1.402	2.207
beta[1]	−0.762	0.4436	0.0020	−1.642	−0.761	0.1022
beta[2]	0.0068	0.0118	<0.00001	−0.016	0.0068	0.0301
beta[3]	−0.0635	0.164	<0.0001	−0.3873	−0.0637	0.2591
beta[4]	−0.0286	0.1621	<0.0001	−0.346	−0.028	0.2893
beta[5]	0.3387	0.2312	0.0011	−0.1135	0.3376	0.7915

credible interval of (−0.016,0.0301) on the logit of false positive occurrence, and the corresponding parameter A, which estimates the ratio of the odds of the false positive rate of a room to that of a booth, therefore regardless of the model used, the implications are much the same, and it appears that the most reasonable model is one that does not take into account the effect of age, location, and type of test. The simulation produced very small MCMC errors for all parameters, that is, the posterior means of all parameters appear to be estimated quite accurately with the above estimates given by Table 5.2.

5.2.4   Diagnostic likelihood ratio

The PDLR is briefly discussed in Section 5.3.4, and is defined as

$$PDLR = TPF/FPF,$$

where TPF is the true positive fraction, and FPF is the false positive fraction. Thus, if

$$\log(TPF) = beta[1] + beta[2]X, \tag{5.3}$$

and

$$\log(FPF) = alpha[1] + alpha[2]X,$$

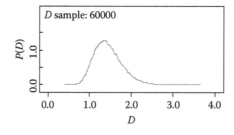

FIGURE 5.1: Posterior density of D.

TABLE 5.3: Posterior distribution of the PDLR—audiology example.

Parameter	Mean	sd	2.5%	Median	97.5%
alpha[1]	−1.547	0.1273	−1.786	−1.557	−1.282
alpha[2]	1.382	0.1296	1.105	1.392	1.621
beta[1]	−1.005	0.052	−1.111	−1.004	−0.9053
beta[2]	0.08506	0.07002	−0.04601	0.0839	0.2214
fphat	0.3826	0.01357	0.3565	0.3828	0.4088
tphat	0.6016	0.01625	0.5727	0.6043	0.6369
pdlr	1.582	0.0709	1.45	1.58	1.725
rpdlrab	3.696	0.5335	2.641	3.688	4.795

where X is the indicator function for test a or b (X = 1 for test a and 0 for test b), then

$$\log(\text{PDLR}) = \text{beta}[1] - \text{alpha}[1] + (\text{beta}[2] - \text{alpha}[2])X, \qquad (5.4)$$

and exp(beta[2]-alpha[2]) is the ratio of the PDLR for test a relative to test b. Note that the TPF is estimated from the diseased patients and the FPF from only the non diseased.

Below is the code for estimating the PDLR and the effect of the test modality (test a or b) on the PDLR. The following program can be downloaded from http://medtestacc.blogspot.com onto the worksheet and executed. There are 529 hearing-impaired ears where there is information on the true positive status (given by the tp vector), along with the matched information on test a or b (given by the x4 vector). With regard to the false positive information, there are 1276 normal ears with false positive information indicated by the vector fp, and the corresponding information on the occurrence of test a or test b, given by the vector x5. There are two parameters of interest, namely, the PDLR and the ratio of the PDLR for test a relative to test b. The results of the Bayesian analysis are given in Table 5.3.

BUGS CODE 5.3

```
Model;
{
for( i in 1 : M ) {
fp[i] ~ dbern(p[i])
p[i] <- exp(beta[1] + beta[2]*x5[i])
}
for(i in 1:N) { tp[i]~dbern(q[i])
q[i]<-exp(alpha[1]+alpha[2]*x4[i])
}
tphat<-mean(q[])
fphat <- mean(p[])
```

```
for (i in 1:2 ){
beta[i] ~ dnorm(0.0,0.0001)
alpha[i] ~ dnorm(0.0,0.0001)
}
A<-exp(alpha[2])
B<-exp(beta[2])
rpdlrab <-exp(alpha[2]-beta[2])
pdlr<-tphat/fphat
}
```

What does this tell us about the effect of test modality on the value of the PDLR? The average value over tests a and b is 1.58, with a 95% credible interval of 1.45–1.725. On the other hand, the ratio of the PDLR for test a relative to test b is given by rpdlrab and has a posterior mean of 3.696, that is, the average value of the positive diagnostic ratio for test a is 3.69 times the PDLR for test b. The analysis is executed with 25,000 observations, with a burn in of 1,000 and a refresh of 100, and all the MCMC errors were <0.001. See Exercise 4 for additional information.

5.3   Receiver Operating Characteristic Area and Patient Covariates

5.3.1   Introduction

When the diagnostic scores are binary, accuracy is measured by sensitivity and specificity, and other classification probabilities, however, when they are continuous, the acknowledged method of determining accuracy is with the area under the ROC curve. If the diagnostic score is continuous, the score could be dichotomized, and the sensitivity and specificity estimated, however, this could result in a loss of information, leading to unreliable estimates of test accuracy. When the continuous score is dichotomized, the threshold value must be chosen with care, as we have seen in previous chapters.

As with binary scores, patient covariate effects on ordinal or continuous diagnostic scores should be taken into account when estimating the area under the ROC curve. The problem of assessing the effect of covariates on the ROC curve was first considered by Tosteson and Begg [14] for ordinal data, but was later generalized to continuous scores by Toledano and Gatsonis [15]. For a good introduction to the subject, see Chapter 6 of Pepe [1] and Chapter 8 of Zhou et al. [2], both of whom present several methods of incorporating patient covariate information into the ROC curve. Pepe, for example, describes non-parametric, semi-parametric, and parametric regression methods to estimate the ROC area. See also Alonzo and Pepe [16] for using distribution-free methods in the estimation of the ROC area.

For the binormal model (assuming the diagnostic scores are normally distributed), the induced ROC curve technique of regressing the diagnostic score on covariates will be adopted for the Bayesian approach taken here.

In what follows, several examples will illustrate the effect of patient covariates on the ROC curve. First, an example is taken from two therapies for lung cancer and was analyzed from a Bayesian viewpoint by Gregurich [7], who developed an ordinal regression technique, based on a generalized least squares concept, to compare the two therapies. The diagnostic score is the response to therapy, measured on a four-point ordinal scale. The area under the ROC curve measures the separation between the two therapy groups of patients. Other examples involving ordinal scores are the staging of lung cancer with ultrasound and the staging of melanoma involving several radiologists. The principal covariate in both studies is the several radiologists, thus the degree of agreement among them is also of interest in estimating the ROC area.

5.3.2 Ordinal regression methods

An ordinal regression model is employed to estimate the ROC area for medial tests with ordinal scores. This particular formulation of regression uses an underlying latent scale assumption. The cumulative odds model is often expressed in terms of an underlying continuous response. The following specification of the ordinal model follows Congdon [8: 102], where the observed response score, Y_i, with possible values $1, 2, \ldots, K$ is taken to reflect an underlying continuous part of the cumulative probability:

$$\gamma_{ij} = \Pr(Y_i \leq j) = F(\theta_j - \mu_i), \tag{5.5}$$

where $i = 1, 2, \ldots, N$ is the number of patients and $j = 1, 2, \ldots, K - 1$.

It is noted that

$$\mu_i = \beta X_i, \tag{5.6}$$

expresses the relationship between the ordinal responses and the covariates, X_i, for the ith patient. F is a distribution function and θ_j are the cut points corresponding to the jth rank. For our purposes, F is usually given a logistic or probit link, where the former leads to a proportional odds model. Suppose p_{ij} is the probability that the ith patients have response j, then

$$\gamma_{ij} = p_{i1} + p_{i2} + \cdots + p_{ij}. \tag{5.7}$$

Of course, Equation 5.7 can also be inverted to give

$$p_{i1} = \gamma_{i1},$$
$$p_{ij} = \gamma_{ij} - \gamma_{i,j-1},$$

and

$$p_{i,K} = 1 - \gamma_{i,K-1}. \tag{5.8}$$

Suppose F is the logistic distribution function, and

$$C_{ij} = \text{logit}(\gamma_{ij})$$
$$= \theta_j - \beta X_i, \tag{5.9}$$

where β, the vector of unknown regression coefficients, is constant across response categories j, then θ_j are the logits of the probabilities of belonging to the categories $1, 2, \ldots j$ as compared to belonging to the categories $j + 1, \ldots, K$, for subjects with $X = 0$. The difference in cumulative logits for different values of X, say X_1 and X_2, is independent of j, which is called the proportional odds assumption, namely,

$$C_{1j} - C_{2j} = \beta(X_1 - X_2). \tag{5.10}$$

Using the above ordinal regression model, the posterior distribution of the individual probabilities, p_{ij}, are determined, as are the probabilities q_j $(j = 1, 2, \ldots, K)$ of the basic ordinal responses.

Once the posterior distribution of the basic responses is known for the diseased and non-diseased groups, the posterior distribution of the area under the ROC curve can also be computed. Several scenarios will be displayed for a given example of ordinal regression: (a) the ROC area induced by all covariates or selected subsets of covariates, and (b) the ROC area conditional on certain values of the covariates or subsets of covariates.

The Gregurich [7] dissertation is based on a lung cancer clinical trial with two therapeutic strategies, where the treatments were compared with respect to tumor response. The sequential therapy method was used in the first group, who were given the same combination of agents, while the second group was given an alternating approach, with three different combinations of agents, alternating from cycle to cycle. The tumor response was assessed at the end of treatment as progressive disease, no change, partial remission, and complete remission. This is an unconventional application of the ROC area in that it is usually computed to measure the accuracy of a medical test between diseased and non-diseased populations, however, it should be remembered that the ROC area measures the separation between two populations and is another way to view the two sample problem. Table 5.4 displays the response to therapy.

TABLE 5.4: Tumor response from Gregurich [7].

Therapy	Gender	Progressive disease	No change	Partial remission	Complete remission	Total
Sequential	Male	28	45	29	26	128
	Female	4	12	5	2	23
Total		32	57	34	28	151
Alternating	Male	41	44	20	20	125
	Female	12	7	3	1	23
Total		53	51	23	21	148

The two treatments were sequential and alternating, and for each treatment both males and females responded to treatment. For example, for sequential therapy, there are 28 out of 151 who have a complete response, compared to 21 out of 148 who received alternating therapy. Based on this evidence, it appears that the two treatments are almost equivalent in the pattern of response to therapy, and one would expect a ROC area close to 0.50. Note the small fraction of females for both treatments—15% for the sequential treatment and 15% for the alternating arm. Thus, an important consideration for the analysis is: does gender have an effect on the ROC area? If gender is important and does have an effect, one would want to estimate the ROC area conditionally, say for males and for females separately.

Our Bayesian analysis is based on the following code, which closely follows the ordinal regression model outlined in Equations 5.5 through 5.10. Note that the remarks denoted by # explain the analysis by citing the important steps in determining the posterior distribution of the ROC area. The program statements are taken from Congdon [8: 102].

BUGS CODE 5.4

```
model;
{
# ROC area
# see Gregurich
# code is from Congdon [8: 102]
# non diseased patients
for(i in 1:151){for(j in 1:4){logit(ndgamma[i,j])<-ndtheta[j]-ndmu[i]}}
for(i in 1:151){ndp[i,1]<-ndgamma[i,1]}
for(i in 1:151){ndp[i,2]<-ndgamma[i,2]-
ndgamma[i,1]}
for(i in 1:151){ndp[i,3]<-ndgamma[i,3]-
ndgamma[i,2]}
for(i in 1:151){ndy[i]~dcat(ndp[i,1:4])}
for(i in 1:151){ndp[i,4]<-1-ndgamma[i,3]}
# intercept depends on y
for(i in 1:151){
ndmu[i]<-ndb0[ndy[i]]+ndx1[i]*ndb[1]
}
# prior distribution for the regression coefficients
ndb[1]~dnorm(0,.0001)
for(i in 1:4){ndb0[i]~dnorm(0,.001)}
# the following give the cut points for the latent variable
ndtheta[1]~dnorm(0,1)
ndtheta[2]~dnorm(0,1)
ndtheta[3]~dnorm(0,1)
ndtheta[4]~dnorm(0,1)I(ndtheta[3],)
# the ndq[i] are the probabilities of the responses for non diseased
```

```
for( i in 1:4){ndq[i]<-mean(ndp[,i])}
# diseased patients
for(i in 1:148){for(j in 1:4){logit(dgamma[i,j])<-dtheta[j]-dmu[i]}}
for(i in 1:148){dp[i,1]<-dgamma[i,1]}
for(i in 1:148){dp[i,2]<-dgamma[i,2]-
dgamma[i,1]}
for(i in 1:148){dp[i,3]<-dgamma[i,3]-
dgamma[i,2]}
for(i in 1:148){dp[i,4]<-1-
dgamma[i,3]}
for(i in 1:148){dy[i]~dcat(dp[i,1:4])}
# intercept depends on y
for(i in 1:148){
dmu[i]<-db0[dy[i]]+dx1[i]*db[1]}
# prior distributions for the regression coefficients
db[1]~dnorm(0,.0001)
for(i in 1:4){db0[i]~dnorm(0,.001)}
# the following dtheta are the cut points for the underlying
# latent variable
dtheta[1]~dnorm(0,1)
dtheta[2]~dnorm(0,1)
dtheta[3]~dnorm(0,1)
dtheta[4]~dnorm(0,1)I(dtheta[3],)
# the dq[i] are the probabilities of the four ordinal responses for diseased
for( i in 1:4){dq[i]<-mean(dp[,i])}
# roc area
area<-a1+a2/2
a1<-dq[2]*ndq[1]+dq[3]*(ndq[1]+ndq[2])+
dq[4]*(ndq[1]+ndq[2]+ndq[3])
# a2 is the probability of a tie
a2<-dq[1]*ndq[1]+dq[2]*ndq[2]+dq[3]*ndq[3]+
dq[4]*ndq[4]
}
list(
# for sequential therapy
ndy=c(1,1,1,1,1,1,1,1,1,1,1,1,
      1,1,1,1,1,1,1,1,1,1,1,1,
      1,1,1,1,
      2,2,2,2,2,2,2,2,2,2,2,2,2,2,2,
      2,2,2,2,2,2,2,2,2,2,2,2,2,2,2,
      2,2,2,2,2,2,2,2,2,2,2,2,2,2,2,
      3,3,3,3,3,3,3,3,3,3,3,3,3,3,3,
      3,3,3,3,3,3,3,3,3,3,3,3,3,3,3,
      4,4,4,4,4,4,4,4,4,4,4,4,4,4,
      4,4,4,4,4,4,4,4,4,4,4,
```

```
        1,1,1,1,
        2,2,2,2,2,2,2,2,2,2,2,2,
        3,3,3,3,3,
        4,4),

ndx1=c(1,1,1,1,1,1,1,1,1,1,1,1,1,1,1,1,1,1,1,1,1,
        1,1,1,1,1,1,1,1,1,1,1,1,1,1,1,1,1,1,1,1,1,1,
        1,1,1,1,1,1,1,1,1,1,1,1,1,1,1,1,1,1,1,1,1,1,
        1,1,1,1,1,1,1,1,1,1,1,1,1,1,1,1,1,1,1,1,1,1,
        1,1,1,1,1,1,1,1,1,1,1,1,1,1,1,1,1,1,1,1,1,1,
        1,1,1,1,1,1,1,1,1,1,1,1,1,1,1,1,1,1,1,1,1,1,
        1,1,1,1,1,1,1,1,
        0,0,0,0,0,0,0,0,0,0,0,0,0,0,0,0,0,0,0,0,0,
        0,0,0),

# data for diseased
# for alternating therapy
dy=c(
        1,1,1,1,1,1,1,1,1,1,1,1,1,1,1,
        1,1,1,1,1,1,1,1,1,1,1,1,1,1,1,
        1,1,1,1,1,1,1,1,1,1,1,
        2,2,2,2,2,2,2,2,2,2,2,2,2,2,2,
        2,2,2,2,2,2,2,2,2,2,2,2,2,2,2,
        2,2,2,2,2,2,2,2,2,2,2,2,2,2,
        3,3,3,3,3,3,3,3,3,3,3,3,3,3,3,
        3,3,3,3,3,
        4,4,4,4,4,4,4,4,4,4,4,4,4,4,4,
        4,4,4,4,4,
        1,1,1,1,1,1,1,1,1,1,1,1,1,
        2,2,2,2,2,2,2,2,
        3,3,3,
        4),
dx1=c(
        1,1,1,1,1,1,1,1,1,1,1,1,1,1,1,1,1,1,1,1,1,
        1,1,1,1,1,1,1,1,1,1,1,1,1,1,1,1,1,1,1,1,1,1,
        1,1,1,1,1,1,1,1,1,1,1,1,1,1,1,1,1,1,1,1,1,1,
        1,1,1,1,1,1,1,1,1,1,1,1,1,1,1,1,1,1,1,1,1,1,
        1,1,1,11,1,1,1,1,1,1,1,1,1,1,1,1,1,1,1,1,1,1,
        1,1,1,1,1,
        0,0,0,0,0,0,0,0,0,0,0,0,0,0,0,0,0,0,0,0,0,
        0,0,0))

# initial values
list(ndtheta=c(0,0,0,0),dtheta=c(0,0,0,0))
# initialize other values
```

The Bayesian analysis is executed with 65,000 observations, with a burn in of 5,000 and a refresh of 100. Note that in each of the two populations (sequential and alternating treatments) there is a separate regression function for each one of the four ordinal responses (1, 2, 3, 4), and the effect of gender is assumed to be uniform for all four responses, thus, there is a separate intercept for the four responses (1, 2, 3, 4). In total, there are eight regression functions for the analysis, namely, four for the sequential therapy and four for the alternating therapy. Also in the above code, non-informative normal prior distributions are placed on the regression coefficients. The primary objective is to estimate the ROC area and to determine the effect of gender on the area.

The non-disease designation refers to sequential therapy and the disease label refers to alternating therapy, thus for Table 5.5, ndb[1] refers to the gender effect of the sequential therapy and db[1] to the gender effect of the alternating therapy. The value of db0[1] is the estimate (posterior mean) of the intercept for alternative therapy when the response is 1, while the corresponding value for the sequential therapy is nb0[1], which has a posterior mean of -31.02. Recall that the four responses are: $1 = $ progressive disease, $2 = $ no change, $3 = $ partial remission, and $4 = $ complete remission. The effect of gender on the logit of a response appears to be the same for both therapies, and the posterior means of the intercepts are quite similar for both treatments, that is to say, the intercepts corresponding to the response for each of the four responses (1, 2, 3, 4) are almost the same for both therapies.

The most important parameter is the ROC area, estimated as 0.4177 (0.0092), which corresponds to a ROC area of 0.582 if the roles of the two therapies are reversed. This implies that sequential therapy is more effective

TABLE 5.5:   Posterior distribution for Gregurich study.

Parameter	Mean	sd	Error	2 1/2	Median	97 1/2
a1	0.286	0.00888	<0.0001	0.2689	0.2859	0.3039
a2	0.2632	0.003984	<0.00001	0.2557	0.2631	0.2713
Area	0.4176	0.009211	<0.0001	0.3998	0.4175	0.436
db[1]	0.0268	0.8505	0.0495	−1.558	−0.00221	1.835
db0[1]	−31.299	17.46	0.1822	−73.87	−27.52	−0.524
db0[2]	−2.059	1.026	0.05745	−4.027	−2.069	−0.09438
db0[3]	2.329	1.059	0.0529	0.2989	2.34	4.508
db0[4]	30.6	18.1	0.2001	8.805	26.58	75.02
dq[1]	0.3863	0.01195	<0.0001	0.3676	0.3848	0.4138
dq[2]	0.3108	0.0187	<0.0001	0.2727	0.3113	0.3466
dq[3]	0.1366	0.0176	<0.0001	0.1031	0.1359	0.1727
dq[4]	0.1663	0.0104	<0.0001	0.1498	0.165	0.1896
ndb[1]	−0.0077	0.642	0.0297	−1.283	−0.0105	1.243
ndb0[1]	−31.02	17.4	0.196	−73.11	−27.16	−9.226
ndb0[2]	−2.259	0.8358	0.0337	−3.934	−2.257	−0.5516
ndb0[3]	2.446	0.888	0.03661	0.771	2.44	4.268
ndb0[4]	30.78	17.69	0.1968	8.84	26.73	74.07
ndq[1]	0.2415	0.01226	<0.0001	0.2224	0.24	0.2693
ndq[2]	0.3443	0.0194	<0.0001	0.3045	0.3447	0.3816
ndq[3]	0.2019	0.0186	<0.0001	0.1657	0.2017	0.2391
ndq[4]	0.2123	0.0111	<0.0001	0.1948	0.2111	0.2382

than alternating therapy, but not by much, which can be deduced by looking at the raw data in Table 5.4.

What is the effect of gender on the ROC area? It appears that gender has little effect on the logit scale for both therapies, thus one would expect the ROC area to be the same or close to 0.582 if gender is not taken into account.

The remaining parameters are dq[i] and ndq[i] for $i = 1, 2, 3, 4$. The latter are the posterior means of the probability of response i for the sequential therapy, while the former are the corresponding values for the alternating therapy, and both are used in the formula for the ROC area given by

$$\text{area} = \text{a1} + \text{a2}/2, \tag{5.11}$$

where

$$\begin{aligned} \text{a1} = &\ \text{dq}[2]^*\text{ndq}[1] + \text{dq}[3]^*(\text{ndq}[1] + \text{ndq}[2]) \\ &+ \text{dq}[4]^*(\text{ndq}[1] + \text{ndq}[2] + \text{ndq}[3]), \end{aligned} \tag{5.12}$$

and

$$\text{a2} = \text{dq}[1]^*\text{ndq}[1] + \text{dq}[2]^*\text{ndq}[2] + \text{dq}[3])^* + \text{ndq}[3] + \text{dq}[4]^*\text{ndq}[4]. \tag{5.13}$$

Equations 5.12 and 5.13 appear as code for the ROC area in BUGS CODE 5.4. Thus, so far, the conclusion for this study is that gender has no effect on a small estimated ROC area of 0.582. One word of caution is that the MCMC errors for db0[1], ndb0[1], and ndb0[4] are large relative to the errors of the other parameters. For the most part, the errors are quite small, but one might want to increase the simulation size to see the effect on the MCMC error for those parameters that have relative large values with 65,000 observations initially generated from the joint posterior distribution.

Using a non-parametric technique, I verified this value with the ROC area in SPSS (version 11.5) with a value of 0.583(0.033).

5.3.3   Staging metastasis for melanoma: Accuracy of four radiologists

As a second example with ordinal scores, a study involving melanoma metastasis to the lymph nodes is considered. A sentinel lymph node biopsy is performed on patients to determine the degree of metastasis, where the diagnosis is made on the basis of the depth of the primary lesion, the Clark level of the primary lesion, and the age and gender of the patient. The procedure involves the cooperation of an oncologist, a surgical team that dissects the primary tumor, pathologists, and radiologists who perform the imaging aspect of the biopsy. A radiologist makes the primary determination of the degree of metastasis on a five-point ordinal scale, where 1 = absolutely no evidence of metastasis, 2 = no evidence of metastasis, 3 = very little evidence of metastasis, 4 = some evidence of metastasis, and 5 = strong evidence of metastasis. For

TABLE 5.6:   Metastasis of melanoma patients—rating of metastasis.

Reader	Metastasis	1	2	3	4	5	Total
1	0	12	10	5	2	1	30
1	1	3	7	11	13	16	50
2	0	15	7	4	3	1	30
2	1	2	8	10	12	18	50
3	0	11	9	2	3	5	30
3	1	8	10	6	10	16	50
4	0	13	8	6	2	1	30
4	1	10	6	8	14	12	50

more information on the procedure, refer to Pawlik and Gershenwald [17] and Gershenwald et al. [18].

The study is paired where each radiologist examines each patient. The results of the hypothetical example are reported in Table 5.6.

The melanoma study has one covariate, namely, the reader, thus the study is analyzed under the following scenarios: (1) using the effect of the radiologists simultaneously, (2) determining if the effect of the four is the same on the ROC area, (3) determining the ROC area separately for each reader, and (4) estimating the ROC area conditionally on a particular reader. The analysis is executed with the following code.

BUGS CODE 5.5

```
model;
{
# 4 readers
# ROC area
# melanoma example
# code is from Congdon [8: 102]
# non diseased
for(i in 1:30){for(j in 1:5){logit(ndgamma[i,j])<-ndtheta[j]-ndmu[i]}}
for(i in 1:30){ndp[i,1]<-ndgamma[i,1]}
for(i in 1:30){ndp[i,2]<-ndgamma[i,2]-
ndgamma[i,1]}
for(i in 1:30){ndp[i,3]<-ndgamma[i,3]-
ndgamma[i,2]}
for(i in 1:30){ndp[i,4]<-ndgamma[i,4]-
ndgamma[i,3]}
for(i in 1:30){ndy[i]~dcat(ndp[i,1:5])}
for(i in 1:30){ndp[i,5]<-1-ndgamma[i,4]}
# intercept depends on y
for(i in 1:30){
ndmu[i]<-ndb0[ndy[i]]+ ndx1[i]*ndb[1]+ndx2[i]*ndb[2]+ndx3[i]*ndb[3]}
```

```
for(i in 1:3){ndb[i]~dnorm(0,.001)}
for(i in 1:5){ndb0[i]~dnorm(0,.001)}
ndtheta[1]~dnorm(0,1)
ndtheta[2]~dnorm(0,1)
ndtheta[3]~dnorm(0,1)
ndtheta[4]~dnorm(0,1)
ndtheta[5]~dnorm(0,1)I(ndtheta[4],)
for(i in 1:5){ndq[i]<-mean(ndp[,i])}
# diseased population
for(i in 1:50){for(j in 1:5){logit(dgamma[i,j])<-dtheta[j]-dmu[i]}}
for(i in 1:50){dp[i,1]<-dgamma[i,1]}
for(i in 1:50){dp[i,2]<-dgamma[i,2]-
dgamma[i,1]}
for(i in 1:50){dp[i,3]<-dgamma[i,3]-
dgamma[i,2]}
for(i in 1:50){dp[i,4]<-dgamma[i,4]-
dgamma[i,3]}
for(i in 1:50){dy[i]~dcat(dp[i,1:5])}
for(i in 1:50){dp[i,5]<-1-dgamma[i,4]}
# intercept depends on y
for(i in 1:50){
dmu[i]<-db0[dy[i]]+ dx1[i]*db[1]+dx2[i]*db[2]+dx3[i]*db[3]}
for(i in 1:3){db[i]~dnorm(0,.001)}
for(i in 1:5){db0[i]~dnorm(0,.001)}
dtheta[1]~dnorm(0,1)
dtheta[2]~dnorm(0,1)
dtheta[3]~dnorm(0,1)
dtheta[4]~dnorm(0,1)
dtheta[5]~dnorm(0,1)I(dtheta[4],)
for( i in 1:5){dq[i]<-mean(dp[,i])}
# roc area
area<-a1+a2/2
a1<-dq[2]*ndq[1]+dq[3]*(ndq[1]+ndq[2])+
dq[4]*(ndq[1]+ndq[2]+ndq[3])+
dq[5]*(ndq[1]+ndq[2]+ndq[3]+ndq[4])
a2<-dq[1]*ndq[1]+dq[2]*ndq[2]+dq[3]*ndq[3]+
dq[4]*ndq[4]+dq[5]*ndq[5]
}
list(
ndy=c(1,1,1,1,1,1,1,1,1,1,1,1,1,
      2,2,2,2,2,2,2,2,2,2,
      3,3,3,3,3,
      4,4,
      5,
```

```
1,1,1,1,1,1,1,1,1,1,1,1,1,1,1,
2,2,2,2,2,2,2,
3,3,3,3,
4,4,4,
5,
1,1,1,1,1,1,1,1,1,1,1,1,
2,2,2,2,2,2,2,2,2,
3,3,
4,4,4,
5,5,5,5,5,
1,1,1,1,1,1,1,1,1,1,1,1,1,
2,2,2,2,2,2,2,2,
3,3,3,3,3,3,
4,4,
5
```

  ),

```
ndx1=c(1,1,1,1,1,1,1,1,1,1,1,1,1,1,1,1,1,1,1,1,
       1,1,1,1,1,1,1,1,1,1,

   0,0,0,0,0,0,0,0,0,0,0,0,0,0,0,0,0,0,0,0,0,
   0,0,0,0,0,0,0,0,0,0,
   0,0,0,0,0,0,0,0,0,0,0,0,0,0,0,0,0,0,0,0,0,
   0,0,0,0,0,0,0,0,0,0,
   0,0,0,0,0,0,0,0,0,0,0,0,0,0,0,0,0,0,0,0,0,
   0,0,0,0,0,0,0,0,0,0
   ),
ndx2=c(0,0,0,0,0,0,0,0,0,0,0,0,0,0,0,0,0,0,0,0,0,
       0,0,0,0,0,0,0,0,0,0,
       1,1,1,1,1,1,1,1,1,1,1,1,1,1,1,1,1,1,1,1,1,
       1,1,1,1,1,1,1,1,1,1,
       0,0,0,0,0,0,0,0,0,0,0,0,0,0,0,0,0,0,0,0,0,
       0,0,0,0,0,0,0,0,0,0,
       0,0,0,0,0,0,0,0,0,0,0,0,0,0,0,0,0,0,0,0,0,
       0,0,0,0,0,0,0,0,0,0
   ),

ndx3=c(0,0,0,0,0,0,0,0,0,0,0,0,0,0,0,0,0,0,0,0,0,
       0,0,0,0,0,0,0,0,0,0,
       0,0,0,0,0,0,0,0,0,0,0,0,0,0,0,0,0,0,0,0,0,
       0,0,0,0,0,0,0,0,0,0,
       1,1,1,1,1,1,1,1,1,1,1,1,1,1,1,1,1,1,1,1,1,
       1,1,1,1,1,1,1,1,1,1,
       0,0,0,0,0,0,0,0,0,0,0,0,0,0,0,0,0,0,0,0,0,
```

```
        0,0,0,0,0,0,0,0,0,0,0
    ),
# data for diseased
dy=c(1,1,1,
     2,2,2,2,2,2,2,
     3,3,3,3,3,3,3,3,3,3,3,
     4,4,4,4,4,4,4,4,4,4,4,4,4,
     5,5,5,5,5,5,5,5,5,5,5,5,5,5,5,5,

     1,1,
     2,2,2,2,2,2,2,2,2,
     3,3,3,3,3,3,3,3,3,3,
     4,4,4,4,4,4,4,4,4,4,4,4,
     5,5,5,5,5,5,5,5,5,5,5,5,5,5,5,5,5,5,
     1,1,1,1,1,1,1,
     2,2,2,2,2,2,2,2,2,2,
     3,3,3,3,3,3,
     4,4,4,4,4,4,4,4,4,4,
     5,5,5,5,5,5,5,5,5,5,5,5,5,5,5,5,
     1,1,1,1,1,1,1,1,1,1,1,
     2,2,2,2,2,2,
     3,3,3,3,3,3,3,3,
     4,4,4,4,4,4,4,4,4,4,4,4,4,4,
     5,5,5,5,5,5,5,5,5,5,5,5
    ),

dx1=c(
     1,1,1,1,1,1,1,1,1,1,1,1,1,1,1,1,1,1,1,1,1,
     1,1,1,1,1,1,1,1,1,1,1,1,1,1,1,1,1,1,1,1,1,
     1,1,1,1,1,1,1,1,1,1,

     0,0,0,0,0,0,0,0,0,0,0,0,0,0,0,0,0,0,0,0,0,
     0,0,0,0,0,0,0,0,0,0,0,0,0,0,0,0,0,0,0,0,0,
     0,0,0,0,0,0,0,0,0,0,
     0,0,0,0,0,0,0,0,0,0,0,0,0,0,0,0,0,0,0,0,0,
     0,0,0,0,0,0,0,0,0,0,0,0,0,0,0,0,0,0,0,0,0,
     0,0,0,0,0,0,0,0,0,0,

     0,0,0,0,0,0,0,0,0,0,0,0,0,0,0,0,0,0,0,0,0,
     0,0,0,0,0,0,0,0,0,0,0,0,0,0,0,0,0,0,0,0,0,
     0,0,0,0,0,0,0,0,0,0
    ),
dx2=c(0,0,0,0,0,0,0,0,0,0,0,0,0,0,0,0,0,0,0,0,0,
     0,0,0,0,0,0,0,0,0,0,0,0,0,0,0,0,0,0,0,0,0,
     0,0,0,0,0,0,0,0,0,0,
```

```
      1,1,1,1,1,1,1,1,1,1,1,1,1,1,1,1,1,1,1,1,1,
      1,1,1,1,1,1,1,1,1,1,1,1,1,1,1,1,1,1,1,1,1,
      1,1,1,1,1,1,1,1,1,1,
      0,0,0,0,0,0,0,0,0,0,0,0,0,0,0,0,0,0,0,0,0,
      0,0,0,0,0,0,0,0,0,0,0,0,0,0,0,0,0,0,0,0,0,
      0,0,0,0,0,0,0,0,0,0,
      0,0,0,0,0,0,0,0,0,0,0,0,0,0,0,0,0,0,0,0,0,
      0,0,0,0,0,0,0,0,0,0,0,0,0,0,0,0,0,0,0,0,0,
      0,0,0,0,0,0,0,0,0,0

   ),

dx3=c(0,0,0,0,0,0,0,0,0,0,0,0,0,0,0,0,0,0,0,0,0,
      0,0,0,0,0,0,0,0,0,0,0,0,0,0,0,0,0,0,0,0,0,
      0,0,0,0,0,0,0,0,0,0,
      0,0,0,0,0,0,0,0,0,0,0,0,0,0,0,0,0,0,0,0,0,
      0,0,0,0,0,0,0,0,0,0,0,0,0,0,0,0,0,0,0,0,0,
      0,0,0,0,0,0,0,0,0,0,
      1,1,1,1,1,1,1,1,1,1,1,1,1,1,1,1,1,1,1,1,1,
      1,1,1,1,1,1,1,1,1,1,1,1,1,1,1,1,1,1,1,1,1,
      1,1,1,1,1,1,1,1,1,1,
      0,0,0,0,0,0,0,0,0,0,0,0,0,0,0,0,0,0,0,0,0,
      0,0,0,0,0,0,0,0,0,0,0,0,0,0,0,0,0,0,0,0,0,
      0,0,0,0,0,0,0,0,0,0
   ))

list(ndtheta=c(0,0,0,0,0),dtheta=c(0,0,0,0,0))
```

The first list statement gives the basic information, where ndy refers to the ratings for those patients without metastasis, ndx1 gives the indicator (1 indicates the corresponding rating in ndy given by radiologist 1 and 0 otherwise) for the first reader for those patients without metastasis, the ndx2 for the second reader, etc. The variable dy refers to the rating for the patients with metastasis, while dx1 is the column of the indicator (a zero indicates the first rater did not give the rating and a 1 indicates reader 1 gives the rating) for the first reader, for those patients with metastasis, etc. Refer to Table 5.6 and the first list statement and the meaning is obvious for coding the data.

A Bayesian analysis is executed with 75,000 observations, a burn in of 5,000 and a refresh of 100. The output is given in Table 5.7 with the following identification for the parameters: area refers to the ROC area, db[1] refers to the effect of reader 1 for the diseased (those with metastasis) patients, while db[3] is the effect for reader 3. In addition, db0[1] is the estimate of the intercept corresponding to the ordinal response 1 for diseased patients, while db0[5] is the intercept corresponding to ordinal response 5. Continuing in a similar fashion, ndb[1] is the effect of reader 1 on the logit scale for

TABLE 5.7:   Bayesian analysis for melanoma study with four radiologists.

Parameter	Mean	sd	Error	2 1/2	Median	97 1/2
a1	0.711	0.0296	<0.0001	0.6501	0.712	0.7663
a2	0.1522	0.01172	<0.0001	0.1301	0.1519	0.1761
area	0.7871	0.0242	<0.0001	0.7373	0.788	0.8321
db[1]	0.7786	5.535	0.3389	−8.559	0.9357	9.274
db[2]	−0.0165	31.63	0.1189	−61.86	0.05461	61.66
db[3]	0.2739	31.76	0.1267	−61.73	0.1926	62.59
db0[1]	−28.85	18.65	0.2988	−73.45	−25.27	−1.922
db0[2]	−3.139	5.594	0.3402	−11.94	−3.407	6.138
db0[3]	−1.006	5.56	0.3391	−9.757	−1.276	8.071
db0[4]	1.478	5.523	0.3367	−7.289	1.309	10.56
db0[5]	29.64	18.66	0.2877	2.925	26.07	74.36
dq[1]	0.1128	0.0226	<0.0001	0.07577	0.1103	0.1628
dq[2]	0.1153	0.0308	<0.0001	0.0603	0.1137	0.1794
dq[3]	0.1812	0.0382	<0.0001	0.1105	0.1797	0.2602
dq[4]	0.2108	0.0370	<0.0001	0.1402	0.2101	0.2861
dq[5]	0.3799	0.0258	<0.0001	0.3387	0.3769	0.4385
ndb[1]	−9.859	12.43	0.7639	−30.82	−6.585	8.275
ndb[2]	−0.0148	31.86	0.1198	−62.59	0.0011	62.66
ndb[3]	0.0700	31.52	0.109	−61.67	−0.0528	61.85
ndb0[1]	−23.55	21.04	0.4894	−70.02	−21.36	14.22
ndb0[2]	8.081	12.45	0.7644	−10.22	4.845	28.97
ndb0[3]	10.22	12.43	0.7628	−8.036	7.048	31.21
ndb0[4]	11.62	12.4	0.7588	−6.845	8.59	32.67
ndb0[5]	33.68	19.24	0.5689	1.399	32.6	76.41
ndq[1]	0.4828	0.0381	<0.0001	0.4229	0.4785	0.5693
ndq[2]	0.244	0.0475	<0.0001	0.1508	0.2438	0.3376
ndq[3]	0.1219	0.0405	<0.0001	0.0531	0.1185	0.2113
ndq[4]	0.0586	0.0288	<0.0001	0.0148	0.0546	0.125
ndq[5]	0.0926	0.0292	<0.0001	0.0468	0.0893	0.1597

non-diseased (those without metastasis) patients and ndb0[3] is the intercept corresponding to ordinal score 3 for non-diseased patients, etc. Also, dq[1] is the probability of ordinal score 1 for diseased patients, while ndq[1] is the corresponding quantity for the non-diseased patients, etc.

Note that with 75,000 observations for the simulation, the MCMC error for db[i] and db0[i] is relatively large and the corresponding posterior distributions are very skewed. Recall that db[i], for i = 1, 2, 3 are the effects of readers 1, 2, and 3, respectively, for the diseased, while db0[i], i = 1, 2, 3, 4, 5 are the intercepts for the five regressions corresponding to the five ordinal responses of those patients where the disease has metastasized. There are five regressions of the cumulative logits on the readers. The same is observed for the non-diseased patients, that is, the MCMC error is fairly large and the posterior distributions skewed for the effects of the three readers and the five intercepts. Thus, the posterior median should be used to estimate the location of the skewed distributions. It does appear though that the effects of the three

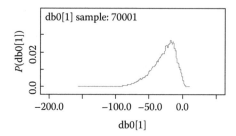

FIGURE 5.2: Posterior density of the intercept for ordinal score 1 for diseased patients.

readers on the cumulative logits are not the same for the diseased and are not the same for the non-diseased patients. It is safe to say that the pattern of the posterior medians implies that the intercepts are the same for the diseased as they are for the non diseased. The skewness of the posterior distribution of db0[1] is exhibited in Figure 5.2.

On the other hand, the MCMC error is quite small for the ROC area, which has a posterior mean of 0.7871(0.0242) and a 95% credible interval of (0.7373,0.8321). The ROC area is "adjusted" for the simultaneous effects of the four readers. I also used SPSS 11.5 to estimate the ROC area and got a value of 0.767(0.027). When the ROC area is estimated with reader 3 information only, the posterior mean is 0.6483(0.0213) and a 95% credible interval of (0.6053,0.6891). I revised BUGS CODE 5.5 and executed the analysis with 75,000 observations, a burn in of 5,000 and a refresh of 100. The MCMC error of the ROC area is <0.0001. Thus, it appears that the readers do not have the same effect on the ROC area. When all four readers are used simultaneously, the area is 0.7871, but with reader 3 it is only 0.6484. Quite a difference!

Two major scenarios have been presented for regression techniques for determining the accuracy of medical tests. Up to this point, regression methods for false positive rates and for the ROC area of ordinal test scores have been considered. The next section is a portrayal of the Bayesian analysis for continuous test scores.

5.4   Regression Methods for Continuous Test Scores

The particular regression approach used for continuous scores will be based on the assumption of normality for the test scores, both for the diseased and non-diseased populations. Such a model is called binormal and is the foundation for defining the ROC curve and its area. We begin with the following assertion.

Suppose

$$Y_d \sim N(\mu_d, \sigma_d^2)$$

and

$$Y_{nd} \sim N(\mu_{nd}, \sigma_{nd}^2),$$

then it can be shown that the ROC curve is given by

$$\text{ROC}(s) = \Phi(a + b\Phi^{-1}(s)), \qquad (5.14)$$

where

$$a = (\mu_d - \mu_{nd})/\sigma_d \qquad (5.15)$$

and

$$b = \sigma_{nd}/\sigma_d, \qquad (5.16)$$

where μ_d and μ_{nd} are the means of the diseased and non-diseased populations, respectively. In addition, σ_d and σ_{nd} are the standard deviations of the diseased and non-diseased populations, respectively.

The above result (Equation 5.14) is shown to be true as follows: For any threshold t, let

$$\text{FPF}(t) = P(Y_{nd} > t) = \Phi((\mu_{nd} - t)/\sigma_{nd})$$

and

$$TPF(t) = P(Y_d > t) = \Phi((\mu_d - t)/\sigma_d).$$

For an FPF s, the corresponding threshold is

$$t = \mu_{nd}\Phi^{-1}(s),$$

thus

$$\begin{aligned}
\text{ROC}(s) &= \text{TPF}(t) \\
&= \Phi((\mu_d - t)/\sigma_d) \\
&= \Phi((\mu_d - \mu_{nd} + \sigma_{nd}\Phi^{-1}(s))/\sigma_d) \\
&= \Phi(a + b\Phi^{-1}(s)), \qquad (5.17)
\end{aligned}$$

and the binormal ROC curve is defined by Equation 5.17. The derivation of Equation 5.15 is given by Pepe [1: 82].

With regard to the area of the binormal ROC curve, it is

$$\text{AUC} = \Phi\left(a/\sqrt{(1 + b^2)}\right), \qquad (5.18)$$

where a and b are given by Equations 5.15 and 5.16, respectively.

From a Bayesian viewpoint, the posterior distribution of the AUC is easily generated, once one knows the posterior distribution of a and b.

To show Equation 5.18, Pepe [1: 84] gives the following proof:

The ROC area is given by

$$\text{AUC} = P(Y_d > Y_{nd})$$
$$= P(W > 0),$$

where

$$W = Y_d - Y_{nd}.$$

Note that

$$W \sim N(\mu_d - \mu_{nd}, \sigma_d^2 + \sigma_{nd}^2),$$

thus

$$P(W > 0) = \Phi\left((\mu_d - \mu_{nd})/\sqrt{(\sigma_d^2 + \sigma_{nd}^2)}\right),$$

and

$$\text{AUC} = \Phi\left(a/\sqrt{(1 + b^2)}\right). \tag{5.19}$$

Another useful property of the ROC curve is that it is invariant under a monotone increasing transformation. That is, if h is such a transformation, and

$$W_d = h(Y_d)$$

and

$$W_{nd} = h(Y_{nd}),$$

then the ROC area based on W_d and W_{nd} is the same as that based on Y_d and Y_{nd}. This is also shown by Pepe [1: 85] and is left as an exercise.

The WinBUGS code, BUGS CODE 5.6, to be used for the following examples is based on Equations 5.17 and 5.18 and presented by O'Malley et al. [9]. Other Bayesian approaches to ROC area estimation with covariates are given by Peng and Hall [19] and Hellmich et al. [20], while non-Bayesian approaches to ROC estimation with covariates are described by Hanley [21] and Gatsonis [22].

5.4.1 Induced receiver operating characteristic curves

Suppose the continuous test scores are normally distributed and follow the regression function:

$$Y = \beta[1] + \beta[2]D + \beta[3]X + \beta[4]XD + \sigma(D)\varepsilon, \qquad (5.20)$$

where D is an indicator for disease, that is, $D = 0$ for non-diseased patients and $D = 1$ for diseased, $\varepsilon \sim N(0,1)$,

$$\sigma(d) = \sigma_d I[D = 1] + \sigma_{nd} I[D = 0],$$

the beta are unknown regression coefficients and X is a covariate, then it can be shown that the induced ROC curve conditional on the covariate value x is

$$\text{ROC}_x(t) = \Phi(\beta[2]/\sigma_d + \beta[4]x/\sigma_d + \sigma_{nd}\Phi^{-1}(1-t)/\sigma_d), \qquad (5.21)$$

where $0 \leq t \leq 1$. The regression function is a linear function of the regression coefficients, where the variance depends on the type of population, whether diseased or not. The main result of Equation 5.21 is a result of Equation 5.17. Remember that the ROC curve is conditional on a particular covariate value, where $X = x$.

5.4.2 Diagnosing prostate cancer with total prostate-specific antigen

In Etzioni et al. [11], of 683 patients who had their PSA levels measured for prostate cancer, 454 did not have the disease and 229 did. (To download this dataset, go to http://www.fhcrc.org/labs/pepe/book and to Chapter 1 of Pepe [1] for details of this study.) Patient covariates included age, where the average age among those with and without the disease was 64.8 years. Among those with and without prostate cancer, the total PSA was 10.31 and 2.02 mg/dL, respectively. The total PSA measurements were highly skewed to the right with mean and median levels of 2.02 and 1.31, respectively, for those without disease, but were 10.31 and 4.39, respectively, for those with cancer. Figure 5.3 shows the skewness via a P–P plot, thus it was decided to take logarithms of the total PSA levels for binomial analysis. From Figure 5.3, it appears that the log transformation did indeed induce approximate normality for the total PSA values. Also, for many subjects, repeated measurements of total PSA levels were taken, but this was not considered a covariate.

The regression of log total PSA levels on disease status ($d = 0, 1$), age, and the age by disease interaction gave the posterior analysis in Table 5.8. The effect of age by disease interaction on PSA is given by beta[4] and is quite small with a posterior mean of -0.0025 and a 95% credible interval of $(-0.0301, 0.0250)$, with a similar result for beta[3], which estimates the effect of age on average total PSA. Thus, it appears that age is not an important

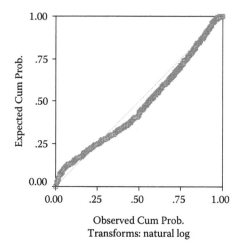

FIGURE 5.3: Total PSA for the Etzioni et al. study.

covariate to consider when estimating the ROC area, and one would expect a similar ROC area if age was not included in the model. An interesting aspect of this analysis is the posterior density of the ROC area, which has a posterior mean of 0.8057(0.1624) but a posterior median 0.8471, and the skewness is evident from Figure 5.4. I would go with the posterior median in order to estimate the accuracy of PSA to discriminate between disease and non disease. In summary, I would say that PSA is an accurate test for prostate cancer, but that age does not influence the accuracy. The analysis should be repeated without age as a covariate, but this is left as an exercise.

The above analysis is based on BUGS CODE 5.6, and 90,000 observations were generated, with a burn in of 10,000 and a refresh of 100. The regression function of PSA on disease indicator and age is given by the code:

$$mu[i] <\text{-}beta[1] + beta[2]*d[i] + beta[3]*age[i] + beta[4]*d[i]*age[i], \qquad (5.22)$$

TABLE 5.8:   Posterior analysis for prostate cancer.

Parameter	Mean	sd	Error	2 1/2	Median	97 1/2
auc	0.8057	0.1624	0.0074	0.3864	0.8471	0.991
beta[1]	−0.327	0.4031	0.01893	−1.122	−0.3195	0.4464
beta[2]	1.414	0.9202	0.0429	−0.3958	1.394	3.228
beta[3]	0.0106	0.0062	<0.001	−0.0012	0.01056	0.0229
beta[4]	−0.0025	0.0140	<0.001	−0.0301	−0.0022	0.0250
la1	1.848	1.203	0.0560	−0.5135	1.821	4.215
la2	2.173	0.2529	0.0010	1.725	2.157	2.713
vary[1]	0.5876	0.0392	<0.001	0.5156	0.5858	0.6691
vary[2]	1.271	0.1208	<0.001	1.057	1.265	1.529

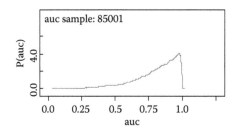

FIGURE 5.4: Posterior density of ROC area for prostate cancer.

and is based on Equations 5.15, 5.16, and 5.18. The reader should refer to the notes indicated by # in the program statements. Note that the parameters la1 and la2 are used in the formula for the ROC area. Refer to the code below and the notes indicated by #.

BUGS CODE 5.6

```
model;
# Binormal model with interaction between the true disease state and the
    covariates.
# Calculates posterior distribution of model parameters and the area under
    curve.
{
# likelihood function
      for(i in 1:N) {

      # log of y is yt
         yt[i]<-log(y[i])

               yt[i]~dnorm(mu[i],precy[d[i]+1]);
#              yt[i] <- log(y[i]); # logarithmic transformation
# the regression function
               mu[i] <- beta[1] + beta[2]*d[i]+beta[3]*age[i]+beta[4]*d[i]*age[i]
                     }
# prior distributions - non-informative prior
      for(i in 1:P) {
            beta[i] ~ dnorm(0, 0.000001);
                  }
      for(i in 1:K) {
            precy[i]~dgamma(0.001, 0.001);
            vary[i] <- 1.0/precy[i];
                  }
```

```
# calculates area under the curve
    la1 <- beta[2]/sqrt(vary[1]);
    la2 <- vary[2]/vary[1];
    auc <- phi(la1/sqrt(1+la2));
}
# psa data from Etzioni et al. [11]
list(K=2, P=4, N=683, y=c(.03,
```
.09,.23,.27,.27,.29,.29,.29,.30,.31,.33,.35,.37,.37,.42,.43,.44,.45,
.45,.46,.46,.47,.47,.48,.49,.49,.50,.50,.50,.51,.51,.55,.55,.56,.57,
.57,.58,.58,.58,.58,.59,.59,.59,.61,.61,.62,.62,.63,.63,.64,.64,.64,
.64,.65,.65,.65,.66,.66,.66,.66,.66,.66,.67,.67,.67,.67,.67,.68,.68,
.69,.69,.69,.69,.69,.70,.71,.72,.72,.73,.74,.74,.75,.75,.75,.75,.75,
.76,.76,.77,.77,.77,.77,.77,.77,.78,.78,.78,.78,.78,.78,.79,.79,.79,
.79,.80,.80,.80,.81,.81,.81,.81,.82,.83,.83,.84,.85,.86,.87,.87,.87,
.87,.87,.88,.89,.89,.89,.89,.89,.92,.92,.92,.93,.93,.93,.93,.93,.93,
.94,.94,.95,.95,.95,.95,.96,.96,.97,.97,.98,.98,.98,.98,.98,.99,1.00,
1.00,1.00,1.01,1.01,1.02,1.03,1.03,1.03,1.03,1.03,1.03,1.04,1.04,
1.04,1.04,1.04,1.05,1.05,1.05,1.05,1.06,1.06,1.06,1.06,1.07,1.07,
1.07,1.08,1.08,1.08,1.11,1.11,1.12,1.12,1.13,1.13,1.13,1.14,1.15,
1.15,1.15,1.15,1.15,1.15,1.15,1.15,1.16,1.16,1.16,1.17,1.17,1.17,
1.17,1.18,1.18,1.18,1.18,1.18,1.19,1.19,1.19,1.20,1.20,1.21,1.22,
1.22,1.22,1.23,1.23,1.24,1.24,1.24,1.25,1.25,1.25,1.25,1.25,1.25,
1.26,1.26,1.26,1.27,1.27,1.27,1.27,1.27,1.27,1.28,1.28,1.29,1.30,
1.30,1.31,1.31,1.32,1.32,1.33,1.34,1.35,1.35,1.35,1.35,1.35,1.35,
1.35,1.36,1.37,1.37,1.37,1.38,1.39,1.39,1.40,1.40,1.40,1.40,1.41,
1.41,1.41,1.41,1.41,1.41,1.43,1.43,1.43,1.43,1.44,1.44,1.45,1.46,
1.46,1.47,1.47,1.47,1.48,1.48,1.49,1.49,1.50,1.50,1.50,1.50,1.51,
1.51,1.51,1.51,1.53,1.54,1.54,1.55,1.55,1.56,1.57,1.57,1.58,1.58,
1.58,1.61,1.62,1.62,1.62,1.62,1.64,1.67,1.67,1.67,1.67,1.67,1.68,
1.69,1.69,1.70,1.70,1.70,1.71,1.71,1.71,1.71,1.71,1.71,1.71,1.71,
1.73,1.73,1.73,1.74,1.79,1.80,1.80,1.83,1.85,1.85,1.88,1.88,1.88,
1.89,1.89,1.89,1.91,1.91,1.91,1.92,1.93,1.93,1.94,1.95,1.96,2.01,
2.01,2.03,2.03,2.03,2.04,2.04,2.05,2.05,2.06,2.07,2.08,2.08,2.10,
2.11,2.13,2.13,2.14,2.16,2.17,2.19,2.19,2.19,2.22,2.22,2.23,2.24,
2.27,2.27,2.27,2.28,2.28,2.29,2.29,2.30,2.30,2.33,2.34,2.34,2.35,
2.36,2.36,2.37,2.40,2.41,2.42,2.43,2.43,2.43,2.43,2.46,2.50,2.50,
2.51,2.51,2.52,2.53,2.55,2.55,2.56,2.56,2.57,2.58,2.61,2.62,2.62,
2.63,2.63,2.63,2.66,2.69,2.70,2.71,2.73,2.77,2.79,2.82,2.82,2.82,
2.83,2.84,2.84,2.85,2.86,2.86,2.87,2.88,2.88,2.90,2.92,2.92,2.93,
2.95,2.96,2.96,2.96,2.97,2.98,3.03,3.03,3.04,3.05,3.05,3.08,3.10,
3.11,3.13,3.17,3.17,3.18,3.20,3.21,3.24,3.25,3.25,3.29,3.30,3.30,
3.32,3.32,3.33,3.34,3.35,3.38,3.41,3.42,3.43,3.45,3.51,3.55,3.57,
3.57,3.58,3.58,3.61,3.65,3.65,3.66,3.68,3.69,3.70,3.73,3.77,3.78,
3.78,3.78,3.80,3.84,3.88,3.89,3.95,3.97,3.97,4.00,4.03,4.03,4.04,

4.05,4.08,4.12,4.15,4.19,4.20,4.20,4.20,4.30,4.33,4.34,4.38,4.39,
4.40,4.41,4.44,4.47,4.47,4.48,4.52,4.54,4.60,4.62,4.64,4.70,4.75,
4.75,4.76,4.78,4.90,4.90,4.93,4.94,4.98,5.02,5.09,5.10,5.11,5.12,
5.13,5.13,5.25,5.28,5.37,5.39,5.44,5.44,5.53,5.54,5.64,5.65,5.67,
5.73,5.75,5.81,5.85,6.07,6.07,6.16,6.18,6.27,6.29,6.31,6.41,6.48,6.48,
6.50,6.52,6.52,6.54,6.54,6.56,6.56,6.77,6.92,6.93,7.09,7.19,7.21,
7.23,7.24,7.28,7.29,7.42,7.43,7.53,7.59,7.64,7.78,7.90,8.04,8.15,
8.31,8.37,8.57,8.62,8.69,9.07,9.11,9.15,9.15,9.17,9.24,9.30,9.33,
9.76,9.94,9.96,9.97,10.11,10.60,10.71,10.92,11.33,11.40,11.54,
11.62,11.65,12.69,12.69,13.61,13.94,14.82,15.41,15.84,15.84,15.89,
16.18,16.48,16.70,16.81,17.10,17.17,17.57,19.35,20.10,20.24,20.47,
20.53,21.48,22.50,23.81,24.63,25.06,26.67,27.68,29.31,31.46,33.02,
35.93,37.63,37.66,38.39,43.30,48.80,49.16,51.72,61.16,72.07,79.21,
90.66,99.97,99.98,99.98,99.98),
# disease status
d=c(.00,1.00,.00,.00,.00,.00,.00,.00,.00,.00,.00,.00,.00,.00,.00,.00,
1.00,.00,.00,.00,.00,.00,.00,.00,.00,.00,.00,.00,.00,.00,.00,.00,.00,
.00,.00,.00,.00,.00,.00,1.00,.00,.00,1.00,.00,.00,.00,.00,.00,.00,.00,
.00,.00,.00,.00,.00,.00,.00,.00,.00,.00,.00,.00,.00,.00,.00,1.00,
.00,1.00,.00,.00,.00,.00,.00,.00,1.00,.00,.00,.00,.00,.00,.00,.00,
.00,.00,.00,.00,.00,.00,.00,.00,.00,.00,.00,.00,.00,.00,.00,.00,.00,
.00,.00,.00,.00,.00,.00,.00,.00,.00,.00,.00,.00,.00,.00,.00,.00,.00,
.00,.00,.00,.00,.00,.00,.00,.00,.00,.00,.00,.00,.00,.00,.00,.00,.00,
.00,.00,1.00,.00,.00,.00,.00,.00,.00,.00,.00,.00,.00,.00,.00,.00,.00,
.00,.00,.00,.00,.00,.00,.00,.00,.00,.00,.00,.00,.00,.00,.00,.00,.00,
.00,.00,.00,.00,.00,.00,.00,.00,1.00,1.00,.00,.00,1.00,.00,.00,.00,
.00,1.00,.00,.00,.00,.00,.00,.00,.00,.00,.00,.00,.00,.00,.00,.00,
.00,.00,1.00,.00,.00,.00,.00,.00,.00,.00,1.00,1.00,.00,.00,1.00,.00,
1.00,1.00,.00,.00,.00,.00,.00,.00,.00,1.00,.00,.00,.00,.00,.00,.00,
.00,.00,1.00,.00,.00,.00,.00,.00,1.00,.00,.00,.00,.00,.00,.00,.00,
.00,.00,.00,.00,.00,.00,.00,.00,.00,.00,1.00,.00,.00,.00,.00,.00,.00,
.00,.00,.00,.00,1.00,.00,.00,.00,.00,1.00,1.00,.00,.00,1.00,1.00,.00,
.00,.00,.00,.00,.00,.00,.00,.00,1.00,.00,.00,.00,.00,1.00,1.00,.00,.00,
.00,.00,.00,.00,.00,.00,.00,.00,.00,1.00,.00,.00,1.00,.00,.00,.00,.00,
.00,.00,.00,.00,.00,.00,1.00,.00,.00,.00,.00,.00,1.00,.00,.00,.00,.00,
.00,.00,1.00,1.00,.00,.00,1.00,.00,.00,.00,.00,1.00,.00,1.00,.00,.00,
.00,.00,1.00,1.00,.00,.00,.00,.00,.00,1.00,.00,1.00,.00,.00,.00,.00,.00,
.00,.00,.00,.00,1.00,1.00,1.00,.00,1.00,.00,.00,.00,1.00,1.00,.00,.00,.00,
.00,1.00,.00,.00,.00,.00,.00,1.00,1.00,.00,1.00,.00,1.00,.00,1.00,.00,
.00,.00,1.00,.00,.00,.00,.00,.00,.00,.00,.00,.00,1.00,1.00,.00,1.00,.00,
1.00,.00,1.00,.00,1.00,.00,.00,1.00,.00,1.00,.00,.00,.00,1.00,1.00,.00,.00,
.00,.00,.00,1.00,1.00,.00,.00,.00,1.00,.00,1.00,1.00,.00,1.00,1.00,.00,
1.00,.00,.00,.00,1.00,.00,.00,.00,.00,1.00,1.00,.00,1.00,.00,.00,.00,
1.00,.00,1.00,.00,.00,1.00,.00,1.00,.00,.00,.00,1.00,.00,1.00,1.00,.00,

1.00,.00,.00,1.00,1.00,.00,.00,1.00,.00,1.00,.00,.00,.00,.00,.00,.00,
.00,1.00,1.00,.00,1.00,1.00,.00,1.00,.00,1.00,1.00,.00,.00,.00,1.00,1.00,
.00,1.00,1.00,1.00,1.00,1.00,1.00,1.00,1.00,1.00,.00,1.00,1.00,1.00,
1.00,.00,1.00,1.00,1.00,1.00,.00,.00,.00,.00,1.00,1.00,1.00,1.00,.00,
1.00,1.00,.00,1.00,.00,1.00,1.00,1.00,.00,1.00,.00,.00,.00,1.00,1.00,
.00,1.00,1.00,1.00,1.00,.00,1.00,.00,1.00,1.00,1.00,.00,1.00,1.00,.00,
.00,.00,1.00,.00,1.00,1.00,1.00,.00,1.00,.00,1.00,.00,1.00,1.00,.00,
1.00,1.00,1.00,.00,1.00,.00,1.00,.00,1.00,.00,.00,1.00,1.00,1.00,.00,
1.00,1.00,1.00,.00,1.00,.00,1.00,1.00,1.00,1.00,.00,1.00,1.00,.00,1.00,
1.00,1.00,1.00,1.00,1.00,1.00,1.00,1.00,1.00,1.00,1.00,1.00,1.00,1.00,
1.00,1.00,1.00,1.00,.00,1.00,1.00,1.00,1.00,1.00,1.00,1.00,1.00,1.00,
1.00,1.00,1.00,1.00,1.00,.00,1.00,1.00,.00,1.00,.00,1.00,1.00,1.00,
1.00,1.00,1.00,1.00,1.00,1.00,1.00,1.00,1.00,1.00,.00,1.00,1.00,
1.00,1.00,1.00,1.00,1.00,1.00,1.00,1.00,1.00,1.00,1.00,1.00),
age=c(68,70,73,74,55,57,58,59,60,61,62,63,64,57,60,62,63,68,
69,70,71,72,72,58,57,59,61,68,71,58,61,62,63,64,65,66,
67,65,69,61,64,65,66,67,58,61,62,63,53,56,58,60,63,64,
62,65,66,67,68,71,72,53,54,58,60,61,63,64,65,64,65,66,
68,70,71,68,61,63,65,67,69,70,72,58,61,62,63,64,65,66,
67,67,70,71,72,74,75,76,63,66,67,68,69,70,71,72,73,58,
62,61,65,66,67,68,70,53,55,57,62,64,66,57,59,60,61,62,
63,64,65,66,48,50,52,57,61,62,63,64,65,66,67,68,70,72,
54,57,59,62,64,65,66,67,68,69,70,71,58,61,62,63,64,65,
66,67,68,64,66,66,69,70,57,63,64,65,66,67,58,62,64,64,
65,66,67,58,62,63,64,65,67,68,61,63,65,56,57,58,60,62,
63,75,77,78,55,58,59,60,62,65,69,70,71,72,73,74,75,54,
56,57,59,61,62,57,59,61,62,65,66,67,63,65,67,55,57,58,
59,60,62,63,65,67,70,65,67,68,69,70,61,64,65,66,67,68,
61,59,61,62,63,65,66,67,68,59,62,63,64,65,58,61,62,63,
64,65,66,67,70,74,75,76,53,55,57,57,58,59,61,63,64,50,
52,53,54,55,56,57,58,59,65,67,67,68,69,70,71,72,54,56,
58,60,56,59,60,61,69,71,73,70,71,72,73,74,63,65,66,57,
60,62,63,61,65,74,78,79,80,81,61,65,66,66,67,68,71,61,
58,62,63,64,65,66,67,68,68,67,68,69,71,72,65,69,70,73,
74,59,64,65,56,60,67,70,72,72,57,68,71,73,75,61,64,68,
69,56,59,60,61,62,64,66,68,50,53,54,55,57,58,60,61,62,
63,65,67,68,65,68,69,63,66,67,68,69,70,71,72,64,68,70,
71,72,73,58,62,63,63,63,64,66,66,68,53,55,56,57,58,65,
67,68,69,70,71,72,73,75,72,76,77,78,80,59,62,63,64,65,
66,49,52,53,54,55,56,57,58,65,67,69,57,58,59,61,63,64,
58,61,62,63,64,65,66,67,68,54,57,59,67,69,71,56,60,61,
62,63,65,55,59,64,64,67,69,70,71,72,73,74,52,54,61,65,
67,70,71,71,72,66,69,70,71,72,73,74,75,76,62,71,72,73,
74,75,76,77,70,71,72,73,61,65,67,57,60,60,62,62,63,64,

65,66,70,71,72,74,75,76,62,64,65,65,68,69,69,70,71,60,
63,64,68,72,74,67,71,72,73,74,75,76,59,62,63,64,65,66,
67,68,69,64,67,68,68,69,70,72,73,74,60,47,51,52,53,54,
55,69,71,66,68,71,72,68,70,60,63,64,65,69,71,73,75,51,
53,55,65,69,70,76,58,64,66,67,65,68,70,62,63,64,65,66,
67,65,67,68,69,70,62,66,67,61,64,65,66,67,69,70,71,66,
69,70,71,72,73,59,63,64,65,66,68,69,69,60,63,65,66,67,
68,69,63,65,66,67,68,70,65,66,67,50,53,55,57,58,71))
# initial values
   list(beta=c(0,0,0,0), precy=c(1,1))

5.4.3   Stover audiology study

Another example of continuous test scores is provided by the Stover et al. [10] audiology study. (To download this dataset with 1848 cases, go to http://www.fhcrc.org/labs/pepe/book, and Chapter 6 of Pepe [1] for details of this study.) For each ear, the distortion-product otoacoustic emission (DPOAE) test is applied under several experimental settings. Each setting is defined by a particular frequency (hertz), intensity (decibel), and threshold of the auditory stimulus. The test score is the negative signal-to-noise ratio, where higher values are associated with hearing impairment. The following code does not include the data for the test response, the disease incidence, the frequency, threshold, and intensity from the study, but the analysis is executed with 75,000 observations, with a burn in of 5,000 and a refresh of 100. Note that the reader must import the data from the address given above (Table 5.9).

BUGS CODE 5.7

```
model;
# Calculates posterior distribution of model parameters and the area under
   curve. y=test
{
# likelihood function
      for(i in 1:N) {

            yt[i]~dnorm(mu[i],precy[d[i]+1]);
            yt[i] <--(y[i]);
# d is the disease indicator
# amt is the threshold of the stimulus
# f is the frequency
# int is the intensity
# this is the regression function
            mu[i] <- beta[1]+beta[2]*d[i]+beta[3]*amt[i]+beta[4]*f[i]
```

TABLE 5.9: Posterior analysis of the Stover audiology study.

Parameter	Mean	sd	Error	2 1/2	Median	97 1/2
auc	0.5033	0.1848	0.0094	0.1324	0.5056	0.8428
beta[1]	−7.439	0.7583	0.0186	−8.915	−7.44	−5.957
beta[2]	0.0679	5.881	0.3022	−12.55	0.1571	11.32
beta[3]	0.405	0.0316	<0.0001	0.3432	0.4048	0.4669
beta[4]	<0.0001	<0.0001	<0.00001	−0.0018	<0.0001	<0.00001
beta[5]	<0.00000001	<0.0000001	<0.0000001	<0.000001	<0.0000001	<0.0000001
beta[6]	−0.122	0.0430	<0.0001	−0.2063	−0.1219	−0.0378
beta[7]	0.0038	0.0010	<0.00001	0.0017	0.0038	0.0059
beta[8]	0.0455	0.0919	0.0047	−0.1254	0.04516	0.2393
la1	0.0091	0.7998	0.0410	−1.707	0.02144	1.538
la2	1.332	0.1001	<0.0001	1.15	1.328	1.54
vary[1]	54.16	2.085	0.0078	50.23	54.1	58.38
vary[2]	72.06	4.659	0.0197	63.5	71.85	81.7

```
+beta[5]*int[i]+beta[6]*d[i]*amt[i]+beta[7]*d[i]*f[i]+beta[8]*d[i]*int[i]
        }
# prior distributions - non-informative prior; similarly for informative priors
    for(i in 1:8){

            beta[i] ~ dnorm(0, 0.0001);
                }
    for(i in 1:2 ){
            precy[i]~dgamma(0.001, 0.001);
            vary[i] <- 1.0/precy[i];
                }
# calculates area under the curve
    la1 <- beta[2]/sqrt(vary[1]); # ROC curve parameters
    la2 <- vary[2]/vary[1];
    auc <- phi(la1/sqrt(1+la2));
}
```

The analysis tentatively implies that all the covariates should be eliminated. If this is the case and all covariates are deleted from the model, the ROC area is estimated as 0.922(0.0070) and a 95% credible interval of (0.9078,0.9355); a plot of the ROC curve is given by Figure 5.5.

This result is left as an exercise. It should be noted that the presence of the three covariates (threshold, intensity, and frequency) attenuate the estimated ROC error by some 45% from an area of 0.92, without covariates, to an area of 0.50, when all three are in the model, however, it appears that all three are not important predictors of the area!

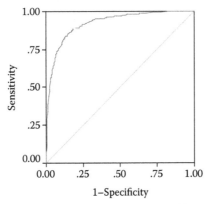

Diagonal segments are produced by ties

FIGURE 5.5: ROC curve for the Stover et al. study.

5.5   Exercises

1. Verify Table 5.1, the analysis for the audiology data. Use BUGS CODE 5.1 with 65,000 observations, a burn in of 5,000 and a refresh of 100. Estimate the effect of age, location of test, and the interaction between location and type of test. Note that the data and code are given in BUGS CODE 5.1.

2. The analysis of Exercise 1 implies that the interaction and other effects are not significant, therefore drop the interaction term in the model and repeat the analysis by revising BUGS CODE 5.1. Do the location and type of test have an effect on the false positive rate? Use 65,000 observations generated from the joint posterior distribution, with a burn in of 5,000 and a refresh of 100. What is the MCMC error for the parameters? What is the posterior mean of A and B? What is the posterior mean of the beta coefficients in the model? What is your final determination for the most appropriate model?

3. Verify Table 5.2, using the same information as stored in BUGS CODE 5.1. This problem assumes a logistic link, where 65,000 observations were generated, with a burn in of 5,000 and a refresh of 100.

4. Verify Table 5.3, that is, perform the Bayesian analysis with a logistic link, but with the true positive occurrence and the false positive occurrences as the dependent variables. See Section 5.2.4 and BUGS CODE 5.3. For this problem, the audiology data should be uploaded from www.fhcrc.org/labs/pepe/book, and note that from this source, the cases (ears) should be extracted as follows. When using false positive as the dependent variable and test type as the independent variable, use those cases where $0 \leq fp \leq 1$ and $0 \leq testab \leq 1$, while when true positive is the dependent variable and test type is the dependent variable, use only those cases when $0 \leq tp \leq 1$ and $0 \leq testab \leq 1$. Verify that the MCMC errors are <0.001 for all parameters and the information in Table 5.3.

5. Devise a strategy for choosing between the log and logistic links to analyze the audiology data. Explain in detail and provide a convincing argument for your choice.

6. Based on Table 5.4 and BUGS CODE 5.4, verify the results of Table 5.5 and perform the posterior analysis with 65,000 observations, with a burn in of 5,000 and a refresh of 100.

7. Based on Table 5.4 and BUGS CODE 5.4, find the posterior distribution of the cut points dtheta[i] and ndtheta[i] for $i = 1, 2, 3, 4$. What are the posterior means and standard deviations of these parameters? Use 65,000 observations, with a burn in of 5,000 and a refresh of 100.

What are the MCMC errors for these parameters? Are you satisfied with the MCMC errors?

8. Refer to Table 5.4 and BUGS CODE 5.4, and estimate the ROC area for the clinical trial, but do not use gender as a covariate. With 75,000 observations generated from the posterior distribution, with a burn in of 5,000 and a refresh of 100, execute the analysis by revising the code, that is, delete the gender effect from the code statements and delete the gender data (labeled dx1 and dx2) from the list statement in BUGS CODE 5.4. Plot the posterior distribution of the ROC area. What is the posterior mean of the ROC area? Is it close to 0.417, which is the value computed for the area when gender is taken into account? Recall that the value 0.417 was computed when gender was taken into account, but that the effect of gender on the logit of the response probabilities was quite small, thus, one would expect the ROC area computed without gender as a factor to be very close to 0.417. What is the MCMC error for the ROC area? Is it sufficiently small for your taste?

9. As a second hypothetical example with ordinal scores, consider the case, where there are four readers interpreting CT images for the metastasis of lung cancer to the lymph nodes. The study is multicenter and reports the accuracy of CT for detecting lymph node invasion in patients with known lung cancer. All patients enrolled were examined preoperatively with CT, and after reading the image, the readers were asked to rate the degree of disease on a five-point scale in order to score the metastasis to the lymph nodes. The gold standard for metastasis was established by pathologic analysis obtained from surgery. If no invasion occurred, the population is referred to as non diseased, but if metastasis occurred, the patient belongs to the diseased population. A subset of the study is reported in Table 5.10, where only CT images are involved and four radiologists scored the degree of invasion. Our objective is to determine

TABLE 5.10:   CT rating data of four radiologists—lung cancer metastasis.

Reader	Disease	Ordinal scale ratings					Total
		1	2	3	4	5	
1	0	7	1	4	9	4	25
1	1	5	1	4	12	6	29
2	0	7	1	5	3	2	18
2	1	8	3	5	20	8	45
3	0	15	0	4	7	1	27
3	1	15	0	6	16	3	41
4	0	3	1	3	3	1	11
4	1	2	2	4	1	8	18

the ROC area accounting for all four readers and to examine the effect of the four readers on the ROC area. A similar example is given in Zhou et al. [2: 246].

Perform the analysis with 75,000 observations, with a burn in of 5,000 and a refresh of 100, using a revision of BUGS CODE 5.4. The information needed for the analysis is given below and should be placed as a list statement. The second list statement contains the initial values for executing the program. In order to accommodate the new data, the code must be revised!

```
# data is from Table 5.10
list(
ndy=c(1,1,1,1,1,1,1,
      2,
      3,3,3,3,
      4,4,4,4,4,4,4,4,4,
      5,5,5,5,

      1,1,1,1,1,1,1,
      2,
      3,3,3,3,3,
      4,4,4,
      5,5,

      1,1,1,1,1,1,1,1,1,1,1,1,1,1,1,
      3,3,3,3,
      4,4,4,4,4,4,4,
      5,

      1,1,1
      2,
      3,3,3,
      4,4,4,
      5),

# reader 1 for non diseased
ndx1=c(1,1,1,1,1,1,1,1,1,1,1,1,1,1,1,1,1,1,1,1,
       1,1,1,1,1,

       0,0,0,0,0,0,0,0,0,0,0,0,0,0,0,0,0,0,

       0,0,0,0,0,0,0,0,0,0,0,0,0,0,0,0,0,0,0,0,0,
       0,0,0,0,0,0,0,
       0,0,0,0,0,0,0,0,0,0,0),
```

```
# reader 2 for non diseased
ndx2=c(0,0,0,0,0,0,0,0,0,0,0,0,0,0,0,0,0,0,0,0,0,0,0,
       0,0,0,0,0,
       1,1,1,1,1,1,1,1,1,1,1,1,1,1,1,1,1,1,

       0,0,0,0,0,0,0,0,0,0,0,0,0,0,0,0,0,0,0,0,0,0,0,
       0,0,0,0,0,0,0,0,

       0,0,0,0,0,0,0,0,0,0,0,0
   ),
# reader 3 non diseased
ndx3=c(0,0,0,0,0,0,0,0,0,0,0,0,0,0,0,0,0,0,0,0,0,0,0,
       0,0,0,0,0,
       0,0,0,0,0,0,0,0,0,0,0,0,0,0,0,0,0,0,0,0,0,0,0,
       1,1,1,1,1,1,1,1,1,1,1,1,1,1,1,1,1,1,1,1,1,1,1,
       1,1,1,1,1,1,1,
       0,0,0,0,0,0,0,0,0,0,0,0),
# data for diseased
# CT rating for diseased
dy=c(1,1,1,1,1,
     2,
     3,3,3,3,
     4,4,4,4,4,4,4,4,4,4,4,4,
     5,5,5,5,5,5,

     1,1,1,1,1,1,1,1,
     2,2,2,
     3,3,3,3,3,
     4,4,4,4,4,4,4,4,4,4,4,4,4,4,4,4,4,4,4,4,
     5,5,5,5,5,5,5,5,
     1,1,1,1,1,1,1,1,1,1,1,1,1,1,1,1,1,
     3,3,3,3,3,3,
     4,4,4,4,4,4,4,4,4,4,4,4,4,4,4,4,
     5,5,5,
     1,1,
     2,2,
     3,3,3,3,
     4),

# for reader 1 diseased
dx1=c(
     1,1,1,1,1,1,1,1,1,1,1,1,1,
     1,1,1,1,1,1,1,1,1,1,1,1,1,1,1,1,1,
```

```
0,0,0,0,0,
0,0,0,0,0,0,0,0,0,0,0,0,0,0,0,0,0,0,0,0,0,
0,0,0,0,0,0,0,0,0,0,0,0,0,0,0,0,0,0,0,0,0,

0,
0,0,0,0,0,0,0,0,0,0,0,0,0,0,0,0,0,0,0,0,0,
0,0,0,0,0,0,0,0,0,0,0,0,0,0,0,0,0,0,0,0,0,

0,0,0,0,0,0,0,0,0,0,0,0,0,0,0,0,0,0),
# reader 2 diseased
dx2=c(
0,0,0,0,0,0,0,0,0,
0,0,0,0,0,0,0,0,0,0,0,0,0,0,0,0,0,0,0,0,0,

1,1,1,1,1,
1,1,1,1,1,1,1,1,1,1,1,1,1,1,1,1,1,1,1,1,1,
1,1,1,1,1,1,1,1,1,1,1,1,1,1,1,1,1,1,1,1,1,
0,0,0,0,0,0,0,0,0,0,0,0,0,0,0,0,0,0,0,0,0,
0,0,0,0,0,0,0,0,0,0,0,0,0,0,0,0,0,0,0,0,0,
0,
0,0,0,0,0,0,0,0,0,0,00,0,0,0,0,0,0),
# reader 3 diseased
dx3=c(0,0,0,0,0,0,0,0,0,0,0,0,0,0,0,0,0,0,0,0,0,
0,0,0,0,0,0,0,0,0,0,

0,0,0,0,0,0,0,0,0,0,0,0,0,0,0,0,0,0,0,0,0,
0,0,0,0,0,0,0,0,0,0,0,0,0,0,0,0,0,0,0,0,0,
0,0,0,0,0

1,1,1,1,1,1,1,1,1,1,1,1,1,1,1,1,1,1,1,1,1,
1,1,1,1,1,1,1,1,1,1,1,1,1,1,1,1,1,1,1,1,1,
1,
0,0,0,0,0,0,0,0,0,0,0,0,0,0,0,0,0,0))
list(ndtheta=c(0,0,0,0,0),dtheta=c(0,0,0,0,0))
```

10. To verify the results of Table 5.7, refer to BUGS CODE 5.5 and execute the analysis with 75,000 observations from the joint posterior distribution of the parameters, with a burn in of 5,000 and a refresh of 100. Also,

(a) Plot the posterior density of the ROC area.

(b) Are the effects of readers 1, 2, and 3 the same for the diseased patients?

(c) Determine the posterior distribution of the cut points of the latent variables for the non-diseased patients. What are their means and 95% credible intervals?

(d) What is the posterior mean of the ROC area for reader 2?

11. Refer to BUGS CODE 5.5, the code for the melanoma metastasis study with four readers. Add the following two columns to the first list statement of BUGS CODE 5.5 and execute the Bayesian analysis with 75,000 observations, a burn in of 5,000 and a refresh of 100.

```
agend=c(45,78,56,68,72,81,45,67,71,69,63,
        71,74,68,80,49,57,59,73,72,68,52,
        69,73,75,62,65,59,80,72,
        5,78,56,68,72,81,45,67,71,69,63,
        71,74,68,80,49,57,59,73,72,68,52,
        69,73,75,62,65,59,80,72,
        5,78,56,68,72,81,45,67,71,69,63,
        71,74,68,80,49,57,59,73,72,68,52,
        69,73,75,62,65,59,80,72,
        5,78,56,68,72,81,45,67,71,69,63,
        71,74,68,80,49,57,59,73,72,68,52,
        69,73,75,62,65,59,80,72),
aged=c(45,78,42,68,72,81,45,67,71,69,63,59,77,78,
       71,80,83,72,74,68,80,49,57,59,73,72,68,52,
       69,49,59,75,80,71,59,73,75,62,65,59,80,49,
       38,80,71,74,80,78,79,59,
       45,78,42,68,72,81,45,67,71,69,63,59,77,78,
       71,80,83,72,74,68,80,49,57,59,73,72,68,52,
       69,49,59,75,80,71,59,73,75,62,65,59,80,49,
       38,80,71,74,80,78,79,59,
       45,78,42,68,72,81,45,67,71,69,63,59,77,78,
       71,80,83,72,74,68,80,49,57,59,73,72,68,52,
       69,49,59,75,80,71,59,73,75,62,65,59,80,49,
       38,80,71,74,80,78,79,59,
       45,78,42,68,72,81,45,67,71,69,63,59,77,78,
       71,80,83,72,74,68,80,49,57,59,73,72,68,52,
       69,49,59,75,80,71,59,73,75,62,65,59,80,49,
       38,80,71,74,80,78,79,59)
```

The agend column contains the ages of the non-diseased patients, where the ages are repeated four times for the four readers, and in the same way, the aged column contains the ages for the diseased patients. Revise the code accordingly and add the age factor to the regression statements for the non-diseased and diseased populations.

(a) What is the posterior mean and standard deviation of the ROC area?

(b) What is the effect of age on the ROC area?

(c) Is the ROC area attenuated by the age factor?

(d) What are the posterior means of the probabilities of the five ordinal responses for the diseased and non-diseased populations?

12. Refer to Section 5.4.1 and derive a formula for the induced ROC curve (Equation 5.21).

13. For the prostate cancer example (see Table 5.8 and BUGS CODE 5.6), estimate the area under the ROC curve without the covariate age. Is there much of a change in the posterior median of the ROC area? Use 90,000 observations for the simulation, with a burn in of 5,000 and a refresh of 1,000. Compare the MCMC error for the ROC area of Table 5.8 and the MCMC error of the present analysis (without using a covariate).

14. For the Stover et al. [10] audiology study, plot the main response y against the three covariates to see if there is a linear response. There are only three values for the frequency and three for the intensity.

15. Refer to BUGS CODE 5.7 and perform a Bayesian analysis for the Stover audiology data without using any covariates (threshold (amt), frequency, and intensity) and estimate the ROC area, the beta parameters, la1, la2, vary[1], and vary[2]. Execute the analysis with 45,000 observations for the simulation, with a burn in of 5,000 and a refresh of 500. Verify the entries in Table 5.11 for the analysis. Note that the code must be revised, including deleting the covariate data in the first list statement, revising the regression function, and revising the initial values in the last list statement.

    (a) Compare these results with Table 5.9. Notice the difference in the posterior means and standard deviations of the two ROC areas!

    (b) Compare the beta[2] coefficient in the two cases. For the case when all covariates are in the model, the posterior standard error of beta[2] is much larger than the corresponding quantity when the covariates are not in the model. The change is from 5.888 to 0.4733.

    (c) In view of (b), why do the covariates in the model induce such a large standard deviation for beta[2]?

16. Refer to BUGS CODE 5.7 and perform a Bayesian analysis for the Stover audiology data without the three interaction terms, but including the

TABLE 5.11:   Stover audiology study without covariates.

Parameter	Mean	sd	Error	2 1/2	Median	97 1/2
auc	0.9222	0.0070	<0.00001	0.9078	0.9227	0.9255
beta[1]	−6.188	0.2104	0.0013	−6.603	−6.188	−5.776
beta[2]	17.29	0.4733	0.0029	16.36	17.29	18.22
la1	2.218	0.0724	<0.0001	2.073	2.218	2.364
la2	1.431	0.1073	<0.0001	1.236	1.426	1.655
vary[1]	60.8	2.347	0.01197	56.37	60.75	65.58
vary[2]	86.9	5.585	0.02575	76.65	86.69	98.45

main effects of the covariates (threshold (amt), frequency, and intensity) and estimate the ROC area, the four beta parameters, la1, la2, vary[1], and vary[2]. Execute the analysis with 75,000 observations for the simulation, with a burn in of 5,000 and a refresh of 500. Verify the entries in Table 5.12 for the analysis. Note that the code must be revised, including deleting the covariate data in the first list statement, revising the regression function, and revising the initial values in the last list statement. The code must be revised by changing the regression function statement. Verify the results shown in Table 5.12.

(a) The effect of the two covariates beta[4] and beta[5], appear to be small, however, the effect of beta[3] on amt (the threshold of the stimulus) does have some effect. Please comment.

(b) Recall without covariates that the ROC area is about 0.922, but including the covariates without their interaction with the disease indicator gives an area of 0.70. Also, recall that with all covariates and their interactions with d, the area is 0.50. Why does the inclusion or exclusion of the covariates and their interactions have such a big effect on the ROC area?

17. The blood glucose test diagnoses type 2 diabetes and consist of taking a sample after the subject fasts for 12 hours. A normal test result is a blood glucose value < 110 mb/dL, while values between 111 and 125 mg/dL indicate problems with glucose metabolism; and values in excess of 125 mg/dL show definite signs of type 2 diabetes. A study was conducted with 217 subjects known not to have type 2 diabetes and 90 subjects who were suspected as having problems metabolizing glucose. Those without diabetes have a mean glucose level of 100(4.622) mg/dL, while the mean glucose level for the other group is 109.3(3.94) mg/dL, and the glucose levels appear to be normally distributed. In addition to the glucose values, the age and gender of each subject is recorded, and the code below contains the data for the study.

BUGS CODE 5.8

```
model;
    # Calculates posterior distribution of model parameters and the area
        under curve.
    {
    # likelihood function
        for(i in 1:N) {

            g[i]~dnorm(mu[i],precy[d[i]+1]);
            mu[i] <- beta[1]+beta[2]*d[i]+beta[3]*age[i]+beta[4]*male[i];
            }
```

TABLE 5.12: Bayesian analysis audiology study without interactions.

Parameter	Mean	sd	Error	2 1/2	Median	97 1/2
auc	0.701	0.0263	<0.0001	0.648	0.7015	0.751
beta[1]	-8.09	0.6843	0.0038	-9.436	-8.09	-6.748
beta[2]	5.991	0.8474	0.0029	4.335	5.993	7.646
beta[3]	0.3413	0.0218	<0.00001	0.2987	0.3414	0.3841
beta[4]	<0.0001	<0.0001	<0.000001	-0.0010	$-1*10^{-4}$	$7*10^{-4}$
beta[5]	<0.00000001	<0.00000001	10^{-10}	<0.0000001	$-1*10^{-8}$	$1*10^{-7}$
vary[1]	54.4	2.086	0.0019	50.48	54.35	58.65
vary[2]	74.03	4.822	0.0057	65.19	73.82	84.06

prior distributions - non-informative prior; similarly for informative priors

```
for(i in 1:P) {

        beta[i] ~ dnorm(0, 0.001);
                }

    for(i in 1:K) {

        precy[i]~dgamma(0.001, 0.001);
        vary[i] <- 1.0/precy[i];
                }
```

calculates area under the curve

```
    la1 <- beta[2]/sqrt(vary[1]); # ROC curve parameters
    la2 <- vary[2]/vary[1];
    auc <- phi(la1/sqrt(1+la2));
}
list(P=4,N=307, K=2,
g=c(96.63,102.98,97.72,97.82,106.94,105.32,92.61,94.99,
105.49,97.34,97.72,96.87,98.53,102.57,99.69,96.46,93.68,
97.46,104.60,98.49,107.34,96.03,105.17,96.87,98.16,
104.14,99.73,94.68,93.85,99.70,95.07,99.74,102.22,98.99,
103.72,101.55,101.55,95.54,97.47,103.37,100.31,100.55,
99.76,103.12,92.16,106.42,102.03,96.97,103.79,96.58,
113.96,100.26,95.07,104.00,101.47,105.84,103.61,98.03,
93.45,92.92,98.48,100.14,97.46,97.88,104.21,92.92,
104.49,95.51,100.49,99.46,105.03,91.78,100.75,105.68,
100.31,91.27,103.92,98.78,92.80,107.75,104.85,104.24,
93.57,100.69,97.11,101.41,84.43,101.88,94.94,94.91,
100.04,104.18,104.81,98.06,107.01,94.13,99.19,98.87,
99.01,96.42,103.20,109.30,97.20,94.74,103.36,103.82,
93.54,97.27,96.29,100.58,102.62,94.51,101.84,98.10,
102.66,99.73,96.50,104.86,100.69,97.57,101.81,98.88,
101.00,100.48,98.99,108.75,105.34,108.13,100.90,105.06,
98.10,106.16,105.64,94.18,104.07,98.64,97.82,98.49,
100.74,100.63,93.91,94.89,103.31,102.42,98.5,196.68,
109.31,95.59,99.23,102.60,104.24,103.14,109.07,103.23,
103.72,98.41,93.53,92.92,101.26,98.75,106.58,94.80,
102.49,101.80,99.97,97.73,106.66,100.91,93.13,105.04,
101.92,91.52,107.76,94.59,97.97,98.59,104.58,107.60,
98.14,101.84,101.41,92.35,99.41,99.63,96.51,100.77,
100.67,93.19,103.83,108.11,96.35,106.37,99.29,
102.72,89.20,101.92,105.87,96.66,101.85,103.92,101.38,
95.23,99.60,98.08,99.64,111.32,108.37,91.69,95.38,98.09,
```

```
        92.05,106.36,93.98,102.26,103.81,98.00,99.20,106.46,
        109.58,113.86,103.72,105.94,114.61,111.08,106.89,
        119.51,110.30,110.00,108.31,108.68,108.98,115.01,
        113.07,114.89,109.79,105.70,114.20,113.53,113.97,
        110.91,110.33,115.78,111.05,108.53,111.56,110.78,
        109.71,112.18,112.05,109.46,103.84,112.23,118.56,
        110.60,109.54,112.31,100.78,114.07,112.14,107.85,
        111.65,105.94,108.63,109.89,107.14,108.76,110.11,
        104.60,107.11,112.49,113.74,103.19,105.07,109.04,
        110.45,105.02,108.27,109.17,110.37,110.92,107.53,
        109.22,113.01,108.74,116.74,112.10,110.88,111.08,
        110.22,111.23,112.94,99.04,113.51,107.26,110.76,
        108.06,97.03,109.14,105.56,111.55,108.85,98.46,110.24,
        112.22,108.57,105.95,106.30),
        d
        =c(0,0,0,0,0,0,0,0,0,0,0,0,0,0,0,0,0,0,0,0,0,0,0,0,0,0,0,0,0,0,0,0,0,0,0,0,0,
        0,0,0,0,0,0,0,0,0,0,0,0,0,0,0,0,0,0,0,0,0,0,0,0,0,0,0,0,0,0,0,0,0,0,0,0,0,0,0,
        0,0,0,0,0,0,0,0,0,0,0,0,0,0,0,0,0,0,0,0,0,0,0,0,0,0,0,0,0,0,0,0,0,0,0,0,0,0,0,
        0,0,0,0,0,0,0,0,0,0,0,0,0,0,0,0,0,0,0,0,0,0,0,0,0,0,0,0,0,0,0,0,0,0,0,0,0,0,0,
        0,0,0,0,0,0,0,0,0,0,0,0,0,0,0,0,0,0,0,0,0,0,0,0,0,0,0,0,0,0,0,0,0,0,0,0,0,0,0,
        0,0,0,0,0,0,0,0,0,0,0,0,0,0,0,0,0,0,0,0,0,0,0,0,0,0,0,0,0,0,0,0,0,0,0,0,0,0,0,
        0,0,0,0,0,0,0,0,0,0,0,0,0,0,0,0,0,0,0,0,0,0,0,1,1,1,1,1,1,1,1,1,1,1,1,1,1,1,
        1,1,1,1,1,1,1,1,1,1,1,1,1,1,1,1,1,1,1,1,1,1,1,1,1,1,1,1,1,1,1,1,1,1,1,1,1,1,1,
        1,1,1,1,1,1,1,1,1,1,1,1,1,1,1,1,1,1,1,1,1,1,1,1,1,1,1,1,1,1,1,1,1,1,1,1,1),
        age=c(41,36,38,46,38,45,42,42,43,43,43,55,47,46,47,47,32,38,45,37,43,36,
        52,48,37,42,41,47,46,46,45,49,44,48,50,48,43,47,44,57,48,49,40,41,45,45,
        52,45,48,45,42,46,48,44,36,48,39,44,51,48,47,39,38,43,39,45,40,35,36,41,
        46,48,55,41,44,35,38,47,45,50,40,44,46,38,38,50,44,40,46,37,43,40,46,43,
        36,44,32,47,40,38,42,46,45,41,53,45,41,40,55,48,44,47,47,45,41,53,41,38,
        43,47,50,45,43,46,42,43,42,47,46,44,37,42,43,44,44,46,36,50,40,39,37,55,
        41,45,43,39,54,37,38,42,44,48,50,33,42,48,40,49,38,47,39,38,47,39,44,49,
        46,48,38,39,38,48,42,42,43,36,34,41,36,49,43,35,40,46,44,41,49,46,42,47,
        42,42,49,43,41,47,47,44,39,42,43,51,43,46,37,44,42,38,35,42,45,49,42,40,
        45,48,42,52,53,49,63,53,62,57,57,64,53,54,55,59,54,53,55,63,52,58,59,57,
        56,56,55,59,59,62,59,56,64,59,56,60,54,60,54,54,53,57,58,54,59,63,55,59,
        51,52,57,60,58,50,62,59,61,53,64,50,55,57,60,58,59,56,55,53,57,53,54,59,
        61,59,56,56,58,57,60,57,59,54,60,51,61,57,53,53,60,64,58,56,63),
        male=c(1,0,1,0,0,0,1,1,1,0,0,0,0,0,0,1,0,1,1,1,0,1,0,1,1,1,1,0,1,0,0,1,1,1,0,
        1,0,1,1,0,1,0,1,0,0,1,0,0,1,0,1,1,0,1,0,0,1,0,0,1,1,1,0,0,1,1,1,0,0,1,1,0,0,0,1,
        1,0,0,0,0,0,0,0,0,1,0,1,0,1,0,0,0,0,0,1,1,1,1,0,1,1,1,1,0,1,1,0,1,0,1,1,0,1,1,
        0,0,1,1,1,0,0,1,1,1,1,1,0,1,0,0,1,1,1,1,0,0,0,0,1,0,0,1,1,1,1,1,1,0,0,0,1,0,0,1,
        0,0,1,0,1,1,1,0,1,0,1,0,1,1,0,0,1,0,0,1,0,0,0,0,1,0,0,1,0,0,1,1,1,0,0,0,1,1,0,1,
        1,0,1,0,0,0,0,1,1,0,1,1,1,1,0,1,1,0,0,1,1,1,0,0,0,0,0,0,1,0,1,1,1,0,1,1,1,0,1,1,
        0,1,1,1,0,0,0,1,0,0,0,0,1,1,1,0,1,0,1,1,0,1,1,1,1,1,1,1,1,0,0,1,1,0,0,1,1,1,1,1,0,
        0,0,1,0,0,1,1,1,1,0,0,1,1,0,1,1,0,0,0,0,0,0,1,0,1,0,1,0,0,1,1,1,0))
    list(beta=c(0,0,0,0), precy=c(1,1))
```

TABLE 5.13: Bayesian analysis of type 2 diabetes study.

Parameter	Mean	sd	Error	2 1/2	Median	97 1/2
auc	0.8262	0.0360	0.0014	0.7494	0.8286	0.8901
beta[1]	0.9798	3.182	0.1603	91.71	98.01	104.2
beta[2]	8.523	1.209	0.0499	6.145	8.525	10.89
beta[3]	0.04966	0.0724	0.0036	−0.0920	0.0488	0.1912
beta[4]	0.6139	0.671	0.0050	−0.7152	0.6151	1.92
vary[1]	65.34	6.35	0.0304	53.99	64.95	78.92
vary[2]	15.86	2.44	0.0123	11.8	15.63	21.31

The Bayesian analysis is executed with 55,000 observations, with a burn in of 5,000 and a refresh of 200. Verify the entries in Table 5.13.

In Table 5.13, beta[3] is the effect of age and beta[4] is the effect of gender(male), and the ROC area has a mean of 0.8262(0.0360) with a 95% credible interval of (0.7494,0.8901), indicating good accuracy for the blood glucose test in differentiating between those with and without type 2 diabetes.

(a) Are age and gender important covariates?

(b) Analyze the study without the covariate.

(c) Is there a difference in the ROC areas? Plot the blood glucose values vs. the ages of the subjects. Is it a linear association?

References

[1] Pepe, M.S. *The Statistical Evaluation of Medical Tests for Classification and Prediction.* Oxford University Press, Oxford, UK, 2003.

[2] Zhou, X.H., McClish, D.K., and Obuchowski, N.A. *Statistical Methods in Diagnostic Medicine.* John Wiley, New York, 2002.

[3] Leisenring, W., Pepe, M.S., and Longton, G. A marginal regression framework for evaluating medical diagnostic tests. *Statistics in Medicine*, 16:1263, 1997.

[4] Armstrong, B. and Sloan, P. Ordinal regression models for epidemiologic data. *American Journal of Epidemiology*, 29:191–204, 2004.

[5] McCullagh, P. Regression models for ordinal data. *Journal of the Royal Statistical Society: Series B*, 42:109–142, 1980.

[6] Rifkin, M.D., Zerhouni, E.A., Gatsonis, C.A., Quint, I.E., Paushter, D.M., Epstein, J.I., Hamper, U., Walsh, P.C., and McNeil, B.J. Comparison of magnetic resonance imaging, and ultrasonagraphy in staging

early prostate cancer. *New England Journal of Medicine*, 323:621–626, 1980.

[7] Gregurich, M. A Bayesian approach to estimating the regression coefficients of a multinomial logit model. PhD dissertation. The University of Texas School of Public Health, Houston, TX, 1992.

[8] Congdon, P. *Applied Bayesian Modeling*. John Wiley, Chichester, 2003.

[9] O'Malley, A.J., Zou, K.H., Fielding, J.R., and Tempany, C.M.C. Bayesian regression methodology for estimating receiver operating characteristic curve with two radiologic applications: Prostate biopsy and spriral CT of ureteral stones. *Academic Radiology*, 8:713–726, 2001.

[10] Stover, L., Gorga, M.P., Neely, S.T., and Montoya, D. Toward optimizing the clinical utility of distortion product otoacoustic emission measurements. *Journal of the Acoustical Society of America*, 100:956–967, 1996.

[11] Etzioni, R., Pepe, M.S., Longton, G., Hu, C., and Goodman, G. Incorporating the time dimension in receiver operating characteristic curves: A case study of prostate cancer. *Medical Decision Making*, 19:242, 1991.

[12] Leisenring, W. and Pepe, M.S. Regression modeling of diagnostic likelihood ratio tests for the evaluation of medical diagnostic tests. *Biometrics*, 54:444, 1998.

[13] Leisenring, W., Alonzo, T., and Pepe, M.S. Comparison of predictive values of binary medical diagnostic tests for paired designs. *Biometrics*, 56:345, 2000.

[14] Tosteson, A.A.N. and Begg, C.B. A general regression methodology for ROC curve estimation. *Medical Decision Making*, 8:204, 1998.

[15] Toledano, A. and Gatsonis, C.A. Regression analysis of correlated receiver operating characteristic curves. *Academic Radiology*, 2:(Suppl. 1)S30, 1995.

[16] Alonzo, T.A. and Pepe, M.S. Distribution-free ROC analysis using binary regression techniques. *Biostatisics*, 3:421–432, 2002.

[17] Pawlik, T.M. and Gershenwald, J.E. Sentinel lymph node biopsy for melanoma. *Contemporary Surgery*, 61:175, 2005.

[18] Gershenwald, J.E., Teng, C.H., Thompson, W., et al. Improved sentinel lymph node localization in patients with primary melanoma with the use of radio labeled colloid. *Surgery*, 124:203, 1998.

[19] Peng, F. and Hall, J. Bayesian analyses of ROC curves using Markov Chain Monte Carlo Methods. *Medical Decision Making*, 16:404–444, 1986.

[20] Hellmich, M., Abrams, K.R., Jones, D.R., and Lambert, P.C. A Bayesian approach to a general regression model for ROC curves. *Medical Decision Making*, 18:438–443, 1986.

[21] Hanley, J.A. The use of the binormal model for parametric ROC analyses of quantitative diagnostic tests. *Statistics in Medicine*, 15:1575–1585, 1996.

[22] Gatsonis, C. Random effects models for diagnostic accuracy data. *Academic Radiology*, 2:S14–S21, 1995.

Chapter 6

Agreement and Test Accuracy

6.1 Introduction

The previous chapters provide the foundation for the Bayesian approach to agreement. Chapter 6 will introduce the reader to estimating the accuracy of a test when several readers are involved in interpreting the test scores. It is important to remember that test accuracy depends not only on the medical device that gives the test scores, but also on the various other elements involved in measuring accuracy. Probably the most important factor to take into account are the various readers who have the main responsibility of reporting the accuracy of the test. Chapters 4 and 5 reported several examples that involved multiple readers.

Recall the melanoma staging example in Chapter 5, where four radiologists used the same images to score the degree of metastasis in melanoma patients. There were two groups determined by the gold standard, those with metastasis to the lymph nodes and those with no metastasis. The area under the receiver operating characteristic (ROC) curve estimates the accuracy of the test, but there are four areas corresponding to the four radiologists. Which areas do we use? All four, or do we report one overall score representing the effect of all four? This is a case of ordinal scores with a gold standard; however, there are cases where no gold standard is available.

This chapter is divided into two scenarios: the case where a gold standard is present and the situation where no gold standard is present. The first case will be considered when a gold standard is present, then the other scenario will be presented. In addition, the chapter broadly deals with two other cases: when the test scores are ordinal and when they are continuous. When a gold standard is present, the main measure of accuracy is the area under the ROC curve and agreement will be estimated by some average of the ROC areas corresponding to the various readers. On the other hand, for ordinal scores and when no gold standard is available, agreement will be measured by the Kappa statistic. When no gold standard is present with continuous scores, the intraclass correlation and other correlations measure the accuracy of a medical test with several readers. For additional information on Kappa and other measures of agreement, the reader should refer to Shoukri [1], Von Eye and Mun [2], and Broemeling [3].

In what follows, first to be presented is the case of ordinal scores when a gold standard is available, to be followed by the case when a gold standard is present but with continuous test scores. Then, the situation of no gold standard when ordinal scores are available is reported, and lastly, the case of no gold standard with continuous scores is described.

6.2   Ordinal Scores with a Gold Standard

The first case to be considered when a gold standard is present is when the medical test is based on ordinal scores. Chapters 4 and 5 considered such cases, and the reader is referred specifically to Section 4.3.5 on the Bayesian analysis of the ROC area. Recall that the Bayesian analysis of the ROC area is based on BUGS CODE 4.2. In addition, Section 5.3.2 presents another way to compute the ROC area, via an ordinal regression model that incorporates covariate information into the calculation. The analysis is based on BUGS CODE 5.5 for the melanoma metastasis example, where four radiologists read the results of a sentinel lymph node biopsy.

A sentinel lymph node biopsy is performed to see if metastasis to the lymph nodes has occurred; the diagnosis is based on the Clark level and depth of the primary tumor. The biopsy is a nuclear medicine procedure that identifies a sentinel lymph node, which is dissected by surgery and sent to a pathologist to determine the degree of metastasis to the lymph node basin. The results are reported in Table 5.6.

In Chapter 5, the analysis consisted of determining the effect of each radiologist on the ROC area, but here the focus will be on reporting the ROC area for each radiologist and determining an overall estimate of test accuracy. A Bayesian analysis is performed based on BUGS CODE 6.1, which is a revision of BUGS CODE 5.5.

BUGS CODE 6.1

```
model;
{
# reader 2
# ROC area
# melanoma example
# code is from Congdon [2003: 1020]
# non diseased
for(i in 1:30){for(j in 1:5){logit(ndgamma[i,j])<-ndtheta[j]-ndmu[i]}}
for(i in 1:30){ndp[i,1]<-ndgamma[i,1]}
for(i in 1:30){ndp[i,2]<-ndgamma[i,2]-
ndgamma[i,1]}
for(i in 1:30){ndp[i,3]<-ndgamma[i,3]-
ndgamma[i,2]}
```

```
for(i in 1:30){ndp[i,4]<-ndgamma[i,4]-
ndgamma[i,3]}
for(i in 1:30){ndy[i]~dcat(ndp[i,1:5])}
for(i in 1:30){ndp[i,5]<-1-ndgamma[i,4]}
# intercept depends on y
for(i in 1:30){
ndmu[i]<-ndb0[ndy[i]]}
for(i in 1:5){ndb0[i]~dnorm(0,.001)}
ndtheta[1]~dnorm(0,1)
ndtheta[2]~dnorm(0,1)
ndtheta[3]~dnorm(0,1)
ndtheta[4]~dnorm(0,1)
ndtheta[5]~dnorm(0,1)I(ndtheta[4],)
for( i in 1:5){ndq[i]<-mean(ndp[,i])}
# diseased population
for(i in 1:50){for(j in 1:5){logit(dgamma[i,j])<-dtheta[j]-dmu[i]}}
for(i in 1:50){dp[i,1]<-dgamma[i,1]}
for(i in 1:50){dp[i,2]<-dgamma[i,2]-
dgamma[i,1]}
for(i in 1:50){dp[i,3]<-dgamma[i,3]-
dgamma[i,2]}
for(i in 1:50){dp[i,4]<-dgamma[i,4]-
dgamma[i,3]}
for(i in 1:50){dy[i]~dcat(dp[i,1:5])}
for(i in 1:50){dp[i,5]<-1-dgamma[i,4]}
# intercept depends on y
for(i in 1:50){
dmu[i]<-db0[dy[i]]}
for(i in 1:5){db0[i]~dnorm(0,.001)}
dtheta[1]~dnorm(0,1)
dtheta[2]~dnorm(0,1)
dtheta[3]~dnorm(0,1)
dtheta[4]~dnorm(0,1)
dtheta[5]~dnorm(0,1)I(dtheta[4],)
for( i in 1:5){dq[i]<-mean(dp[,i])}
# roc area
area<-a1+a2/2
a1<-dq[2]*ndq[1]+dq[3]*(ndq[1]+ndq[2])+
dq[4]*(ndq[1]+ndq[2]+ndq[3])+
dq[5]*(ndq[1]+ndq[2]+ndq[3]+ndq[4])
a2<-dq[1]*ndq[1]+dq[2]*ndq[2]+dq[3]*ndq[3]+
dq[4]*ndq[4]+dq[5]*ndq[5]
}
list( ndy=c(1,1,1,1,1,1,1,1,1,1,1,1,1,1,1,1,
    2,2,2,2,2,2,2,
```

```
3,3,3,3,
4,4,4,
5 ),
```

```
# data for diseased
dy=c(1,1,2,2,2,2,2,2,2,2,
    3,3,3,3,3,3,3,3,3,3,
    4,4,4,4,4,4,4,4,4,4,4,4,
    5,5,5,5,5,5,5,5,5,5,5,5,5,5,5,5,5,5))
list(ndtheta=c(0,0,0,0,0),dtheta=c(0,0,0,0,0))
```

The above code contains the information for reader 2 (see Table 5.6) in the first list statement, and the second list statement gives the initial values for the Markov Chain Monte Carlo (MCMC) simulation. In order to perform the analysis, revise the first list statement by inserting the relevant information for that reader, using the information in Table 5.6. A Bayesian analysis is executed with 65,000 observations, a burn in of 5,000 and a refresh of 5,000, resulting in MCMC errors <0.0001 for the four ROC areas (Table 6.1).

The ROC areas vary from a low of 0.6483 for reader 3 to a high of 0.8085 for reader 2! This example presents a more realistic assessment of test accuracy, because the variation between readers is evident and presents a dilemma. It could be true that the reader with the lowest score is the most accurate (compared to the true unknown accuracy). What is the accuracy of this medical test? There are several ways to come up with one measure of overall accuracy: (a) drop the high and low areas and average the two middle areas, 0.7872 and 0.7057; (b) use a weighted average of the four areas; and (c) consider the four readers as one reader and recompute the ROC area.

By joining the code for readers 1, 2, 3, and 4 and adding statements for the differences in the ROC areas of the four readers, and statements for the simple and weighted averages of the four readers, the following tables give the Bayesian analysis, which further elucidates the accuracy of the sentinel lymph node biopsy as measured by the four radiologists.

For Table 6.2, dij is the difference in the ROC areas of reader i minus reader j, sarea is the simple average of four ROC areas, and warea is the weighted mean of the four, where the weight of area$i$ is the inverse of the posterior variance (see Table 6.1) of the ROC area of reader i. It appears

TABLE 6.1:   Posterior analysis for ROC area of four readers—melanoma example.

Parameter	Mean	sd	Error	2 1/2	Median	97 1/2
Area 1	0.7872	0.0241	<0.0001	0.6502	0.7122	0.7633
Area 2	0.8085	0.0227	<0.0001	0.7619	0.8092	0.8509
Area 3	0.6483	0.0213	<0.0001	0.6053	0.6488	0.689
Area 4	0.7056	0.0229	<0.0001	0.6586	0.7062	0.7487

TABLE 6.2: Posterior analysis for the difference in the four ROC areas.

Parameter	Mean	sd	Error	2 1/2	Median	97 1/2
d12	−0.0214	0.0225	<0.0001	−0.0873	−0.0213	0.0444
d13	0.1386	0.0325	<0.0001	0.0739	0.139	0.0212
d14	0.0814	0.0335	<0.0001	0.0514	0.1601	0.1473
d23	0.16	0.0313	<0.0001	0.0982	0.2212	0.2212
d24	0.1028	0.0321	<0.0001	0.0398	0.1026	0.1659
d34	−0.0571	0.0312	<0.0001	−0.1183	−0.0572	0.0042
sarea	0.7371	0.1148	<0.0001	0.7141	0.7374	0.7591
warea	0.7321	0.0114	<0.0001	0.7092	0.7323	0.754

that the following pairs of readers have different ROC areas: 1 and 3; 1 and 4; 2 and 3; and 2 and 4. This implies that it is appropriate to use some sort of average to represent the overall accuracy of the test for melanoma metastasis. Note that the simple and weighted averages are almost the same, because the posterior standard deviations are the same. It should be stressed that the most representative analysis is to report the four ROC areas, then if appropriate, use a weighted average as an overall estimate of the accuracy. The analysis that produced Table 6.2 is based on 65,000 observations, with a burn in of 5,000 and a refresh of 100. Note the small MCMC error <0.0001 for all the parameters in the analysis.

The analysis for the case of ordinal scores with a gold standard is continued with the first three exercises at the end of the chapter. The following section will consider the case of continuous scores with a gold standard.

6.3 Continuous Scores with a Gold Standard

The presentation of the problem of agreement with several readers and a continuous test score is considered. Recall that the ROC area for continuous scores is analyzed from a Bayesian approach in Chapter 5, where the area is based on

$$\text{AUC} = \Phi\left(a/\sqrt{(1+b^2)}\right), \tag{5.18}$$

where

$$a = (\mu_d - \mu_{nd})/\sigma_d$$

and

$$b = \sigma_{nd}/\sigma_d.$$

TABLE 6.3:   Blood glucose levels (mg/dL) for three readers.

Reader	d = 0		d = 1	
	Mean	sd	Mean	sd
1	100.01	4.62	109.66	3.94
2	98.01	4.52	113.66	3.40
3	105.01	4.69	106.66	3.88

The means and standard deviations are for the diseased and non-diseased populations, respectively, and the analysis is executed using BUGS CODE 5.6, which is based on the O'Malley et al. [4] study.

The first example to be analyzed is the type 2 diabetes example of Chapter 5, described in Exercise 17 of that chapter. The study has three readers, each measuring the blood glucose levels (milligram per deciliter) of two types of patients, namely, those without the disease and those patients having problems with glucose metabolism. There are a total of 307 subjects, of which 217 do not have type 2 diabetes and 90 that have problems metabolizing glucose. The descriptive statistics for the three readers are shown in Table 6.3.

Based on Table 6.3, one would expect the ROC area of reader 2 to be the smallest, and the largest ROC area to correspond to reader 3, while the ROC area of reader 1 will be between that of readers 2 and 3. BUGS CODE 6.2 will provide the Bayesian analysis for the ROC areas of the three readers, and the analysis is executed with 65,000 observations, a burn in of 5,000 and a refresh of 100. The information for this study includes two covariates, the age and gender of each patient, and the analysis includes this additional information (Table 6.4).

BUGS CODE 6.2

```
model;
    # Calculates posterior distribution of model parameters and the area
        under curve.
    # type 2 diabetes with three readers
    {
    # likelihood function
        for(i in 1:N) {

            g1[i]~dnorm(mu1[i],precy1[d[i]+1]);
            mu1[i] <- beta1[1]+beta1[2]*d[i]+beta1[3]*age[i]+beta1[4]*male[i];

            g2[i]~dnorm(mu2[i],precy2[d[i]+1]);
            mu2[i] <- beta2[1]+beta2[2]*d[i]+beta2[3]*age[i]+beta2[4]*male[i];
            g3[i]~dnorm(mu3[i],precy3[d[i]+1]);
```

```
        mu3[i] <- beta3[1]+beta3[2]*d[i]+beta3[3]*age[i]+beta3[4]*male[i];
                }

# prior distributions - non-informative prior; similarly for informative priors
        for(i in 1:P) {
                beta1[i] ~ dnorm(0, 0.001);
                beta2[i] ~ dnorm(0, 0.001);
                beta3[i] ~ dnorm(0, 0.001);
                        }

        for(i in 1:K) {
                precy1[i]~dgamma(0.001, 0.001);
                vary1[i] <- 1.0/precy1[i];

                precy2[i]~dgamma(0.001, 0.001);
                vary2[i] <- 1.0/precy2[i];

                precy3[i]~dgamma(0.001, 0.001);
                vary3[i] <- 1.0/precy3[i];
                        }
# calculates area under the curve
# reader 1
        la11 <- beta1[2]/sqrt(vary1[1]); # ROC curve parameters
        la21 <- vary1[2]/vary1[1];
        auc1 <- phi(la11/sqrt(1+la21));
#reader 2
        la12 <- beta2[2]/sqrt(vary2[1]); # ROC curve parameters
        la22 <- vary2[2]/vary2[1];
        auc2 <- phi(la12/sqrt(1+la22));
# reader 3
        la13 <- beta3[2]/sqrt(vary3[1]); # ROC curve parameters
        la23 <- vary3[2]/vary3[1];
        auc3 <- phi(la13/sqrt(1+la23));


}
list(P=4,N=307, K=2,
# g1 is blood glucose levels for reader 1
g1=c(96.63,102.98,97.72,97.82,106.94,105.32,92.61,94.99,
105.49,97.34,97.72,96.87,98.53,102.57,99.69,96.46,93.68,
97.46,104.60,98.49,107.34,96.03,105.17,96.87,98.16,
104.14,99.73,94.68,93.85,99.70,95.07,99.74,102.22,98.99,
103.72,101.55,101.55,95.54,97.47,103.37,100.31,100.55,
99.76,103.12,92.16,106.42,102.03,96.97,103.79,96.58,
113.96,100.26,95.07,104.00,101.47,105.84,103.61,98.03,
93.45,92.92,98.48,100.14,97.46,97.88,104.21,92.92,
```

104.49,95.51,100.49,99.46,105.03,91.78,100.75,105.68,
100.31,91.27,103.92,98.78,92.80,107.75,104.85,104.24,
93.57,100.69,97.11,101.41,84.43,101.88,94.94,94.91,
100.04,104.18,104.81,98.06,107.01,94.13,99.19,98.87,
99.01,96.42,103.26,109.30,97.20,94.74,103.36,103.82,
93.54,97.27,96.29,100.58,102.62,94.51,101.84,98.10,
102.66,99.73,96.50,104.86,100.69,97.57,101.81,98.88,
101.00,100.48,98.99,108.75,105.34,108.13,100.90,105.06,
98.10,106.16,105.64,94.18,104.07,98.64,97.82,98.49,
100.74,100.63,93.91,94.89,103.31,102.42,98.5,196.68,
109.31,95.59,99.23,102.60,104.24,103.14,109.07,103.23,
103.72,98.41,93.53,92.92,101.26,98.75,106.58,94.80,
102.49,101.80,99.97,97.73,106.66,100.91,93.13,105.04,
101.92,91.52,107.76,94.59,97.97,98.59,104.58,107.60,
98.14,101.84,101.41,92.35,99.41,99.63,96.51,100.77,
100.67,93.19,103.83,108.11,96.35,106.37,99.29,
102.72,89.20,101.92,105.87,96.66,101.85,103.92,101.38,
95.23,99.60,98.08,99.64,111.32,108.37,91.69,95.38,98.09,
92.05,106.36,93.98,102.26,103.81,98.00,99.20,106.46,
109.58,113.86,103.72,105.94,114.61,111.08,106.89,
119.51,110.30,110.00,108.31,108.68,108.98,115.01,
113.07,114.89,109.79,105.70,114.20,113.53,113.97,
110.91,110.33,115.78,111.05,108.53,111.56,110.78,
109.71,112.18,112.05,109.46,103.84,112.23,118.56,
110.60,109.54,112.31,100.78,114.07,112.14,107.85,
111.65,105.94,108.63,109.89,107.14,108.76,110.11,
104.60,107.11,112.49,113.74,103.19,105.07,109.04,
110.45,105.02,108.27,109.17,110.37,110.92,107.53,
109.22,113.01,108.74,116.74,112.10,110.88,111.08,
110.22,111.23,112.94,99.04,113.51,107.26,110.76,
108.06,97.03,109.14,105.56,111.55,108.85,98.46,110.24,
112.22,108.57,105.95,106.30),
d
=c(0,0,0,0,0,0,0,0,0,0,0,0,0,0,0,0,0,0,0,0,0,0,0,0,0,0,0,0,0,0,0,0,0,0,0,0,0,0,
0,0,0,0,0,0,0,0,0,0,0,0,0,0,0,0,0,0,0,0,0,0,0,0,0,0,0,0,0,0,0,0,0,0,0,0,0,0,0,0,0,0,0,
0,0,0,0,0,0,0,0,0,0,0,0,0,0,0,0,0,0,0,0,0,0,0,0,0,0,0,0,0,0,0,0,0,0,0,0,0,0,0,0,0,0,0,
0,0,0,0,0,0,0,0,0,0,0,0,0,0,0,0,0,0,0,0,0,0,0,0,0,0,0,0,0,0,0,0,0,0,0,0,0,0,0,0,0,0,0,
0,0,0,0,0,0,0,0,0,0,0,0,0,0,0,0,0,0,0,0,0,0,0,0,0,0,0,0,0,0,0,0,0,0,0,0,0,0,0,0,0,0,0,
0,0,0,0,0,0,0,0,0,0,0,0,0,1,1,1,1,1,1,1,1,1,1,1,1,1,1,1,1,1,1,1,1,1,1,1,1,1,1,1,1,
1,1,1,1,1,1,1,1,1,1,1,1,1,1,1,1,1,1,1,1,1,1,1,1,1,1,1,1,1,1,1,1,1,1,1,1,1,1,1,1,1,1,
1,1,1,1,1,1,1,1,1,1,1,1,1,1,1,1,1,1,1,1,1,1),
age=c(41,36,38,46,38,45,42,42,43,43,43,55,47,46,47,47,32,38,45,37,43,36,52,
48,37,42,41,47,46,46,45,49,44,48,50,48,43,47,44,57,48,49,40,41,45,45,52,45,
48,45,42,46,48,44,36,48,39,44,51,48,47,39,38,43,39,45,40,35,36,41,46,48,55,
41,44,35,38,47,45,50,40,44,46,38,38,50,44,40,46,37,43,40,46,43,36,44,32,47,

40,38,42,46,45,41,53,45,41,40,55,48,44,47,47,45,41,53,41,38,43,47,50,45,43,
46,42,43,42,47,46,44,37,42,43,44,44,46,36,50,40,39,37,55,41,45,43,39,54,37,
38,42,44,48,50,33,42,48,40,49,38,47,39,38,47,39,44,49,46,48,38,39,38,48,42,
42,43,36,34,41,36,49,43,35,40,46,44,41,49,46,42,47,42,42,49,43,41,47,47,44,
39,42,43,51,43,46,37,44,42,38,35,42,45,49,42,40,45,48,42,52,53,49,63,53,62,
57,57,64,53,54,55,59,54,53,55,63,52,58,59,57,56,56,55,59,59,62,59,56,64,59,
56,60,54,60,54,54,53,57,58,54,59,63,55,59,51,52,57,60,58,50,62,59,61,53,64,
50,55,57,60,58,59,56,55,53,57,53,54,59,61,59,56,56,58,57,60,57,59,54,60,51,
61,57,53,53,60,64,58,56,63),
male=c(1,0,1,0,0,0,1,1,1,0,0,0,0,0,0,1,0,1,1,1,0,1,0,1,1,1,1,0,1,0,0,1,1,1,0,1,0,
1,1,0,1,0,1,0,0,1,0,0,1,0,1,1,0,1,0,0,1,0,0,1,1,1,0,0,1,1,1,0,0,1,1,0,0,0,1,1,0,0,
0,0,0,0,0,0,1,0,1,0,1,0,0,0,0,0,0,1,1,1,1,0,1,1,1,1,0,1,1,0,1,0,1,1,0,1,1,0,0,1,1,
1,0,0,1,1,1,1,1,0,1,0,0,1,1,1,1,0,0,0,0,1,0,0,1,1,1,1,1,1,0,0,0,1,0,0,1,0,0,1,0,1,
1,1,0,1,0,1,0,1,1,0,0,1,0,0,1,0,0,0,0,1,0,0,1,0,0,1,1,1,0,0,0,1,1,0,1,1,0,1,0,0,0,
0,1,1,0,1,1,1,1,0,1,1,0,0,1,1,1,0,0,0,0,0,0,1,0,1,1,1,0,1,1,1,0,1,1,0,1,1,1,0,0,0,
1,0,0,0,0,1,1,1,0,1,0,1,1,0,1,1,1,1,1,1,1,0,0,1,1,0,0,1,1,1,1,1,0,0,0,1,0,0,1,1,1,
1,0,0,1,1,0,1,1,0,0,0,0,0,0,1,0,1,0,1,0,0,0,1,1,1,0),
# glucose values of reader 2
g2=c(94.63,100.98,95.72,95.82,104.94,103.32,90.61,92.99,103.49,95.34,
95.72,94.87,96.53,100.57,97.69,94.46,91.68,95.46,102.60,96.49,105.34,
94.03,103.17,94.87,96.16,102.14,97.73,92.68,91.85,97.70,93.07,97.74,100.22,
96.99,101.72,99.55,99.55,93.54,95.47,101.37,98.31,98.55,97.76,101.12,90.16,
104.42,100.03,94.97,101.79,94.58,111.96,98.26,93.07,102.00,99.47,103.84,
101.61,96.03,91.45,90.92,96.48,98.14,95.46,95.88,102.21,90.92,102.49,93.51,
98.49,97.46,103.03,89.78,98.75,103.68,98.31,89.27,101.92,96.78,90.80,
105.75,102.85,102.24,91.57,98.69,95.11,99.41,82.43,99.88,92.94,92.91,98.04,
102.18,102.81,96.06,105.01,92.13,97.19,96.87,97.01,94.42,101.26,107.30,
95.20,92.74,101.36,101.82,91.54,95.27,94.29,98.58,100.62,92.51,99.84,96.10,
100.66,97.73,94.50,102.86,98.69,95.57,99.81,96.88,99.00,98.48,96.99,106.75,
103.34,106.13,98.90,103.06,96.10,104.16,103.64,92.18,102.07,96.64,95.82,
96.49,98.74,98.63,91.91,92.89,101.31,100.42,96.51,94.68,107.31,93.59,97.23,
100.60,102.24,101.14,107.07,101.23,101.72,96.41,91.53,90.92,99.26,96.75,
104.58,92.80,100.49,99.80,97.97,95.73,104.66,98.91,91.13,103.04,99.92,
89.52,105.76,92.59,95.97,96.59,102.58,105.60,96.14,99.84,99.41,90.35,97.41,
97.63,94.51,98.77,98.67,91.19,101.83,106.11,94.35,104.37,97.29,100.72,
87.20,99.92,103.87,94.66,99.85,101.92,99.38,93.23,97.60,96.08,97.64,109.32,
106.37,89.69,93.38,96.09,90.05,104.36,91.98,100.26,101.81,96.00,97.20,
110.46,113.58,117.86,107.72,109.94,118.61,115.08,110.89,123.51,114.30,
114.00,112.31,112.68,112.98,119.01,117.07,118.89,113.79,109.70,118.20,
117.53,117.97,114.91,114.33,119.78,115.05,112.53,115.56,114.78,113.71,
116.18,116.05,113.46,107.84,116.23,122.56,114.60,113.54,116.31,104.78,
118.07,116.14,111.85,115.65,109.94,112.63,113.89,111.14,112.76,114.11,
108.60,111.11,116.49,117.74,107.19,109.07,113.04,114.45,109.02,112.27,
113.17,114.37,114.92,111.53,113.22,117.01,112.74,120.74,116.10,114.88,
115.08,114.22,115.23,116.94,103.04,117.51,111.26,114.76,112.06,101.03,
113.14,109.56,115.55,112.85,102.46,114.24,116.22,112.57,109.95,110.30),

```
# glucose levels of reader 3
g3=c(101.63,107.98,102.72,102.82,111.94,110.32,97.61,99.99,110.49,102.34,
102.72,101.87,103.53,107.57,104.69,101.46,98.68,102.46,109.60,103.49,
112.34,101.03,110.17,101.87,103.16,109.14,104.73,99.68,98.85,104.70,
100.07,104.74,107.22,103.99,108.72,106.55,106.55,100.54,102.47,108.37,
105.31,105.55,104.76,108.12,97.16,111.42,107.03,101.97,108.79,101.58,
118.96,105.26,100.07,109.00,106.47,110.84,108.61,103.03,98.45,97.92,
103.48,105.14,102.46,102.88,109.21,97.92,109.49,100.51,105.49,104.46,
110.03,96.78,105.75,110.68,105.31,96.27,108.92,103.78,97.80,112.75,109.85,
109.24,98.57,105.69,102.11,106.41,89.43,106.88,99.94,99.91,105.04,109.18,
109.81,103.06,112.01,99.13,104.19,103.87,104.01,101.42,108.26,114.30,
102.20,99.74,108.36,108.82,98.54,102.27,101.29,105.58,107.62,99.51,106.84,
103.10,107.66,104.73,101.50,109.86,105.69,102.57,106.81,103.88,106.00,
105.48,103.99,113.75,110.34,113.13,105.90,110.06,103.10,111.16,110.64,
99.18,109.07,103.64,102.82,103.49,105.74,105.63,98.91,99.89,108.31,107.42,
103.51,101.68,114.31,100.59,104.23,107.60,109.24,108.14,114.07,108.23,
108.72,103.41,98.53,97.92,106.26,103.75,111.58,99.80,107.49,106.80,
104.97,102.73,111.66,105.91,98.13,110.04,106.92,96.52,112.76,99.59,102.97,
103.59,109.58,112.60,103.14,106.84,106.41,97.35,104.41,104.63,101.51,
105.77,105.67,98.19,108.83,113.11,101.35,111.37,104.29,107.72,94.20,
106.92,110.87,101.66,106.85,108.92,106.38,100.23,104.60,103.08,104.64,
116.32,113.37,96.69,100.38,103.09,97.05,111.36,98.98,107.26,108.81,103.00,
104.20,103.46,106.58,110.86,100.72,102.94,111.61,108.08,103.89,116.51,
107.30,107.00,105.31,105.68,105.98,112.01,110.07,111.89,106.79,102.70,
111.20,110.53,110.97,107.91,107.33,112.78,108.05,105.53,108.56,107.78,
106.71,109.18,109.05,106.46,100.84,109.23,115.56,107.60,106.54,109.31,
97.78,111.07,109.14,104.85,108.65,102.94,105.63,106.89,104.14,105.76,107.11,
101.60,104.11,109.49,110.74,100.19,102.07,106.04,107.45,102.02,105.27,
106.17,107.37,107.92,104.53,106.22,110.01,105.74,113.74,109.10,107.88,
108.08,107.22,108.23,109.94,96.04,110.51,104.26,107.76,105.06,94.03,
106.14,102.56,108.55,105.85,95.46,107.24,109.22,105.57,102.95,103.30))
# initial values for simulation
list(beta1=c(0,0,0,0), beta2=c(0,0,0,0),beta3=c(0,0,0,0),
precy1=c(1,1), precy2=c(1,1), precy3=c(1,1))
```

Note that beta$i[j]$ for $i = 1, 2, 3$ and $j = 1, 2, 3, 4$ is the jth beta coefficient for reader i. The intercept for reader i is beta$i[1]$, while the effect of the disease indicator d for reader i is given by beta$i[2]$. The effect of age on the blood glucose level for reader i is beta$i[3]$, while the effect of gender for reader i is given by beta$i[4]$. It should be noted that by including age and gender as covariates, the estimated ROC area might be attenuated. It appears from this table that age and gender are not important covariates and should be dropped from the analysis and the ROC areas re-estimated. The disparity between the ROC areas of the three readers presents a dilemma. Why the large range in the three areas? Is reader 3 inexperienced? When age and gender are eliminated, the posterior analysis is shown in Table 6.5.

TABLE 6.4: Posterior analysis for type 2 diabetes study—three readers with age and gender.

Parameter	Mean	sd	Error	2 1/2	Median	97 1/2
auc1	0.8298	0.0374	0.0014	0.7495	0.8232	0.8958
auc2	0.9907	0.0046	<0.0001	0.9793	0.9916	0.9969
auc3	0.5371	0.061	0.0024	0.4163	0.5376	0.6555
beta1[1]	98.45	3.52	0.1697	91.67	98.52	105.4
beta1[2]	8.66	1.284	0.0516	6.148	8.666	11.19
beta1[3]	0.03915	0.0800	0.0038	−0.1191	0.0373	0.1942
beta1[4]	0.6082	0.6689	0.0046	−0.7142	0.6112	1.907
beta2[1]	94.72	2.235	0.1054	89.99	94.76	99.13
beta2[2]	14.61	0.8885	0.0322	12.84	14.62	16.33
beta2[3]	0.0772	0.0528	0.0024	−0.0231	0.0758	0.1845
beta2[4]	−0.1322	0.5069	0.0037	−1.128	−0.1317	0.8657
beta3[1]	101.5	2.584	0.1241	96.48	101.6	106.7
beta3[2]	0.5756	0.9525	0.0377	−1.298	0.5765	2.453
beta3[3]	0.0810	0.0587	0.0028	−0.0362	0.0806	0.1966
beta3[4]	−0.1314	0.5044	0.0035	−1.124	−0.1302	0.8501
vary1[1]	65.34	6.367	0.0294	53.99	64.96	78.87
vary1[2]	15.91	2.46	0.0110	11.82	15.66	21.42
vary2[1]	21.59	2.108	0.0100	17.85	21.45	26.12
vary2[2]	16	2.476	0.0114	11.881	15.76	21.53
vary3[1]	21.58	2.088	0.0104	17.86	21.45	26.06
vary3[2]	16.01	2.463	0.0111	11.88	15.78	21.51

A comparison of Tables 6.4 and 6.5 reveals that age and gender attenuate the ROC area for readers 1 and 3, however, the area for reader 2 appears much too high. Reader 2 is almost perfect in differentiating the diseased from the non-diseased subject! Also, the ROC area for reader 3 now increases to 0.607 from the previous value of 0.53, when age and gender are taken into account. I have confidence in Table 6.5 because the posterior analysis with covariates, reported in Table 6.4, imply that they are not needed in the estimation of the ROC area. What is the accuracy of the blood glucose test for diagnosing type 2 diabetes? One possibility is to use the simple average or a weighted average (see Table 6.6).

The analysis is executed with 65,000 observations, with a burn in of 5,000 and a refresh of 100, and the resulting MCMC error is quite small. It is interesting that the weighted area is 0.991(0.0022), still large because the posterior

TABLE 6.5: Posterior analysis for type 2 diabetes study—three readers and no covariates.

Parameter	Mean	sd	Error	2 1/2	Median	97 1/2
auc1	0.847	0.0206	<0.0001	0.8038	0.848	0.8849
auc2	0.9945	0.0022	<0.0001	0.989	0.9948	0.9977
auc3	0.607	0.0333	<0.0001	0.5409	0.6073	0.6711

TABLE 6.6:   Simple and weighted ROC areas for diabetes study.

Parameter	Mean	sd	Error	2 1/2	Median	97 1/2
sarea	0.8162	0.0130	<0.00001	0.7902	0.8164	0.8412
warea	0.991	0.0022	<0.00001	0.9856	0.9914	0.9943

standard deviation of reader 2 is 0.0022. The weighted average is computed by weighting the ith area by the inverse of the corresponding variance of the posterior distribution of the ROC area of the ith reader. The ROC area for reader 2 is much too high and hard to believe, thus I would go with the simple average of 0.8162(0.0130) as my best guess of the accuracy of the blood glucose test.

6.4   Agreement with Ordinal Scores and No Gold Standard

When no gold standard is available, other methods must be used to measure agreement. An introduction to the subject begins with describing various measures of agreement that preceded the Kappa coefficient, then Kappa is described for two readers when the medical test scores are binary (i.e., positive or negative indicators of disease). The Kappa coefficient is the accepted measure to estimate agreement in many areas of science, including medicine, sociology, psychology, and psychiatry. When first introduced, it applied to two raters, but was later extended to more than two raters and to nominal and ordinal (more than binary) test scores. These developments will be described from a Bayesian approach and illustrated from various areas of medicine. For an introduction to the subject, two non-Bayesian sources are Shoukri [1], Von Eye and Mun [2], and for a Bayesian viewpoint, Broemeling [3] presents elementary and advanced methodology.

6.4.1   Precursors of Kappa

Our study of agreement begins with some early work before Kappa, where Kappa is introduced with a 2×2 table (Table 6.7) giving a binary score to n subjects. Each subject is classified as either positive, denoted by $X = 1$, or negative, denoted by $X = 0$, by reader 1 and $Y = 0$ or 1, by rater 2.

Let θ_{ij} be the probability that rater 1 gives a score of i and rater 2 a score of j, where i, $j = 0$ or 1, and let n_{ij} be the corresponding number of subjects. The experimental results have the structure of a multinomial distribution. Obviously, the probability of agreement is the sum of the diagonal probabilities $\theta_{00} + \theta_{11}$, however, this measure is usually not used. Instead, the Kappa parameter is often employed as an overall measure, and is defined as

$$\kappa = [(\theta_{00} + \theta_{11}) - (\theta_{0.}\theta_{.0} + \theta_{1.}\theta_{.1})]/[1 - (\theta_{0.}\theta_{.0} + \theta_{1.}\theta_{.1})], \qquad (6.1)$$

TABLE 6.7: Classification table.

Rater 1	Rater 2		
	$Y = 0$	$Y = 1$	
$X = 0$	(n_{00}, θ_{00})	(n_{01}, θ_{01})	$(n_{0.}, \theta_{0.})$
$X = 1$	(n_{10}, θ_{10})	(n_{11}, θ_{11})	$(n_{1.}, \theta_{1.})$
Total	$(n_{.0}, \theta_{.0})$	$(n_{.1}, \theta_{.1})$	

where the dot indicates summation of θ_{ij} over the missing subscript, and the probability of a positive response for rater 2 is $\theta_{0.1}$. Kappa is a so-called chance corrected measure of agreement.

Before Kappa, there were some attempts to quantify the degree of agreement between two raters, and what follows is based on the Fleiss [5] account of the early history of the subject, beginning with Goodman and Kruskal [6]. They assert that the raw agreement measure $\theta_{00} + \theta_{11}$ is the only sensible measure, however, as Fleiss stresses, other logical measures have been adopted. This presentation closely follows Fleiss, who summarizes the early publications of indices of agreement and the two origins of Kappa, then presents his own version of chance corrected early indices and how they relate to Kappa.

One of the earliest measures is from Dice [7], who formulated

$$S_D = \theta_{11}/[(\theta_{.1} + \theta_{1.})/2], \tag{6.2}$$

as an index of agreement. First select a rater at random, then S_D is the conditional probability that judge has assigned a positive rating to a subject, given that the other judge has assigned a positive rating. This seems reasonable if the probability of a negative rating is greater than that of a positive rating. In a similar way, define $S'_D = \theta_{00}/[(\theta_{.0} + \theta_{0.})/2]$, then based on the Dice approach, Rogot and Goldberg [6] defined

$$A_2 = \theta_{11}/(\theta_{1.} + \theta_{.1}) + \theta_{00}/(\theta_{0.} + \theta_{.0}), \tag{6.3}$$

as a measure of agreement, with the desirable property that $A_2 = 0$ if there is complete disagreement and $A_2 = 1$ if there is complete agreement.

According to Fleiss [5], Rogot and Goldberg [8] define another measure of agreement as

$$A_1 = (\theta_{00}/\theta_{.0} + \theta_{00}/\theta_{0.} + \theta_{11}/\theta_{1.} + \theta_{11}/\theta_{.1})/4, \tag{6.4}$$

which is an average of the four conditional probabilities. This index has the same properties as A_1, namely, $0 \leq A_2 \leq 1$, but in addition, $A_2 = 1/2$ when the raters are giving independent scores, that is, when $\theta_{ij} = \theta_{i.}\theta_{.j}$ for all i and j.

Suppose two raters assign scores to 100 subjects, as shown in Table 6.8. Then, what are the estimated values of A_1 and A_2? It can be verified that $\tilde{A}_1 = (22/55 + 22/37 + 30/45 + 30/63)/4 = (0.4 + 0.59 + 0.68 + 0.47)/4 = 0.53$, and $\tilde{A}_2 = 30/(63 + 45) + 22/(55 + 37) = 0.52$, where both indicate fair agreement; remember that the maximum value of both is 1. The first index is

TABLE 6.8: Hypothetical example of 100 subjects.

	Rater 2		
Rater 1	$Y = 0$	$Y = 1$	
$X = 0$	22	15	37
$X = 1$	33	30	63
Total	55	45	100

very close to 1/2; does this indicate independence between the two readers? Does $22/100 = 22/55 * 22/37$? Or does $0.22 = 0.23$? This is evidence that the raters are acting independently in their assignment of scores to the 100 subjects.

What is the Bayesian analysis? Some assumptions are in order, therefore suppose that the 100 subjects are selected at random from some well-defined population, that the structure of the experiment is multinomial, and that the assignment of scores to subjects is such that the probability of each of the four outcomes stays constant from subject to subject. Initially, a uniform prior is assigned to the four mutually exclusive outcomes, and the posterior distribution of $(\theta_{00}, \theta_{01}, \theta_{10}, \theta_{11})$ is Dirichlet (23, 16, 34, 31). What is the posterior distribution of A_1 and A_2?

Using 25,000 observations generated, with a burn in of 1,000 observations and a refresh of 100, the posterior analysis is displayed in Table 6.9. Good accuracy is achieved with an MCMC error <0.0001 for all parameters, and the parameter g is

$$g(\theta) = \sum_{i,j=0}^{i,j=1} (\theta_{ij} - \theta_{i.}\theta_{.j})^2,$$

which can be used to investigate the independence of the two raters. From Table 6.9, it appears that independence is a tenable assertion. Note that g is non negative and has a median of 0.0015 and a mean of 0.00308, and that independence is also implied by the posterior mean of $A_2 = 0.5138$. The role of the independence of the two raters is important for chance corrected measures such as Kappa. Fleiss [5] reports two additional indices proposed by Armitage [9] and Goodman and Kruskal [6], but these will not be presented here.

TABLE 6.9: Posterior distribution of A_1 and A_2.

Parameter	Mean	sd	2 1/2	Median	97 1/2
A_1	0.5324	0.0482	0.4362	0.5326	0.6254
A_2	0.5138	0.0487	0.4180	0.5138	0.6088
g	0.00308	0.0040	$3.437*10^{-6}$	0.0015	0.01457

6.4.2   Chance corrected measures of agreement

Fleiss [5] continues with a review of the chance corrected indices of agreement, which have a general form of

$$M = (I_r - I_c)/(1 - I_c), \tag{6.5}$$

where I_r is a measure of raw agreement and I_c is a measure of agreement by chance. Scott [10] first introduced such a measure called Kappa, which was followed by Cohen [11], who gave a related version. Recall Equation 6.1 for Kappa, where Scott assumed both raters have the same marginal distribution; however, Cohen did not make such an assumption. The Cohen version of Kappa will be used in this book.

From the information in Table 6.8 and the WinBUGS code used to produce Table 6.9, the posterior distribution of Kappa and its components are given in Table 6.10. The value of Kappa indicates very poor agreement, because the raw agreement estimate of 0.5196 and the estimated chance agreement of 0.4882 are quite close. This demonstrates the effect of chance agreement on Kappa.

Fleiss [5] continues the presentation by correcting the two precursors of Kappa, A_1 and A_2, for chance agreement using Equation 2.4, thus the chance corrected A_2 index is

$$M(A_2) = 2(\theta_{11}\theta_{00} - \theta_{01}\theta_{10})/(\theta_{1.}\theta_{.0} + \theta_{.1}\theta_{0.})$$
$$= \kappa, \tag{6.6}$$

where $I_c = E(A_2)$, assuming independent raters.

In similar fashion, the chance corrected value of S_D can be shown to be κ, however, this is not true for A_1, that is, the chance corrected value of A_1 is not Kappa.

Kappa has been embraced as the most popular index of agreement and has been extended to ordinal scores and to multiple raters and scores, and its use is ubiquitous in the social and medical sciences. Kappa gives an idea of the overall agreement between two raters with nominal scores, but once the value is estimated, it is important to know why Kappa has that particular value, and this will entail investigating the marginal and joint distribution of the raters. Performing an agreement analysis is much like doing an analysis of variance (ANOVA), where if at the first stage the null hypothesis of equal means is rejected, it is followed by a multiple comparison procedure. In the

TABLE 6.10:   Posterior distribution of Kappa.

Parameter	Mean	sd	2 1/2	Median	97 1/2
Kappa	−0.0905	0.1668	−0.4131	−0.0916	0.2437
I_r	0.5196	0.0486	0.4239	0.5199	0.6148
I_c	0.4882	0.0135	0.458	0.4896	0.5126

case that no agreement is indicated by Kappa, a look at the disagreement between the two readers at the off-diagonal cells would be informative. On the other hand, if Kappa indicates strong agreement, a comparison between the two raters at the on-diagonal cells can be fruitful.

6.4.3   Conditional Kappa

The conditional Kappa for rating category i is

$$\kappa_i = (\theta_{ii} - \theta_{i.}\theta_{.i})/(\theta_{i.} - \theta_{i.}\theta_{.i})$$
$$= (\theta_{ii}/\theta_{i.} - \theta_{i.}\theta_{.i}/\theta_{i.})/(1 - \theta_{i.}\theta_{.i}/\theta_{i.}) \qquad (6.7)$$

and is the conditional probability that the second rater assigns a score of i to a subject, given that the first rater scores an i, adjusted for conditional independence between the raters. Since

$$P[Y = i \mid X = i] = \theta_{ii}/\theta_{i.}, \quad i = 0, 1, \qquad (6.8)$$

and assuming conditional independence between the two raters, Equation 6.7 is apparent.

Thus, if overall Kappa gives an indication of strong agreement, then conditional Kappa will identify the degree of agreement between the two raters for each possible category, 0 or 1. The Bayesian approach allows one to determine the posterior distribution of conditional Kappa. For the example from Table 6.8, WinBUGS gives 0.0925 for the posterior mean of κ_0 and a posterior standard deviation of 0.1391. Also, a 95% credible interval for κ_0 is $(-0.1837, 0.3646)$. This is not surprising since overall Kappa is only -0.0905. One would expect a value near 0 for the posterior mean of κ_1. Conditional Kappa should be computed after the value of overall Kappa is known, because such estimates provide additional information about the strength of agreement. The concept of conditional Kappa (sometimes referred to as partial Kappa) was first introduced by Coleman [12] and later by Light [13] and is also described by Von Eye and Mun [2] and Liebetrau [14].

For k nominal scores, Kappa is defined as

$$\kappa = \left[\sum_{i=1}^{i=k} \theta_{ii} - \sum_{i=1}^{i=k} \theta_{i.}\theta_{.i}\right] \Big/ \left[1 - \sum_{i=1}^{i=k} \theta_{i.}\theta_{.i}\right], \qquad (6.9)$$

where θ_{ij} is the probability that raters 1 and 2 assign scores i and j, respectively, to a subject, where $i, j = 1, 2, \ldots, k$, and k is an integer of at least 3. With more than two scores, the agreement between raters becomes more complex. There are more ways for the two to agree or disagree. Von Eye and Mun [2: 12] examine the agreement between two psychiatrists who are assigning degrees of depression to 129 patients. Consider Table 6.11, where they report Kappa as 0.375, and the scores are interpreted as: 1 = "not depressed," 2 = "mildly depressed," and 3 = "clinically depressed."

TABLE 6.11: Agreement for depression.

Psychiatrist 1	Psychiatrist 2			Total
	1	**2**	**3**	**Total**
1	11	2	19	32
2	1	3	3	7
3	0	8	82	90
Total	12	13	104	129

Source: From Von Eye, A. and Mun, E.Y. *Analyzing Rater Agreement, Manifest Variable Methods.* Lawrence Erlbaum, Mahwah, NJ, and London, 2005, P. 12, Table 1.3, with permission of T&F.

If one adopts a Bayesian approach with a uniform prior density for θ_{ij}, i and $j = 1, 2, 3$, then the parameters have a Dirichlet $(12, 3, 20, 2, 4, 4, 1, 9, 83)$ posterior distribution; the resulting posterior analysis is shown in Table 6.12.

The above description of the posterior analysis is based on 25,000 observations, with a burn in of 1,000 observations, a refresh of 100, and the MCMC error is <0.0001 for all parameters. It shows that the raw agreement probability is estimated as 0.7174 with the posterior mean. The chance agreement probability has a posterior mean of 0.5595, which determines the posterior mean of Kappa as 0.3579, and the difference d between the probability of raw agreement and the probability of chance agreement has a posterior mean of 0.1578. Overall agreement is fair but not considered strong.

A plot of the posterior density of conditional Kappa for the score of 1 (not depressed) has a mean of 0.2639 with a standard deviation of 0.074, again implying only fair agreement for the not depressed category (see Figure 6.1). Von Eye and Mun [2: 9] report an estimated κ_1 of 0.276 and an estimated standard deviation of 0.0447, thus the conventional and Bayesian are about the same.

How does the Bayesian analysis for overall Kappa compare to Von Eye and Mun [2]? Their estimate of Kappa is 0.375 with a standard error of 0.079, compared to a posterior mean of 0.357, where the standard deviation of the posterior distribution of Kappa is 0.0721. I used SPSS, which gave 0.375 (0.079 is the asymptotic standard error) as an estimate of Kappa, which confirms the Von Eye and Mun analysis.

The code for these calculations and graphics follows. It is of interest to compare the posterior mean and standard deviation of θ_{11} with its known mean

TABLE 6.12: Posterior analysis of depression.

Parameter	Mean	sd	2 1/2	Median	97 1/2
Kappa	0.3579	0.0721	0.219	0.3577	0.4995
Kappa1	0.2639	0.0740	0.1326	0.26	0.4194
agree	0.7174	0.0382	0.62	0.7184	0.7889
cagree	0.5595	0.03704	0.489	0.559	0.6335
D	0.1578	0.0358	0.0914	0.1566	0.232

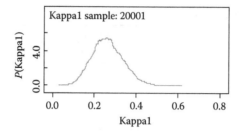

FIGURE 6.1: Posterior density of conditional Kappa for non depressed.

and standard deviation, which are 0.086956 and 0.0238995, respectively. The posterior mean and standard deviation based on 25,000 observations generated by WinBUGS are 0.08699 and 0.02398, respectively, thus the simulation is accurate to four decimal places (truncated) for estimating the mean and to three places for the standard deviation of θ_{11}.

BUGS CODE 6.3

```
model
  {
  g[1,1]~dgamma(12,2)
  g[1,2]~dgamma(3,2)
  g[1,3]~dgamma(20,2)
  g[2,1]~dgamma(2,2)
  g[2,2]~dgamma(4,2)
  g[2,3]~dgamma(4,2)
  g[3,1]~dgamma(1,2)
  g[3,2]~dgamma(9,2)
  g[3,3]~dgamma(83,2)
  h<-sum(g[,])
    for( i in 1 : 3 ) {for( j in 1 :3 ){ theta[i,j]<-g[i,j]/h}}
  theta1.<-sum(theta[1,])
  theta.1<-sum(theta[,1])
  theta2.<-sum(theta[2,])
  theta.2<-sum(theta[,2])
  theta3.<-sum(theta[3,])
  theta.3<-sum(theta[,3])
  kappa<-(agree-cagree)/(1-cagree)
  agree<-theta[1,1]+theta[2,2]+theta[3,3]
  cagree<-theta1.*theta.1+theta2.*theta.2+theta3.*theta.3
  d<-agree-cagree
  Kappa1<-(theta[1,1]-theta1.*theta.1)/(theta1.-theta1.*theta.1)
  }
  list( g = structure(.Data = c(2,2,2,2,2,2,2,2,2),.Dim = c(3,3)))
```

The raw agreement by itself indicates a strong association between psychiatrists, however, the chance agreement is also fairly strong, which reduces the overall agreement to only fair agreement, estimated as 0.357 via a Bayesian approach, and as 0.375 via maximum likelihood by Von Eye and Mun [2]. The Bayesian and conventional analyses of Kappa and conditional Kappa (for the not depressed category) agree quite well, but it is difficult to assign a Kappa value.

6.4.4 Kappa and stratification

Stratification is a way to take into account additional experimental information that will be useful in estimating agreement between observers. Up to this point, the only information brought to bear on estimating Kappa are the scores of the two raters, however, other information is often available, including age, sex, and other subject demographics. In medical studies, in addition to patient demographics, other information on previous and present medical history is available and might have an influence on the way raters disagree. Stratification is the first stage on the way to utilizing other experimental information that influences Kappa. Later, this will be expanded to model-based approaches that will incorporate covariate information in a more efficient way.

Consider a hypothetical national trial that compares x-ray and computed tomography (CT) for detecting lung cancer in high-risk patients. The study is to be conducted at three different sites, under the same protocol. One is primarily interested in the agreement between x-ray and CT, where both images are taken of each patient. The design calls for a total enrollment of 2500 patients from two major cancer centers enrolling 1000 patients each, with a smaller cancer center enrolling 500. The idea of estimating Kappa between x-ray and CT is one of the major issues being studied, along with the usual one of estimating and comparing the diagnostic test accuracies of the two images. For the time being, this information and patient covariate information, as well as other experimental factors, will not be considered. The major objective is to estimate an overall Kappa and to compare the individual Kappas of the three sites. It is assumed that the patients are enrolled according to strict eligibility and ineligibility criteria, which describes a well-defined population of subjects.

The scores assigned are 1, 2, 3, and 4, where 1 indicates "no evidence of disease," 2 indicates "most likely no evidence of disease," 3 implies it is "likely there is evidence of disease," and 4 indicates "evidence of a malignant lesion." Obviously, a team of radiologists assign the scores, however, the details are not considered relevant. Having both images of the same person is somewhat unique in this type of trial. Usually a patient receives only one type of image (Tables 6.13 through 6.15).

A posterior analysis based on WinBUGS with 25,000 observations generated from the joint posterior distribution, with a burn in of 1,000 observations and a refresh of 100 are reported in Table 6.16. A uniform prior density is used for the parameters, and the MCMC error is <0.0001 for all parameters.

TABLE 6.13:   CT vs. x-ray: site 1.

X-ray	CT 1	2	3	4	Total
1	320	57	22	15	414
2	44	197	18	10	269
3	20	26	148	5	199
4	11	7	3	73	94
Total	395	287	191	103	976

The parameter K12 is the difference between κ_1 and κ_2, where κ_1 is the Kappa for site 1, and the overall Kappa is denoted by κ, where

$$\kappa = (\kappa_1\omega_1 + \kappa_2\omega_2 + \kappa_3\omega_3)/(\omega_1 + \omega_2 + \omega_3), \tag{6.10}$$

and ω_i is the inverse of the variance of the posterior distribution of κ_i, $i = 1, 2, 3$. This particular average is one of many that could have been computed, but this issue will be explored later.

D1, D2, and D3 are the difference between the raw agreement probability and the chance agreement probability for sites 1, 2 and 3, respectively. The characteristics of the posterior distribution across the three sites are very consistent, as portrayed by the posterior mean and median of K12 and K13. The posterior analysis for the partial Kappas of the four scores at each site were not computed, but are left as an exercise.

As seen, the posterior analysis for overall Kappa was based on the weighted average, where the weights were the reciprocals of the posterior standard deviations of the individual site Kappas. This is a somewhat arbitrary choice, and Shoukri [1: 32,33] states three alternatives for the weighted average:

A. Equal weights
B. $\omega_i = n_i$, the number of observations in stratum i
C. $\omega_i = [\text{var}(\kappa_i/\text{data})]^{-1}$

The var(κ_i/data) are variances of the posterior distribution of κ_i, and now there is the problem of choosing a particular weighting scheme. Which one should be used for a particular problem? From a frequentist viewpoint, the

TABLE 6.14:   CT vs. x-ray: site 2.

X-ray	CT 1	2	3	4	Total
1	297	44	26	12	379
2	37	210	16	2	265
3	20	26	155	7	208
4	12	6	1	65	84
Total	366	286	198	86	936

TABLE 6.15: CT vs. x-ray: site 3.

X-ray	CT 1	2	3	4	Total
1	185	20	11	4	220
2	15	101	16	2	134
3	8	20	80	7	115
4	8	1	1	15	25
Total	216	142	108	28	494

bias as well as the variance of the estimated (by the posterior mean) should be taken into account (for more on this, see Barlow [15]).

Shoukri [1], using hypothetical data, analyzes the association between magnetic resonance imaging (MRI) and ultrasound as a function of lesion size. The usual approach is to use a weighted Kappa parameter with weights chosen according to several schemes. The weighted Kappa parameter is given by Equation 6.10, where the sum of the weights is unity, κ_i is the usual Kappa parameter for the ith stratum, and ω_i is the weight assigned to the ith stratum.

6.4.5 Weighted Kappa

A weighted Kappa was introduced earlier to perform a stratified analysis, however, a weighted Kappa is important when ordinal scores are being assigned to subjects. Recall that ordinal scores are used for computing the ROC area when a gold standard is available. The basic idea is to assign a weight to each cell of the table where the weight signifies the importance attached to the degree of difference assigned by the raters. Consider the example of the two psychiatrists assigning a degree of depression score to patients (Table 6.17).

When the two agree, a larger weight is assigned to that cell compared to a cell where they disagree, that is, the Kappa value should account for the

TABLE 6.16: Posterior analysis for x-ray and CT.

Parameter	Mean	sd	2 1/2	Median	97 1/2
K12	−0.0296	0.02761	−0.0836	−0.0297	0.0244
K13	0.0008812	0.03367	−0.0643	0.000677	0.0676
κ	0.6508	0.01245	0.6266	0.6509	0.675
κ_1	0.6391	0.0197	0.6004	0.6393	0.677
κ_2	0.6688	0.0194	0.6298	0.6691	0.7059
κ_3	0.6382	0.0274	0.5828	0.6385	0.6908
D1	0.4464	0.0146	0.4175	0.4466	0.4748
D2	0.4689	0.0144	0.44	0.4691	0.4965
D3	0.4323	0.0191	0.3941	0.4324	0.4964

TABLE 6.17: Agreement for depression.

Psychiatrist 1	Psychiatrist 2			Total
	1	2	3	
1	11 (1)	2 (0.5)	19 (0)	32
2	1 (0.5)	3 (1)	3 (0.5)	7
3	0 (0)	8 (0.5)	82 (1)	90
Total	12	13	104	129

Source: From Von Eye, A. and Mun, E.Y. *Analyzing Rater Agreement, Manifest Variable Methods*. Lawrence Erlbaum, Mahwah, NJ, and London, 2005, P. 50, Table 2.8, with permission of T&F.

weights assigned to the cells as:

$$\kappa_w = \sum_{i=1}^{i=k}\sum_{j=1}^{j=k}\omega_{ij}(\theta_{ij} - \theta_{i.}\theta_{.j}) \Bigg/ \left(1 - \sum_{i=1}^{i=k}\sum_{j=1}^{j=k}\omega_{ij}\theta_{i.}\theta_{.j}\right), \qquad (6.11)$$

where ω_{ij} is the weight assigned to the ijth cell of the table. When Cohen [11] introduced this index, he required: (a) $0 \leq \omega_{ij} \leq 1$, and (b) ω_{ij} be a ratio. By (b) is meant that if one weight is given a value of 1, then a weight of 1/2 is a value that is half that of (a) in value. Also, note that when $\omega_{ii} = 1$ for all i, and $\omega_{ii} = 0$ otherwise, the weighted Kappa reduces to the usual Kappa parameter.

Von Eye and Mun [2] assign the weights enclosed in parentheses in Table 6.17. Thus, if the two psychiatrists agree that the assigned weight is 1, while disagreeing by exactly one unit, the assigned weight is 0.5, and for all other disagreements, the assigned weight is 0.

In order to execute a Bayesian analysis, I used a burn in of 5,000 observations, generated 50,000 observations from the posterior distribution of the parameters, and used a refresh of 100. The posterior analysis is given in Table 6.18.

The posterior mean of 0.3866 differs from the estimated Kappa of 0.402 stated by Von Eye and Mun, however, they are still very close. Also, the posterior mean of the weighted Kappa differs from a value of 0.375 for the unweighted Kappa of Table 3.5 from Von Eye and Mun. The following code will perform the posterior analysis reported in Table 6.18.

TABLE 6.18: Posterior analysis for weighted Kappa.

Parameter	Mean	sd	Error	2 1/2	Median	97 1/2
agree	0.7826	0.0314	<0.0001	0.7177	0.7838	0.8405
cagree	0.6451	0.0310	<0.0001	0.5854	0.6447	0.7068
d	0.1375	0.0310	<0.0001	0.0799	0.1363	0.2014
κ_w	0.3866	0.0764	<0.0001	0.2366	0.387	0.5354

BUGS CODE 6.4

Model;
```
{
# generates Dirichlet Distribution
  g[1,1]~dgamma(12,2);  g[1,2]~dgamma(3,2);  g[1,3]~dgamma(20,2);
  g[2,1] ~dgamma(2,2);  g[2,2] ~dgamma(4,2);  g[2,3]~dgamma(4,2;
  g[3,1]~dgamma(1,2);  g[3,2]~dgamma(9,2);  g[3,3]~dgamma(83,2);
  h<- sum(g[,]) ;
  for( i in 1 : 3 ) {for( j in 1 :3 ){ theta[i,j]<-g[i,j]/h }}
  theta1.<-sum(theta[1,]);  theta.1<-sum(theta[,1]);
  theta2.<-sum(theta[2,]);  theta.2<-sum(theta[,2]);
  theta3.<-sum(theta[3,]);  theta.3<- sum(theta[,3]);
# kappa is weighted kappa
  kappa<- (agree-cagree)/(1-cagree);
  agree<-
w[1,1]*theta[1,1]+w[1,2]*theta[1,2]+w[1,3]*theta[1,3]+w[2,1]*theta[2,1]
+w[2,2]*theta[2,2]+w[2,3]*theta[2,3]+w[3,1]*theta[3,1]+w[3,2]*theta[3,2]
+w[3,3]*theta[3,3];
  cagree<-
w[1,1]*theta1.*theta.1+w[1,2]*theta1.*theta.2+w[1,3]*theta1.*theta.3
+w[2,1]*theta2.*theta.1+w[2,2]*theta2.*theta.2+w[2,3]*theta2.*theta.3
+w[3,1]*theta3.*theta.1+w[3,2]*theta3.*theta.2 +w[3,3]*theta3.*theta.3;

  d<-agree-cagree ;
}
list(g = structure(.Data = c(2,2,2,2,2,2,2,2,2),.Dim = c(3,3)))
list(w = structure(.Data = c(1,.5,0,.5,1,.5,0,.5,1),.Dim = c(3,3)))
```

Note that the first list statement contains the initial values of the *g* vector of gamma values, while the second includes the data for the vector of weights, *w*. Obviously, the assignment of weights is somewhat arbitrary, resulting in many possible estimates of weighted Kappa for the same dataset.

One set of weights advocated by Fleiss and Cohen [16] is

$$\omega_{ij} = 1 - (i-j)^2/(k-1)^2, \quad i,j = 1, 2, \ldots, k, \qquad (6.12)$$

where k is the number of scores assigned by both raters. With this assignment of weights to the cells, weighted Kappa becomes an intraclass Kappa, a measure of agreement that will be explained in the following section, along with some additional assumptions that are needed to define the index.

6.4.6 Intraclass Kappa

The intraclass correlation for the one-way random model is well known as an index of agreement between readers assigning continuous scores to subjects, however, the intraclass Kappa for binary observations is not as well known. The intraclass Kappa is based on the following probability model. Let

$$P(X_{ij} = 1) = \pi, \tag{6.13}$$

for all $i = 1, 2, \ldots, c$ and $j = 1, 2, \ldots, n_i$, $i \neq j$, where X_{ij} is the response of the jth subject in the ith group, and suppose the responses from different groups are independent; assume the correlation between two distinct observations in the same group is ρ.

Thus, the intraclass correlation measures a degree of association. Other characteristics of the joint distribution are

$$\begin{aligned}
P(X_{ij} = 1, X_{il} = 1) &= \theta_{11} = \pi^2 + \rho\pi(1-\pi), \\
P(X_{ij} = 0, X_{il} = 0) &= \theta_{00} = (1-\pi)[(1-\pi) + \rho\pi], \\
P(X_{ij} = 0, X_{il} = 1) &= P(X_{ij} = 1, X_{il} = 0) = \pi(1-\pi)(1-\rho) = \theta_{01} = \theta_{10},
\end{aligned} \tag{6.14}$$

where $j \neq l$, $i = 1, 2, \ldots, c$, and $j = 1, 2, \ldots, n_i$.

Note that this model induces some constraints on the joint distribution, where the two probabilities of disagreement are the same, and the marginal probabilities of all raters are the same. For this model, the subjects play the role of raters, there may be several raters, and the raters may differ from group to group. Also, all pairs of raters for all groups have the same correlation! Table 6.19 shows the relation between the θ_{ij} parameters and π and ρ determined by Equation 6.14.

Under independence $\rho = 0$, the probability of chance agreement is $\pi^2 + (1-\pi)^2 = 1 - 2\pi(1-\pi)$, thus a Kappa type index is given by

$$\begin{aligned}
\kappa_I &= \{[1 - 2\pi(1-\pi)(1-\rho)] - [1 - 2\pi(1-\pi)]\}/\{1 - [1 - 2\pi(1-\pi)]\} \\
&= \rho. \tag{6.15}
\end{aligned}$$

The likelihood function as a function of $\theta = (\theta_{00}, \theta_{01}, \theta_{11})$ is

$$L(\theta) \propto \theta_{00}^{n_{00}} \theta_{01}^{(n_{01}+n_{10})} \theta_{11}^{n_{11}}, \tag{6.16}$$

where n_{ij} are the observed cell frequencies. This is the form of a multinomial experiment with four cells, but with three cell probabilities, because $\theta_{01} = \theta_{10}$.

As an example, consider the study shown in Table 6.20, with five groups and five subjects per group. What is the common correlation between the subjects? The corresponding frequency table is Table 6.21. This table is constructed by counting the number of pairs within a group where $x_{ij} = k$,

TABLE 6.19: Cell probabilities for intraclass Kappa.

	0	1
0	$(1-\pi)^2 + \rho\pi(1-\pi) = \theta_{00}$	$\pi(1-\pi)(1-\rho) = \theta_{01}$
1	$\pi(1-\pi)(1-\rho) = \theta_{10}$	$\pi^2 + \rho\pi(1-\pi) = \theta_{11}$

TABLE 6.20: Five
groups and 15 subjects.

Group	Response
1	1, 1, 1, 1, 1
2	0, 1, 0, 1, 0
3	1, 1, 0, 0, 1
4	1, 1, 1, 0, 0
5	0, 0, 1, 1, 1

$x_{ij} = m$, where $k, m = 0$ or 1, $i = 1, 2, 3, 4, 5$, $j, l = 1, 2, 3, 4, 5$, and $j \neq l$. By definition,

$$\kappa_I = \text{cov}(X_{ij}, X_{il})/\sqrt{\text{var}(X_{ij})\text{var}(X_{il})}, \tag{6.17}$$

$$\text{Cov}(X_{ij}, X_{il}) = E(X_{ij} X_{il}) - E(X_{ij})E(X_{il})$$
$$= \theta_{11} - \pi^2.$$

Also, $\text{var}(X_{ij}) = \pi(1 - \pi)$, thus

$$\kappa_I = (\theta_{11} - \pi^2)/\pi(1 - \pi). \tag{6.18}$$

Using a uniform prior for the cell probabilities, the following BUGS CODE 6.5 presents the posterior distribution of the interclass Kappa, and the statement for Kappa is from Equation 6.18.

BUGS CODE 6.5

```
model
 {
 phi~dbeta(1,1)
 for( i in 1 : 5) {
 for( j in 1 : 5 ) {
 Y[i , j] ~ dbern(phi)}}
 theta11~dbeta(20,30)
 kappa<-(theta11-phi*phi)/phi *(1-phi)


 }
 list(Y= structure( .Data=c(1,1,1,1,1,
 0,1,0,1,0,1,1,0,0,1,1,1,0,0,
 0,0,1,1,1), .Dim=c(5,5)))
 list(theta11=.5 )
```

The posterior mean of the intraclass Kappa is 0.02546(0.0995), median -0.00137, and a 95% credible interval of $(-0.0853, 0.2883)$, indicating a weak agreement between pairs of subjects of the groups. The posterior mean(sd)

TABLE 6.21:   Cell frequencies.

	Intraclass Kappa		
	$X = 0$	$X = 1$	Total
$X = 0$	6	11	17
$X = 1$	13	20	33
Total	19	31	50

of π is 0.6295(0.0911). The first list statement is the data from Table 6.20, while the second list statement defines the starting values of theta11. Note that theta11 is given a beta(20, 30) posterior density, which implies the prior for theta11 is

$$g(\theta_{11}) = \theta_{11}^{-1}(1 - \theta_{11})^{-1}, \tag{6.19}$$

an improper distribution.

Twenty-five thousand observations are generated from the posterior distribution, with a burn in of 1000 and a refresh of 100. Note that the MCMC error is <0.0001 for all parameters in the model.

This is a hypothetical example with equal group sizes. Often, however, the groups differ in size, such as in heritability studies, where the groups correspond to litters and the groups consist of littermate offspring that share common traits. The binary outcomes designate the presence or absence of the trait.

Development of the intraclass Kappa is based on the constant correlation model described by Mak [17]. The constant correlation model is a special case of the beta-binomial model, which will be used for the following example. (See Kupper and Haseman [18] and Crowder [19] for some non-Bayesian approaches using the beta-binomial model as the foundation. Other sources for information about intraclass Kappa type indices are Banerjee et al. [20] and Bloch and Kraemer [21].)

Paul [22] presents an example where the intraclass Kappa can be used to estimate the agreement between pairs of fetuses affected by treatment. There are four treatment groups, and several litters are exposed to the same treatment, where each litter consists of live Dutch rabbit fetuses, among which a number respond to the treatment. What is the correlation between the binary responses (yes or no) to treatment? For each treatment group, an intraclass Kappa can be estimated, then the four compared. In order to do this, two models will be the basis for the analysis: (a) the constant correlation model seen earlier, and (b) the beta-binomial model, which is a generalization of the constant correlation model. For this study, Paul reported the data given in Table 6.22, where $1 = $ control group, $2 = $ low dose, $3 = $ medium dose, and $4 = $ high dose. In the high-dose group, there are 17 litters, with the first litter having 9 live fetuses and 1 responding to treatment, while the second litter of 10 has 0 responding. For this treatment group: $n_{00} = 135$, $(n_{01} + n_{10}) = 78$, $n_{11} = 17$, for a total of 230.

TABLE 6.22: Response to therapy of Dutch rabbit fetuses.

1	x	1	1	4	0	0	0	0	1	0	2	2	1	2	0	0	1	0	0	0	3	2	4	0
	n	12	7	6	6	7	8	10	8	6	11	9	7	8	9	7	11	10	8	10	12	8	7	8
2	x	0	1	1	0	2	0	0	1	0	0	1	5	9	5	0	3							
	n	5	11	7	9	12	8	6	6	7	9	5	9	1	6	3	9		3	6				
3	x	2	3	2	1	0	0	0	4	0	6	6	5	0	3	6			0	6				
	n	4	9	8	9	8	4	7	0	4	6	8	11	17	6	6			6	10	6			

Source: From Paul, S.R. *Biometrics*, 38, 361, 1982. With permission.

TABLE 6.23: Posterior analysis for Dutch rabbit fetuses.

Parameter	Mean	sd	2 1/2	Median	97 1/2
κ_I	0.09075	0.1063	−0.0752	0.0766	0.3429
π	0.2262	0.0403	0.1521	0.2245	0.3103
θ_{11}	0.074	0.0172	0.0435	0.0727	0.1110

BUGS CODE 6.3 was revised and executed to give Table 6.23, which provides the posterior analysis of the intraclass Kappa for the high-dose treatment group. The program is executed assuming a uniform prior and using 25,000 observations for the simulation, with a burn in of 5,000 and a refresh of 100. The MCMC error is very small, <0.0001 for all parameters (Table 6.23).

Thus, there is a low correlation between responses (yes or no) of live fetuses in the same litter, with a 95% credible interval of $(−0.0752, 0.3429)$, and the posterior mean of π estimates the probability that a fetus will respond to a high dose as 0.22(0.04). The Bayesian results can be compared to the conventional values calculated from formula 2.4 in Shoukri [1: 27], and is 0.081525 for the estimated intraclass correlation, with an estimated standard deviation of 0.071726. Also, the conventional estimate of π is 0.2434, thus the Bayesian results differ, but there is very little difference. In fact, the posterior median of 0.0766 is very close to the conventional estimate.

6.5 Other Measures of Agreement

The two standard references on measures of agreement are Shoukri [1] and Von Eye and Mun [2]. Their presentations have a lot in common, but they also differ is some respects. Shoukri gives a very complete account of the many measures of agreement in Chapter 3, while Von Eye and Mun place heavy emphasis on the log-linear model, which is barely mentioned by Shoukri. The majority of the material in this section is based mostly on Shoukri [1] and Von Eye and Mun [2], but their developments are not Bayesian, and they give many examples, from which we draw to illustrate the Bayesian counterpart.

The G coefficient and the Jacquard index will be the subject of this section. The G coefficient and the Jacquard index are not well known, and are very rarely used in agreement studies, but do have some attractive properties. Very little is known about the Jacquard coefficient, but it is given by Shoukri [1] as

$$J = \theta_{11}/(\theta_{11} + \theta_{01} + \theta_{10}), \qquad (6.20)$$

an index that ignores the (0,0) outcome, and is the proportion of (1,1) outcomes among the three cell probabilities. I cannot find any references to this coefficient, but it has the property that $0 \leq J \leq 1$, and if there is perfect agreement $J = 1$, but if there is perfect disagreement $J = 0$.

TABLE 6.24: Posterior analysis of indices of agreement.

Parameter	Mean	sd	Error	2 1/2	Median	97 1/2
G	0.5715	0.0558	<0.0001	0.459	0.5728	0.6759
J	0.3845	0.0550	<0.0001	0.2802	0.3828	0.4955
Kappa	0.0636	0.0919	<0.0001	−0.1199	0.0643	0.2415

Shoukri reports that J is estimated by

$$\tilde{J} = n_{11}/(n_{11} + n_{01} + n_{10}), \tag{6.21}$$

with an estimated variance of

$$\text{var}(\tilde{J}) = \tilde{J}^2(1 - \tilde{J})/n_{11}. \tag{6.22}$$

Another adjusted index is the G coefficient defined as

$$G = (\theta_{00} + \theta_{11}) - (\theta_{01} + \theta_{10}), \tag{6.23}$$

and is estimated by

$$\tilde{G} = [(n_{00} + n_{11}) - (n_{01} + n_{10})]/n, \quad -1 \leq G \leq 1, \tag{6.24}$$

with an estimated variance of $(1 - \tilde{G}^2)/n$. Also note that $G = 1$ if the agreement is perfect, and if there is perfect disagreement, $G = -1$.

The estimates and their estimated variances depend on the sampling plan of the 2×2 table, which is assumed to be multinomial, that is, the n subjects are selected at random from a well-defined population and fall into the four categories with the probabilities indicated in Table 5.1.

Suppose the example of Table 6.8 is reconsidered, and the agreement between the two raters is estimated with Kappa, the Jacquard, the G coefficient, and their Bayesian counterparts. The posterior analyses for the three measures of agreement are reported in Table 6.24. These can be compared with the non-Bayesian estimates: for J, the estimate is 0.38 with a standard error of 0.055 via Equations 6.22 and 6.24, while 0.57 is the estimate of the G coefficient with a standard error of 0.0558, via Equation 6.24. The Bayesian and conventional estimates are approximately the same, but they differ because the three measure different properties of agreement. Kappa indicates very poor agreement, but Kappa is adjusted for chance agreement, whereas G adjusts raw agreement by non agreement. The Bayesian analysis is based on:

BUGS CODE 6.6

```
Model;
{
  g00~dgamma(a00,b00);  g01~dgamma(a01,b01);
  g10~dgamma(a10,b10);  g11 ~dgamma(a11,b11);
  h<- g00+g01+ g10 +g11;  theta00<-g00/h;
```

```
theta01<-g01/h; theta10<-g10/h; theta11<-g11/h;
theta0.<-theta00+theta01; theta.0<-theta00+theta10;
theta1.<-theta10+theta11; theta.1<-theta01+theta11;
kappa<- (agree-cagree)/(1-cagree); g<-agree-cagree;
c<- (1/2)*(theta11/(theta11+theta01) +theta11/(theta11+theta10));
J<-theta11/(theta11+theta10+theta01); agree <-theta11+theta00;
                        cagree<-theta1.*theta.1+theta0.*theta.0;
}
list( a00 = 22, b00 = 2, a01 = 15, b01=2, a10= 33 , b10=2, a11=30 , b11=2)
list( g00 = 2, g01 =2, g10 =2, g11=2)
```

which is executed with 25,000 observations, 1,000 for the burn in, and a refresh of 100. An improper prior

$$g(\theta) \propto \prod_{i,j=1}^{i,j=2} \theta_{ij}^{-1}, \tag{6.25}$$

is also assumed for the Bayesian analysis, which implies that the Bayesian and conventional analyses will provide very similar estimates and standard errors. Note that an improper prior like Equation 6.5 can be used if there are no cells with a zero count, and if some of the cells have zero counts, a uniform prior may be used when very little prior information is available.

6.6　Agreement and Test Accuracy

When a gold standard is present, two or more raters are compared with reference to the gold standard, not to each other, as demonstrated in Sections 6.2 and 6.3. If the readers are judging the severity of disease, the gold standard provides an estimate of disease prevalence, and the effect of prevalence on measures of agreement can be explored, but there are situations where raters are judging disease and no gold standard is available. Judges and readers need to be trained, and frequently the gold standard provides the way to do this efficiently. However, in other cases, such as in gymnastics, judges are trained with reference to more experienced people, because a gold standard is not available.

Consider a case where two radiologists are being trained with ultrasound to diagnose prostate cancer, where they either identify the presence of disease or of no disease. The lesion tissue from the patient is sent to pathology where the disease is either detected or declared not present. The pathology report serves as the gold standard. If the same set of 235 lesions is imaged by both radiologists, then how should the two be compared? One approach is to estimate a measure of agreement between the radiologist and pathology for both radiologists and compare them.

TABLE 6.25a: Radiologist 1: prostate cancer with ultrasound.

Radiologist 1	Pathological stage		
	No disease = 0	Disease = 1	Total
No disease = 0	85	25	110
Disease = 1	15	110	125
Total	100	135	235

TABLE 6.25b: Radiologist 2: prostate cancer with ultrasound.

Radiologist 2	Pathological stage		
	No disease = 0	Disease = 1	Total
No disease = 0	2	85	87
Disease = 1	98	50	148
Total	100	135	235

With the improper prior density (Equation 6.25), the Bayesian analysis based on Tables 6.25a and b is executed with WinBUGS and the output is reported in Table 6.26.

There is an obvious difference in agreement between the two radiologists relative to pathology based on Kappa and the G coefficient. The second radiologist is a beginner and is a very inexperienced resident in diagnostic radiology.

Another approach is to estimate the diagnostic accuracy of each radiologist with the sensitivity, specificity, or ROC area of ultrasound, where pathology is the gold standard. If disease detected is the threshold for diagnosis, the sensitivity of ultrasound with the first radiologist is 0.814 and 0.370 with the second, and one would conclude that the first radiologist is doing a much better job than the second radiologist. In fact, the first radiologist is a very experienced diagnostician. What is the effect of disease incidence on Kappa?

Shoukri [1: 35] provides a value of Kappa, the agreement between the two radiologists, based on disease incidence, sensitivity, and specificity. The dependence of Kappa on these measures of test accuracy was formulated by

TABLE 6.26: Posterior analysis for two radiologists.

Parameter	Mean	sd	2 1/2	Median	97 1/2
κ_1	0.6555	0.0492	0.5541	0.6575	0.7458
κ_2	−0.6156	0.04101	−0.3389	−0.2971	−0.2527
G_1	0.3237	0.0246	0.2731	0.3245	0.3692
G_2	−0.2968	−0.0221	−0.3389	−0.2972	−0.2527

Kraemer [23] and later by Thompson and Walter [24] and is given by

$$\kappa_{tw} = 2\theta(1-\theta)(1-\alpha_1-\beta_1)(1-\alpha_2-\beta_2)/[\pi_1(1-\pi_1)+\pi_2(1-\pi_2)], \quad (6.26)$$

where θ, as determined by the gold standard, is the proportion with disease, π_i is the proportion having disease according to rater i, $1-\alpha_i$ is the sensitivity of rater i, and $1-\beta_i$ is the specificity of rater i. Based on Tables 6.25a and b, a quick calculation for an estimate of κ_{tw} is -0.65667, thus the two radiologists have very poor agreement (relative to each other), which in view of Table 6.26 (they had poor agreement relative to the gold standard) is not a surprise. It should be noted that κ_{tw} is often not relevant because a gold standard is not available. When a gold standard is available, the agreement between the readers should be assessed by comparing the readers' measures of accuracy, as demonstrated in Sections 6.2 and 6.3.

The dependence of Kappa on disease incidence shows that the two raters can have high sensitivity and specificity, nevertheless Kappa can be small. It all depends on the disease incidence, which if small, can produce a low value of κ_{tw}. A similar phenomenon occurs in diagnostic testing, when a test can have high sensitivity and specificity, but because of low disease incidence, the positive predictive value is low.

How should the Bayesian analysis for estimating κ_{tw} be performed? One approach is to treat the scores of the two radiologists as independent and find their posterior distribution based on Equation 6.26, where θ is the incidence of disease estimated from either table, π_i the proportion having disease, according to rater i, and is estimated from Table 6.25a. Also, $1-\alpha_1$, the sensitivity of rater 1, is estimated by $\theta_{11}/\theta_{.1}$, (see Table 6.25b), and the specificity of rater 2, $1-\beta_2$, is estimated by $\phi_{10}/\phi_{.0}$, etc. The posterior distribution of all the terms in Equation 6.26 and consequently for κ_{tw}, is determined from the following:

BUGS CODE 6.7

```
model;
  {
  g00~dgamma(a00,b00); g01~dgamma(a01,b01); g10~dgamma(a10,b10);
  g11~dgamma(a11,b11);
  h<- g00+g01+ g10 +g11 ; theta00<-g00/h;
  theta01<-g01/h; theta10<-g10/h; theta11<-g11/h;
  theta0.<-theta00+theta01; theta.0<-theta00+theta10;
  theta1.<-theta10+theta11; theta.1<-theta01+theta11;
  p00~dgamma(c00,d00); p01~dgamma(c01,d01); p10~dgamma(c10,d10);
  p11~dgamma(c11,d11);
  q<-p00+p01+p10+p11; phi00<-p00/q;
  phi01<-p01/q; phi10<-p10/q; phi11<-p11/q;
  phi0.<-phi00+phi01; phi.0<-phi00+phi10;
  phi1.<-phi10+phi11; phi.1<-phi01+phi11 ;
```

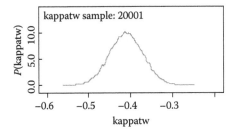

FIGURE 6.2: Posterior density of κ_{tw}.

kappatw<-k11/k12; k11<-2*(theta.1)*(1-theta.1)*(1-theta01/theta.1-theta10/theta.0)*(1-phi01/phi.1-phi10/phi.0);
  k12<-theta1.*(1-theta1.)+ phi1.*(1-phi1.);
  }
list(a00 =85, b00 = 2, a01 = 25, b01=2, a10= 15, b10=2, a11=110 , b11=2,
    c00 =2, d00 = 2,c01 = 85, d01=2, c10= 98, d10=2, c11=50 , d11=2)
list(g00 = 2, g01 =2, g10 =2, g11=2, p00 = 2, p01 =2, p10 =2, p11=2)

The theta parameters refer to Table 6.25a, while the phi parameters correspond to the entries of Table 6.25b, and the Kappa parameter κ_{tw} is referred to as kappatw in the above statements. I computed the posterior mean of κ_{tw} as -0.4106 (0.04013), a median of -0.4105, and a 95% credible interval of $(-0.49, -0.3325)$, indicating very poor agreement between the two radiologists. Utilizing 25,000 observations generated from the posterior distribution of κ_{tw}, with a burn in of 1,000, a refresh of 100, and the MCMC error is <0.0001 produces Figure 6.2, a graph of the posterior density of κ_{tw}.

6.7 Kappa and Association

Obviously, agreement is a form of association between raters, and the word association is ubiquitous in the statistical literature. It can be argued that the goal of most, if not all, statistical methods is to demonstrate an association, or lack thereof, between various experimental variables. By itself, the term association is often used in an informal way. In the context of a 2×2 table, the way to test for no association is to test the hypothesis that the marginal probabilities of the raters are the same. Our last version of Kappa is in the context of association, when the marginal distributions of the two raters are not the same as represented in Table 6.27, where $\theta_1 = P(X_1 = 1)$, $\theta_2 = P(X_2 = 1)$, and $\phi_i = 1 - \theta_i$, $i = 1, 2$. It can be shown that

$$\gamma = \rho\sqrt{\theta_1\theta_2\phi_1\phi_2}, \tag{6.27}$$

TABLE 6.27:　Cell probabilities of two raters.

	Rater 2	
Rater 1	$X_2 = 0$	$X_2 = 1$
$X_1 = 0$	$\phi_1\phi_2 + \gamma = \lambda_{00}$	$\phi_1\theta_2 - \gamma = \lambda_{01}$
$X_1 = 1$	$\phi_2\theta_1 - \gamma = \lambda_{10}$	$\theta_1\theta_2 + \gamma = \lambda_{11}$

where ρ is the correlation between X_1 and X_2, and the chance corrected Kappa under the above distribution for the two raters is

$$\kappa_a = 2\rho\sqrt{\theta_1\theta_2\phi_1\phi_2}/[\theta_1\phi_2 + \theta_2\phi_1]. \tag{6.28}$$

This derivation of κ_a is found in Shoukri [1: 30], who references Bloch and Kraemer [21], who give a complete account of measures of agreement and association.

Note that

$$\rho = (\lambda_{11} - \theta_1\theta_2)/\sqrt{\theta_1\phi_1 + \theta_2\phi_2}, \tag{6.29}$$

where $\theta_1 = \lambda_{1.}$ and $\theta_2 = \lambda_{.1}$.

Under a multinomial sampling scheme for the 2×2 table of outcomes, the cell probabilities have a Dirichlet $(n_{00} + 1, n_{01} + 1, n_{10} + 1, n_{11} + 1)$ distribution under a uniform prior density. This induces a prior distribution for θ_i and hence for the correlation coefficient ρ and also for γ, thus, these distributions should be determined and examined to see if they are compatible with the investigator's prior information about these particular parameters.

Suppose that the outcomes for two raters are represented by Table 6.28, where the probability of success is estimated as $25/125 = 0.2$ for rater 2 and $90/125 = 0.72$ for rater 1, an obvious case of no association between the two raters. What is the posterior distribution of κ_a? I generated 15,500 observations from the joint posterior distribution of κ_a and ρ, and a burn in of 500 with a refresh of 500; the results are shown in Table 6.29.

In order to determine if there is an association between the two raters, the posterior distribution of $d = \lambda_{1.} - \lambda_{.1}$ should be determined. From Table 6.29, the posterior means of κ_a and ρ imply a weak agreement between the two

TABLE 6.28:　Kappa in the context of association.

	Rater 2		
Rater 1	**0**	**1**	**Total**
0	30	5	35
1	70	20	90
Total	100	25	125

TABLE 6.29: Posterior analysis for κ_a.

Parameter	Mean	sd	2 1/2	Median	97 1/2
κ_a	0.0503	0.0472	−0.0476	0.0516	0.1406
ρ	0.0883	0.0799	−0.0839	0.0932	0.2294
d	0.5205	0.0509	0.4183	0.522	0.6168

readers, and the posterior analysis reveals that there is no, or very little, association between the two. It can be verified that the posterior distribution of the usual Kappa is the same as κ_a.

Shoukri [1] also states that the maximum likelihood estimation (MLE) of κ_a is

$$\tilde{\kappa}_a = 2(n_{11}n_{00} - n_{10}n_{01})/[(n_{11} + n_{10})(n_{00} + n_{10}) + (n_{11} + n_{10})(n_{00} + n_{10})],$$

which, using the information from Table 6.28, determines an estimate of 0.0506, which is almost the same as the posterior mean.

6.8 Consensus

Many agreement situations involve a method to come to a conclusion. For example, in boxing, the rules of the contest determine a definite conclusion about the outcome of the contest. If the bout goes the limit without a knockout or technical knockout, one of three things can occur: (1) fighter A wins or (2) fighter B wins or (3) the contest is ruled a draw. Or, as in a jury trial, each juror has an opinion as to guilt or innocence, and one of three outcomes will happen: (a) the defendant is guilty or (b) the defendant is innocent or (c) there is a hung jury.

Still another example is figure skating, where skaters perform at three events (compulsory figures, the short program, and the long), and for each event the skater receives a score from a panel of judges. Finally, each receives a final score and the winner is the skater with the highest total score. How they are scored is governed by the rules of the skating association. The rules can be quite involved, however, there is a mechanism to come to a consensus.

Reviewing and editing papers submitted to a scientific journal involves coming to a consensus about the acceptance of the paper for publication. The consensus revolves around the editor who receives input from a panel of reviewers and, sometimes, other editors, but somehow a final decision is made. The paper can be accepted with or without revision, rejected, etc., depending on the rules set down by the journal's editorial board.

In Phase II clinical trials, a panel of radiologists declares for each patient that the response to therapy is in one of four categories: a complete response, a partial response, progressive disease, or stable disease. How does this panel resolve the individual differences between the radiologists? That is a good question!

Some statistical approaches study the way that people come to a consensus. For example, DeGroot [25] addresses this issue, which is briefly described as follows. A group of individuals must work jointly and each has their own subjective probability distribution about an unknown parameter. DeGroot describes how the group might come to a decision about a common probability distribution for the parameter by combining their individual opinions. The process is iterative, where each individual revises their own subjective probability distribution by taking into account the opinions of the other judges. The author describes how the iterative process might converge, how a consensus is reached, and how the process is related to the Delphi technique. (For additional information about statistical consensus methods, see Winkler [26].)

Is the DeGroot method applicable to the Phase II trial example given above? If applied, apparently a common probability distribution among the panel of radiologists would be agreed on. This distribution would place a probability on each of the four possible outcomes: a complete response, a partial response, progressive disease, or stable disease. This could be very helpful, because if one outcome had a high probability, presumably the panel would agree to label that particular outcome to the patient. On the other hand, if the consensus distribution is uniform, another mechanism is necessary in order to label the patient's outcome.

6.9   Agreement with Multiple Raters and Ordinal Scores—No Gold Standard

With more than two raters, the determination of agreement becomes much more complicated. How should one measure agreement among three or more raters? One could use an overall measure, like Kappa, in order to estimate perfect agreement, corrected for chance, or, one could estimate all pairwise agreements with Kappa, then take an average of them. And, as seen in the last chapter, one could estimate a common intraclass correlation between all raters. Yet another possibility with multiple raters is to introduce the idea of partial agreement, say among six raters, and measure the agreement between five, four, and three raters. What is the role of conditional type Kappa measures with multiple raters? When the scores are ordinal, how should the weights be defined for a weighted Kappa? In some situations, it is obvious how to generalize the Kappa for a 2×2 table to multiple scores and raters, but not so obvious in other cases.

The Bayesian approach to measuring agreement is continued with several interesting examples. For example, six pathologists studying Crohn's disease are asked to identify (yes or no) lesions in 68 biopsy specimens. This example illustrates the complexity involved when there are multiple raters. Several issues not addressed earlier are considered with this example, including: (1) Is there an overall general measurement of agreement for six readers? (2) How is the problem of partial agreement to be met? For example, how does one measure partial agreement between five of the six pathologists? (3) Can one use a weighted Kappa by averaging all pairwise Kappas, and what is the optimal weighting scheme? (4) Should a modeling approach be taken? The example is based on a study by Rogel, Boelle, and Mary [27], who cite Theodossi et al. [28] with an earlier analysis of the data.

Another example with several raters and a binary response serves to exemplify the complexity of analyzing such information. Shoukri [1: 50], in his Table 4.4, presents an example of four veterinary students from a college in Ontario, who are asked to identify foals that have a cervical vertebral malformation. They each examine 20 x-rays and score each as either affected, designated by 1, or not affected, scored as a 0. Shoukri's analysis consists of testing for inter rater bias via Cochran's test and estimating the intraclass correlation coefficient with the one-way ANOVA. The Bayesian approach will be compared to his conventional analysis and the approach extended in order to answer questions about partial agreement and Kappa in the context of association.

A third example, from Williams [29], involves multiple raters and outcomes by the College of American Pathologists, who conducted a reliability test with four laboratories and three classifications for each pathology specimen. How reliable is the testing of these laboratories? The laboratories are the raters, and the analysis is presented by Shoukri [1: 56], who based his examination on Fleiss [5], who developed a Kappa type index, κ_{mc}, which was further studied by Landis and Koch [30] for its relation to intraclass Kappa.

Chapter 6 continues with presentations about weighted Kappa, conditional Kappa, agreement with many raters and ordinal scores, stratified Kappa, the intraclass Kappa, a generalization of the G coefficient, and agreement indices in the context of association. Thus, Chapter 6 is seen as a generalization of Chapter 5, which greatly expands the scope of the Bayesian approach to the subject.

Finally, it should be pointed out that it is just as important to focus on disagreement between raters. Often, more emphasis is placed on agreement, however, in any given real problem, one will usually find more disagreement than agreement, especially so when multiple raters and multiple outcomes are encountered. What is a good index of disagreement?

6.9.1 Kappa with many raters

There are several ways to define a measure of overall agreement between multiple raters, and two approaches will be taken: (1) a generalization of Kappa from 2×2 tables, and (2) an average of the Kappas for all pairs of 2×2 tables. The details of both approaches and the issues that are raised follow.

For the first approach, suppose each m rater assigns scores to all n subjects using a nominal response with c outcomes labeled $1, 2, \ldots, c$, then an overall Kappa can be defined for $m = c = 3$ as

$$\kappa(3,3) = \left[\sum_{i=1}^{i=3} \theta_{iii} - \sum_{i=1}^{i=3} \theta_{i..} \theta_{.i.} \theta_{..i} \right] \bigg/ \left[1 - \sum_{i=1}^{i=3} \theta_{i..} \theta_{.i.} \theta_{..i} \right], \qquad (6.30)$$

which is a chance corrected index for all three agreeing simultaneously at the three possible outcomes. Of course, it is obvious how to extend Kappa to $\kappa(m, c)$ in general for m raters and c scores, where m is a positive integer of at least 3, and c is a positive integer greater than or equal to 2. It should be pointed out that this is just one of the possible ways to define an agreement measure for multiple raters, and this will be addressed in a later section. Note that θ_{ijk} is the probability that rater 1 assigns a score of i, rater 2 a score of j, and rater 3 a score of k, where $i, j, k = 1, 2, 3$. Assume for the moment that the n subjects are selected at random from a population so that the study has the structure of a multinomial experiment.

The first example taken from Shoukri [1: 50] presents the assessment of cervical vertebral malformation of foals detected by four veterinary students who, on the basis of 20 x-rays, report either a 1 for not affected or a 2 for affected. These responses are reported in Table 3.1 of Shoukri et al., and the appropriate Kappa is

$$\kappa(4,2) = \left[\sum_{i=1}^{i=2} \theta_{iiii} - \sum_{i=1}^{i=2} \theta_{i...} \theta_{.i..} \theta_{..i.} \theta_{...i} \right] \bigg/ \left[1 - \sum_{i=1}^{i=2} \theta_{i...} \theta_{.i..} \theta_{..i.} \theta_{...i} \right]. \quad (6.31)$$

The frequencies that are contained in the 16 categories are presented in Table 6.30.

Assuming a uniform prior for θ_{ijkl}, the probability that student A assigns an i, B a j, C a k, and D an l, the posterior distribution of these parameters is Dirichlet (4, 1, 3, 1, 4, 1, 1, 1, 1, 2, 2, 1, 3, 2, 1, 8), and the posterior analysis for $\kappa(4,2)$ is given in (Table 6.31).

The program is executed with BUGS CODE 6.8 using 45,000 observations, with a burn in of 1,000 and a refresh of 100. The analysis estimates $\kappa(4,2)$ with a posterior mean(sd) of 0.2322(0.0836) and an upper 97.5 percentile of 0.4056. The raw agreement of the four students, agreeing on the outcomes (1,1,1,1) or (2,2,2,2), is 0.333(0.077) with the posterior mean(sd). A quick calculation based on Table 4.2 gives a raw agreement of $10/20 = 0.5$. The only way they can perfectly agree with the same score is at those two outcomes. The posterior mean(sd) of theta$(1, 1, 1, 1) = 0.1109(0.0512)$ and of theta$(2, 2, 2, 2) = 0.2227(0.0685)$.

The parameter d is the difference between the probability of agreement and the probability of agreement by chance (assuming the four raters are independent) and implies that there is a difference in the two probabilities that comprise Kappa42. This parameter is defined as (agree $-$ cagree)$/$(1 $-$ cagree).

TABLE 6.30: Agreement of four students.

Students A B C D	Frequencies	Case
1 1 1 1	3	1
1 1 1 2	0	2
1 1 2 1	2	3
1 2 1 1	0	4
2 1 1 1	3	5
1 1 2 2	0	6
1 2 1 2	0	7
2 1 1 2	0	8
1 2 2 1	0	9
2 1 2 1	1	10
2 2 1 1	1	11
2 2 2 1	0	12
2 2 1 2	2	13
2 1 2 2	1	14
1 2 2 2	0	15
2 2 2 2	7	16

Source: From Shoukri, M.M. *Measures of Interobserver Agreement*. Chapman & Hall/CRC, Boca Raton, 2004, P. 50, Table 4.4, with permission of T&F.

The following code does not follow the notation of the above derivation. Note that theta[1] of the code corresponds to θ_{1111} and theta[16] corresponds to θ_{2222} of the derivation (see Table 6.30).

BUGS CODE 6.8

```
{
# the following assumes a uniform prior
# Uses information in Table 6.30
# The 16 observations are labeled 1 to 16 from top to bottom
g[1]~dgamma(4,2); g[2]~dgamma(1,2);
g[3]~dgamma(3,2); g[4] ~dgamma(1,2);
g[5] ~dgamma(4,2); g[6]~dgamma(1,2);
g[7]~dgamma(1,2); g[8]~dgamma(1,2);
g[9]~dgamma(1,2); g[10]~dgamma(2,2);
g[11]~dgamma(2,2); g[12]~dgamma(1,2);
```

TABLE 6.31: Posterior analysis for cervical vertebral malformation.

Parameter	Mean	sd	Error	2 1/2	Median	97 1/2
agree	0.3334	0.0770	<0.0001	0.1924	0.3307	0.4914
cagree	0.1322	0.0198	<0.0001	0.1087	0.126	0.1865
$\kappa(4,2)$	0.2322	0.0836	<0.0001	0.0794	0.2289	0.4056

```
g[13]~dgamma(3,2); g[14]~dgamma(2,2);
g[15]~dgamma(1,2); g[16]~dgamma(8,2);
h<- sum(g[])
for( i in 1:16){ theta[i]<- g[i]/h}
# The following are the terms for chance agreement
atheta2...<-theta[5]+theta[8]+theta[10]+theta[11]+theta[12]+theta[13]+
         theta[14]+theta[16];
atheta.2..<-theta[4]+theta[7]+theta[9]+theta[11]+theta[12]+theta[13]+
         theta[15]+theta[16];
atheta..2.<-theta[3]+theta[6]+theta[9]+theta[10]+theta[12]+theta[14]+
         theta[15]+theta[16];
atheta...2<-theta[2]+theta[6]+theta[7]+theta[8]+theta[13]+theta[14]+
         theta[15]+theta[16];
atheta1...<-theta[1]+theta[2]+theta[3]+theta[4]+theta[6]+theta[7]+
         theta[9]+theta[15];
atheta.1..<-theta[1]+theta[2]+theta[3]+theta[5]+theta[6]+theta[8]+
         theta[10]+theta[14];
atheta..1.<-theta[1]+theta[2]+theta[4]+theta[5]+theta[7]+theta[8]+
         theta[11]+theta[13];
atheta...1<-theta[1]+theta[3]+theta[4]+theta[5]+theta[9]+theta[10]+
         theta[11]+theta[12];
kappa42<- (agree-cagree)/(1-cagree)
agree<-theta[1]+theta[16]
cagree<-atheta1...*atheta.1..*atheta..1.*atheta...1
        +atheta2...*atheta.2..*atheta..2.*atheta...2
d<- agree-cagree
}
# initail values
list(g = c(2,2,2,2,2,2,2,2,2,2,2,2,2,2,2,2))
```

6.9.2 Partial agreement

Another multi rater example with binary scores is presented to demonstrate the Bayesian approach to estimating partial agreement between six pathologists, who examined the presence of epithelioid granuloma in 68 intestinal biospy lesions of patients with Crohn's disease. The study was carried out by Rogel, Boelle, and Mary [27], using a log-linear model approach. For the time being, a more basic descriptive approach is proposed. Consider Table 6.32, where 1 designates the presence and 2 the absence of epithelioid granuloma. An overall measure of agreement for the six pathologists is

$$\kappa(6,2) = (\text{agree} - \text{cagree})/(1 - \text{cagree}), \tag{6.32}$$

where

$$\text{agree} = \sum_{i=1}^{i=2} \theta_{iiiiii}. \tag{6.33}$$

TABLE 6.32: Observed frequencies of epithelioid granuloma.

	Pathologist						
Case	A	B	C	D	E	F	Number
1	1	1	1	1	1	1	15
2	1	1	1	1	1	2	0
3	1	1	1	1	2	1	2
4	1	1	1	1	2	2	0
5	1	1	1	2	1	1	2
6	1	1	2	2	2	1	1
7	1	1	2	1	1	1	0
8	1	1	2	1	2	1	0
9	1	1	2	1	2	2	0
10	1	1	2	2	1	1	1
11	1	2	1	1	1	1	0
12	1	2	1	1	1	2	0
13	1	2	1	1	2	1	0
14	1	2	1	1	2	2	1
15	1	2	1	2	2	1	0
16	1	2	2	2	2	2	1
17	1	2	2	1	1	1	0
18	1	2	2	1	1	2	1
19	1	2	2	1	2	1	0
20	1	2	2	1	2	2	0
21	1	2	1	2	2	2	0
22	2	1	1	1	1	1	1
23	2	1	1	1	1	2	0
24	2	1	1	1	2	1	0
25	2	1	1	2	1	1	1
26	2	1	1	2	2	1	1
27	2	1	2	2	2	2	1
28	2	1	2	1	1	1	0
29	2	1	2	1	2	1	0
30	2	1	2	1	2	2	0
31	2	1	1	2	2	2	0
32	2	2	1	1	1	2	0
33	2	2	1	1	2	1	0
34	2	2	1	1	2	2	0
35	2	2	1	2	1	1	3
36	2	2	1	2	2	1	2
37	2	2	1	2	2	2	2
38	2	2	2	1	1	2	0
39	2	2	2	1	2	1	0
40	2	2	2	1	2	2	0
41	2	2	2	2	1	2	2
42	2	2	2	2	2	1	1
43	2	2	2	2	2	2	30

Source: From Rogel, A., Boelle, P.Y., and Mary, J.Y. Global and partial agreement among several observers. *Statistics in Medicine*. 1998, 17, 489. Copyright Wiley-VCH Verlag GmbH & Co. KGaA. Reproduced with permission.

TABLE 6.33:   Agreement of six pathologists.

Parameter	Mean	sd	Error	2 1/2	Median	97 1/2
θ_{111111}	0.2206	0.0498	<0.0001	0.1307	0.2178	0.3257
θ_{222222}	0.4414	0.0597	<0.0001	0.3267	0.4408	0.5595
agree	0.6615	0.0571	<0.0001	0.5461	0.6627	0.7696
cagree	0.0631	0.0268	<0.0001	0.0308	0.0587	0.1316
$\kappa(6,2)$	0.6382	0.0598	<0.0001	0.5183	0.6392	0.7521
agree5	0.8088	0.0473	<0.0001	0.7076	0.8118	0.8918
cagree5	0.2072	0.0502	<0.0001	0.1389	0.1984	0.327
$\kappa_5(6,2)$	0.758	0.0611	<0.0001	0.6266	0.7623	0.8644
GC	0.3241	0.1143	<0.0001	0.0918	0.3269	0.5382
GC5	0.6178	0.0948	<0.0001	0.4146	0.6235	0.7854

is the probability of agreement among the six pathologists, and the probability of agreement by chance is

$$\text{cagree} = \sum_{i=1}^{i=2} \theta_{i.....} \theta_{.i....} \theta_{..i...} \theta_{...i..} \theta_{....i.} \theta_{.....i}, \tag{6.34}$$

where θ_{ijklmn} is the probability that pathologist 1 assigns a score of i, and pathologist 6 assigns a score of n, etc., where $i, j, k, l, m, n = 1, 2$. It can be verified that the posterior distribution of a Kappa type index $\kappa(6,2)$ of overall agreement is given in Table 6.33.

The analysis is based on BUGS CODE 6.9 and executed with 75,000 observations, with a burn in of 5,000 and a refresh of 100, resulting in MCMC errors <0.0001 for all parameters.

BUGS CODE 6.9

```
model
  {
# g1 corresponds to the first case
  g1~dgamma(15,2) g3~dgamma(2,2)
  g5~dgamma(2,2) g6 ~dgamma(1,2)
  g10~dgamma(1,2)
  g14~dgamma(1,2) g16~dgamma(1,2)
  g18~dgamma(1,2) g22~dgamma(1,2)
  g25~dgamma(1,2) g26~dgamma(1,2)
  g27~dgamma(1,2) g35~dgamma(3,2)
  g36~dgamma(2,2) g37~dgamma(2,2)
  g41~dgamma(2,2) g42~dgamma(1,2)
  g43~dgamma(30,2)
h<- g1+ g3+ g5+ g6 + g10+ g14+ g16+ g18+ g22+ g25+g26+g27+g35+g36
     +g37+ g41+g42+g43
```

```
# a1 is the probability of the outcome for case 1
a1<- g1/h; a3<- g3/h; a5<- g5/h; a6<- g6/h; a10<- g10/h;
a14<- g14/h ; a16<- g16/h; a18<- g18/h; a22<- g22/h;
a25<- g25/h ; a26<- g26/h; a27<- g27/h; a35<- g35/h;
a36<- g36/h; a37<- g37/h; a41<- g41/h; a42<- g42/h;
a43<- g43/h ;
agree<- a1+a43
GC<-agree -(1-agree)
# a1..... is the probability that pathologist A assigns a 1 to a lesion
a1.....<- a1+a3+a5+a6+a10+a14+a16+a18
a.1....<- a1+a3+a5+a6+ a10+a22+a25+a26+a27
a..1...<- a1+a3+a5+a14+a22+a25+a26+a35+a36+a37
a...1..<- a1+a3+a14+a18+a22
a....1.<- a1+a5+a10+a18+a22+a25+a35+a41
a.....1<- a1+a3+a5+a6+a10+a22+a25+a26+a35+a36+a42
a2.....<- a22+a25+a26+a27+a35+a36+a37+a41+a42+a43
a.2....<- a14+a16+a18+a35+a36+a37+a41+a42+a43
a..2...<- a6+a10+a16+a18+a27+a41+a42+a43
a...2..<- a5+a6 +a10+a16+a25+a26+a27+a35+a36+a37+a41+a42+a43
a....2.<- a3+a6+a14+a16+a26+a27+a36+a37+a42+a43
a.....2<- a14+a16+a18+a27+a37+a41+a43
# a.....2 is the probability that reader F assigns a score of 2 to a lesion
cagree<- (a1.....*a.1....*a..1...*a...1..*a....1.*a.....1)+
         (a2.....*a.2....*a..2...*a...2..*a....2.*a.....2)
# kappa is the index that all six agree
kappa<- (agree-cagree)/(1-cagree)
al5<- a1+a3+a5+a22+a37+a41+a42+a43
#al5 is the probability at least five agree
cal5<- a1.....*a.1....*a..1...*a...1..*a....1.*a.....1+
       a1.....*a.1....*a..1...*a...1..*a....2.*a.....1+
       a1.....*a.1....*a..1...*a...2..*a....1.*a.....1+
       a2.....*a.1....*a..1...*a...1..*a....1.*a.....1 |
       a2.....*a.2....*a..1...*a...2..*a....2.*a.....2+
       a2.....*a.2....*a..2...*a...2..*a....1.*a.....2+
       a2.....*a.2....*a..2...*a...2..*a....2.*a.....1+
       a2.....*a.2....*a..2...*a...2..*a....2.*a.....2
# cal5 is the probability at least 5 agree by chance
# kappa5 is the index that at least 5 agree
kappa5<-(al5-cal5)/(1-cal5)
GC5<- al5 -(1-al5)
  }
list(g1=2, g3=2, g5 =2, g6=2, g14=2, g16=2, g18=2, g22=2, g25=2, g26=2,
  g27 =2, g35=2, g36 =2,g37 =2, g41=2, g42=2,g43=2).
```

In the above code, a3 denotes the probability of the outcome corresponding to Case 3 of Table 6.32.

The analysis show a fairly good overall agreement between the six pathologists based on the Kappa $\kappa(6,2)$. Most agreement is when all six concur 30 times on the absence of an epithelioid granuloma, which has a posterior mean of 0.441 and a 95% credible interval of (0.326, 0.559), while they concur 15 times that an epithelioid granuloma is present, with a posterior mean of 0.2206 and a 95% credible interval of (0.1307, 0.3257). Also, there is very little evidence that the pathologists are presenting independent scores.

As the number of raters increase, the chances of overall agreement decrease and it is important to measure partial agreement. For example, when do at least five of the pathologists agree on either the absence or presence of the condition?

Suppose the agreement of at least five of the pathologists is measured by a chance type Kappa parameter $\kappa_5(6,2) = (a15 - cal5)/(1 - cal5)$, where

$$a15 = a1 + a3 + a5 + a10 + a22 + a37 + a41 + a42 + a43$$

and

$$
\begin{aligned}
cal5 = &\; a1.....*a.1....*a..1...*a...1..*a....1.*a.....1+ \\
&\; a1.....*a.1....*a..1...*a...1..*a....2.*a.....1+ \\
&\; a1.....*a.1....*a..1...*a...2..*a....1.*a.....1+ \\
&\; a2.....*a.1....*a..1...*a...1..*a....1.*a.....1+ \\
&\; a2.....*a.2....*a..1...*a...2..*a....2.*a.....2+ \\
&\; a2.....*a.2....*a..2...*a...2..*a....1.*a.....2+ \\
&\; a2.....*a.2....*a..2...*a...2..*a....2.*a.....1+ \\
&\; a2.....*a.2....*a..2...*a...2..*a....2.*a.....2.
\end{aligned}
$$

Note, a1..... is the probability that pathologist A assigns a score of 1 to a lesion, while a.....2 is the probability that pathologist F assigns a score of 2 to a lesion.

The a15 parameter is the raw probability that at least five agree and cal5 is the probability that at least five agree, assuming all six are giving independent scores. Table 6.34 also shows the posterior analysis for the agreement of at least five pathologists. Exactly five agree (at either 1 or 2) for cases 3, 5, 10, 22, 41, and 42 of Table 6.32, with frequencies 2, 2, 1, 1, 2, and 1, respectively. All six agree for cases 1 and 43 with frequencies 15 and 30,

TABLE 6.34:   Posterior analysis for split agreement.

Parameter	Mean	sd	2 1/2	Median	97 1/2
GCS	−0.7942	0.0729	−0.9142	−0.8019	−0.6311
agrees	0.1029	0.0364	0.0428	0.0990	0.1844
cagrees	0.0723	0.0114	0.0473	0.0734	0.0915
Kappas	0.03292	0.0389	−0.0331	0.0294	0.1178

respectively. A quick calculation shows a raw probability of $8/68 = 0.1117$ for the probability that exactly five agree and a probability of 0.794 that at least 5 agree. These results agree very well with the corresponding posterior means of Table 6.34.

Another approach to partial agreement is to ask: when do the pathologists disagree the most? Of course, when they split 3–3, that is when three pathologists agree for the absence of epithelioid granuloma and when the other three agree for the presence of granuloma or vice versa. This occurs for cases 6, 14, 18, 26, and 35, of Table 6.32, with frequencies 1, 1, 1, 1, and 3, respectively. See Table 6.32, where a quick calculation reveals that the probability of such a split is $7/68 = 0.1029$, while the posterior analysis for a Kappa-like index for this event appears in Table 6.34.

As one would expect, the analysis shows very poor agreement between the six pathologists when they split 3–3. The probability of a split agreement is 0.1029 and the corresponding chance agreement is 0.0723. When discussing Kappa, the probability of agreement and of chance agreement of the event should always be stated.

Of course, the analysis of partial agreement can be expanded by determining Kappa in the context of association and the intraclass Kappa. This will be considered in the exercise section. One could examine the following situations for partial agreement: (a) when exactly five of the six agree, or a 5–1 split (either way for absence or presence of granuloma); and (b) when exactly four agree (a 4–2 split). The case where exactly three agree with a 3–3 split and pairwise agreement have already been analyzed. (See Table 6.33 for the latter case, and Table 6.34 for the former.)

6.9.3 Stratified Kappa

The example of a national lung cancer screening trial is continued, but another modality is added, namely, MRI, that is, each subject receives three images: one from x-ray, one from CT, and one from MRI, and three hospital sites are involved. The study was designed to have 1000 subjects in the first two sites, and 500 in the third, but the final assignment of patients was 976 and 936 for sites 1 and 2, respectively, and 494 for the third site.

For the first site, the outcomes are displayed in Table 6.35a, where the cell entries are interpreted as follows: 1 indicates no evidence of disease, 2 indicates that most likely no disease is present, 3 denotes that it is likely that disease is present, and 4 that there is a malignant lesion. Thus, for case 64 of site 1, the three images agree 50 times that there is a malignant lesion.

Obviously, this is the most complex example introduced so far. It involves multiple raters, the three imaging modalities, four ordinal outcomes, and three sites. Such complexity introduces a myriad of ways that agreement is to be approached. The overall Kappa for agreement between the three modalities can be determined, as well as many indices for partial agreement. How should the various indices of agreement of the three strata (sites) be combined and compared?

TABLE 6.35a: X-ray, CT, and MRI for site 1.

Case	X-ray	CT	MRI	Total
1	1	1	1	200
2	1	1	2	55
3	1	1	3	40
4	1	1	4	25
5	1	2	1	25
6	1	2	2	15
7	1	2	3	10
8	1	2	4	7
9	1	3	1	9
10	1	3	2	6
11	1	3	3	4
12	1	3	4	3
13	1	4	1	3
14	1	4	2	1
15	1	4	3	1
16	1	4	4	10
17	2	1	1	10
18	2	1	2	20
19	2	1	3	8
20	2	1	4	6
21	2	2	1	100
22	2	2	2	40
23	2	2	3	35
24	2	2	4	22
25	2	3	1	2
26	2	3	2	3
27	2	3	3	7
28	2	3	4	6
29	2	4	1	1
30	2	4	2	1
31	2	4	3	1
32	2	4	4	7
33	3	1	1	10
34	3	1	2	5
35	3	1	3	3
36	3	1	4	2
37	3	2	1	1
38	3	2	2	3
39	3	2	3	20
40	3	2	4	2
41	3	3	1	15
42	3	3	2	10
43	3	3	3	100
44	3	3	4	23
45	3	4	1	1
46	3	4	2	1
47	3	4	3	2

TABLE 6.35a (continued): X-ray, CT, and MRI for site 1.

Case	X-ray	CT	MRI	Total
48	3	4	4	1
49	4	1	1	1
50	4	1	2	1
51	4	1	3	2
52	4	1	4	7
53	4	2	1	1
54	4	2	2	3
55	4	2	3	1
56	4	2	4	2
57	4	3	1	0
58	4	3	2	0
59	4	3	3	2
60	4	3	4	1
61	4	4	1	8
62	4	4	2	6
63	4	4	3	9
64	4	4	4	50

TABLE 6.35b: X-ray, CT, and MRI for site 2.

Case	X-ray	CT	MRI	Total
1	1	1	1	150
2	1	1	2	50
3	1	1	3	50
4	1	1	4	47
5	1	2	1	5
6	1	2	2	25
7	1	2	3	10
8	1	2	4	4
9	1	3	1	5
10	1	3	2	5
11	1	3	3	15
12	1	3	4	1
13	1	4	1	3
14	1	4	2	2
15	1	4	3	1
16	1	4	4	6
17	2	1	1	20
18	2	1	2	8
19	2	1	3	6
20	2	1	4	3
21	2	2	1	50

(continued)

TABLE 3.35b (continued):   X-ray, CT, and MRI for site 2.

Case	X-ray	CT	MRI	Total
22	2	2	2	100
23	2	2	3	40
24	2	2	4	20
25	2	3	1	3
26	2	3	2	2
27	2	3	3	10
28	2	3	4	1
29	2	4	1	0
30	2	4	2	0
31	2	4	3	0
32	2	4	4	2
33	3	1	1	10
34	3	1	2	5
35	3	1	3	4
36	3	1	4	1
37	3	2	1	2
38	3	2	2	18
39	3	2	3	3
40	3	2	4	3
41	3	3	1	30
42	3	3	2	20
43	3	3	3	90
44	3	3	4	15
45	3	4	1	2
46	3	4	2	1
47	3	4	3	0
48	3	4	4	4
49	4	1	1	6
50	4	1	2	3
51	4	1	3	2
52	4	1	4	1
53	4	2	1	1
54	4	2	2	3
55	4	2	3	1
56	4	2	4	1
57	4	3	1	0
58	4	3	2	0
59	4	3	3	1
60	4	3	4	0
61	4	4	1	7
62	4	4	2	3
63	4	4	3	50
64	4	4	4	5

TABLE 6.35c: X-ray, CT, and MRI for site 3.

Case	X-ray	CT	MRI	Total
1	1	1	1	100
2	1	1	2	40
3	1	1	3	22
4	1	1	4	23
5	1	2	1	4
6	1	2	2	10
7	1	2	3	3
8	1	2	4	3
9	1	3	1	2
10	1	3	2	2
11	1	3	3	5
12	1	3	4	2
13	1	4	1	1
14	1	4	2	1
15	1	4	3	0
16	1	4	4	2
17	2	1	1	10
18	2	1	2	2
19	2	1	3	2
20	2	1	4	1
21	2	2	1	20
22	2	2	2	60
23	2	2	3	11
24	2	2	4	10
25	2	3	1	3
26	2	3	2	2
27	2	3	3	10
28	2	3	4	1
29	2	4	1	0
30	2	4	2	0
31	2	4	3	1
32	2	4	4	1
33	3	1	1	4
34	3	1	2	2
35	3	1	3	1
36	3	1	4	1
37	3	2	1	2
38	3	2	2	10
39	3	2	3	4
40	3	2	4	4
41	3	3	1	5
42	3	3	2	5
43	3	3	3	60
44	3	3	4	10
45	3	4	1	2
46	3	4	2	1

(continued)

TABLE 3.35c (continued): X-ray, CT, and MRI for site 3.

Case	X-ray	CT	MRI	Total
47	3	4	3	1
48	3	4	4	3
49	4	1	1	4
50	4	1	2	1
51	4	1	3	1
52	4	1	4	2
53	4	2	1	0
54	4	2	2	1
55	4	2	3	0
56	4	2	4	0
57	4	3	1	0
58	4	3	2	0
59	4	3	3	1
60	4	3	4	0
61	4	4	1	3
62	4	4	2	2
63	4	4	3	2
64	4	4	4	8

Consider combining the three Kappas for the three sites into one, by a weighted average, where the weights depend on the inverse of the variance of the posterior distribution of Kappa of Table 6.36.

BUGS CODE 6.10 is for the posterior analysis of overall Kappa for site 3 and was executed with 45,000 observations generated from the joint posterior distribution of Kappa3, agree3, and cagree3, with a burn in of 5,000 observations and a refresh of 100. A similar program can be executed for the other two sites, giving Kappas for the three sites (strata) and the corresponding standard deviations, and the stratified Kappa is easily computed. This is left as an exercise, and note a uniform prior is used for prior information.

TABLE 6.36: Stratified Kappa for screening trial.

Parameter	Mean	sd	Error	2 1/2	Median	97 1/2
agree1	?	?		?	?	?
cagree1	?	?		?	?	?
Kappa1	?	?		?	?	?
agree2	?	?		?	?	?
cagree2	?	?		?	?	?
Kappa2	?	?		?	?	?
agree3	0.3348	0.0196	<0.0001	0.2966	0.3348	0.374
cagree3	0.0892	0.0042	<0.0001	0.0817	0.0890	0.09835
Kappa3	0.2697	0.0206	<0.0001	0.2295	0.2696	0.3107
Kappa stratified	?	?		?	?	?

BUGS CODE 6.10

```
model
  {
# for site 3
# g[i] corresponds to case i of Table 6.35c.
# a uniform prior is assumed
  g[1]~dgamma(101,2) g[2]~dgamma(41,2) g[3]~dgamma(23,2)
  g[4]~dgamma(24,2) g[5]~dgamma(5,2) g[6]~dgamma(11,2)
  g[7]~dgamma(4,2) g[8]~dgamma(4,2) g[9]~dgamma(4,2)
  g[10]~dgamma(3,2) g[11]~dgamma(6,2) g[12]~dgamma(3,2)
  g[13]~dgamma(2,2) g[14]~dgamma(2,2) g[15]~dgamma(.1,2)
  g[16]~dgamma(3,2) g[17]~dgamma(11,2) g[18]~dgamma(3,2)
  g[19]~dgamma(3,2) g[20]~dgamma(2,2) g[21]~dgamma(21,2)
  g[22]~dgamma(61,2) g[23]~dgamma(12,2) g[24]~dgamma(11,2)
  g[25]~dgamma(4,2) g[26]~dgamma(3,2) g[27]~dgamma(11,2)
  g[28]~dgamma(2,2) g[29]~dgamma(1,2) g[30]~dgamma(11,2)
  g[31]~dgamma(2,2) g[32]~dgamma(2,2) g[33]~dgamma(5,2)
  g[34]~dgamma(3,2) g[35]~dgamma(2,2) g[36]~dgamma(2,2)
  g[37]~dgamma(3,2) g[38]~dgamma(11,2) g[39]~dgamma(5,2)
  g[40]~dgamma(5,2) g[41]~dgamma(6,2) g[42]~dgamma(6,2)
  g[43]~dgamma(61,2) g[44]~dgamma(11,2) g[45]~dgamma(3,2)
  g[46]~dgamma(2,2) g[47]~dgamma(2,2) g[48]~dgamma(4,2)
  g[49]~dgamma(5,2) g[50]~dgamma(2,2) g[51]~dgamma(2,2)
  g[52]~dgamma(3,2) g[53]~dgamma(.1,2) g[54]~dgamma(2,2)
  g[55]~dgamma(11,2) g[56]~dgamma(1,2) g[57]~dgamma(1,2)
  g[58]~dgamma(.01,2) g[59]~dgamma(1,2) g[60]~dgamma(.01,2)
  g[61]~dgamma(3,2) g[62]~dgamma(2,2) g[63]~dgamma(2,2)
  g[64]~dgamma(8,2)
  h<- sum(g[])
  for(i in 1:64){a[i]<- g[i]/h}
agree3<- a[1]+a[21]+a[43]+a[64]
b1..<-a[1]+a[2]+a[3]+a[4]+a[5]+a[6]+a[7]+a[8]+a[9]+a[10]+a[11]+a[12]+
      a[13]+a[14]+a[15]+a[16]
b.1.<-a[1]+a[2]+a[3]+a[4]+a[17]+a[18]+a[19]+a[20]+a[33]+a[34]+a[35]+
      a[36]+a[49]+a[50]+a[51]+a[52]
b..1 <-a[1]+a[5]+a[9]+a[13]+a[17]+a[21]+a[25]+a[29]+a[33]+a[37]+a[41]+
      a[45]+a[49]+a[53]+a[57]+a[61]
b2..<-a[17]+a[18]+a[19]+a[20]+a[21]+a[22]+a[23]+a[24]+a[25]+a[26]+a[27]+
      a[28]+a[29]+a[30]+a[31]+a[32]
b.2.<-a[5]+a[6]+a[7]+a[8]+a[21]+a[22]+a[23]+a[24]+a[37]+a[38]+a[39]+
      a[40]+a[53]+a[54]+a[55]+a[56]
b..2<-a[2]+a[6]+a[10]+a[14]+a[18]+a[22]+a[26]+a[30]+a[34]+a[38]+a[42]+
      a[46]+a[50]+a[54]+a[58]+a[62]
b3..<-a[33]+a[34]+a[35]+a[36]+a[37]+a[38]+a[39]+a[40]+a[41]+a[42]+a[43]+
      a[44]+a[45]+a[46]+a[47]+a[48]
```

b.3.<-a[9]+a[10]+a[11]+a[12]+a[25]+a[26]+a[27]+a[28]+a[41]+a[42]+a[43]+
    a[44]+a[57]+a[58]+a[59]+a[60]
b..3<-a[3]+a[7]+a[11]+a[15]+a[19]+a[23]+a[27]+a[31]+a[35]+a[39]+a[43]+
    a[47]+a[51]+a[55]+a[59]+a[63]
b4..<-a[49]+a[50]+a[51]+a[52]+a[53]+a[54]+a[55]+a[56]+a[57]+a[58]+a[59]+
    a[60]+a[61]+a[62]+a[63]+a[64]
b.4.<-a[13]+a[14]+a[15]+a[16]+a[29]+a[30]+a[31]+a[32]+a[45]+a[46]+a[47]+
    a[48]+a[61]+a[62]+a[63]+a[64]
b..4<-a[4]+a[8]+a[12]+a[16]+a[20]+a[24]+a[28]+a[32]+a[36]+a[40]+a[44]+
    a[48]+a[52]+a[56]+a[60]+a[64]
cagree3<- b1..*b.1.*b..1+b2..*b.2.*b..2+b3..*b.3.*b..3+b4..*b.4.*b..4
kappa3<- (agree3-cagree3)/(1-cagree3) }
list(g=c(2,2,2,2,2,2,2,2,2,2,2,2,2,2,2,2,2,2,2,2,2,2,2,2,2,2,2,2,2,2,2,2,2,2,2,2,2,2,2,
    2,2,2,2,2,2,2,2,2,2,2,2,2,2,2,2,2,2,2,2,2,2,2,2,2,2,2,2,2))

6.10   Conclusions for Agreement and Accuracy

This chapter introduces ideas that allow one to assess the accuracy of a medical device when there are several readers involved in the diagnosis of the disease. In the first three sections, methodology that addresses agreement when a gold standard is present, is described and illustrated with many examples from the medical field. When a gold standard is present, one can compare the accuracy of several readers, and if there is wide variation in the stated accuracies, an overall assessment of accuracy can be devised. The case where the test scores were ordinal and continuous were both considered.

When no gold standard is present, the main focus of the chapter is on estimating the degree of agreement between the various readers. The main parameter for estimating agreement is the posterior distribution of the Kappa coefficient. Several scenarios are presented, when there are two readers and binary scores, two readers with multiple nominal and ordinal scores, and finally for multiple readers and multiple ordinal scores. Generalizations of Kappa are introduced, namely, for the situation when stratified designs are appropriate and a weighted Kappa is estimated via Bayesian techniques. All examples use WinBUGS, where the MCMC errors of the parameters are controlled to be small, frequently <0.0001.

The case where the medical test scores are continuous is not considered, but is an important topic and the reader is referred to Chapter 6 of Broemeling [6], who used the intraclass correlation and other correlation type measures to estimate the agreement between and among readers.

It should be noted that when no gold standard is available, the focus on accuracy is somewhat blunted, because agreement is not the same as accuracy. When a gold standard is present, the accuracies of the readers can be compared, but when it is not present, this cannot be done. The situation where some patients are subject to the gold standard and other patients are not is often done in practice, and under certain conditions, accuracy can be assessed. This important scenario will be considered in Chapters 7 and 8, one addressing verification bias, and the other chapter addressing the situation when no gold standard is present, but prior information about the disease rate is available. In the latter scenario, Bayesian approaches are employed to estimate the test accuracies.

Even if there is no gold standard, each patient either has or does not have the disease, and if the "present" study does not have a gold standard, but there is reliable information about the disease rate from previous related studies, then mathematically it is possible to build models that estimate the test accuracy. This interesting situation will be addressed in later chapters.

6.11 Exercises

1. Verify Table 6.1 for the four ROC areas of the four readers of the sentinel lymph node biopsy. Execute the analysis with 65,000 observations, with a burn in of 5,000 and a refresh of 100 based on BUGS CODE 6.1.

2. Verify Table 6.2 by joining the code of the four readers and add statements that allow for computation of the differences in the ROC areas of all pairs of readers. Also, add statements that allow for computation of the simple and weighted averages. Execute the analysis with 65,000 observations, with a burn in of 5,000 and a refresh of 100. State your conclusion about the overall accuracy of the test and justify your answer.

3. As a second example with ordinal scores, consider Exercise 9 of Chapter 5, where four readers are interpreting the metastasis of lung cancer to the lymph nodes using CT. The study is multicenter and reports the accuracy of CT in detecting lymph node invasion in patients with known lung cancer. All patients enrolled were examined preoperatively with CT, and after reading the image readers were asked to rate the degree of disease on a five-point scale to rate the metastasis to the lymph nodes. The gold standard for metastasis was established by pathologic analysis obtained from surgery. If no invasion occurred, the population is referred to as non diseased, but if metastasis occurred, the patient belongs to the diseased population. A subset of the study is reported in Table 5.10, where only CT images are involved and four radiologists scored the degree of invasion. Our objective is to determine the ROC area accounting for all four readers and to examine the effect of the four readers on the ROC area.

These data were analyzed in Exercise 9 of Chapter 5, based on BUGS CODE 5.4, where the effect of the four readers on the ROC area was determined.

(a) Similar to Exercises 1 and 2 above, using BUGS CODE 6.1, perform a Bayesian analysis that determines the ROC area of the four readers.

(b) As with Exercises 1 and 2 above, revise the code by adding statements that will provide the posterior analysis for the difference in the ROC areas of the four readers.

(c) Based on the ROC areas of the four readers, what is your conclusion about the overall accuracy of the medical test for periprostatic invasion?

(d) Find the posterior distribution of the simple average of the four ROC areas.

(e) Determine the posterior distribution of a weighted average of the four ROC areas, where the weight of a particular ROC area is the inverse of the variance of the posterior distribution of that ROC area.

(f) Based on (d) and (e), what is your conclusion about the overall accuracy of this medical test for lung cancer metastasis to the lymph nodes?

4. Verify Table 6.4 using BUGS CODE 6.2. Use 65,000 observations, with a burn in of 5,000 and a refresh of 100.

5. Verify Table 6.5 by revising BUGS CODE 6.2. The code is revised by eliminating age and gender as independent variables in the three regressions.

Again use 65,000 observations, with a burn in of 5,000 and a refresh of 100. You should get exactly the same results as reported in Table 6.5.

6. Verify Table 6.6 by revising BUGS CODE 6.2 to include statements that allow computation of the simple and weighted averages of the three ROC areas. I used 65,000 observations for the simulation, with a burn in of 5,000 and a refresh of 100. You should get the same results as I did for Table 6.6.

7. Refer to Table 6.8 and suppose the results of a previous related experiment are given in Table 6.37. Combining this prior information and using a Bayes theorem with the data given in Table 6.8, perform the appropriate analysis and compare with the posterior analysis of Table 6.9. Use 35,000 observations, with a burn in of 5,000 and a refresh of 100. Also, plot the posterior densities of A_1 and A_2.

8. The following code is based on the beta-binomial model. With 25,000 observations generated from the joint posterior distribution of theta11, π, and κ_I, with a burn in of 1,000 and a refresh of 100, execute the

TABLE 6.37: Two raters.

Rater 1	Rater 2		
	$Y = 0$	$Y = 1$	
$X = 0$	2	1	3
$X = 1$	3	22	25
Total	5	23	28

statements below and estimate the three parameters. This is the Dutch rabbit fetus study. How do your results compare with those based on the constant correlation model? (See Table 6.23.)

BUGS CODE 6.11

```
model
{ for (i in 1:17) {phi[i]~dbeta(1,1)}
for( i in 1 : 17) {
for( j in 1 : 10) {
Y[i , j] ~ dbern(phi[i])}}
theta11~dbeta(17,213)
kappa<-(theta11-pi*pi)/pi*(1-pi)
 pi<-mean(phi[]) }
list(Y= structure( .Data=
c(1,0,0,0,0,0,0,0,0,NA,0,0,0,0,0,0,0,0,0,0,
1,0,0,0,0,0,0,NA,NA,NA,0,0,0,0,0,NA,NA,NA,NA, NA,
1,0,0,0,NA,NA,NA,NA, NA,NA,0,0,0,0,0,0,NA,NA, NA,NA,
1,0,0,NA,NA,NA,NA, NA,NA,NA,1,0,0,0,0,0,0,0,NA,NA,
1,1,0,0,0,0,0,0,0,NA,NA,0,0,0,0,NA,NA,NA, NA,NA,NA,
1,1,1,1,NA,NA,NA, NA,NA,NA,1,0,0,0,0,NA,NA, NA,NA,NA,
1,0,0,NA,NA,NA, NA,NA,NA,NA,1,1,1,1,0,0,0,0,NA,NA,
1,1,0,0,0,0,NA,NA,NA,NA,1,1,1,0,0,0,0,0,NA,NA,
1,0,0,0,0,0,NA,NA,NA,NA),.Dim=c(17,10)))
list(theta11=.5 )
```

9. Using the above program, estimate the intraclass Kappa for the four treatment groups of Table 6.21. Do the four groups differ with respect to the correlation between fetuses across groups? Explain in detail how the WinBUGS code is executed and carefully explain the posterior analysis for comparing the four groups. If there is no difference in the four Kappas, how should they be combined in order to estimate the overall intraclass correlation?

10. Refer to Table 6.24 and give an interpretation of the degree of agreement between the three indices. What is your overall conclusion about the degree of agreement? Should only one index be used or should all four be reported? Kappa is the usual choice as an index of agreement.

11. Verify Table 6.26 and determine the posterior distribution of the difference in the two Kappas. Is there a difference between them? Based on the G coefficient, what is your interpretation of the degree of agreement between the radiologists and the gold standard? Test for a difference in G_1 and G_2.

12. In view of Equation 6.26, show that two raters can have high sensitivity and specificity, but κ_{tw} can be small.

13. Verify Equation 6.26, and if necessary, refer to Kraemer [22] and Thompson and Walter [23].

14. Verify Equations 6.27 and 6.28.

15. If $\theta_1 = \theta_2$, what is the value of κ_a (Equation 6.28)?

16. Verify Table 6.29 with your own WinBUGS code. Verify that the posterior distribution of the usual Kappa (1.1) is the same as that of κ_a, given by Table 6.29.

17. Using BUGS CODE 6.8, verify the posterior analysis for Table 6.31, and produce a plot of the posterior distribution of $\kappa(4,2)$.

18. Amend BUGS CODE 6.9, verify Table 6.33, and perform the posterior analysis for a split agreement between the six pathologists. Generate 75,000 observations from the joint posterior distribution of the parameters, with a burn in of 5,000 and a refresh of 100.

19. Amend BUGS CODE 6.9 and determine the posterior distribution of a Kappa type index for the following events: (a) exactly five agree, or a 5–1 split between pathologists; and (b) exactly four agree for a 4–2 split. For each of the two events above, find the posterior probability of the event and the posterior probability of the same event, but assuming that the six pathologists are acting independently in their assignment of scores to the lesions. Use 125,000 observations, with a burn in of 10,000 observations and a refresh of 100, and employ a uniform prior density for the parameters. Provide a plot of the posterior distribution of the Kappa type index for both events (a) and (b).

20. Using BUGS CODE 6.10, verify the posterior analysis of Table 6.36. What is the posterior mean and standard deviation of the stratified Kappa? Provide the posterior density of stratified Kappa and give a 95% credible interval for the same parameter. What is your overall conclusion about the agreement between the three imaging modalities? Recall that 30,000 observations are generated from the posterior distribution with a uniform prior.

References

[1] Shoukri, M.M. *Measures of Interobserver Agreement.* Chapman & Hall/CRC, Boca Raton, 2004.

[2] Von Eye, A. and Mun, E.Y. *Analyzing Rater Agreement, Manifest Variable Methods.* Lawrence Erlbaum, Mahwah, NJ, and London, 2005.

[3] Broemeling, L.D. *Bayesian Methods of Measures of Agreement.* Chapman & Hall/CRC, Boca Raton, 2009.

[4] O'Malley, J.A., Zou, K.H., Fielding, J.R., and Tampany, C.M.C. Bayesian regression methodology for estimating a receiver operating characteristic curve with two radiologic applications: Prostate biopsy and spiral CT of ureteral stones. *Academic Radiology,* 8:713, 2001.

[5] Fleiss, J.L. Measuring agreement between two judges on the presence or absence of a trait. *Biometrics,* 31(3):641, 1975.

[6] Goodman, L.A. and Kruskal, W.H. Measures of association for cross classification. *Journal of the American Statistical Association,* 49:732, 1954.

[7] Dice, L.R. Measures of the amount of ecologic association and disagreement. *Ecology,* 26:297, 1945.

[8] Rogot, E. and Goldberg, I.D. A proposed idea for measuring agreement in test re-test studies. *Journal of Chronic Diseases,* 19:991, 1966.

[9] Armitage, P., Blendis, L.M., and Smyllie, H.C. The measurement of observer disagreement in the recording of signs. *Journal of the Royal Statistical Society: Series A,* 129:98, 1966.

[10] Scott, W.A. Reliability of content analysis: The cases of nominal scale coding. *Public Opinion Quarterly,* 19:321, 1955.

[11] Cohen, J.A. Coefficient of agreement for nominal scales. *Educational and Psychological Measurement,* 20:37, 1960.

[12] Coleman, J.S. Measuring concordance in attitudes. Unpublished manuscript. Johns Hopkins University, Baltimore, MA, 1966.

[13] Light, R.J. Analysis of variance for categorical data with applications for agreement and association. Unpublished dissertation. Harvard University, Cambridge, MA, 1969.

[14] Liebetrau, A.M. *Measures of Association.* Sage, Newbury Park, 1983.

[15] Barlow, W., Lai, M.Y., and Azen, S.P. A comparison of methods for calculating a stratified Kappa. *Statistics in Medicine,* 10:1465, 1991.

[16] Fleiss, J.L. and Cohen, J. The equivalence of the weighted Kappa and the interclass Kappa. *Psychological Bulletin*, 72:232, 1973.

[17] Mak, T.K. Analyzing intraclass correlation for dichotomous variables. *Applied Statistics*, 37:344, 1988.

[18] Kupper, L.L. and Haseman, J.K. The use of correlated binomial model for the analysis of certain toxicological experiments. *Biometrics*, 25:281, 1978.

[19] Crowder, M.J. Inferences about the intraclass correlation coefficient in the beta-binomial ANOVA for proportions. *Journal of the Royal Statistical Society: Series B*, 41:230, 1979.

[20] Banerjee, M., Capozzoli, M., McSweeney, L., and Sinha, D. Beyond Kappa: A review of interrater agreement measures. *Canadian Journal of Statistics*, 27:3, 1999.

[21] Bloch, D.A. and Kraemer, H.C. 2 × 2 Kappa coefficients: Measures of agreement or association. *Biometrics*, 45:269, 1989.

[22] Paul, S.R. Analysis of proportions of affected fetuses in teratological experiments. *Biometrics*, 38:361, 1982.

[23] Kraemer, H.C. Ramifications of a population model for Kappa as a coefficient of reliability. *Psychometrika*, 44:461, 1979.

[24] Thompson, W.D. and Walter, S.D. A reappraisal of the Kappa coefficient. *Journal of Clinical Epidemiology*, 41(10):949, 1988.

[25] DeGroot, M.H. Reaching a consensus. *Journal of the American Statistical Association*, 69:118, 1974.

[26] Winkler, R.L. The consensus of subjective probability distributions. *Management Science*, 15:B61, 1969.

[27] Rogel, A., Boelle, P.Y., and Mary, J.Y. Global and partial agreement among several observers. *Statistics in Medicine*, 17:489, 1998.

[28] Theodossi, A., Spieglehalter, D.J., Jass, J., Firth, J., Dixon, M., Leader, M., Levison, D.A., et al. 1994. Observer variation and discriminatory value of biopsy features of inflammatory bowel disease. *Gut* 35:961, 1994.

[29] Williams, G.W. Comparing the joint agreement of several raters with one rater. *Biometrics*, 32:619, 1976.

[30] Landis, J.R. and Kock, G.C. The measurement of observer agreement for categorical data. *Biometrics*, 33:159, 1977.

Chapter 7

Estimating Test Accuracy with an Imperfect Reference Standard

7.1 Introduction

Suppose that a gold standard does not exist, but the accuracy of a new test will be assessed with an imperfect gold standard. Many cases exist where there is no perfect gold standard. For example, depression is usually determined by a series of questions and observing the behavior of the patient, but such assessments are highly subjective, and there is no one test that provides a perfect diagnosis. For infectious diseases, a perfect diagnosis can be elusive, where a culture is taken, however, the culture may not contain the infective agent or if the agent is present, it may not grow in the culture. Pepe [1] gives other examples, including tests for diagnosing cancer and hearing loss. Zhou, McClish, and Obuchowski [2] also present various studies, including the diagnosis of a bacterial infection with the stool and serology tests. The method of analysis is maximum likelihood, while the approach taken here is Bayesian. Other examples presented by Zhou, McClish, and Obuchowski include two tests for tuberculosis, with the Tine and Mantour tests, at two different sites, while a third example for detecting pleural thickening is performed by x-ray with three readers. Pepe describes another interesting example of multiple tests, where *Chlamydia* bacterial infection is diagnosed with a blood culture, polymerase chain reaction (PCR), and enzyme linked immunosorbent assay (ELISA).

Previous work has focused on maximum likelihood estimation (MLE) and Bayesian. Zhou, McClish, and Obuchowski emphasize maximum likelihood and Bayesian. The Bayesian method is based on earlier work by Joseph, Gyorkos, and Coupal [3], who employ an augmented data approach. The augmented data approach views the missing data (the disease status D of a patient) as an unobservable random variable that can be modeled in such a way as to provide the posterior density of the measures of disease accuracy (true and false positive rates). Such an approach will be used here, because the Bayesian method has the advantage of using prior information and being able to separate the parameters of interest from nuisance parameters. Fortunately, prior information is available for diagnostic tests, especially the disease rates and the accuracy assessments of medical tests, and can be used as part of the posterior analysis.

With the Bayesian approach of Joseph, Gyorkos, and Coupal [3] and Dendukuri and Joseph [4], the various tests are assumed to be conditionally independent, an assumption that will be used in the present approach; however, the assumption will be relaxed in some cases and the two ways compared in estimating test accuracy.

Pepe [1: 195] presents an example of using an imperfect reference standard R to assess the accuracy of a new test T (see Table 7.1). The new test T has a "true" sensitivity of 0.80 (80/100) and a specificity of 0.70 (70/100), but of course this is actually not known because there is no gold standard. Relative to the reference test R, the estimated sensitivity is also 0.8 (64/80) but has a specificity of 0.61(74/120), thus the new test is assessed to be less specific than it actually is. Also, with respect to the gold standard, the prevalence of disease is 50%, but is estimated to be 40% with regard to R. Remember, the gold standard is not present, we do not know the "true" measures of accuracy, only those with regard to the reference standard can be estimated, and can be misleading!

The two tests are said to be conditionally independent if

$$P[T, R \mid D] = P[T \mid D]P[R \mid D], \tag{7.1}$$

and, but is usually employed with both the conventional and Bayesian approaches. Using this assumption, Pepe [1: 195] states that it is likely that both the observed (relative to the reference test R) sensitivity and specificity will be decreased.

Are there methods that will improve on the measures of accuracy provided by the imperfect standard test? Using primarily the Bayesian approach, this question will be explored in this chapter. In what follows, the subject is introduced with two binary tests, one is the reference test R and the other is a new test, T, whose accuracy is to be assessed. Note, none of the patients will have their true disease status (D) measured, instead each patient will be given a positive or negative score on both tests. A Bayesian approach is taken, where based on the likelihood function, the posterior distribution of the sensitivity and specificity is determined. The likelihood function is presented in two forms. The first is presented without augmented variables and the second assumes that the missing disease status is modeled by augmented or latent variables. In the first form of the likelihood function, conditional independence is not assumed and the posterior distribution of disease prevalence is isolated as a product of four functions, each of which is a mixture of beta

TABLE 7.1: Hypothetical example—imperfect reference.

	$D = 0$	$D = 1$	$R = 0$	$R = 1$
$T = 0$	70	20	74	16
$T = 1$	30	80	46	64
Total	100	100	120	80

random variables. In the second form of the likelihood function with latent variables, and assuming conditional independence, the posterior distribution of the sensitivity, specificity, and disease prevalence is determined. An example earlier analyzed by Joseph, Gyorkos, and Coupal [3] and presented by Zhou, McClish, and Obuchowski [2] involves a bacterial infection of immigrants to Canada and employs the augmented data method to estimate the sensitivity and specificity of the reference test R (a serology test) and another test T, the stool examination.

Presented next is an extension of two binary tests to several populations, using the augmented data method. An example from Zhou, McClish, and Obuchowski taken from Hui and Walter [5] involves two tests to diagnose tuberculosis. The reference test is the Mantour test and the new test is the Tine test. Both tests are used on two different populations, one with a low prevalence of tuberculosis and the other with a very high prevalence rate. Of interest here is the comparison of the sensitivity and specificity between the two populations. In all cases, the accuracy of the new test relative to the reference is compared to the accuracy computed by the Bayesian approach.

Multiple tests are also of interest in diagnostic medicine, and the Bayesian approach with augmented variables is easily extended. An interesting example of this is comparing the accuracies of three readers who are using x-ray to diagnose pleural thickening of the lungs of South African asbestos workers. In this case, there is no reference test (reader) and the sensitivities and specificities of the three are estimated using the latent variable method with and without the conditional independence assumption (CIA). As for agreement, see Broemeling [6] for Bayesian methods of estimating agreement between readers diagnosing a disease. Of course, of interest is an extension of using multiple tests in various populations. For example, different sets of three readers could be assessing pleural thickening in two locations, say among asbestos workers in South Africa and among those in Montana asbestos mines. Of interest is comparing the combined sensitivity among the three readers in South Africa with the combined sensitivity of the three readers in Montana (see Megibow et al. [7]).

The last topic to be presented in this chapter, is estimating the area under the receiver operating characteristic (ROC) curve of a new test compared to the area of a reference standard test. Both tests provide ordinal scores to patients, and an example of this is staging the disease of, say, cancer patients. Staging involves assigning a stage to each patient, e.g., using magnetic resonance imaging (MRI) and computed tomography (CT) to stage breast cancer. The stage reflects the state of the disease, where local disease is designated 1, disease that has spread to the lymph nodes is assigned a 2, and 3 is assigned to those where the disease has spread beyond the lymphatic system to other organs. Knowing the accuracy of the staging system is crucial in patient management.

Much of the work in the area of medical test accuracy when there is no gold standard is presented in Pepe [1] and Zhou, McClish, and Obuchowski [2].

What is presented here follows the same general topics and theme, however, the emphasis is inclusively Bayesian. With Bayesian inferences, prior information is used in an optimal fashion and is of great advantage for these types of problems when there is no gold standard. When determining the accuracy of a "new" medical test, prior information is usually plentiful, because often there have been many previous related studies that are highly related to the study at hand. When studying a "new" test, its accuracy is usually better than the reference test. Of course, this information is valuable to a Bayesian! Of special concern is prior information about the disease rate or prevalence, because accurate estimates of the disease rate are crucial in order to estimate efficiently the sensitivity and specificity of the new test. Since no gold standard is available, the prevalence rate is not measured in the present study; therefore, prior studies must be used for prior information about disease rates.

What is the scientific interest in estimating test accuracy when no gold standard is available? The radiology literature contains many studies where no gold standard is available, thus it is important to have methods that allow one to determine accuracy. Also, many tests are routinely used when no gold standard is present, and based on the outcome of the test, the patient is referred to another test procedure, which is not necessarily the gold standard. For example, a patient's symptoms might indicate the possibility of heart disease, and the patient is referred for an exercise stress test. The diagnosis based on the symptoms is preliminary, but it is of interest to know the accuracy of the diagnosis based only on the symptoms, where the reference test is the exercise stress test. A gold standard is not involved up to this point. Using a series of tests or multiple tests is an area of statistical and clinical interest and will be examined in more detail in Chapter 10.

The present chapter is somewhat related to the next, when the test is carried out but the disease status of some patients is not verified. The bias of measures of accuracy is corrected by statistical techniques based on missing value methodology. In this chapter, a patient is not referred to a gold standard, but the accuracy of the test at hand is based on a reference test. Statistically, missing values techniques based on augmented data lead to methods that correct the bias of accuracy measures given by the reference test.

7.2   Two Binary Tests

Consider the experiment shown in Table 7.2 with a reference test R and a new test T.

The likelihood function is

$$L(\theta, \phi) \propto [\theta_{11}p + \phi_{11}(1-p)]^{x_{11}} [\theta_{10}p + \phi_{10}(1-p)]^{x_{10}}$$
$$\theta_{01}p + \phi_{01}(1-p)]^{x_{01}} [\theta_{00}p + \phi_{00}(1-p)]^{x_{00}}, \tag{7.2}$$

TABLE 7.2: Two binary tests.

Reference test	Test T	
	$T = 1$	$T = 0$
$R = 1$	x_{11}	x_{10}
$R = 0$	x_{01}	x_{11}
Total	n_1	n_0

where p is the disease rate,

$$\theta_{ij} = P[R = i, T = j \mid D = 1], \quad i, j = 0, 1,$$

and

$$\phi_{ij} = P[R = i, T = j \mid D = 0].$$

Also note that

$$\sum_{i=0}^{i=1} \sum_{j=0}^{j=1} \theta_{ij} = 1$$

and

$$\sum_{i=0}^{i=1} \sum_{j=0}^{j=1} \phi_{ij} = 1.$$

The likelihood function is a product of four functions, corresponding to the four cells of Table 7.2. Each function gives the joint probability of R and T conditional on $D = 1$, followed by the same probability conditional on $D = 0$, and the likelihood function is a highly nonlinear function of the parameters. Since x_{ij} are positive integers, the binomial theorem can be used to expand each of the four functions into a mixture, e.g., consider the first factor, then

$$[\theta_{11} p + \phi_{11}(1-p)]^{x_{11}} = \sum_{i=0}^{i=x_{11}} BC(x_{11}, i)\theta_{11}^i \phi_{11}^{x_{11}-i} p^i (1-p)^{x_{11}-i},$$

where $BC(x_{11}, i) = x_{11}!/i!(x_{11} - i)!$.

A similar expansion can be used for the other three factors of the likelihood function, but in general:

$$[\theta_{ij} p + \phi_{ij}(1-p)]^{x_{ij}} = \sum_{l=0}^{l=x_{ij}} BC(x_{ij}, l)\theta_{ij}^l \phi_{ij}^{x_{ij}-l} p^l (1-p)^{x_{ij}-l}, \tag{7.3}$$

for $i, j = 0, 1$, thus the likelihood function is

$$L(\theta, \phi) = \prod_{i,j=0}^{i,j=1} \sum_{l=0}^{l=x_{ij}} BC(x_{ij}, l)\theta_{ij}^l \phi_{ij}^{x_{ij}-l} p^l (1-p)^{x_{ij}-l}, \tag{7.4}$$

and, when combined with a uniform prior for the parameters, gives an interesting and complex posterior distribution. It can be seen that the thetas and phis can be eliminated by integration using the properties of the beta distribution, leaving

$$g(p/\text{data}) = \prod_{i=0}^{i=1} \prod_{j=0}^{j=1} \sum_{l=0}^{l=x_{ij}} w_{ij} p^l (1-p)^{x_{ij}-l}, \qquad (7.5)$$

as the marginal posterior distribution of p, expressed as the product of four mixtures of beta distributions. Note that the weights are

$$w_{ij} = [\Gamma(x_{ij}+2)/\Gamma(l+1)\Gamma(x_{ij}-l+1)]/x_{ij},$$

thus, each of the four factors of the marginal distribution of p is a mixture of beta distributions. It is difficult to work with this form of the likelihood function, therefore an augmented data form (latent variables) of the likelihood function will be employed.

7.3 Posterior Distribution for Two Binary Tests

Consider an alternative layout for the experiment with the two tests R and T. When $D = 1$, the results of the study are shown in Table 7.3a, and when $D = 0$, the results are shown in Table 7.3b, where the augmented data is represented by y_{ij} and the observations by the corresponding n_{ij}. As above,

$$\theta_{ij} = P[R = i, T = j \mid D = 1], \quad i, j = 0, 1,$$

and

$$\phi_{ij} = P[R = i, T = j \mid D = 0].$$

Thus, the likelihood function is

$$L(\theta, \phi/\text{data}) \propto p^{y..}(1-p)^{n..-y..} \prod_{i=0}^{i=1} \prod_{j=0}^{j=1} \theta_{ij}^{y_{ij}} \prod_{i=0}^{i=1} \prod_{j=0}^{j=1} \phi_{ij}^{n_{ij}-y_{ij}}, \qquad (7.6)$$

TABLE 7.3a: Augmented data for reference R and test T when $D = 1$.

	Test T	
Reference test	$T = 1$	$T = 0$
$R = 1$	y_{11}	y_{10}
$R = 0$	y_{10}	y_{00}
Total		

TABLE 7.3b: Augmented data for R and T when $D = 0$.

Reference test	Test T	
	$T = 1$	$T = 0$
$R = 1$	$n_{11} - y_{11}$	$n_{10} - y_{10}$
$R = 0$	$n_{01} - y_{10}$	$n_{00} - y_{00}$
Total		

and assuming a uniform prior, the posterior distribution of the parameters p, θ_{ij}, and ϕ_{ij} can be determined in terms of all the conditional distributions as follows.

7.4 Posterior Distribution without Conditional Independence

The conditional distribution of θ_{ij} $(i, j = 0, 1)$, given the other parameters (including the augmented data y_{ij}) is Dirichlet with parameter vector:

$$(y_{11} + 1, y_{10} + 1, y_{01} + 1, y_{00} + 1). \tag{7.7}$$

The conditional distribution of ϕ_{ij} $(i, j = 0, 1)$, given the other parameters (including the augmented data y_{ij}) is Dirichlet with parameter vector:

$$(n_{11} - y_{11} + 1, n_{10} - y_{10} + 1, n_{01} - y_{01} + 1, n_{00} - y_{00} + 1). \tag{7.8}$$

The conditional distribution of p given the other parameters is beta with parameters:

$$\text{alpha}p = y_{..} + 1$$

and

$$\text{beta}p = n_{..} - y_{..} + 1. \tag{7.9}$$

The conditional distribution of y_{ij} given the other parameters is binomial with hyperparameters $\theta_{ij} p / [\theta_{ij} p + \phi_{ij}(1 - p)]$ for the probability parameter and n_{ij} for the second parameter, where

$$i, j = 0, 1. \tag{7.10}$$

Note that the assumption of conditional independence does not hold and the sensitivity of T is

$$P[T = 1 \mid D = 1],$$

or

$$seT = \theta_{11} + \theta_{01}, \tag{7.11}$$

while that for R is

$$seR = \theta_{11} + \theta_{10}. \tag{7.12}$$

With regard to the specificity, that of T is

$$spT = \phi_{10} + \phi_{00}, \tag{7.13}$$

and for R is

$$spR = \phi_{01} + \phi_{00}. \tag{7.14}$$

7.5 Posterior Distribution Assuming Conditional Independence

If one assumes conditional independence, the likelihood function is expressed directly in terms of sensitivity and specificity as

$$L(p, s_1, s_2, c_1, c_2) \propto s_1^{y_{11}+y_{01}} (1-s_1)^{y_{10}+y_{00}} s_2^{y_{11}+y_{10}} (1-s_2)^{y_{01}+y_{00}},$$
$$c_1^{n_{10}+n_{00}-y_{10}-y_{00}} (1-c_1)^{n_{11}+n_{01}-y_{11}-y_{01}},$$
$$c_2^{n_{01}+n_{00}-y_{01}-y_{00}} (1-c_2)^{n_{11}+n_{10}-y_{11}-y_{10}},$$
$$p^{y_{..}} (1-p)^{n_{..}-y_{..}}, \tag{7.15}$$

where $y_{..}$ is the sum of y_{ij} and $n_{..}$ is the sum of the four cell frequencies. The notation has been changed to denote s_1 and c_1 as the sensitivity and specificity of T, respectively, while s_2 and c_2 denote the corresponding quantities for reference R.

For computational purposes and assuming a uniform prior, it is obvious from the above likelihood function (Equation 7.15) that the conditional distribution of the unknown parameters is as follows.

The marginal distribution of p is beta with parameters:

$$ap = y_{..} + 1$$

and

$$bp = n_{..} - y_{..} + 1.. \tag{7.16}$$

The conditional distribution of s_1, given the other parameters, is beta with parameters $as1$ and $bs1$, where

$$as1 = y_{11} + y_{01} + 1$$

and

$$bs1 = y_{10} + y_{00} + 1. \tag{7.17}$$

The conditional distribution of s_2, given the other parameters, is beta with hyperparameters:

$$as2 = y_{11} + y_{10} + 1$$

and

$$bs2 = y_{01} + y_{00} + 1. \tag{7.18}$$

The conditional distribution of c_1 is beta with hyperparameters:

$$ac1 = n_{10} + n_{00} - y_{10} - y_{00} + 1$$

and

$$bc1 = n_{11} + n_{01} - y_{11} - y_{01} + 1. \tag{7.19}$$

The conditional distribution of c_2 is beta with parameters:

$$ac2 = n_{01} + n_{00} - y_{01} - y_{00} + 1$$

and

$$bc2 = n_{11} + n_{10} - y_{11} - y_{10} + 1. \tag{7.20}$$

In addition, the posterior distribution of the latent variables is as follows.

The conditional distribution of y_{11} given the other variables is binomial with parameters:

$$m11 = ps_1s_2/[ps_1s_2 + (1-p)(1-c_1)(1-c_2)] \quad \text{(the probability parameter)}$$

and

$$q11 = n_{11}. \tag{7.21}$$

The conditional distribution of y_{10}, given the other parameters, is binomial with parameters:

$$m10 = p(1-s_1)s_2/[p(1-s_1)s_2 + (1-p)c_1(1-c_2)]$$

and

$$q10 = n_{10}. \tag{7.22}$$

The conditional distribution of y_{01}, given the other parameters, is binomial with hyperparameters:

$$m01 = ps_1(1-s_2)/[ps_1(1-s_2) + (1-p)(1-c_1)c_2]$$

and

$$q01 = n_{01}. \tag{7.23}$$

Lastly, the conditional distribution of y_{00}, given the other parameters, is binomial with hyperparameters:

$$m00 = p(1 - s_1)(1 - s_2)/[p(1 - s_1)(1 - s_2) + (1 - p)c_1 c_2]$$

and

$$q00 = n_{00}. \tag{7.24}$$

It is important to know that the above posterior distributions for the accuracy of two binary tests assume a uniform prior for p, s_1, s_2, c_1, and c_2 and the assumption of conditional independence between R and T! In Section 7.6, conjugate-type prior distributions will be used for these quantities, namely, informative beta distributions for the disease prevalence and accuracy parameters.

7.6   Example of Accuracy for Diagnosing a Bacterial Infection

This section examines three alternative Bayesian analyses for the study of a bacterial infection. The three analyses are: (a) assuming a uniform prior and conditional independence between an imperfect reference test R and a new test T, (b) assuming an informative prior and conditional independence, and (c) assuming a uniform prior, but not assuming conditional independence.

Consider the diagnosis of a bacterial infection with *Strongyloides* in 162 Cambodian refugees in Canada. They entered Canada from July 1982 to February 1983 and were tested with a stool examination, which serves as the "new" test T, and a serologic reference test R; the results as reported by Zhou, McClish, and Obuchowski [2: 366] are given in Table 7.4.

A number of people, including Joseph, Gyorkos, and Coupal [3] and Dendukuri and Joseph [4], have analyzed this information, but earlier studies involving the microbiology of the infective agent *Strongyloides* were done by Genta [8] and Bailey [9], both of whom studied the ELISA to detect the pathogen. The observed sensitivity and specificity of the stool examination relative to the serology examination are $38/125 = 0.304$ and $35/37 = 0.945$, respectively. The main focus is to correct the actual sensitivity and specificity of the stool examination via the methodology derived in the previous section. In order to estimate the sensitivity and specificity of the two tests, two Bayesian analyses will be provided: (a) assuming conditional independence between the stool and serology examinations, and (b) not assuming conditional independence.

TABLE 7.4: Results of a stool examination T and a serologic reference examination R.

	Stool examination T		
Serology test R	$T = 1$	$T = 0$	Total
$R = 1$	38	87	125
$R = 0$	2	35	37
Total	40	122	162

Source: From Joseph, L., Gyorkos, T.W., and Coupal, L. *American Journal of Epidemiology*, 1995, by permission of Oxford University Press.

First, assume conditional independence between T and R, then the posterior distribution of the relevant parameters is given by the conditional distributions of each parameter given the others, and these are identified in Equations 7.14 through 7.24. The computations are executed with the following code.

BUGS CODE 7.1

```
model;
{
# conditional distribution of p
p~dbeta(alp,bep)
alp<- y11+y10+y01+y00+ap
bep<- sn -ap+bp
sn<- n11+n10+n01+n00
# conditional distribution of s1
s1~dbeta(als1,bes1)
als1<-y11+y01+as1
bes1<-y10+y00+bs1
#conditional distribution of s2
s2~dbeta(als2,bes2)
als2<-y11+y10+as2
bes2<-y01+y00+bs2
# conditional distribution of c1
c1~dbeta(alc1,bec1)
alc1<-n10+n00-y10-y00+ac1
bec1<-n11+n01-y11-y01+bc1
# conditional distribution of c2
c2~dbeta(alc2,bec2)
alc2<- n01+n00-y01-y00+ac2
bec2<- n11+n10-y11-y10+bc2
# conditional distribution of y11
y11~dbin( py11,n11)
py11<-p*s1*s2/(p*s1*s2+(1-c1)*(1-c2)*(1-p))
```

```
# conditional distribution of y10
y10~dbin(py10,n10)
py10<-p*(1-s1)*s2/(p*(1-s1)*s2+c1*(1-c2)*(1-p))
# conditional distribution of y01
y01~dbin(py01,n01)
py01<-p*s1*(1-s2)/(p*s1*(1-s2)+c2*(1-c1)*(1-p))
# conditional distribution of y00
y00~dbin(py00,n00)
py00<-p*(1-s1)*(1-s2)/(p*(1-s1)*(1-s2)+c1*c2*(1-p))
}
# hyperparameters for uniform prior Joseph
list(ap=1,bp=1,as1=1,bs1=1,as2=1,bs2=1,
ac1=1,bc1=1,ac2=1,bc2=1, n11=38,n10=87,n01=2,n00=35)
# hyperparameters for informative prior joseph
list(ap=80,bp=80,as1=4.44,bs1=13.31,as2=21.96,
bs2=5.49,ac1=71.25,bc1=3.75,ac2=4.1,bc2=1.76, n11=38,n10=87,n01=2,n00=35)
# initial values
list(p=.5,c1=.95,c2=.9,s1=.9,s2=.9,y11=1,y10=1,
y01=1,y00=00)
```

This code is very similar to those appearing in Equations 7.14 through 7.24 and are self-evident, but are also described by the comment statements labeled by #. There are three list statements: (a) the hyperparameters for a uniform prior, (b) the hyperparameters for an informative prior, and (c) a list of starting values for the algorithm. The first analysis will be using a uniform prior for all parameters, thus, the first list statement of BUGS CODE 7.1 is the data. For a uniform prior, the posterior analysis is shown in Table 7.5.

The analysis is performed with 125,000 observations generated from the posterior distribution of the parameters, with a burn in of 5,000 and a refresh of 100. As seen from Table 7.5, the standard deviation for the two sensitivities is almost as large as the mean, indicating uncertainty for these measures of accuracy, and the Markov Chain Monte Carlo (MCMC) errors are relatively large for all parameters. Also of interest is that the prevalence is estimated at of 0.4986, whereas prior information stated in earlier studies believed the prior mean was 0.5, however, much uncertainty was expressed about that

TABLE 7.5:   Posterior analysis of the stool (T) and serology (R) examinations—uniform prior.

Parameter	Mean	sd	Error	2 1/2	Median	97 1/2
p	0.4986	0.2011	0.0053	0.1604	0.4991	0.8352
c_1	0.6977	0.2586	0.0109	0.0922	0.7046	0.9942
c_2	0.2504	0.2552	0.0109	0.0042	0.1237	0.8862
s_1	0.2517	0.2503	0.0107	0.0046	0.1308	0.8826
s_2	0.7014	0.2623	0.0109	0.0889	0.7192	0.9945

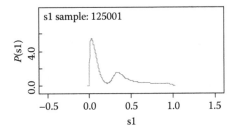

FIGURE 7.1: Posterior density of the sensitivity of the stool examination with a uniform prior.

value. A plot of the posterior density of s_1 (the stool test) also reveals much uncertainty (see Figure 7.1).

A plot of the last 200 observations generated from the posterior density of s_1 also shows the variation in values for this parameter (see Figure 7.2). On the other hand, the analyses of Zhou, McClish, and Obuchowski [2: 367] utilized informative prior information about the parameters (Table 7.6).

The prior information was elicited from a panel of experts and the ranges of the parameter values were converted to hyperparameters of the corresponding beta prior distribution, that is, a beta prior was used for each parameter with the above values for the parameters of that variable. Note the uncertainty for p, expressed as a range $(0,1)$ and a uniform prior for the prevalence. For example, the prior mean for c_1 the specificity of the stool examination is 0.95, while that for the sensitivity s_2 of the serology test is believed to be 0.80, compared to a prior mean of 0.74 for the sensitivity of the stool examination. Of course, the accuracy of serology is supposed to be better than that compared to the stool examination, and this is reflected in the prior values of Table 7.6.

A Bayesian analysis is performed utilizing the prior information in Table 7.6. Again, 125,000 observations are generated from the joint posterior distribution, with a burn in of 5,000 and a refresh of 100 (see Table 7.7).

Comparing Tables 7.5 and 7.7 reveals less uncertainty in the estimates (posterior means) using informative beta priors for the accuracy parameters,

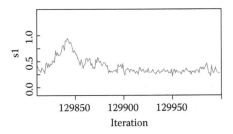

FIGURE 7.2: The "trace" of the last 200 observations generated from the posterior distribution of s_1.

TABLE 7.6:   Prior information about stool and serology tests.

Parameter	Range	Alpha	Beta
p	0–100	1	1
c_1	90–100	71.25	3.75
c_2	35–100	4.1	1.76
s_1	5–45	4.44	13.31
s_2	65–95	21.96	5.49

and the MCMC errors are much smaller when the informative prior is used. For the accuracy parameters (sensitivity and specificity), the posterior standard deviations are less across the board. This is also seen when comparing Figures 7.1 and 7.3, where the latter depicts the posterior density of the stool examination using informative prior information shown in Table 7.7.

The above two analyses show the importance of prior information when estimating the accuracy parameters of a medical test. In the former situation, with little prior information when a uniform prior is used, there was much uncertainty in the posterior distributions, compared to the situation when there is more informative information, such as that expressed in Table 7.6. With "more" prior information, the standard deviations of the posterior distributions are smaller compared to their counterparts when the prior information is vague. More structure is imposed with the informative prior information, resulting in more stability in estimating accuracy. The situation is familiar to the non-Bayesian approach using MLE, where "more" parameters than observations cause instability in the likelihood function, which causes problems with the estimation algorithm. Constraints on the parameters are usually imposed to facilitate convergence of the maximum likelihood procedure, and both Pepe [1] and Zhou, McClish, and Obuchowski [2] discuss the problem in some detail. Of course, for the Bayesian, inducing stability with informative prior information must be done with caution, because one needs to be sure that the informative prior information is indeed reliable!

The *Strongyloides* study will be analyzed in another way, where the CIA between R (serology) and T (the stool examination) does not hold. Recall that Equations 7.7 through 7.14 state the various conditional posterior

TABLE 7.7:   Posterior analysis of the stool (T) and serology (R) examinations—informative prior.

Parameter	Mean	sd	Error	2 1/2	Median	97 1/2
p	0.7618	0.1007	0.0014	0.5236	0.7755	0.9286
c_1	0.957	0.0214	<0.0001	0.9065	0.9603	0.9885
c_2	0.6901	0.1605	0.0020	0.3727	0.7006	0.9558
s_1	0.3093	0.0518	<0.0001	0.2224	0.3043	0.4269
s_2	0.8831	0.0423	<0.0001	0.7892	0.8874	0.9535

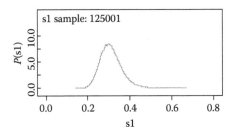

FIGURE 7.3: Posterior density of the stool examination with informative prior.

distributions of all the parameters, assuming a uniform prior distribution, and the following code.

BUGS CODE 7.2

```
# accuracy parameters for two binary tests R and T
# the sensitivity and specificity of R and T
# T is the new and R is the reference
model;
{
p~dbeta(y..,m)
# p and d are the disease rate.
d<- p
m<-n-y..+1
y..<-y11+y10+y01+y00+ 1
y11~dbin(a11,b11)
a11<- theta11*d/(theta11*d+ph11*(1-d))
y10~dbin(a10,b10)
a10<-theta10*d/(theta10*d | ph10*(1 d))
y01~dbin(a01,b01)
a01<- theta01*d/(theta01*d+ph01*(1-d))
y00~dbin(a00,b00)
a00<-theta00*d/(theta00*d+ph00*(1-d))
# Dirichlet distribution for the thetas
g11~dgamma(r11,2)
g10~dgamma(r10,2)
g01~dgamma(r01,2)
g00~dgamma(r00,2)
r11<-y11+1
r10<-y10+1
r01<-y01+1
r00<-y00+1
sg<-g11+g10+g01+g00
```

```
theta11<-g11/sg
theta10<-g10/sg
theta01<-g01/sg
theta00<-g00/sg
# Dirichlet distribution for the phs
h11~dgamma(s11,2)
h10~dgamma(s10,2)
h01~dgamma(s01,2)
h00~dgamma(s00,2)
s11<-z11+1
s10<-z10+1
s01<-z01+1
s00<-z00+1
sh<-h11+h10+h01+h00
ph11<-h11/sh
ph10<-h10/sh
ph01<-h01/sh
ph00<-h00/sh
z11<-b11-y11
z10<-b10-y10
z01<-b01 -y01
z00<-b00-y00
# tests for conditional independence of the thetas
theta11c<- (theta11+theta10)*(theta11+theta01)-theta11
theta10c<- (theta11+theta10)*(theta10+theta00)-theta10
theta01c<- (theta11+theta01)*(theta01+theta00)-theta01
theta00c<- (theta10+theta00)*(theta01+theta00)-theta00
# tests for conditional independence for phs
ph11c<- (ph11+ph10)*(ph11+ph01)-ph11
ph10c<- (ph11+ph10)*(ph10+ph00)-ph10
ph01c<- (ph11+ph01)*(ph01+ph00)-ph01
ph00c<- (ph10+ph00)*(ph01+ph00)-ph00
#sensitivity and specificity
# sensitivity and specificity of the stool exam
s1<- (theta11+ theta01)
c1<- ph10+ph00
# sensitivity and specificity of the serology exam
s2<-theta11+theta10
c2<- ph01+ph00
}
#  Strongyloides example taken from Joseph
list(n=163, b11=38,b10=87,b01=2,b00=35)
# starting values
list(g11=1,g10=1,g01=1, g00=1, p=.5, h11=1,h10=1,h01=1,h00=1, y11=1,
    y10=1,y01=1, y00=1))
```

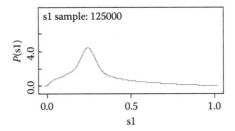

FIGURE 7.4: Posterior density of s_1 with uniform prior and without CIA.

An analysis without assuming conditional independence between stool and serology is performed with a uniform prior and appears in Table 7.7. The results should be compared with Table 7.5, which reports the posterior analysis with a uniform prior when the CIA between R and T is imposed. It is difficult to make a general overall comparison, but it appears that the results of the two tables are somewhat similar, suggesting that the CIA is true. The posterior density of s_1 for the two scenarios is depicted in Figures 7.1 and 7.4, and the posterior distributions of s_1 are reported in Tables 7.5 and 7.7 and reveal the similarity between the two.

The posterior analysis shown in Table 7.8 used 130,000 observations generated for the simulation, with a burn in of 5,000 and a refresh of 100. Compare the results below with the corresponding results of Table 7.5.

Since the two analyses are somewhat similar, one might believe that the CIA is valid, but this can be tested by referring to BUGS CODE 7.2, which provides the relevant posterior distributions. Recall that the CIA is defined by

$$P[R = i, T = j \mid D = k] = P[R = i \mid D = k]P[T = j \mid D = k],$$

for $i, j, k = 0, 1$, and eight conditions need to be checked. Refer to the relevant code of BUGS CODE 7.2, and Table 7.9, which describes the various posterior distributions for testing conditional independence.

The first row is the posterior distribution of the difference:

$$P[R = 0, T = 0 \mid D = 0] - P[R = 0 \mid D = 0]P[T = 0 \mid D = 0],$$

TABLE 7.8: Posterior analysis of *Strongyloides*—without CIA uniform prior.

Parameter	Mean	sd	Error	2 1/2	Median	97 1/2
p	0.4736	0.3271	0.0128	0.0113	0.4412	0.987
c_1	0.7036	0.1731	0.0038	0.2358	0.7428	0.9526
c_2	0.2906	0.176	0.0042	0.0481	0.2442	0.7671
s_1	0.3183	0.1847	0.0030	0.0537	0.2702	0.7913
s_2	0.7006	0.1852	0.0040	0.2183	0.7513	0.9541

TABLE 7.9: Posterior distributions for conditional independence.

Parameter	Mean	sd	Error	2 1/2	Median	97 1/2
ph00	0.2425	0.1593	0.0027	0.0167	0.2138	0.6711
ph01	0.0549	0.092	0.0020	0.0010	0.0225	0.3450
ph10	0.4453	0.2113	0.0058	0.0267	0.5052	0.7978
ph11	0.2572	0.1599	0.0025	0.0188	0.2333	0.6772
theta00	?					
theta01	?					
theta10	?					
theta11	?					

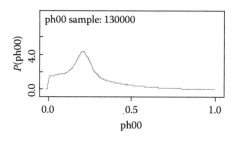

FIGURE 7.5: Posterior density of the conditional independence of ph00.

and if the 95% credible interval contains zero, the implication is that the CIA is valid, at least for this condition. Of course, there are seven other conditions that need to be checked, and the last four of Table 7.9 are left as an exercise. For the first row, the 95% credible interval for ph00 is (0.0167,0.6711) implying that for this restriction, the CIA is not justified, which is also demonstrated with a plot of the corresponding posterior density of Figure 7.5.

Note that the MCMC errors are relatively large, thus a larger simulation size should be adopted. This is left as an exercise.

7.7 Accuracies of Two Binary Tests for Several Populations with Conditional Independence

The previous analysis of two binary tests is extended to several populations, and a good example of this is presented by Zhou, McClish, and Obuchowski [2: 371], with two tests that diagnose tuberculosis at two different sites. The Zhou approach is MLE and is based on the clinical study of Hui and Walter [5]. The two diagnostic tests are the Tine test (T) and the Mantour test (R), while the two populations of patients are quite different in the prevalence of the disease, as the first is a southern school district, while the second is a tuberculosis sanatorium in Missouri. Of interest for this type

TABLE 7.10a: Mantour and Tine tests for tuberculosis at site 1—a southern school district.

Mantour test R	Tine test T		
	$T = 1$	$T = 0$	
$R = 1$	14	4	19
$R = 0$	9	528	537
Total	23	532	556

Source: From Zhou, X.H., McClish, D.K., and Obuchowski, N.A. *Statistical Methods for Diagnostic Medicine.* 2002; P. 371, Table 11.8. Copyright Wiley-VCH Verlag GmbH & Co. KGaA. Reproduced with permission.

TABLE 7.10b: Mantour and Tine tests for tuberculosis at site 2—a tuberculosis sanatorium.

Mantour test R	Tine test T		
	$T = 1$	$T = 0$	
$R = 1$	887	31	918
$R = 0$	37	367	404
Total	924	398	1322

Source: From Zhou, X.H., McClish, D.K., and Obuchowski, N.A. *Statistical Methods for Diagnostic Medicine.* 2002; P. 371, Table 11.8. Copyright Wiley-VCH Verlag GmbH & Co. KGaA. Reproduced with permission.

of study is a comparison of the accuracies of both tests between the two populations. Also for this problem, the prevalence rates of tuberculosis are known with some confidence and such information can be used to specify an informative prior distribution. Since the two locations are quite different and one would expect the clinicians of the two sites to not be the same people, the results from the two sites are statistically independent.

Results from the two sites are presented in Tables 7.10a and b. Note the observed sensitivity for T, relative to R, at site 1 is $14/19 = 0.736$, while the specificity is $528/537 = 0.983$, also the sensitivity of T, relative to R, at site 2 is $887/918 = 0.966$, with a specificity of $367/404 = 0.908$. What will be the change in the accuracies of the two tests at the two sites when corrected by the Bayesian methodology presented earlier? Some of the questions pertinent to the Bayesian approach are: What prior information should be employed and should conditional independence between R and T be imposed? Reasonable guesses about the prior information for the prevalence rates of tuberculosis are possible. For the first site, a low rate of, say, 4%, while 95% for the second site will be specified. One would expect the sensitivity of T for the first site to be low because the prevalence is low compared to the corresponding sensitivity of T for the second site, where the prevalence rate is extremely high. On the other hand, for the specificity of T, one would expect it to be higher for the first site compared to the second, because the prevalence rate is lower for the first

TABLE 7.11a: Accuracies of the Tine (test 1) and Mantour (test 2) tuberculosis tests for site 1—a southern school district.

Parameter	Mean	sd	Error	2 1/2	Median	97 1/2
p	0.0304	0.0049	<0.00001	0.0216	0.0302	0.0409
c_1	0.9862	0.0066	<0.00001	0.972	0.9866	0.9979
c_2	0.9934	0.0043	<0.00001	0.9833	0.9948	0.9997
s_1	0.8595	0.1016	<0.0001	0.6227	0.8775	0.9945
s_2	0.8149	0.1269	<0.0001	0.536	0.8339	0.9918

TABLE 7.11b: Accuracies of the Tine (test 1) and Mantour (test 2) tuberculosis tests for site 2—a tuberculosis sanatorium.

Parameter	Mean	sd	Error	2 1/2	Median	97 1/2
p	0.9896	0.0028	<0.0001	0.9832	0.9899	0.9944
c_1	0.0392	0.0397	<0.0001	0.00099	0.0270	0.1471
c_2	0.0367	0.0028	<0.0001	0.00091	0.0255	0.1365
s_1	0.6946	0.0128	<0.0001	0.6693	0.6947	0.7194
s_2	0.69	0.0128	<0.0001	0.6646	0.6902	0.715

compared to the second! This is borne out, to some extent, by inspection of the observed accuracies for T (relative to R) from Table 7.10.

A Bayesian approach to estimating test accuracies will be done with three scenarios, where the first is assuming an informative prior for the disease rate, namely, a beta(30,970) distribution and a uniform prior for the accuracies; the results shown in Table 7.11a are for site 1.

These are interesting results because the reported sensitivities and specificities are quite close to the observed accuracies relative to the reference, that is to say, the Tine test has a sensitivity and specificity of 0.777 and 0.98, respectively, relative to the Mantour test, while the Mantour test, has a sensitivity and specificity of 0.60 and 0.99, relative to the Tine test. Thus, the observed accuracies appear to be less than those estimated by the Bayesian analysis. Recall from Pepe [1: 195,196], that if the two tests are conditionally independent, the observed accuracies (sensitivities and specificities) are less than the accuracies relative to the gold standard. Note, a uniform prior is used for all parameters, except the disease rate, thus the estimated accuracies are based on the data only, and if the estimated accuracies are indeed accurate, the implication is that the two tests are conditionally independent, but this will be tested later in this section.

As for the second site, if a uniform prior is used for all parameters, the Bayesian analysis gives the results shown in Table 7.11b. These are interesting results, recall the prior distribution of the disease rate is given as beta(990,10) and the specificities are quite small, as they should be, because the disease is seen almost all the time. I am surprised that the sensitivities are no larger, both of which are about 70%. Note that the observed accuracies for the second site are 0.969 and 0.908 for the sensitivity and specificity of the Tine test relative to the Mantour test, while the sensitivity and specificity of the Mantour test relative to the Tine test are 0.95 and 0.922, respectively. I would expect the observed accuracies to be less than those estimated above by Bayesian techniques. To compute the observed accuracies, see Tables 7.10a and b.

These results should be compared to the MLEs of Zhou, McClish, and Obuchowski [2: 371]. They assume that the sensitivities and specificities of the Tine and Mantour tests are the same in the two populations, but assume that the prevalence rates are not the same for the two sites, and compute the sensitivity and specificity of the Tine test as 0.9841 and 0.9688, respectively, with associated standard errors 0.01279 and 0.00623. Of course, this is quite different for the present Bayesian analysis. They imposed a restriction (equality) on the sensitivities and specificities of the two tests because of the number of parameters relative to the number of cell frequencies.

On the other hand, the Bayesian approach did not impose those types of restrictions. It is important to remember that the accuracies of the two tests will surely depend on the prevalence rates of tuberculosis, which differ dramatically between the students of a school district and the patients at a tuberculosis sanatorium. Thus, the two populations were considered to be independent and a uniform prior was assumed for the sensitivities and specificities of the two populations. The following code generated the output for the tuberculosis example at the two sites. The first list statement is for site 1 with a uniform prior, while the second list statement provides the input for the second site, again with a uniform prior distribution.

BUGS CODE 7.3

```
model;
{
# conditional distribution of p
# p0 is an informative prior for p
# note that the distribution for p0 has be deactivated by the comment sign #!
# p0~dbeta(ap0,bp0)
# the following are hyperparameters for the disease rate of the second site
ap0<- 990
bp0<-10
# p1 depends on the data only
p1~dbeta(alp,bep)
# p can be expressed as a mixture of p0 and p1
P<-.90*p0+.10*p1
alp<- y11+y10+y01+y00+ap
bep<- sn -alp+bp
sn<- n11+n10+n01+n00+1
# conditional distribution of s1
s1~dbeta(als1,bes1)
als1<-y11+y01+as1
bes1<-y10+y00+bs1
```

```
#conditional distribution of s2
s2~dbeta(als2,bes2)
als2<-y11+y10+as2
bes2<-y01+y00+bs2
# conditional distribution of c1
c1~dbeta(alc1,bec1)
alc1<-n10+n00-y10-y00+ac1
bec1<-n11+n01-y11-y01+bc1
# conditional distribution of c2
c2~dbeta(alc2,bec2)
alc2<- n01+n00-y01-y00+ac2
bec2<- n11+n10-y11-y10+bc2
# conditional distribution of y11
y11~dbin( py11,n11)
py11<-p*s1*s2/(p*s1*s2+(1-c1)*(1-c2)*(1-p))
# conditional distribution of y10
y10~dbin(py10,n10)
py10<-p*(1-s1)*s2/(p*(1-s1)*s2+c1*(1-c2)*(1-p))
# conditional distribution of y01
y01~dbin(py01,n01)
py01<-p*s1*(1-s2)/(p*s1*(1-s2)+c2*(1-c1)*(1-p))
# conditional distribution of y00
y00~dbin(py00,n00)
py00<-p*(1-s1)*(1-s2)/(p*(1-s1)*(1-s2)+c1*c2*(1-p))
}
# hyperparameters for Tuberculosis example
# Page 371 of Zhou site 1
# Test 1 is Tine Test
# Test 2 is Mantour test
# uniform prior for accuracy parameters
list(ap= 1, bp=1, as1=1,bs1=1,as2=1,bs2=1,
ac1=1,bc1=1,ac2=1,bc2=1, n11=14,n10=4,n01=9,n00=528)
# hyper parameters for Tuberculosis example
# Page 371 of Zhou site 2
# Test 1 is Tine Test
# Test 2 is Mantour test
# uniform prior for accuracy parameters
list(ap=1,bp=1,as1=1,bs1=1,as2=1,bs2=1,
ac1=1,bc1=1,ac2=1,bc2=1, n11=887,n10=31,n01=37,n00=367)
# initial values
list(c1=.5,c2=.5,s1=.5,s2=.5,y11=1,y10=1,
y01=1,y00=1)
```

In the above examples for Tables 7.10a and b, the distribution of p is expressed as a mixture. The posterior distribution of p may be determined by the prior distribution $p0$ for p and the posterior distribution for p, labeled $p1$.

The distribution of p is specified as a mixture between $p0$ and $p1$, and the hyperparameters for $p0$ are $ap0$ and $bp0$, which may be specified in the first two list statements of BUGS CODE 7.3. The first list statement gives the data for site 1, the second list statement gives the data for site 2, while the third list statement provides the initial values for the MCMC computations.

7.8   Accuracies of Two Binary Tests without Conditional Independence: Two Populations

Tables 7.10a and b present the results for two diagnostic tests for tuberculosis at two sites, and the previous section presented the Bayesian analysis based on the CIA, assuming a uniform prior for all parameters, except disease prevalence. For now, conditional independence will not be imposed. How will the posterior analysis differ between the two scenarios?

The first analysis is for site 1, the population of a southern school district, where the prevalence is assumed to be 4%. A uniform prior is assumed for the sensitivities and specificities of the Tine and Mantour tests and the analysis is based on the study results (shown in Table 7.10). The computations are executed with 130,000 observations generated from the posterior distribution, with a burn in of 5,000 and a refresh of 100. BUGS CODE 7.2 is slightly revised as follows: the first few statements give the posterior distribution of d, the disease rate as a mixture of the posterior distribution of p, the disease rate with a uniform prior, and the prior distribution of the disease rate expressed as beta distribution.

BUGS CODE 7.4

```
p~dbeta(y..,m)
p0~dbeta(ap0,bp0)
# d is a mixture of the uniform prior and an
# informative prior p0
# d is the posterior distribution of disease rate
d<-.10*p+.90*p0
m<-n-y..+1
y..<-y11+y10+y01+y00+ 1
```

The first list statement, among those below gives the data for the first site:

```
# Tine test & Mantour test site 1
list(n=556,ap0=17,bp0=538, b11=14,b10=4,b01=9,b00=528)
# Tine test & Mantour test site 2
list( n=1323,ap0=1296,bp0=26, b11=887,b10=31,b01=37,b00=367)
# starting values
list(g11=1,g10=1,g01=1, g00=1, p=.5, h11=1,h10=1,h01=1,h00=1, y11=1,
    y10=1,y01=1,y00=1))
```

TABLE 7.12a: Posterior analysis of the Tine and Mantour tests without conditional independence—site 1.

Parameter	Mean	sd	Error	2 1/2	Median	97 1/2
d	0.0299	0.0049	<0.0001	0.0189	0.0300	0.0449
c_1	0.9695	0.0106	<0.0001	0.9473	0.97	0.9885
c_2	0.9753	0.0097	<0.0001	0.9554	0.9757	0.9929
s_1	0.5345	0.2131	0.0018	0.1158	0.5458	0.9038
s_2	0.4416	0.2081	0.0017	0.0840	0.4304	0.8559

TABLE 7.12b: Posterior analysis of the Mantour and Tine tests without conditional independence—site 2.

Parameter	Mean	sd	Error	2 1/2	Median	97 1/2
d	0.9802	0.0034	<0.0001	0.9728	0.9805	0.9864
c_1	0.5006	0.2241	0.0025	0.0945	0.5006	0.9069
c_2	0.5057	0.2235	0.0024	0.0982	0.5065	0.9099
s_1	0.7024	0.0137	<0.0001	0.6753	0.7024	0.729
s_2	0.697	0.0135	<0.0001	0.6707	0.6979	0.7243

For site 1, the posterior analysis is given by Table 7.12a. It is interesting to compare Table 7.11a with Table 7.12a, where the former assumes conditional independence and the latter does not. The sensitivities of the two tests are higher under the assumption of conditional independence, and the specificities appear to be the same. When the prevalence rate is low, I would expect higher specificities compared to sensitivities. In summary, comparing the two tables implies that the assumption of conditional independence may not be valid.

The posterior distribution of the two sensitivities, s_1 and s_2, appear to have more uncertainty than the other parameters in Table 7.11a, and this is demonstrated by a plot of the posterior density as shown in Figure 7.6, and a plot of the posterior density of s_2 is similar.

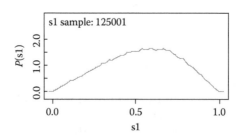

FIGURE 7.6: Posterior density of s_1 with no conditional independence.

With a similar approach, not assuming conditional independence, the posterior analysis of location 2 is displayed by Table 7.12b. Recall that the prevalence for tuberculosis is assumed to be 98%, expressed as a beta(980,20) distribution, a prior.

Note that the two specificities are the same, as are the two sensitivities, and that the two tests have higher sensitivity than specificity, which is to be expected with such a high prevalence rate. Looking at Tables 7.10a and b for the basic information on the two tests at sites 1 and 2, respectively, one sees symmetry (the marginal totals are similar) in the test results at both sites, suggesting that the specificities of the two would be similar, as would the specificities, at both sites.

The tuberculosis example has been analyzed by assuming conditional independence, resulting in Tables 7.11a and b for the Bayesian results. Without assuming conditional independence, the analyses are reported in Tables 7.12a and b. It cannot be over emphasized that an informative prior distribution is placed on the disease rate, with a value of approximately 4% for site 1 (a southern school district) and a rate of 98% for site 2 (a tuberculosis sanatorium). For the former site the prevalence is expressed as the beta(40,960) distribution and as beta(980,20) for the latter site. It should also be noted that if the prior distributions for the disease rate and for the accuracies are changed, the results will be quite different than that reported in Tables 7.11 and 7.12. This sensitivity to the prior will be explored in the exercises at the end of the chapter.

7.9   Multiple Tests in a Single Population

What is to be presented is for three tests, where the generalization to more than three tests is obvious and will be left as an exercise at the end of this chapter. Label the tests T_1, T_2, and T_3, and when $T_i = 1$, the ith test is positive, and when $T_i = 0$, the test is negative, where $i = 1, 2, 3$. Suppose p denotes the prevalence of the disease and that the three tests are scored on the same units.

A Bayesian approach requires a likelihood function, which, if conditional independence is assumed, will have seven parameters, three sensitivities denoted by s_i ($i = 1, 2, 3$), three specificities denoted by c_i, and the disease prevalence p. A typical layout for the study results for three tests is given by Tables 7.13a and b.

There are eight cell frequencies, and the posterior distribution is based on the marginal distribution of p, the conditional distribution of each of the other six parameters, given the other parameters.

As before, let the marginal distribution of p be

$$p \sim \text{beta}(\text{alp}, \text{bep}), \tag{7.25}$$

TABLE 7.13a: Three
binary tests results when $T_1 = 1$.

	T_2	
T_3	$T_2 = 1$	$T_2 = 0$
$T_3 = 1$	n_{111}	n_{101}
$T_3 = 1$	n_{110}	n_{100}

TABLE 7.13b: Three
binary tests results when $T_1 = 0$.

	T_2	
T_3	$T_2 = 1$	$T_2 = 0$
$T_3 = 1$	n_{011}	n_{001}
$T_3 = 1$	n_{010}	n_{000}

where

$$\text{alp} = \sum_{i=0}^{i=1} \sum_{j=0}^{j=1} y_{ij} + 1$$

and

$$\text{bep} = n_{..} - \text{alp},$$

where

$$n_{..} = \sum_{i=0}^{i=1} \sum_{j=1}^{j=1} n_{ij},$$

where y_{ij} are augmented variables, which will be defined in a moment. The following sensitivities will have the following conditional beta posterior distributions.

$$s_1 \sim \text{beta}(\text{als1}, \text{bes1}), \tag{7.26}$$

where

$$\text{als1} = y_{111} + y_{101} + y_{110} + y_{100} + \text{as1}$$

and

$$\text{bes1} = y_{011} + y_{001} + y_{010} + y_{000} + \text{bs1}.$$

$$s_2 \sim \text{beta}(\text{al2}, \text{bes2}), \tag{7.27}$$

where

$$\text{als2} = y_{111} + y_{110} + y_{011} + y_{010} + \text{as2}$$

and

$$bes2 = y_{101} + y_{100} + y_{001} + y_{000} + bs2.$$
$$s_3 \sim beta(al3, bes3), \tag{7.28}$$

where

$$als3 = y_{111} + y_{101} + y_{011} + y_{001} + as3$$

and

$$bes3 = y_{110} + y_{100} + y_{010} + y_{000} + bs3.$$

As for the specificities,

$$c_1 \sim beta(alc1, bec1), \tag{7.29}$$
$$alc1 = n_{010} + n_{001} + n_{010} + n_{000} - y_{010} - y_{001} - y_{010} - y_{000} + ac1,$$
$$bec1 = n_{111} + n_{101} + n_{110} + n_{100} - y_{111} - y_{101} - y_{101} - y_{100} + bc1.$$
$$c_2 \sim beta(alc2, bec2), \tag{7.30}$$
$$alc2 = n_{101} + n_{100} + n_{101} + n_{000} - y_{101} - y_{100} - y_{101} - y_{000} + ac2,$$
$$bec2 = n_{111} + n_{110} + n_{011} + n_{010} - y_{111} - y_{110} - y_{011} - y_{010} + bc2.$$
$$c_3 \sim beta(alc3, bec3), \tag{7.31}$$
$$alc3 = n_{110} + n_{100} + n_{010} + n_{000} - y_{110} - y_{100} - y_{010} - y_{000} + ac3,$$
$$bec3 = n_{111} + n_{101} + n_{011} + n_{001} - y_{111} - y_{101} - y_{011} - y_{001} + bc3.$$

The augmented variables are given binomial distributions as follows.

$$y_{111} \sim binomial(py_{111}, n_{111}), \tag{7.32}$$

where

$$py_{111} = ps_1 s_2 s_3 / [ps_1 s_s s_3 + (1 - p)(1 - c_1)(1 - c_2)(1 - c_3)].$$
$$y_{101} \sim binomial(py_{101}, n_{101}), \tag{7.33}$$

where

$$py_{101} = ps_1(1 - s_2)s_3 / [ps_1(1 - s_2)s_3 + (1 - p)(1 - c_1)c_2(1 - c_3)].$$
$$y_{110} \sim binomial(py_{110}, n_{110}), \tag{7.34}$$

where

$$py_{110} = ps_1 s_2(1 - s_3) / [ps_1 s_2(1 - s_3) + (1 - p)(1 - c_1)(1 - c_2)c_3].$$
$$y_{100} \sim binomial(py_{100}, n_{100}), \tag{7.35}$$

where

$$py_{100} = ps_1(1 - s_2)(1 - s_3) / [ps_1(1 - s_2)(1 - s_3) / [ps_1(1 - s_2)(1 - s_3)$$
$$+ (1 - p)(1 - c_1)c_2 c_3].$$
$$y_{011} \sim binomial(py_{011}, n_{011}), \tag{7.36}$$

where

$$py_{011} = p(1 - s_1)s_2s_3/[p(1 - s_1)s_2s_3 + (1 - p)c_1(1 - c_2)(1 - c_3)].$$
$$y_{001} \sim \text{binomial}(py_{001}, n_{001}), \tag{7.37}$$

where

$$py_{001} = p(1 - s_1)(1 - s_2)s_3/[p(1 - s_1)(1 - s_2)s_3 + (1 - p)c_1c_2(1 - c_3)].$$
$$y_{010} \sim \text{binomial}(py_{010}, n_{010}), \tag{7.38}$$

where

$$py_{010} = p(1 - s_1)s_2(1 - s_3)/[p(1 - s_1)s_2(1 - s_3) + (1 - p)c_1(1 - c_2)c_3].$$
$$y_{000} \sim \text{binomial}(py_{000}, n_{000}), \tag{7.39}$$

where

$$py_{000} = p(1 - s_1)(1 - s_2(1 - s_3)/[p(1 - s_1)(1 - s_2)(1 - s_3) + (1 - p)c_1c_2c_3].$$

Equations 7.25 through 7.39 comprise the posterior distribution of the seven parameters of the problem and will be executed with BUGS CODE 7.5. A good example of multiple tests is the study by Irwig et al. [10], presented as an example of multiple tests and consisting of three radiologists who are conducting a study on the accuracy of x-ray in detecting pleural thickening of asbestos miners in South Africa. The example presented in Tables 7.14a and b is similar in that there are three readers who are studying the enlargement of the prostate gland via MRI images, where a positive reading indicates enlargement and a negative reading implies the reader believes the prostate was not enlarged (Tables 7.14a and b).

TABLE 7.14a: Three radiologists when radiologist 1 scores positive $T_1 = 1$.

Prostate enlargement	Radiologist 2	
Radiologist 3	$T_2 = 1$	$T_2 = 0$
$T_3 = 1$	$n_{111} = 46$	$n_{101} = 27$
$T_3 = 0$	$n_{110} = 20$	$n_{100} = 31$

TABLE 7.14b: Three radiologists when radiologist 1 scores negative $T_1 = 0$.

Prostate enlargement	Radiologist 2	
Radiologist 3	$T_2 = 1$	$T_2 = 0$
$T_3 = 1$	$n_{011} = 31$	$n_{001} = 41$
$T_3 = 1$	$n_{010} = 79$	$n_{000} = 1533$

TABLE 7.15: Posterior analysis of three radiologists for prostate enlargement.

Parameter	Mean	sd	Error	2 1/2	Median	97 1/2
p	0.02674	0.0035	<0.0001	0.0200	0.0266	0.0341
c_1	0.9709	0.0052	<0.0001	0.9598	0.9711	0.9804
c_2	0.9387	0.0067	<0.0001	0.9249	0.9388	0.9513
c_3	0.9267	0.0069	<0.0001	0.9126	0.9269	0.9399
s_1	0.8427	0.0676	<0.0001	0.7062	0.841	0.9702
s_2	0.8024	0.0623	<0.0001	0.6771	0.8035	0.9211
s_3	0.821	0.0541	<0.0001	0.7115	0.8223	0.9235

The posterior analysis assumes conditional independence between the three radiologists and a uniform beta prior for each of the seven parameters, and reveals some interesting results. The disease prevalence is estimated at almost 3%, with high specificity for the three radiologists and good sensitivity ranging from a low of 80% for radiologist 2 to a high of 84% for the first radiologist. These results appear to be reasonable because of the low prevalence of prostate enlargement.

The Bayesian analysis assumes a uniform prior for all parameters, including the disease rate. The readers show good agreement with regard to specificity, but there is less agreement between them with regard to sensitivity, but it is still good, and as it should be, the specificities are greater than the sensitivities, which is to be expected with the disease prevalence so low (Table 7.15).

Stability is evident in the Bayesian analysis with "small" posterior standard deviations and plots of the posterior densities that appear well behaved. The analysis is executed with BUGS CODE 7.5, using 135,000 observations generated from the posterior distribution, with a burn in of 5,000 and a refresh of 100, which gives very small MCMC errors, all of which are <0.0001.

BUGS CODE 7.5

```
model;
{
p1~dbeta(alphap1,betap1)
p0~dbeta(100,900)
# note p is not expressed as a mixture for this example
p<-p1
alphap1<-sy+1
betap1<-sn -alphap1
sn<-n111+n101+n110+n000+n011+n001+n010+n000
sy<-y111+y101+y110+y000+y011+y001+y010+y000
s1~dbeta(alphas1,betas1)
alphas1<-y111+y101+y110+y100+as1
betas1<- y011+y001+y010+y000+ bs1
```

```
s2~dbeta(alphas2,betas2)
alphas2<-y111+y110+y011+y010+as2
betas2<- y101+y100+y001+y000+bs2
s3~dbeta(alphas3,betas3)
alphas3<-y111+y101+y011+y001+as3
betas3<-y110+y100+y010+y000+bs3
c1~dbeta(alphac1,betac1)
alphac1<-n011+n001+n010+n000-y011-y001
-y010-y000+ac1
betac1<-n111+n101+n110+n100-y111-y101
-y110-y100+bc1
c2~dbeta(alphac2,betac2)
alphac2<-n101+n100+n001+n000-y101-y100-
y001-y000 +ac2
betac2<- n111+n110+n011+n010-y111-y110
-y011-y010+bc2
c3~dbeta(alphac2,betac3)
alphac3<-n110+n100+n010+n000-y110+y100+y010+y000+ac3
betac3<-n111+n101+n011+n001-y111+y101+y011+y001+bc3
y111~dbin( py111,n111)
py111<-p*s1*s2*s3/(p*s1*s2*s3+(1-c1)*(1-c2)*(1-c3)*(1-p))
y101~dbin(py101,n101)
py101<-p*s1*(1-s2)*s3/(p*s1*(1-s2)*s3+(1-c1)*c2*(1-c3*(1-p)))
y110~dbin(py110,n110)
py110<-p*s1*s2*(1-s3)/(p*s1*s2*(1-s3)+c3*(1-c2)*(1-c1)*(1-p))
y100~dbin(py100,n100)
py100<-p*s1*(1-s2)*(1-s3)/(p*s1*(1-s2)*(1-s3)+(1-c1)*c2*c3*(1-p))
y011~dbin(py011,n011)
py011<-p*(1-s1)*s2*s3/(p*(1-s1)*s2*s3+(1-p)*c1*(1-c2)*(1-c3))
y001~dbin(py001,n001)
py001<-p*(1-s1)*(1-s2)*s3/(p*(1-s1)*(1-s2)*s3+(1-p)*c1*c2*(1-c3))
y010~dbin(py010,n010)
py010<- p*(1-s1)*s2*(1-s3)/(p*(1-s1)*s2*(1-s3)+(1-p)*c1*(1-c2)*c3)
y000~dbin(py000,n000)
py000<- p*(1-s1)*(1-s2)*(1-s3)/(p*(1-s1)*(1-s2)*(1-s3)+(1-p)*c1*c2*c3)
}
# hyperparameters for Pepe example
# Test1 is culture, 2 is Elisa, 3 is PCR
list(n111=20,n101=4,n110=2,n100=2,
n011=4,n001=3,n010=2,n000=292,
as1=1,bs1=1,as2=1,bs2=1,
ac1=1,bc1=1,ac2=1,bc2=1,
as3=1,bs3=1,ac3=1,bc3=1)
# hyperparameters for prostate enlargement
list(n111=46,n101=27,n110=20,n100=31,n011=31,
```

n001=41,n010=79,n000=1533,
as1=1,bs1=1,as2=1,bs2=1,
ac1=1,bc1=1,ac2=1,bc2=1,
as3=1,bs3=1,ac3=1,bc3=1)
# hyperparameters Pepe table 7.14
# Page 204
list(n111=70,n101=25,n110=5,n100=10,
n011=110,n001=150,n010=100,n000=530,
as1=1,bs1=1,as2=1,bs2=1,
ac1=1,bc1=1,ac2=1,bc2=1,
as3=1,bs3=1,ac3=1,bc3=1)
# initial values
list(c1=.5,c2=.5,s1=.5,s2=.5,s3=.5,c3=.5,
y111=1,y101=1,y110=1,y100=1,y011=1,y001=1,
y010=1,y000=1)

7.10 Multiple Tests without Conditional Independence

Recall that the above analysis for three binary tests depends on conditional independence between them. Will the accuracies change if this assumption is not imposed? Our next challenge is to develop an analysis that not only does not include this assumption, but also provides reasonable estimates of test accuracy. Recall also that the Bayesian analysis without conditional independence was based on the likelihood function (Equation 7.6) and developed in Section 7.6, and that BUGS CODE 7.2 was executed to analyze two diagnostic tests for *Strongyloides*, where the assumption of conditional independence was tested. A similar approach is now taken, where Equations 7.7 through 7.14 are generalized to three binary tests.

For three tests with latent variables y_{ijk}, let

$$\theta_{ijk} = P[T_1 = i, T_2 = j, T_3 = k \mid D = 1], \quad i, j, k = 0, 1,$$

and

$$\phi_{ijk} = P[T_1 = i, T_2 = j, T_3 = k \mid D = 0].$$

Then, the likelihood function is

$$L(\theta, \phi/\text{data}) \propto p^{y\cdots} (1 - p)^{n\cdots - y\cdots} \prod_{i=0}^{i=1} \prod_{j=0}^{j=1} \prod_{k=0}^{k=1} \theta_{ijk}^{y_{ijk}} \prod_{i=0}^{i=1} \prod_{j=0}^{j=1} \prod_{k=0}^{k=1} \phi_{ijk}^{n_{ijk} - y_{ijk}}, \quad (7.40)$$

and assuming a uniform prior, the posterior distribution of the parameters p, θ_{ijk}, and ϕ_{ijk} can be determined in terms of all the conditional distributions as follows.

The conditional distribution of θ_{ijk} $(i, j, k = 0, 1)$, given the other parameters (including the augmented data y_{ijk}), is Dirichlet with parameter vector $(y_{111} + 1, y_{110} + 1, y_{101} + 1, y_{100} + 1, y_{011} + 1, y_{010} + 1, y_{001} + 1, y_{000} + 1)$.

The conditional distribution of ϕ_{ijk} $(i, j = 0, 1)$, given the other parameters (including the augmented data y_{ij}), is Dirichlet with parameter vector $(n_{111} - y_{111} + 1, n_{110} - y_{110} + 1, n_{101} - y_{101} + 1, n_{100} - y_{100} + 1, n_{011} - y_{011}, n_{010} - y_{010}, n_{001} - y_{001}, n_{000} - y_{000})$.

The conditional distribution of p, given the other parameters, is beta with parameters

$$\text{alpha} p = y_{...} + 1$$

and

$$\text{beta} p = n_{...} - y_{...} + 1,$$

where

$$y_{...} = \sum_{i=0}^{i=1} \sum_{j=0}^{j=1} \sum_{k=0}^{k=1} y_{ijk}$$

and

$$n_{...} = \sum_{i=0}^{i=1} \sum_{j=0}^{j=1} \sum_{k=0}^{k=1} n_{ijk}.$$

The conditional distribution of y_{ijk}, given the other parameters, is binomial with hyperparameters $\theta_{ijk} p/[\theta_{ijk} p + \phi_{ijk}(1 - p)]$ for the probability parameter and n_{ijk} for the second parameter, where $i, j, k = 0, 1$.

This is sufficient to determine the joint posterior distribution of all the parameters, assuming a uniform prior for all. WinBUGS uses Gibbs sampling from all the conditional distributions to determine the joint posterior distribution of all the parameters.

The analysis shown in Table 7.16 estimates the disease (prostate enlargement) prevalence at 4.1% using a uniform prior. The disease rate d can be defined as a mixture of two distributions for p and $p0$. In effect, the prior distribution of d is uniform. One may adjust the prior distribution for d by varying the parameters of the beta prior for $p0$ and the mixing weights.

It is seen that the specificities are all high and very similar among the three radiologists, while the sensitivities are quite low, varying from 0.46 for radiologist 1 to a high of 0.49 for radiologist 2. Are these results reasonable?

It is interesting to compare the results of Table 7.15 with Table 7.16, where the former reports the analysis when the conditional independence is imposed,

TABLE 7.16:   Posterior analysis of three binary tests without conditional independence—prostate enlargement with three radiologists.

Parameter	Mean	sd	Error	2 1/2	Median	97 1/2
d	0.0414	0.0285	<0.0001	0.0024	0.0369	0.1102
c_1	0.9447	0.0119	<0.0001	0.9243	0.9435	0.9708
c_2	0.9173	0.0148	<0.0001	0.8939	0.9152	0.9523
c_3	0.9344	0.0134	<0.0001	0.9122	0.9329	0.964
s_1	0.4602	0.1604	0.0023	0.1694	0.4542	0.7826
s_2	0.4924	0.1593	0.0021	0.1901	0.4916	0.8001
s_3	0.487	0.1586	0.0022	0.1877	0.4854	0.7961

while the latter reveals the analysis when the conditional independence is not imposed. The most obvious difference is that the sensitivities of the three readers for the former case are much higher than those for the latter case. It is interesting to observe that the MCMC errors for the sensitivities are larger than that for the other parameters, which is also evident from the plots of the posterior densities of all the posterior distributions. The analysis is executed with BUGS CODE 7.6 using 130,000 observations generated from the joint posterior distribution, with a burn in of 5,000 and a refresh of 100.

BUGS CODE 7.6

```
model;
# three tests wo cia
{
p~dbeta(y...,m)
# p0~dbeta(ap0,bp0)
# d is a mixture of the uniform prior and an
# informative prior p0
# d is the posterior distribution of disease rate
d<- p
m<-n...-y...+1
y...<- y111+y101+y110+y100+y011+y001+y010+y000+1
n...<- n111+n101+n110+n100+n011+n001+n010+n000
y111~dbin(a111,n111)
a111<- theta111*d/(theta111*d+ph111*(1-d))
y110~dbin(a110,n110)
a110<-theta110*d/(theta110*d+ph110*(1-d))
y101~dbin(a101,n101)
a101<- theta101*d/(theta101*d+ph101*(1-d))
y100~dbin(a100,n100)
a100<-theta100*d/(theta100*d+ph100*(1-d))
y011~dbin(a011,n011)
a011<-theta011*d/(theta011*d+ph011*(1-d))
```

```
y001~dbin(a001,n001)
a001<-theta001*d/(theta001*d+ph001*(1-d))
y010~dbin(a010,n010)
a010<-theta010*d/(theta010*d+ph010*(1-d))
y000~dbin(a000,n000)
a000<-theta000*d/(theta000*d+ph000*(1-d))
g111~dgamma(r111,2)
g110~dgamma(r110,2)
g101~dgamma(r101,2)
g100~dgamma(r100,2)
g011~dgamma(r011,2)
g010~dgamma(r010,2)
g001~dgamma(r001,2)
g000~dgamma(r000,2)
r111<-y111+1
r110<-y110+1
r101<-y101+1
r100<-y100+1
r011<-y011+1
r010<-y010+1
r001<-y001+1
r000<-y000+1
sg<-g111+g110+g101+g100+g011+g010+g001+g000
theta111<-g111/sg
theta110<-g110/sg
theta101<-g101/sg
theta100<-g100/sg
theta011<-g011/sg
theta010<-g010/sg
theta001<-g001/sg
theta000<-g000/sg
h111~dgamma(s111,2)
h110~dgamma(s110,2)
h101~dgamma(s101,2)
h100~dgamma(s100,2)
h011~dgamma(s011,2)
h010~dgamma(s010,2)
h001~dgamma(s001,2)
h000~dgamma(s000,2)
s111<-z111+1
s110<-z110+1
s101<-z101+1
s100<-z100+1
s011<-z011+1
s010<-z010+1
```

```
s001<-z001+1
s000<-z000+1
sh<-h111+h110+h101+h100 + h011+h010+h001+h000
ph111<-h111/sh
ph110<-h110/sh
ph101<-h101/sh
ph100<-h100/sh
ph011<-h011/sh
ph010<-h010/sh
ph001<-h001/sh
ph000<-h000/sh
z111<-n111-y111
z110<-n110-y110
z101<-n101 -y101
z100<-n100-y100
z011<-n011-y011
z010<-n010-y010
z001<-n001 -y001
z000<-n000-y000
# test for conditional independence
theta111c<- (theta111+theta110)*(theta111+theta101)-
theta111
theta110c<- (theta111+theta110)*(theta110+theta100)-
theta110
theta101c<- (theta111+theta101)*(theta101+theta100)-
theta101
theta100c<- (theta110+theta100)*(theta101+theta100)-
theta100
theta011c<- (theta011+theta010)*(theta011+theta001)-
theta011
theta010c<- (theta011+theta010)*(theta010+theta000)-
theta010
theta001c<- (theta011+theta001)*(theta001+theta000)-
theta001
theta000c<- (theta010+theta000)*(theta001+theta100)-
theta000
# test for conditional independence for phs
ph111c<- (ph111+ph110)*(ph111+ph101)-ph111
ph110c<- (ph111+ph110)*(ph110+ph100)-ph110
ph101c<- (ph111+ph101)*(ph101+ph100)-ph101
ph100c<- (ph110+ph100)*(ph101+ph100)-ph100
ph011c<- (ph011+ph010)*(ph011+ph001)-ph011
ph010c<- (ph011+ph010)*(ph010+ph000)-ph010
ph001c<- (ph011+ph001)*(ph001+ph000)-ph001
ph000c<- (ph010+ph000)*(ph001+ph000)-ph000
```

```
#sensitivity and specificity
s1<- theta111+ theta101+theta110+theta100
c1<- ph011+ph001+ph010+ph000
s2<-theta111+theta110+theta011+theta010
c2<- ph101+ph100+ph001+ph000
s3<-theta111+theta101+theta011+theta001
c3<-ph110+ph100+ph010+ph000
}
# prostate enlargement with 3 readers
list( n111=46,n110=20,n101=27,n100=31,
n011=31,n010=79,n001=41,n000=1533)
# hyperparameters for pepe example
# Test1 is culture, 2 is Elisa, 3 is PCR
list(ap0=50,bp0=1000,n111=20,n101=4,n110=2,
n100=2,n011=4,n001=3,n010=2,n000=292)
# starting values
list(p=.5, h111=1,h110=1,h101=1,h100=1,
h011=1,h010=1,h001=1,h000=1,
y111=1,y110=1,y101=1,y100=1,
y011=1,y010=1,y001=1,y000=1)
```

It is obvious how to generalize multiple tests to several populations with and without conditional independence and this will not be presented here, but will be considered as an exercise.

7.11   Two Ordinal Tests and the Receiver Operating Characteristic Area

The central theme of this chapter is estimating test accuracy when there is no gold standard, and at best an imperfect reference test is available. Thus far, two ordinal tests have been considered for a variety of scenarios. First, two tests for one population of patients is considered under two conditions: with and without conditional independence between the two, then the situation is generalized to two binary tests and several populations, with and without conditional independence. Lastly, multiple binary tests are considered, with and without conditional independence, and all scenarios were illustrated with a variety of real-life examples of estimating medical test accuracy. Latent variables were used in each case, which allows one to envision what can happen, both when the disease is present and when it is absent.

Now the goal is to estimate the area under the ROC curve for two ordinal tests, when there is no gold standard, but one of the tests serves as an imperfect reference test.

Suppose the two tests, T_1 and T_2, have ordinal scores $1, 2, \ldots, c$ and that

$$\theta_{ij} = P[T_1 = i, T_2 = j \mid D = 1], \tag{7.41}$$

and

$$\phi_{ij} = P[T_1 = i, T_2 = j \mid D = 0], \tag{7.42}$$

where $i, j = 1, 2, \ldots, c$. Then

$$\theta_{i.} = P[T_1 = i \mid D = 1], \tag{7.43}$$

and

$$\theta_{.j} = P[T_2 = j \mid D = 1]. \tag{7.44}$$

Also,

$$\phi_{i.} = P[T_1 = i \mid D = 0], \tag{7.45}$$

and

$$\phi_{.j} = P[T_2 = j \mid D = 0]. \tag{7.46}$$

In addition, suppose that the latent variables are distributed

$$y_{ij} \sim \text{beta}(a_{ij}, n_{ij}), \tag{7.47}$$

where

$$a_{ij} = p\theta_{ij} / [p\theta_{ij} + (1-p)\phi_{ij}], \tag{7.48}$$

and p is the disease rate and the observations are n_{ij} for $i, j = 1, 2, \ldots, c$.
The area under the ROC curve for T_1 is defined as follows:

$$AT_1 = AT_{11} + AT_{12}/2, \tag{7.49}$$

where

$$AT_{11} = \sum_{i=2}^{i=c} \theta_{i.} \sum_{j=1}^{j=i-1} \phi_{j.},$$

and

$$AT_{12} = \sum_{i=1}^{i=c} \sum_{j=1}^{j=c} \theta_{i.} \phi_{i.}.$$

In a similar fashion, the area under the ROC curve for T_2 is

$$AT_2 = AT_{21} + AT_{22}/2, \tag{7.50}$$

where

$$AT_{21} = \sum_{i=2}^{i=c} \theta_{.i} \sum_{j=1}^{j=i-1} \phi_{.j},$$

and

$$AT_{22} = \sum_{i=1}^{i=c} \sum_{j=1}^{j=c} \theta_{.i} \phi_{.i}.$$

As an example of finding the ROC area of two tests, consider the example of staging melanoma with two readers, T_1 and T_2, where the ordinal scores are 1, 2, 3, indicating the stage of the disease, and the results are given by Table 7.17.

On examination of the staging results of Table 7.17, the following observations are made: assuming T_2 is the reference test and that a score of 3 identifies disease (stage 3), the ROC area for T_1 is approximately 0.58; on the other hand, assuming T_1 (reader 1) is the reference with a T_1 score of 3 indicating disease, it can be shown that the ROC area of T_2 (reader 2) is approximately 0.58. Also, if a score of 3 from either test indicates disease, the disease prevalence is approximately 60%; on the other hand, if disease is indicated only if both tests score 3, the disease rate is estimated as only $46/286 = 16\%$. Notice the symmetry in the two tests, the marginal distributions are each u-shaped, implying that the accuracy of both tests will be similar.

Because there is no gold standard, a Bayesian analysis is performed based on the derivations above, consisting of Equations 7.41 through 7.50. First, a prevalence rate of 20% is imposed with uniform priors for the thetas and phis and an analysis is executed with 135,000 observations generated from the posterior distribution of all parameters, with a burn in of 5,000 and a refresh of 100. See BUGS CODE 7.7, where the list statement gives the data for the MRI and CT results of Table 7.18.

It is evident that the ROC area for reader 1 is greater than that for reader 2. The prevalence rate was set at 20%, however, different values of the rate

TABLE 7.17:   Two readers staging melanoma.

		T_1		
T_2	$T_1 = 1$	$T_1 = 2$	$T_1 = 3$	Total
$T_2 = 1$	45	26	28	99
$T_2 = 2$	25	20	20	65
$T_2 = 3$	46	30	46	122
Total	116	76	94	286

TABLE 7.18: ROC areas for T_1 and T_2 for staging melanoma.

Parameter	Mean	sd	Error	2 1/2	Median	97 1/2
d	0.2004	0.0362	<0.0001	0.1344	0.1987	0.2765
T_1	0.5272	0.101	0.0017	0.331	0.5272	0.7218
T_2	0.4721	0.1004	0.0016	0.2814	0.4712	0.6663

should be tried to see the effect on the ROC areas of the two modalities. It is imperative that one knows the disease prevalence with a high degree of confidence in order to have reliable estimates of the test accuracies.

BUGS CODE 7.7

```
model;
{
# Two tests T1 and T2 with ordinal values
# Dirichlet for theta
for(i in 1:3){ for (j in 1:3){h[i,j]~dgamma( zh[i,j],2)}}
sh<-sum(h[,])
for(i in 1:3){ for (j in 1:3){theta[i,j]<-h[i,j]/sh}}
for(i in 1:3){ for (j in 1:3){zh[i,j]<-y[i,j]+1}}
# Dirichlet for ph
for(i in 1:3){ for (j in 1:3){g[i,j]~dgamma( zg[i,j],2)}}
sg<-sum(g[,])
for(i in 1:3){ for (j in 1:3){ph[i,j]<-g[i,j]/sg}}
for(i in 1:3){ for (j in 1:3){zg[i,j]<- n[i,j] - y[i,j]+1}}
# dist of augmented data y[i,j]
for(i in 1:3){ for (j in 1:3){y[i,j]~dbin(w[i,j],n[i,j])}}
for(i in 1:3){ for (j in 1:3){w[i,j]<-d*theta[i,j]/(d*theta[i,j]+(1-d)*ph[i,j])}}
# posterior distribution of p
# the disease rate d is a mixture of p and p0
# p0 is the prior distribution of the disease rate
# p is the disease rate with a uniform prior
d<-.1*p+.9*p0
p0~dbeta(ap,bp)
p~dbeta(alphap,betap)
sy<-sum(y[,])
sn<-sum(n[,])
alphap<- sy+1
betap<- sn-sy+1
# ROC area T1
# Prob T1 = 1,2,3 given D=1
theta.1<-
 theta[1,1]+theta[2,1]+theta[3,1]
theta.2<-
```

```
 theta[1,2]+theta[2,2]+theta[3,2]
theta.3<-
 theta[1,3]+theta[2,3]+theta[3,3]
# Prob T1=1,2,3 given D=0
ph.2<- ph[1,2]+ph[2,2]+ph[3,2]
ph.3<- ph[1,3]+ph[2,3]+ph[3,3]
ph.1<- ph[1,1]+ph[2,1]+ph[3,1]
T1<-T11+T12/2
T11<- theta.2*ph.1+theta.3*(ph.1+ph.2)
# T12 is the prob of a tie
T12<- theta.1*ph.1+theta.2*ph.2+theta.3*ph.3
# ROC area T2
# Prob T2=1,2,3 given D=1
theta1.<- theta[1,1]+theta[1,2]+theta[1,3]
theta2.<-
 theta[2,1]+theta[2,2]+theta[2,3]
theta3.<-
 theta[3,1]+theta[3,2]+theta[3,3]
# Prob T2=1,2,3 given D=0
ph2.<-
 ph[2,1]+ph[2,2]+ph[2,3]
ph3.<-
 ph[3,1]+ph[3,2]+ph[3,3]
ph1.<-
 ph[1,1]+ph[1,2]+ph[1,3]
T2<- T21+T22/2
T21<-theta2.*ph1.+theta3.*(ph1.+ph2.)
# T22 is the Prob of a tie
T22<- theta1.*ph1.+theta2.*ph2.+theta3.*ph3.
}
# staging for melanoma
list(ap=20,bp=80, n=structure(.Data=c(45,26,28,25,20,20,46,30,46),
.Dim=c(3,3)))
# hypothetical
list(ap=64,bp=36, n=structure(.Data=c(30,40,50,1,30,60,1,1,70),
.Dim=c(3,3)))
# initial values
list(p=.5,
 g=structure(.Data=c(1,1,1,1,1,1,1,1,1),.Dim=c(3,3)),
 h=structure(.Data=c(1,1,1,1,1,1,1,1,1),.Dim=c(3,3)))
```

Note for the example that the disease rate is expressed as a mixture of two random variables, p and $p0$, where p is the posterior distribution of the disease rate assuming a uniform prior, and $p0$ is the prior distribution of the disease rate, with parameters ap and bp, which were set at 20 and 80, respectively.

The mixture weights were chosen at 0.20 and 0.80 for p and $p0$, respectively. See the appropriate statements in BUGS CODE 7.7.

7.12   Exercises

1. Show that if two tests are conditionally independent

$$P[T, R \mid D] = P[T \mid D]P[R \mid D], \tag{7.1}$$

that it is likely that both the observed sensitivity and specificity are decreased.

Assume that $P[T = 1 \mid D = 0] < P[T = 1 \mid D = 1]$, which is reasonable if test T has reasonable accuracy. That is, the test is better at detecting disease when it is present compared to when disease is absent.

2. This problem is due to Pepe [1: 196]. Show that if the reference test R is 100% specific, but less that 100% sensitive, then the observed sensitivity of the "new" test T is increased if $P[T = 1 \mid R = 1, D = 1] > P[T = 1 \mid D = 1]$.

The latter inequality implies that additional information given by the reference test that disease is present (in addition to knowing disease is actually present) increases the chance of detecting disease with the new test.

3. Refer to Section 7.4 on the posterior distribution of the parameters without assuming conditional independence between R and T.
   (a) Verify Equations 7.7 through 7.14.
   (b) Derive the likelihood function Equation 7.6.

4. Refer to Section 7.5 on the posterior distribution of the parameters assuming conditional independence between R and T.
   (a) Verify Equations 7.15 through 7.24.
   (b) In particular, derive the likelihood function Equation 7.15.

5. Verify the posterior analysis of Table 7.5, with BUGS CODE 7.1 and a uniform prior distribution. Use as I did, 125,000 observations generated from the posterior distribution of the parameters, with a burn in of 5,000 and a refresh of 100. Note that the first list statement provides the data and the third list statement the starting values. What are the MCMC errors for the simulation?

6. Verify the posterior analysis of Table 7.7 with BUGS CODE 7.1 and an informative prior distribution. Use as I did, 125,000 observations generated from the posterior distribution of the parameters, with a burn

in of 5,000 and a refresh of 100. What are the MCMC errors for all the parameters? Note the second list statement provides the data and the third list statement the starting values.

7. Complete the last four rows of Table 7.9, which is a check on the conditional independence for the *Strongyloides* study when $D = 1$.
    (a) Use BUGS CODE 7.2 and generate 130,000 observations, a burn in of 5,000 and a refresh of 100 from the posterior distribution of the parameters. The section for checking the CIA is labeled by # tests for conditional independence of the thetas.
    (b) What is your overall conclusion about the CIA assumption? Does it hold for this study?
    (c) Are 130,000 observations sufficiently large for the simulation?

8. The *Strongyloides* study was analyzed three ways (see Tables 7.5, 7.7 and 7.8). Which analysis is the most appropriate? Provide convincing reasons for your conclusion.

9. Verify the posterior analysis of Table 7.5, using 130,000 observations generated from the posterior distribution of the parameters, with a burn in of 5,000 and a refresh of 100. Recall this is for the case of the CIA and a uniform prior.

10. Verify the posterior analysis of Table 7.7, using 130,000 observations generated from the posterior distribution of the parameters, with a burn in of 5,000 and a refresh of 100. Recall this is for the case of imposing the CIA and an informative prior.

11. Refer to Table 7.4 and compute the observed sensitivity and specificity of the stool examination using serology as a reference test.
    (a) Explain how the Bayesian analyses corrected the observed accuracy of the stool examination.
    (b) Did the Bayesian analyses increase or decrease the sensitivity of the stool examination?
    (c) Did the Bayesian analyses increase or decrease the specificity of the stool examination?

12. Two dermatologists, R and T, are diagnosing melanoma, where R is an experienced clinician in skin cancer, but T is a first year resident in dermatology (Table 7.19).
    (a) What is the observed sensitivity and specificity of T using R as a reference?
    (b) Perform a Bayesian analysis assuming a uniform prior and with conditional independence, estimate the prevalence rate p, as well as the sensitivities and specificities of both dermatologists. Use BUGS CODE 7.1 and generate 150,000 observations from the joint posterior distribution of the parameters, with a burn in of 5,000

TABLE 7.19:   Two dermatologists
diagnosing melanoma.

	Dermatologist T	
Dermatologist	$T = 1$	$T = 0$
$R = 1$	17	42
$R = 0$	13	200

and a refresh of 100. Calculate the posterior mean, median, standard deviation, and the lower and upper 2 1/2 percentiles. Plot the posterior density of the sensitivity for dermatologist T. Do the corrected accuracies of resident T differ from the observed accuracies relative to R?

(c) Perform a Bayesian analysis assuming an informative prior and with conditional independence, estimate the prevalence rate p, as well as the sensitivities and specificities of both dermatologists. For informative prior information, assume the sensitivity of R and T are approximately 0.70 and 0.50, respectively, and assume the specificities of R and T are approximately 0.9 and 0.8, respectively. Also assume the prevalence rate of melanoma is approximately 0.30.

Using BUGS CODE 7.1, generate 150,000 observations from the joint posterior distribution of the parameters, with a burn in of 5,000 and a refresh of 100. Calculate the posterior mean, median, standard deviation, and the lower and upper 2 1/2 percentiles. Plot the posterior density of the sensitivity for dermatologist T.

(d) Perform a Bayesian analysis assuming a uniform prior, but not assuming conditional independence, estimate the prevalence rate p, as well as the sensitivities and specificities of both dermatologists. Using BUGS CODE 7.2, generate 150,000 observations from the joint posterior distribution of the parameters, with a burn in of 5,000 and a refresh of 100. Calculate the posterior mean, median, standard deviation, and the lower and upper 2 1/2 percentiles. Plot the posterior density of the sensitivity for dermatologist T.

(e) What are the MCMC errors for the parameters?

13. Refer to Tables 7.11a and b, which display the posterior analysis of the sensitivities and specificities of the Tine and Mantour tests for two sites. Test the hypothesis that the sensitivity of the Tine test is the same for both sites. The posterior mean(sd) of s_1 is 0.8595(0.1016) for site 1, and 0.6941(0.0128) for site 2.

14. Repeat the posterior analysis of the accuracies of the Tine and Mantour tests for the two sites, but use a prevalence rate of 10% for site 1 and 85% for site 2. Refer to BUGS CODE 7.3, and use the following parameters for the prior distribution of p for site 1:

In the first list statement for site 1, let $ap0 = 55$ and $bp0 = 500$. This puts a prior mean of 0.10 for p. Also, let the weights in the mixture distribution of p (the 7th statement in the code) be 0.10 for $p1$ and 0.90 for $p0$.

For site 2, in the second list statement, let $ap0 = 1124$ and $bp0 = 198$. This puts a prior mean of 0.85 for p of site 2. Use the same weights as above, namely, 0.90 for $p0$ and 0.10 for $p1$.

Use 130,000 observations generated from the posterior distribution of the parameters, with a burn in of 5,000 and a refresh of 100, and assume a uniform prior for the sensitivities and specificities of both sites.

15. Refer to Tables 7.11a and b. Zhou, McClish, and Obuchowski [2: 371] assume the sensitivities and specificities of the two tests are the same for the two locations. On inspection of the two tables, does this assumption appear valid? Based on the Bayesian analysis, it appears to me that the assumption is not unreasonable. Do you agree?

16. Refer to Tables 7.10a and b.
    (a) What is the observed sensitivity and specificity of the Tine test relative to the Mantour test?
    (b) What is the observed sensitivity and specificity of the Mantour test relative to the Tine test?
    (c) Does the Bayesian analysis reported in Tables 7.11a and b change these observed accuracies of the two tests? How? Explain your conclusions.

17. Refer to Figure 7.6, the plot of the posterior density of s_1 for site 1, and plot the posterior densities of the other accuracy parameters: s_2, c_1, and c_2. The posterior analysis is based on an informative beta prior for the disease rate of 4% for d (expressed as beta(40,960)), a uniform prior for the other parameters, 130,000 generated from the posterior distribution, with a burn in of 5,000 and a refresh of 100. BUGS CODE 7.2 is revised according to the description in Section 7.8.

18. Refer to Table 7.12a and test the hypothesis that conditional independence does not hold for site 1. Use BUGS CODE 7.4, which is a revision of BUGS CODE 7.2, which contains the code (it is labeled with a comment # that identifies the test for independence of the thetas and phis) that implements the test for conditional independence. Use 130,000 observations generated from the posterior distribution, with a burn in of 5,000 and a refresh of 100. Plot the posterior density of theta00c. Does the plot suggest conditional independence? What is the 95% credible interval for theta00c? What are the MCMC errors for the parameters? Are the errors sufficiently small to trust the analysis with 130,000 observations?

19. Two pathologists are classifying prostate biopsies as either positive or negative at a New York hospital, with the results shown in Table 7.20.

TABLE 7.20:   Two New York pathologists and prostate biopsies.

	Pathologist 1		
Pathologist 2	$T = 1$	$T = 0$	
$R = 1$	51	23	74
$R = 0$	16	212	228
Total	67	235	302

TABLE 7.21:   Two Houston pathologists and prostate biopsies.

	Pathologist 1		
Pathologist 2	$T = 1$	$T = 0$	
$R = 1$	123	45	168
$R = 0$	51	36	87
Total	174	81	255

At another hospital in Houston, two pathologists are performing the same tests with the results shown in Table 7.21. The disease rate for this particular clinical population of prostate cancer in the New York hospital is approximately 20% (with a high degree of certainty), and is believed to be 52% (also with a high degree of certainty) for the Houston hospital.

(a) Assuming conditional independence and a uniform prior for the sensitivities and specificities for the New York hospital, perform a Bayesian analysis and estimate the test accuracy of the two pathologists by generating 150,000 observations from the joint posterior distribution, with a burn in of 5,000 and a refresh of 100. Do the two pathologists have similar accuracy for classifying prostate tissue biopsies? Plot the posterior density of the specificity of pathologist 2.

(b) Repeat (a) above for the Houston hospital and compare the results for the two hospitals.

(c) Repeat (a) and (b) above, not assuming conditional independence.

(d) Test the hypothesis that there is no conditional independence between the two pathologists in New York.

(e) Test the hypothesis that there is no conditional independence between the two pathologists in Houston.

(f) What are the MCMC errors for the above parameters?

20. Refer to Section 7.8 for multiple tests with one population, assuming conditional independence, and consider an example presented by Pepe [1: 199] involving three diagnostic tests for *Chlamydia*. PCR, ELISA,

TABLE 7.22a: Three binary tests results when $T_1 = 1$.

	T_2	
T_3	$T_2 = 1$	$T_2 = 0$
$T_3 = 1$	$n_{111} = 20$	$n_{101} = 4$
$T_3 = 1$	$n_{110} = 2$	$n_{100} = 2$

Source: From Pepe, M.S. *The Statistical Evaluation of Medical Tests for Classification and Prediction*, 2003, P. 199, Table 7.8, by permission of Oxford University Press.

TABLE 7.22b: Three binary tests results when $T_1 = 0$.

	T_2	
T_3	$T_2 = 1$	$T_2 = 0$
$T_3 = 1$	$n_{011} = 4$	$n_{011} = 3$
$T_3 = 1$	$n_{010} = 2$	$n_{000} = 292$

Source: From Pepe, M.S. *The Statistical Evaluation of Medical Tests for Classification and Prediction*, 2003, P. 199, Table 7.8, by permission of Oxford University Press.

and a bacterial culture were scored as either positive or negative on 324 specimens from two clinics in China. The results are given in Tables 7.22a and b, where test 1 is culture, 2 is ELISA, and 3 is PCR.

Perform a Bayesian analysis with a uniform beta prior for all seven parameters, use 155,000 observations generated from the posterior distribution, with a burn in of 5,000 and a refresh of 100. Assume a uniform prior for all parameters and report a complete posterior analysis by citing the posterior mean, sd, lower and upper 2 1/2 percentiles, and the median of each parameter. Execute the calculations with BUGS CODE 7.5, using the first list statement for the data and the third list statement as initial values.

Verify that the posterior analysis is given by Table 7.23.

(a) Which test is most accurate?

(b) Plot the posterior densities of all seven parameters.

(c) Why are the specificities larger than the sensitivities?

(d) What is the observed sensitivity of test 1 relative to test 2?

(e) Report the MCMC error for each parameter of Table 7.22.

21. Repeat Exercise 20, but use the following parameters for the mixture p: $ap0 = 200$, $bp0 = 800$ with weights 0.90 for $p0$ and 0.10 for $p1$.

TABLE 7.23:   Posterior analysis of three tests for *Chlamydia* with conditional independence.

Parameter	Mean	sd	Error	2 1/2	Median	97 1/2
p	0.0468	0.0089	<0.0001	0.0311	0.0462	0.0658
c_1	0.9865	0.0078	<0.0001	0.9679	0.9878	0.9977
c_2	0.9764	0.0087	<0.0001	0.9565	0.9774	0.9909
c_3	0.9423	0.0140	<0.0001	0.9126	0.9434	0.9677
s_1	0.864	0.0695	<0.0001	0.7089	0.8721	0.9742
s_2	0.8595	0.0720	<0.0001	0.6999	0.8686	0.9757
s_3	0.9005	0.0587	<0.0001	0.7598	0.9113	0.9857

22. Compare Tables 7.15 and 7.16, which give the results for the Bayesian analysis of three radiologists who assess enlargement of the prostate. The former analysis assumes conditional independence between the three readers and a uniform prior for all parameters, while the latter analysis does not assume independence. Verify the results of Table 7.16 and explain the difference between the two tables. Test for no conditional independence using BUGS CODE 7.6 (the relevant statements are easily found) with 130,000 observations generated from the posterior distribution, with a burn in of 5,000 and a refresh of 100. Plot the posterior density of theta111c. Does the plot suggest conditional independence?

23. Perform a Bayesian analysis of the accuracies of three tests for the diagnosis of *Chlamydia*, where the data are given in problem 20. Do the analysis, but do not assume conditional independence and use uniform priors for all parameters. See the first statements in BUGS CODE 7.6. Note the second list statement of the code lists the data for this problem. Execute the analysis with BUGS CODE 7.6 using 130,000 observations generated from the joint posterior distribution, with a burn in of 5,000 and a refresh of 100. Verify the analysis as given in Table 7.24.

    (a) Compare the two tables in Exercises 20 and 23. What is your overall impression of the difference?

TABLE 7.24:   Posterior analysis of three tests for *Chlamydia* without conditional independence.

Parameter	Mean	sd	2 1/2	Median	97 1/2
p	0.023	0.0202	0.00089	0.018	0.0741
c_1	0.9201	0.0176	0.8836	0.9207	0.9531
c_2	0.9202	0.0176	0.8837	0.9208	0.9531
c_3	0.9142	0.0182	0.8766	0.9148	0.9482
s_1	0.4924	0.1592	0.1906	0.4916	0.7993
s_2	0.4927	0.1598	0.1912	0.4920	0.7996
s_3	0.5438	0.1579	0.2276	0.55	0.8305

(b) Plot the posterior densities of all parameters.

(c) Why is there more uncertainty in the sensitivities compared to the specificities?

(d) Do the posterior plots of the densities reveal any instability in the posterior distributions?

(e) Is conditional independence justified for the *Chlamydia* data?

(f) Vary the disease prevalence p and study the effect on the accuracies of the three tests.

(g) Report the MCMC error for each parameter.

24. Refer to Exercises 20 and 21 and repeat the analysis of each one using uniform beta priors for all parameters. You will notice a big change in the estimate (posterior mean) of the disease rate d. Use weights 1 and 0 for p and $p0$, this will impose a uniform prior for the prevalence of *Chlamydia*. Recall that Exercise 20 imposes conditional independence but Exercise 21 does not! Compare the sensitivities and specificities between the two analyses. Use 150,000 observations, with a burn in of 10,000 and a refresh of 500, and plot the posterior densities of the seven parameters.

25. (a) Verify Table 7.18, the posterior analysis for estimating the ROC areas for T_1 and T_2. Use BUGS CODE 7.7 and generate 45,000 observations from the joint posterior distribution, with a burn in of 5,000 and a refresh of 100.

(b) Repeat the analysis with a disease rate d set at 60%. Let $ap = 60$ and $bp = 40$ in the first list statement of BUGS CODE 7.7 and express d as a mixture with weights 0.2 and 0.8 for p and $p0$, respectively. See the relevant statements and set the appropriate values.

(c) Compare the ROC areas for T_1 and T_2. Why the difference?

(d) What is the MCMC error for each parameter? Is it small enough? Why or why not?

26. Consider two tests, T_1 and T_2, with ordinal scores given in Table 7.25. Perform a Bayesian analysis using a prevalence rate d of approximately 60%, setting $ap = 64$ and $bp = 36$ in the second list statement of BUGS CODE 7.7. In the statement for d, set d as a mixture of p and $p0$ with weights 0.2 and 0.8, respectively, and execute the analysis with 45,000

TABLE 7.25: Two tests—T_1 and T_2.

	T_1			
T_2	$T_1 = 1$	$T_1 = 2$	$T_1 = 3$	Total
$T_2 = 1$	30	40	50	120
$T_2 = 2$	1	30	60	91
$T_2 = 3$	1	1	70	72
Total	32	71	180	283

TABLE 7.26: ROC areas for T_1 and T_2.

Parameter	Mean	sd	2 1/2	Median	97 1/2
d	0.6397	0.0396	0.56	0.6403	0.7151
T_1	0.5475	0.1093	0.3413	0.5459	0.7622
T_2	0.5343	0.1383	0.2502	0.5427	0.7729

observations generated from the posterior distribution of all parameters, with a burn in of 5,000 and a refresh of 500.

(a) Verify Table 7.26 for the posterior distribution of the two ROC curves.

(b) Plot the posterior densities of the ROC areas.

(c) Vary the prevalence rate d and determine the effect on the two ROC areas.

(d) What is the MCMC error for the two ROC areas?

References

[1] Pepe, M.S. *The Statistical Evaluation of Medical Tests for Classification and Prediction*. Oxford University Press, Oxford, UK, 2003.

[2] Zhou, X.H., McClish, D.K., and Obuchowski, N.A. *Statistical Methods for Diagnostic Medicine*. John Wiley, New York, 2002.

[3] Joseph, L., Gyorkos, T.W., and Coupal, L. Bayesian estimation of disease prevalence and the parameters of diagnostic tests in the absence of a gold standard. *American Journal of Epidemiology*, 3:263, 1995.

[4] Dendukuri, N. and Joseph, L. Bayesian approaches to modeling the conditional dependence between multiple diagnostics tests. *Biometrics*, 57:158, 2001.

[5] Hui, S.L. and Walter, S. Estimating the error rates of diagnostic tests. *Biometrics*, 36:167, 1980.

[6] Broemeling, L.D. *Bayesian Methods for Measures of Agreement*. Taylor & Francis, Boca Raton, FL, 2009.

[7] Megibow, A.J., Zhou, X.H., Rotterdam, H., Francis, I., Zerhouni, E.A., Balfe, D.M., Weinreb, J.C., Aisen, A., Kuhlman, J., and Heiken, J.P. Pancreatic adenocarcinoma: CT versus MR imaging in the evaluation of resectability—report of the Radiology Diagnostic Oncology Group. *Radiology*, 195:327, 1995.

[8] Genta, R. Global prevalence of *Strongyloides*: Critical review with epidemiologic insights into the prevention of disseminated disease. *Reviews of Infectious Diseases*, 2:755, 1989.

[9] Bailey, J.W. A serological test for the diagnosis of *Strongyloides* in the ex-Far East prisoners of war. *Annals of Tropical Medicine and Parasitology*, 83:241, 1989.

[10] Irwig, L.M., du Troit, R.S., Sluis-Cremer, G.K., Soloman, A., Thomas, R.G., Hamel, P.P., Webster, I., and Hastie, T. Risk of asbestos in crocidolite and amosite mines in South Africa. *Annals of the New York Academy of Sciences*, 330:357, 1979.

Chapter 8

Verification Bias and Test Accuracy

8.1 Introduction

Consider a standard diagnostic test, say the blood glucose test for type 2 diabetes, where after fasting for at least 8 hours, the patient is declared positive if the measured level is in excess of 126 mg/dL. If this is the case, the person is often given an oral glucose tolerance test, which is considered a gold standard for the disease. For those with levels between 111 and 125 mg/dL, problems with glucose metabolism are suspected, while for those with levels below 111 mg/dL, type 2 diabetes is not suspected. For this latter group, the patient would not ordinarily be given the oral glucose tolerance test, while for those between 111 and 125 mg/dL, a follow-up test might be appropriate if other factors point to diabetes. In this example, the diagnostic test value Y falls into one of three categories: (1) those with values below 111 mg/dL, (2) those between 111–125 mg/dL, and (3) those with values of $Y > 125$ mg/dL. Suppose the oral glucose tolerance test is considered a gold standard, where all those in the third group undergo the gold standard for disease verification, say 23% are verified for the second group, and none are verified for the first group. This is a typical case of verification bias because the usual estimates of test accuracy are biased if they are estimated using only the verified cases. For simplicity, suppose the test Y is positive if $Y > 125$ mg/dL, otherwise the test is declared negative. If only the validated cases are used to estimate, say sensitivity, all those in the third group would be verified with the gold standard, while only a subset of those with values $Y \leq 125$ are verified by the gold standard. These usual estimates would be biased, that is to say, the sensitivity would be too high, compared to those estimated by referring all patients for disease verification.

Actually, verification bias is present in many medical test accuracy studies, but often the investigator is unaware that bias is present. According to Zhou, McClish, and Obuchowski [1], Greenes and Begg [2] reviewed 145 investigations that took place over the period 1976–1980 and found that 26% had verification bias that was not recognized by the authors. In addition, Bates, Margolis, and Evans [3] reported that at least one-third of 54 pediatric studies had unrecognized verification bias. There are many more such studies, including those reported by Philbrick, Horwitz, and Feinstein [4], who found that of 33 diagnostic studies for coronary artery disease, 31 had verification bias. In a major review of verification bias, which reviewed 112 studies in major medical journals, Reid, Lachs, and Feinstein [5] reported finding that 54% had verification bias!

It is important to remember that verification bias is present in the routine use of diagnostic procedures, but these are usually not part of a study for accuracy. For example, in mammography, for those women whose image is negative, it would be unethical to refer them for a biopsy. As another example, take the exercise stress test for coronary artery disease. Those that test negative would not usually be referred for a coronary angiography (an invasive nuclear medicine procedure), the gold standard. However, if the main purpose of the study is to assess diagnostic accuracy, the study should be designed in order to avoid verification bias if at all possible.

All is not lost, even if verification bias is unavoidable, statistical techniques are available that correct for verification bias, and this chapter will introduce such methodologies. The concept is first introduced with a binary test that indicates either a positive or a negative result for the patient. Among those that test positive, not all will be referred to the gold standard, and the same applies to those that test negative. In order to correct for verification bias, the missing at random (MAR) assumption is made, that is, for those persons who did not undergo the follow-up gold standard, that observation is considered MAR, that is to say, the decision to verify the disease status depends only on the result, Y, of the diagnostic test. Using the MAR assumption, estimates of sensitivity and specificity are derived based on the likelihood function of the observed results. The likelihood is based on the conditional distribution of the disease status, $D = 1$, given $Y = 0$ or 1, and the marginal distribution for Y. The probability that $Y = 1$, given $D = 1$, is then found via Bayes theorem, and the result is the estimated sensitivity of the test. This method of using Bayes theorem for correction is attributed to Begg and Greenes [6].

Of course, a prior for the unknown parameters must be specified. The specified prior depends on the design of the study and the layout of the observations. Usually, a uniform prior or an improper Jeffrey's type prior is employed, however, in the case where prior information is available based on previous related studies, a conjugate type prior is appropriate.

The above approach is extended to two correlated binary tests. The examples in this case use two readers who are testing the same patients. This is a paired design, where it is important to know the degree to which the two readers agree in their diagnosis. An obvious generalization is to a test with ordinal outcomes. A good example of this is mammography, where the radiologist assigns scores 1, 2, 3, 4, and 5 to each mammogram, where 1 indicates a high confidence of no disease, and 5 signifies a very high confidence that a malignant lesion is present. For ordinal tests, the main emphasis is on the area under the receiver operating characteristic (ROC) curve, which will be estimated in the presence of verification bias.

Again, the methodology is based on the likelihood of the observed result and a prior distribution for the parameters. The actual correction is based on Bayes theorem applied to the probability that $D = 1$, given $Y = i$ (the ith outcome of the test), then the ROC area is computed using a formula in Broemeling [7: 72]. This formula is the familiar rule that the ROC area for ordinal scores is given by $P[Y$ given $D = 1$ is greater than Y given $D = 0] + (1/2)\ P[Y$ given $D = 1$ equals Y given $D = 0]$.

An interesting alternative for computing the corrected estimates of accuracy is to use the inverse probability weighting (IPW) method of estimating the original data from the observed results, where verification bias is present. For a particular value of Y, each cell in the table is multiplied by the inverse of the proportion of patients who have been referred to the gold standard and is repeated for each value of Y; then the accuracy is estimated the usual way. Pepe [8] illustrates the IPW method and states that this method is the same as the correction method of Begg and Greenes [6]. In this chapter, their method will be given a Bayesian flavor.

The latter part of the chapter focuses on two correlated ordinal tests for two readers with and without covariates. Lastly, the case of extreme verification bias is considered. In the simplest case of a binary test Y, extreme bias occurs when all those that test positive are verified for disease status, but among those that test negative, none are verified. Of course, in such cases it is not possible to estimate the standard measures of accuracy, including true and false positive fractions, however, other measures such as detection probabilities and false referral probabilities can be estimated. When the test results are ordinal, it is interesting to observe that if extreme bias does occur for some values of Y, it is still possible to estimate the ROC area. This is an interesting generalization. The interested reader should refer to Pepe [8] for more information on the extreme verification bias, and to Zhou, McClish, and Obuchowski [1] for a good introduction to the maximum likelihood approach to correcting for verification bias. Throughout the chapter, various examples illustrate the Bayesian methods and the exercises at the end of the chapter provide the student with many interesting extensions of the basic concepts.

8.2   Verification Bias and Binary Tests

Consider Table 8.1 for one binary test $Y = 0, 1$, where verification bias is present. In the table, $V = 1$ indicates the patient is verified and the disease status is known, and $V = 0$ indicates a patient has not been verified, thus, there are u_1 individuals who are not verified when $Y = 1$. The total number of patients in the study is $m_1 + m_0$, while the number who tested positive and had the disease is s_1. Let

$$\phi_i = P[D = 1 \mid Y = i], \tag{8.1}$$

and

$$\theta_i = P[Y = i], \tag{8.2}$$

TABLE 8.1: One binary test.

Y	1 (Positive)	0 (Negative)
$V = 1$		
$\qquad D = 1$	s_1	s_0
$\qquad D = 0$	r_1	r_0
$V = 0$	u_1	u_0
Total	m_1	m_0

where $i = 0, 1$, then the likelihood function for the parameters is

$$L(\theta, \phi) \propto \phi_1^{s_1}(1 - \phi_1)^{r_1}\phi_0^{s_0}(1 - \phi_0)^{r_0}\theta_1^{m_1}\theta_0^{m_0}, \tag{8.3}$$

where all parameters are between 0 and 1, and $\theta_0 + \theta_1 = 1$. With a uniform prior for all parameters, the posterior distribution of the parameters is as follows:

$$\phi_i \sim \text{beta}(s_i + 1, r_i + 1), \tag{8.4}$$

for $i = 0, 1$, and (θ_0, θ_1) has a Dirichlet with parameters $(m_0 + 1, m_1 + 1)$. On the other hand, with an improper prior distribution

$$f(\theta, \phi) \propto 1/\theta_0\theta_1\phi_0\phi_1, \tag{8.5}$$

the posterior distribution of the parameters is

$$\phi_i \sim \text{beta}(s_i, r_i), \tag{8.6}$$

and

$$(\theta_0, \theta_1) \sim \text{Dirichlet}(m_0, m_1). \tag{8.7}$$

The approach to correcting for bias is to use Bayes theorem to compute

$$P[Y = 1 \mid D = 1] = P[D = 1 \mid Y = 1]/P[D = 1], \tag{8.8}$$

where

$$P[D = 1] = \phi_1\theta_1 + \phi_0\theta_0. \tag{8.9}$$

Let

$$\alpha_1 = \phi_1\theta_1/(\phi_1\theta_1 + \phi_0\theta_0), \tag{8.10}$$

then α_1 is the sensitivity of the test. On the other hand, let

$$\beta_1 = (1 - \phi_1)\theta_1/(1 - \phi_1\theta_1 - \phi_0\theta_0), \tag{8.11}$$

then β_1 is the false positive fraction, that is, the probability that $Y = 1$, given $D = 0$.

Once the posterior distribution of the parameters is determined, the posterior distribution of the true and false positive fractions is also determined.

TABLE 8.2: Diabetes study for verification bias.

Y		1 (Positive)	0 (Negative)
$V = 1$			
	$D = 1$	$s_1 = 298$	$s_0 = 31$
	$D = 0$	$r_1 = 26$	$r_0 = 48$
$V = 0$		$u_1 = 150$	$u_0 = 117$
Total		$m_1 = 474$	$m_0 = 196$

A good example of verification bias is the study by Drum and Christa-copoulos [9]. The example presented in Table 8.2 is a diabetes study where patients are tested for the disease based on the blood glucose test, but where some of the patients are not further tested by the gold standard, the glucose tolerance test.

This test had two results, $Y = 0$ or 1, where 1 indicates a positive result for diabetes. Note that the total number of subjects was 670, with 474 who tested positive, and among those, 150 were not verified for the disease. Among those who tested negative, 79 were examined by the gold standard, with 31 of those having the disease. The estimated sensitivity based on the selected table is $298/329 = 0.905$, and the estimated false positive rate is $26/74 = 0.35$.

The following is the WinBUGS code for estimating the sensitivity and false positive fraction. The notation in the statements is quite similar to that in the above presentation.

BUGS CODE 8.1

```
model;
{
th1~dbeta(m1,m0)
ph0~dbeta(s0,r0)
ph1~dbeta(s1,r1)
th0<-1-th1
tpf<- ph1*th1/(ph1*th1+ph0*th0)
fpf<- (1-ph1)*th1/(1-ph1*th1-ph0*th0)
}
# diabetes example
# the following values assume an improper prior
list(m0= 196,m1=474,s0=31,r0=48,r1=26,s1=298)
```

Note that the above values used in the list statement assume an improper prior distribution for the parameters in the likelihood (Equation 8.3). I used 45,000 values generated from the posterior distribution, with a burn in of 5,000 and a refresh of 100; the analysis is presented in Table 8.3.

The true positive fraction or sensitivity has a posterior mean of 0.8503, while the false positive fraction has a mean of 0.2424. The 95% posterior credible interval for the true positive fraction is (0.8081,0.8894). Note, the mean

TABLE 8.3:   Posterior distribution for diabetes study.

Parameter	Mean	sd	Error	2 1/2	Median	97 1/2
fpf	0.2424	0.0412	<0.0001	0.167	0.2407	0.3277
tpf	0.8503	0.0208	<0.0001	0.8081	0.8507	0.8894

and median are almost the same, indicating little skewness in the posterior distributions, and the symmetry is also evident from the graph of the posterior distribution. Compare the Bayesian estimates with the naïve estimates (based on the selected data, i.e., those not corrected for bias) of 0.905 and 0.35 for the true and false positive fractions, respectively, thus verifying the theory that suggests the true and false positive fractions will be less than the corresponding naïve estimates based on the verified data only. It is also noted that the Markov Chain Monte Carlo (MCMC) error for all parameters is <0.0001, implying that the 45,000 observations generated for the simulation are sufficient in order to estimate the "true" posterior characteristics (Figure 8.1).

The MAR assumption is formally expressed by

$$P[V = 1 \mid Y, D] = P[V = 1 \mid Y]. \tag{8.12}$$

In other words, the probability of referring a subject to the gold standard depends only on the results of the diagnostic test. If the decision to refer depends on additional factors, such as symptoms or family history, the MAR assumption is not valid; however, a later section considers such an eventuality.

8.3   Two Binary Tests

When assessing the accuracy of two tests, the design in many cases is paired. For example, two imaging devices (e.g., computed tomography [CT] and magnetic resonance imaging [MRI]) are procuring information from the same patients and the two images would be expected to be quite similar.

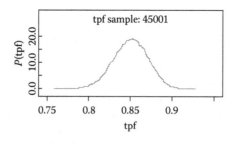

FIGURE 8.1: Posterior density of the true positive fraction.

TABLE 8.4: Two binary scores with verification bias.

$Y_1 =$		1		0	
$V = 1$	$Y_2 =$	1	0	1	0
$D = 1$		s_{11}	s_{10}	s_{01}	s_{00}
$D = 0$		r_{11}	r_{10}	r_{01}	r_{00}
$V = 0$		u_{11}	u_{10}	u_{01}	u_{00}
Total		m_{11}	m_{10}	m_{01}	m_{00}

Another case of a paired design is two readers who are imaging the same set of patients with the same imaging device. One expects the information gained from the two paired sources to be highly correlated, and in the case of two paired readers, agreement between the two is also of interest. The results for a paired design with two binary scores are given in Table 8.4.

Suppose the unknown parameters are

$$\phi_{ij} = P[D = 1 \mid Y_1 = i, Y_2 = j] \tag{8.13}$$

and

$$\theta_{ij} = P[Y_1 = i, Y_2 = j], \tag{8.14}$$

for $i, j = 0, 1$. Also let

$$\phi_{i.} = P[D = 1 \mid Y_1 = i] \tag{8.15}$$

and

$$\theta_{.j} = P[D = 1 \mid Y_2 = j], \tag{8.16}$$

where $i, j = 0, 1$.

The likelihood for the parameters is

$$L(\theta, \phi) \propto \prod_{i=0}^{i=1} \prod_{j=0}^{j=1} \phi_{ij}^{s_{ij}} (1 - \phi_{ij})^{r_{ij}} \prod_{i=0}^{i=1} \prod_{j=0}^{j=1} \theta_{ij}^{m_{ij}}. \tag{8.17}$$

Assuming an improper prior distribution for the parameters, the posterior distributions are

$$\phi_{ij} \sim \text{beta}(s_{ij}, r_{ij}), \tag{8.18}$$

for $i, j = 0, 1$, and θ_{ij} are distributed Dirichlet $(m_{00}, m_{01}, m_{10}, m_{11})$.

Note that

$$\phi_{1.} \sim \text{beta}(s_{1.}, r_{1.}) \tag{8.19}$$

and

$$\phi_{.1} \sim \text{beta}(s_{.1}, r_{.1}), \tag{8.20}$$

where

$$s_1 = s_{11} + s_{10}$$

and

$$r_1 = r_{11} + r_{10}.$$

The main parameters of interest are the true positive fraction and the false positive fraction for the two tests, thus for the first test:

$$\text{tpf}_1 = P[Y_1 = 1 \mid D = 1]$$

and is given by Bayes theorem as

$$\text{tpf}_1 = \phi_{1.}\theta_{1.}/(\phi_{1.}\theta_{1.} + \phi_{0.}\theta_{0.}), \tag{8.21}$$

where $\phi_{i.}$ is given by Equation 8.19 and

$$\theta_{1.} = \theta_{11} + \theta_{10}.$$

As for test 1, the false positive fraction is given by

$$\text{fpf}_1 = (1 - \phi_{1.})\theta_{1.}/(1 - \phi_{1.}\theta_{1.} - \phi_{0.}\theta_{0.}).$$

As a first example of two binary tests, consider two tests for detecting metastasis of colon cancer to the liver, where the first test, Y_1, is MRI, and the second test, Y_2, is a nuclear medicine image provided by single photon emission tomography (SPECT) (Table 8.5).

There are very few (149/855) referred to the gold standard when both readers give a negative score, but when both observers give a positive score, 221/234 are referred to the gold standard. What are the true positive fractions for both observers? What are the false positive fractions? Of the 1301 subjects, 750 received a negative assessment by both images, while 234 received a positive assessment (judged as having the disease) from tests for metastasis. When using two tests, it is of interest to estimate the accuracy of the combined tests, and this will be accomplished in Chapter 10, but this is not done here, and instead the accuracy of the two tests are estimated individually.

The following WinBUGS code follows the notation in the above section and performs the Bayesian analysis for the colon cancer metastasis study.

TABLE 8.5:   Two binary tests for metastasis of colon cancer.

	$Y_1 =$	1		0	
$V = 1$	$Y_2 =$	1	0	1	0
$D = 1$		$s_{11} = 210$	$s_{10} = 20$	$s_{01} = 65$	$s_{00} = 8$
$D = 0$		$r_{11} = 11$	$r_{10} = 29$	$r_{01} = 89$	$r_{00} = 141$
$V = 0$		$u_{11} = 13$	$u_{10} = 11$	$u_{01} = 103$	$u_{00} = 601$
Total		$m_{11} = 234$	$m_{10} = 60$	$m_{01} = 257$	$m_{00} = 750$

BUGS CODE 8.2

```
model;
{
g00~dgamma(m00,2)
g01~dgamma(m01,2)
g10~dgamma(m10,2)
g11~dgamma(m11,2)
h<-g00+g01+g10+g11
th00<-g00/h
th01<-g01/h
th10<-g10/h
th11<-g11/h
phi00~dbeta(s00,r00)
phi01~dbeta(s01,r01)
phi10~dbeta(s10,r10)
phi11~dbeta(s11,r11)
s1.<-s11+s10
r1.<-r11+r10
s.1<-s01+s11
r.1<- r01+r11
r0.<- r00+r01
s0.<-s00+s01
s.0<-s00+s10
r.0<-r00+r10
# for test 1=1 pd=1
ph1.~dbeta(s1.,r1.)
# for test 1=1 d=0
ph.1~dbeta(s.1,r.1)
# for test 1 = 0 d=1
ph0.~dbeta(s0.,r0.)
# for test 2=1 d=1
ph.0~dbeta(s.0,r.0)
th1.<-th11+th10
th.1<-th01+th11
th0.<-th01+th00
th.0<-th00+th10
# accuracy for test 1
tpf1<-ph1.*th1./pd1
fpf1<-(1-ph1.)*th1./(1-pd1)
pd1<-ph1.*th1.+ph0.*th0.
# accuracy for test 2
tpf2<-ph.1*th.1/pd2
fpf2<-(1-ph.1)*th.1/(1-pd2)
pd2<-ph.1*th.1+ph.0*th.0
}
```

TABLE 8.6:   Posterior distribution colon cancer and liver metastasis.

Parameter	Mean	sd	Error	2 1/2	Median	97 1/2
tpf1	0.509	0.0310	<0.0001	0.4492	0.5084	0.5707
fpf1	0.0539	0.0083	<0.0001	0.03882	0.05355	0.0714
tpf2	0.7594	0.0304	<0.0001	0.691	0.7599	0.8241
fpf2	0.1586	0.0142	<0.0001	0.1315	0.1582	0.1876

```
# metastasis study
list(s00=8,r00=141,s01=65,r01=89,s10=20,r10=29,s11=210,r11=11,
m00=750,m01=257,m10=60,m11=234)
# initial values
list( g00=1, g01=1, g10=1, g11=1)
```

BUGS CODE 8.2 closely follows the notation in Section 8.3 and performs the Bayesian analysis for the colon cancer study, and the analysis is executed with 65,000 observations generated for the simulation, with a burn in of 5,000 and a refresh of 100, accordingly the results are reported in Table 8.6.

Simulation errors are quite small, and the sensitivity of the first test (MRI) is estimated as 0.509(0.0310) with the posterior mean, while the false positive fraction is estimated as 0.0539(0.0083), and the two distributions appear to be symmetric. The 95% credible intervals for the two accuracy parameters are (0.4492,0.5707) and (0.0388,0.0714) for the true and false positive fractions, respectively, and it is seen that the MRI test has poor sensitivity but a very small false positive rate. On the other hand, the accuracy estimates for SPECT are much better with a true positive fraction estimated with the posterior mean as 0.759 and a false positive fraction of 0.158. I would prefer the nuclear medicine procedure, which has fair sensitivity and a small false positive fraction. It should be remembered that the above analysis assumed an improper prior (similar to Equation 8.5) for the parameters.

One could study the agreement in the two observers, which is treated somewhat differently from our approach to correcting for verification bias. See Chapter 2 of Broemeling [10] for measuring agreement between two observers with the Kappa statistic.

8.4   Ordinal Tests and Verification Bias

The next step in our study of verification bias is to extend the previous treatment to tests with ordinal scores. Mammography is a good example of such a test, where the scores $Y = 1, 2, 3, 4, 5$ indicate the degree of confidence of the observer in their belief that a lesion is present in the mammogram. For example, a score of 1 indicates a high degree of belief that a lesion is not

TABLE 8.7: Verification bias and one ordinal test.

$Y =$	1	2	$\cdots$	k
$V = 1$				
$D = 1$	s_1	s_2		s_k
$D = 0$	r_1	r_2		r_k
$V = 0$	u_1	u_2		u_k
Total	m_1	m_2		m_k

present, while 5 denotes a high degree of belief that the lesion is present in the breast as indicated by the mammogram. On the other hand, a score of 2 implies a moderate degree of belief that a lesion is not present, while 4 indicates a moderate degree of belief that a lesion is present. Finally, a score of 3 indicates that one is ambivalent as to the presence of a lesion. When verification bias is present, for each level Y of the test, a certain number is subject to verification of the disease, but all will not necessarily be verified by the gold standard. When the test has ordinal outcomes, the best overall measure of accuracy is the area under the ROC curve.

The typical layout for such a test Y with possible values $1, 2, \ldots, k$, is reported with familiar notation as in Table 8.7.

The analysis is likelihood based, where the likelihood function is determined by the conditional distribution of $D = 1$, given $Y = i$ and the marginal distribution of Y, where $i = 1, 2, \ldots, k$. Let

$$\phi_i = P[D = 1 \mid Y = i] \tag{8.22}$$

and

$$\theta_i = P[Y = i],$$

thus, the likelihood function is

$$L(\phi, \theta) \propto \prod_{i=1}^{i=k} \phi_i^{s_i}(1 - \phi_i)^{r_i} \prod_{i=1}^{i=k} \theta_i^{m_i}. \tag{8.23}$$

If an improper prior is used for the parameters, the posterior distribution of ϕ_i is beta with parameters s_i and r_i, and that for θ_i is Dirichlet with parameter $(m_1, m_2, \ldots, m_k)$.

On the other hand, if a uniform prior distribution is deemed appropriate, the posterior distribution of ϕ_i is beta with parameters $s_i + 1$ and $r_i + 1$, and that for θ_i is Dirichlet with parameter $(m_1 + 1, m_2 + 1, \ldots, m_k + 1)$.

In order to compute the area under the ROC, one must compute $P[Y = i \mid D = 1]$ and $P[Y = i \mid D = 0]$ for all $i = 1, 2, \ldots, k$, where the first component is represented by Bayes theorem as

$$P[Y = i \mid D = 1] = P[D = 1 \mid Y = i]P[Y = i]/P[D = 1]$$
$$= \phi_i\theta_i/P[D = 1], \tag{8.24}$$

where

$$P[D = 1] = \sum_{i=1}^{i=k} \phi_i \theta_i. \tag{8.25}$$

On the other hand, the second component is computed as

$$P[Y = i \mid D = 0] = (1 - \phi_i)\theta_i / P[D = 0], \tag{8.26}$$

where

$$P[D = 0] = 1 - P[D = 1].$$

We are now in a position to compute the area under the ROC curve.
    Let

$$\alpha_i = P[T = i \mid D = 1] \tag{8.27}$$

and

$$\beta_i = P[T = i \mid D = 0], \tag{8.28}$$

for $i = 1, 2, \ldots, k$, then the area under the ROC is given by

$$A = A_1 + A_2/2, \tag{8.29}$$

where

$$A_1 = \alpha_2 \beta_1 + \alpha_3 (\beta_1 + \beta_2) + \cdots + \alpha_k (\beta_1 + \beta_2 + \cdots + \beta_{k-1}) \tag{8.30}$$

and

$$A_2 = \sum_{i=1}^{i=k} \alpha_i \beta_i. \tag{8.31}$$

Equation 8.31 for the ROC area is given in Broemeling [7: 72].
    The example for ordinal test scores is taken from a mammography study with 1509 subjects, where each patient is given a score of Y, where $Y = 1, 2, 3, 4, 5$ (see Table 8.8).

TABLE 8.8:   Ordinal results for mammography.

$Y =$	1	2	3	4	5
$V = 1$					
$D = 1$	$s_1 = 72$	$s_2 = 54$	$s_3 = 121$	$s_4 = 145$	$s_5 = 245$
$D = 0$	$r_1 = 308$	$r_2 = 127$	$r_3 = 78$	$r_4 = 33$	$r_5 = 77$
$V = 0$	$u_1 = 92$	$u_2 = 66$	$u_3 = 76$	$u_4 = 10$	$u_5 = 5$
Total	$m_1 = 472$	$m_2 = 247$	$m_3 = 275$	$m_4 = 188$	$m_5 = 327$

BUGS CODE 8.3

```
# uses Bayes theorem on terms in Broemeling formula page 72
# Mammography example
Model;
{
for ( i in 1:5){ ph[i]~dbeta(s[i],r[i])}
for ( i in 1:5){ g[i]~dgamma(m[i],2)}
h<-sum(g[])
for ( i in 1:5){ theta[i]<-g[i]/h}
A<- A1+A2
A2<- (alpha1*beta1+alpha2*beta2+alpha3*beta3+alpha4*beta4+
alpha5*beta5)/2
A1<-alpha2*beta1+alpha3*(beta1+beta2)+alpha4*(beta1+beta2+beta3)
+alpha5*(beta1+beta2+beta3+beta4)
alpha2<-ph[2]*theta[2]/pd
pd<-ph[1]*theta[1]+ph[2]*theta[2]+ph[3]*theta[3]+ph[4]*theta[4]
+ph[5]*theta[5]
alpha1<-ph[1]*theta[1]/pd
beta4<-(1-ph[4])*theta[4]/(1-pd)
beta5<-(1-ph[5])*theta[5]/(1-pd)
beta1<-(1-ph[1])*theta[1]/(1-pd)
alpha3<-ph[3]*theta[3]/pd
beta2<-(1-ph[2])*theta[2]/(1-pd)
alpha4<-ph[4]*theta[4]/pd
alpha5<-ph[5]*theta[5]/pd
beta3<-(1-ph[3])*theta[3]/(1-pd)
}
# mammography example
list(s=c(72,54,121,145,245),r=c(308,127,78,33,77),
m=c( 472,247,275,188,327))
```

The posterior analysis with BUGS CODE 8.3 is performed with 55,000 observations, a burn in of 5,000 and a refresh of 100, and has MCMC errors <0.0001 and is presented in Table 8.9.

The estimated area A under the ROC curve is 0.7762 with a 95% credible interval of (0.7509,0.8005), indicating reasonable accuracy for mammography.

TABLE 8.9: Posterior analysis for mammography study.

Parameter	Mean	sd	Lower 2 1/2	Median	Upper 2 1/2
A	0.7762	0.0126	0.7509	0.7764	0.8005
A_1	0.6972	0.0154	0.6665	0.6974	0.7272
A_2	0.0789	0.00303	0.0729	0.0789	0.0848

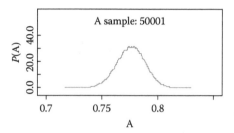

FIGURE 8.2: Posterior density of the ROC area.

The median and mean are identical, indicating symmetry for the posterior distribution of the area, and this is also evident from Figure 8.2. Recall that the ROC area is composed of two parts, where A_2 is the component measuring the ties between α_i and β_i, $i = 1, 2, 3, 4, 5$, therefore, this component of the area is quite small, relative to A_1.

Suppose the mammography study is repeated at another site with 1563 patients, under the same protocol, then how should the results be combined in order to provide an overall estimate of the accuracy of the imaging modality? Assume that the two sites do not share the same patients, but otherwise the way the study is conducted is the same for both hospitals. The radiologists of the two sites have similar training and are guided by the same protocol. In particular, the inclusion and exclusion criteria are the same for both sites, implying similar study populations. Then, one would expect "similar" estimates of the ROC estimates, but how similar will they, in fact, be? See Table 8.10.

The analysis was repeated using BUGS CODE 8.3. The burn in is 5,000 with a refresh of 100, and 55,000 observations are generated from the joint posterior distribution to give the results shown in Table 8.11.

The posterior mean of the ROC area from site 2 is 0.8085, compared to 0.7762 for site 1. How should these two estimates be combined? One way is to use a weighted estimate, with weights proportional to the inverse of the posterior variance, or one could weight by the number of patients in each study. This analysis is left as an exercise.

TABLE 8.10:   Ordinal test for mammography of site 2.

$Y =$	1	2	3	4	5
$V = 1$					
$D = 1$	$s_1 = 65$	$s_2 = 77$	$s_3 = 118$	$s_4 = 162$	$s_5 = 233$
$D = 0$	$r_1 = 331$	$r_2 = 202$	$r_3 = 43$	$r_4 = 22$	$r_5 = 66$
$V = 0$	$u_1 = 88$	$u_2 = 58$	$u_3 = 82$	$u_4 = 13$	$u_5 = 3$
Total	$m_1 = 484$	$m_2 = 337$	$m_3 = 243$	$m_4 = 197$	$m_5 = 302$

TABLE 8.11: Posterior analysis for mammography site 2.

Parameter	Mean	sd	Lower 2 1/2	Median	Upper 2 1/2
A	0.8085	0.01175	0.7848	0.8087	0.8308
A_1	0.7379	0.0146	0.7086	0.7381	0.7659
A_2	0.0705	0.00314	0.0644	0.0705	0.0767

8.5 Two Ordinal Tests and Verification Bias

With two ordinal tests, the design is often paired, where two readers are examining the same patients and assigning a score to each. For example, consider the following study for staging melanoma. A dermatologist and a surgeon are assigning a score to each melanoma patient where the score is the reader's estimate of the stage of the disease. The stages for melanoma are: 1 indicates primary lesion is localized and no metastasis to the lymph nodes; 2 signifies metastasis to the lymph nodes; and 3 indicates an advanced stage, where the disease has metastasized beyond the lymph nodes. The general schematic for two correlated tests is shown in Table 8.12.

As before, let

$$\phi_{i.} = P[D = 1 \mid Y_1 = i]$$

and

$$\phi_{.i} = P[D = 1 \mid Y_2 = i], \tag{8.32}$$

for $i = 1, 2, 3$, then assuming an improper prior distribution,

$$\phi_{i.} \sim \text{beta}(s_{i.}, r_{i.}),$$

and

$$\phi_{.i} \sim \text{beta}(s_{.i}, r_{.i}).$$

TABLE 8.12: Two ordinal tests.

$Y_1 =$	1			2			3		
Y_2	1	2	3	1	2	3	1	2	3
$D = 1$	s_1	s_2	s_3	s_4	s_5	s_6	s_7	s_8	s_9
$D = 0$	r_1	r_2	r_3	r_4	r_5	r_6	r_7	r_8	r_9
$V = 0$	u_1	u_2	u_3	u_4	u_5	u_6	u_7	u_8	u_9
Total	m_1	m_2	m_3	m_4	m_5	m_6	m_7	m_8	m_9

Note that

$$s_{1.} = s_1 + s_2 + s_3,$$
$$s_{.1} = s_1 + s_4 + s_7,$$

and

$$r_{1.} = r_1 + r_2 + r_3, \text{ etc.}$$

Also, let

$$\theta_1 = P[Y_1 = 1, Y_2 = 1]$$

and

$$\theta_9 = P[Y_1 = 3, Y_2 = 3].$$

In addition, let

$$\theta_{1.} = \theta_1 + \theta_2 + \theta_3,$$

thus

$$\theta_{1.} = P[Y_1 = 1]. \tag{8.33}$$

Also let

$$\theta_{.3} = \theta_3 + \theta_6 + \theta_9,$$

then

$$\theta_{.3} = P[Y_2 = 3], \text{ etc.} \tag{8.34}$$

In order to compute the area under the ROC, Bayes theorem is used to compute

$$\alpha 1_i = \phi_{i.} \theta_{i.} / t\alpha 1,$$

where $i = 1, 2, 3$, and

$$t\alpha 1 = \sum_{i=1}^{i=3} \phi_{i.} \theta_{i.}.$$

Note that

$$\alpha 1_i = P[Y_1 = i \mid D = 1], \tag{8.35}$$

and in a similar manner

$$\beta 1_i = (1 - \phi_{i.}) \theta_{i.} / t\beta 1,$$

where

$$t\beta1 = \sum_{i=1}^{i=3}(1-\phi_{i.})\theta_{i.},$$

thus

$$\beta1_i = P[Y_1 = i \mid D = 0]. \tag{8.36}$$

Also,

$$\alpha2_i = P[Y_2 = i \mid D = 1] \tag{8.37}$$

and

$$\beta2_i = P[Y_2 = i \mid D = 0] \tag{8.38}$$

can be defined. The ROC area for the first test is

$$A_1 = A_{11} + A_{12}, \tag{8.39}$$

where

$$A_{11} = \alpha1_2\beta1_1 + \alpha1_3(\beta1_1 + \beta1_2) \tag{8.40}$$

and

$$2A_{12} = \sum_{i=1}^{i=3}\alpha1_i\beta1_i. \tag{8.41}$$

$$A_2 = A_{21} + A_{22} \tag{8.42}$$

Of course, a similar expression holds for the ROC area of test 2. For the melanoma staging study, the first test corresponds to a melanoma surgeon, while the second corresponds to a dermatologist. Both assign a stage to each patient, and we expect the two tests to be correlated (see Table 8.13).

The likelihood function for the parameters is

$$L(\phi, \theta) \propto \prod_{i=1}^{i=9}\phi_i^{s_i}(1-\phi_i)^{r_i}\prod_{i=1}^{i=9}\theta_i^{m_i}, \tag{8.43}$$

and all inferences about the ROC area of the two readers are based on the likelihood and the prior distribution. If an improper prior distribution is appropriate,

$$\phi_i \sim \text{beta}(s_i, r_i) \tag{8.44}$$

and θ_i follow a Dirichlet with parameter m_i. On the other hand, if a uniform prior distribution is used,

$$\phi_i \sim \text{beta}(s_i + 1, r_i + 1) \tag{8.45}$$

and θ_i follow a Dirichlet with parameter $m_1 + 1$.

TABLE 8.13: Staging melanoma by a dermatologist and a surgeon.

$Y_1 =$	1			2			3		
Y_2	1	2	3	1	2	3	1	2	3
$D=1$	$s_1 = 8$	$s_2 = 26$	$s_3 = 51$	$s_4 = 43$	$s_5 = 81$	$s_6 = 94$	$s_7 = 117$	$s_8 = 140$	$s_9 = 208$
$D=0$	$r_1 = 101$	$r_2 = 105$	$r_3 = 83$	$r_4 = 67$	$r_5 = 72$	$r_6 = 40$	$r_7 = 41$	$r_8 = 30$	$r_9 = 4$
$V=0$	$u_1 = 2$	$u_2 = 18$	$u_3 = 62$	$u_4 = 14$	$u_5 = 83$	$u_6 = 67$	$u_7 = 63$	$u_8 = 40$	$u_9 = 108$
Total	$m_1 = 111$	$m_2 = 149$	$m_3 = 196$	$m_4 = 124$	$m_5 = 236$	$m_6 = 201$	$m_7 = 221$	$m_8 = 210$	$m_9 = 320$

TABLE 8.14: Posterior analysis for melanoma staging—two readers.

Parameters	Mean	sd	Lower 2 1/2	Median	Upper 2 1/2
A_1 (surgeon)	0.7867	0.01192	0.763	0.7869	0.8095
A_{11}	0.6656	0.01621	0.6335	0.6658	0.6967
A_{12}	0.1211	0.004355	0.1126	0.1211	0.1296
A_2 (dermatologist)	0.6351	0.01452	0.6061	0.6352	0.6631
A_{21}	0.4762	0.01706	0.440	0.4763	0.5093
A_{22}	0.1589	0.002748	0.1534	0.1589	0.1641
d	0.1517	0.01882	0.1144	0.1517	0.1883

Based on the staging data of Table 8.12, an analysis is performed assuming an improper prior distribution. Forty-five thousand observations were generated from the joint posterior distribution. The MCMC error of estimation was <0.0001 for all seven parameters shown in Table 8.14. The d parameter is the difference in the two areas.

The ROC area based on the surgeon is 0.7867 compared to an area of 0.6351 for the dermatologist, and the 95% credible interval for the difference is (0.1144,0.1883), indicating that there is a real difference in the two assessments of staging accuracy. See the exercises for additional information about the melanoma study. It should be noted that the staging for melanoma is much more complex than presented here, and for additional information refer to the *AJCC Cancer Staging Handbook* [11]. The analysis is performed using BUGS CODE 8.4.

BUGS CODE 8.4

```
model;
# hypothetical data set
# two tests for staging melanoma
# one rater is a surgeon the other a dermatologist
# ratings are: stage 1, stage 2, stage 3
# similar to Zhou page 347 on CT and MRI
{
for (i in 1:9) {ph[i]~dbeta(s[i],r[i])}
for (i in 1:9) {g[i]~dgamma(m[i],2)}
ms<-sum(g[])
for (i in 1:9) {theta[i]<-g[i]/ms }
theta1.<- theta[1]+theta[2]+theta[3]
theta2.<- theta[4]+theta[5]+theta[6]
theta3.<- theta[7]+theta[8]+theta[9]
theta.1<- theta[1]+theta[4]+theta[7]
theta.2<- theta[2]+theta[5]+theta[8]
theta.3<- theta[3]+theta[6]+theta[9]
```

```
s1.<-s[1]+s[2]+s[3]
s2.<-s[4]+s[5]+s[6]
s3.<-s[7]+s[8]+s[9]
s.1<-s[1]+s[4]+s[7]
s.2<-s[2]+s[5]+s[8]
s.3<-s[3]+s[6]+s[9]
r1.<-r[1]+r[2]+r[3]
r2.<-r[4]+r[5]+r[6]
r3.<-r[7]+r[8]+r[9]
r.1<-r[1]+r[4]+r[7]
r.2<-r[2]+r[5]+r[8]
r.3<-r[3]+r[6]+r[9]
# the prob D=1 given Y1=1
ph1.~dbeta(s1.,r1.)
ph2.~dbeta(s2.,r2.)
ph3.~dbeta(s3.,r3.)
ph.1~dbeta(s.1,r.1)
ph.2~dbeta(s.2,r.2)
ph.3~dbeta(s.3,r.3)
# the prob the first test =1 given d=1
alpha1[1]<- ph1.*theta1./dalpha1
alpha1[2]<- ph2.*theta2./dalpha1
alpha1[3]<- ph3.*theta3./dalpha1
dalpha1<- ph1.*theta1.+ph2.*theta2.+ph3.*theta3.
# the prob the first test =1 given D=0
beta1[1]<-((1-ph1.)*theta1.)/dbeta1
beta1[2]<-((1-ph2.)*theta2.)/dbeta1
beta1[3]<-((1-ph3.)*theta3.)/dbeta1
dbeta1<-(1-ph1.)*theta1.+(1-ph2.)*theta2.+(1-ph3.)*theta3.
# the prob that the second test =1 given d=1
alpha2[1]<- ph.1*theta.1/dalpha2
alpha2[2]<- ph.2*theta.2/dalpha2
alpha2[3]<- ph.3*theta.3/dalpha2
dalpha2<- ph.1*theta.1+ph.2*theta.2+ph.3*theta.3
beta2[1]<-((1-ph.1)*theta.1)/dbeta2
beta2[2]<-((1-ph.2)*theta.2)/dbeta2
beta2[3]<-((1-ph.3)*theta.3)/dbeta2
dbeta2<-(1-ph.1)*theta.1+(1-ph.2)*theta.2+(1-ph.3)*theta.3
# area of test 1
  A1<- A11+A12
  A11<- alpha1[2]*beta1[1]+alpha1[3]*( beta1[1]+beta1[2])
  A12<- (alpha1[1]*beta1[1]+alpha1[2]*beta1[2]+alpha1[3]*beta1[3])/2
# area of test 2
  A2<- A21+A22
  A21<- alpha2[2]*beta2[1]+alpha2[3]*( beta2[1]+beta2[2])
```

A22<- (alpha2[1]*beta2[1]+alpha2[2]*beta2[2]+alpha2[3]*beta2[3])/2
d<-A1-A2
    }
# melanoma staging two readers
list(r = c(101,105,83,67,72,40,41,30,4), s = c(8,26,51,43,81,94,117,140,208), m =
c(111,149,196,124,236,201,221,210,320))

8.6   Two Ordinal Tests and Covariates

Consider the melanoma staging study reported in Table 8.13. Suppose the gender of the subjects is taken into account in order to estimate the ROC area. Would one expect gender to make a difference? The approach here is to estimate the ROC areas separately, then combine the estimates. First consider the results for males given in Table 8.15a.

There are four ROC areas to compute, two for the dermatologist and two for the surgeon. The Bayesian analysis is performed using BUGS CODE 8.4 with 45,000 observations generated from the joint posterior distribution of the 18 parameters appearing in the likelihood function, using a burn in of 5,000 and a refresh of 100. An improper prior distribution is employed (Table 8.16).

The estimated area for the surgeon using the posterior mean is 0.8045 with a 95% credible interval of (0.7715,0.8348), while that for the dermatologist is 0.6401, and it appears that the surgeon has more accuracy in staging melanoma patients. Bayesian inferences for females are left as an exercise. Notice that this study is paired by readers, but not by gender.

Refer to Exercise 8 for additional information about comparing the ROC areas for two correlated ordinal tests.

8.7   Inverse Probability Weighting

Pepe [8] describes an interesting variation in estimating test accuracy with verification bias by the IPW technique. Briefly, this method involves constructing an imputed data table from the observed data table. The latter is the actual table observed that has verification bias. Consider the selected data in Table 8.17a.

For the selected data, all subjects were verified when they tested positive, and for those 860 that tested negative, 95/917 were verified for disease status. The inverse probability approach is to multiply each cell frequency by the inverse of the verification rate.

TABLE 8.15a: Staging melanoma by a dermatologist and surgeon—males.

$Y_1 =$	1			2			3		
Y_2	1	2	3	1	2	3	1	2	3
$D=1$	$s_1 = 4$	$s_2 = 12$	$s_3 = 23$	$s_4 = 21$	$s_5 = 45$	$s_6 = 51$	$s_7 = 56$	$s_8 = 78$	$s_9 = 100$
$D=0$	$r_1 = 56$	$r_2 = 51$	$r_3 = 40$	$r_4 = 33$	$s_5 = 35$	$r_6 = 18$	$r_7 = 16$	$r_8 = 13$	$r_9 = 2$
$V=0$	$u_1 = 0$	$u_2 = 6$	$u_3 = 31$	$u_4 = 8$	$u_5 = 40$	$u_6 = 31$	$u_7 = 28$	$u_8 = 22$	$u_9 = 55$
Total	$m_1 = 60$	$m_2 = 69$	$m_3 = 94$	$m_4 = 62$	$m_5 = 120$	$m_6 = 100$	$m_7 = 100$	$m_8 = 113$	$m_9 = 157$

TABLE 8.15b: Staging melanoma by a dermatologist and a surgeon—females.

$Y_1 =$	1			2			3		
Y_2	1	2	3	1	2	3	1	2	3
$D=1$	$s_1 = 4$	$s_2 = 14$	$s_3 = 28$	$s_4 = 22$	$s_5 = 36$	$s_6 = 43$	$s_7 = 61$	$s_8 = 62$	$s_9 = 108$
$D=0$	$r_1 = 45$	$r_2 = 54$	$r_3 = 43$	$r_4 = 34$	$r_5 = 37$	$r_6 = 022$	$r_7 = 25$	$r_8 = 17$	$r_9 = 2$
$V=0$	$u_1 = 2$	$u_2 = 12$	$u_3 = 31$	$u_4 = 6$	$u_5 = 43$	$u_6 = 36$	$u_7 = 35$	$u_8 = 018$	$u_9 = 53$
Total	$m_1 = 51$	$m_2 = 80$	$m_3 = 102$	$m_4 = 62$	$m_5 = 116$	$m_6 = 101$	$m_7 = 121$	$m_8 = 97$	$m_9 = 163$

TABLE 8.16: Posterior analysis for melanoma staging study—staging by a surgeon and a dermatologist for males.

Parameter	Mean	sd	Lower 2 1/2	Median	Upper 2 1/2
A_1 (surgeon)	0.8045	0.0162	0.7715	0.805	0.8348
A_2 (dermatologist)	0.6401	0.02089	0.5986	0.6403	0.6810

The test accuracy, namely, the true and false positive fractions, is estimated in the usual way. The true positive rate is estimated as $280/348 = 0.804$, and the false positive rate as $190/1039 = 0.1828$. Of course, the sensitivity and false positive rate can also be estimated by using the methods described earlier and applied to Table 8.17a. Will the Bayesian approach give the same results when applied to both tables?

First, consider the selected data from Table 8.17a. Then using an improper prior with BUGS CODE 8.1, a burn in of 5,000, a refresh of 100, and generating 25,000 observations from the posterior distribution of the parameters, gives the results shown in Table 8.18.

Additional information about the IPW approach is continued with Exercise 9, but for now the presentation is continued by considering diagnostic tests with verification bias and ordinal outcomes. Recall the mammography study given in Table 8.10, where the results are given in Table 8.7. The inverse probability method is applied by multiplying the cell probabilities by the appropriate inverse verification rate, yielding Table 8.19.

TABLE 8.17a: Selected data with verification bias.

$Y =$	1	0
$V = 1$		
$D = 1$	280	7
$D = 0$	190	88
$V = 0$	0	822
Total	470	917

TABLE 8.17b: Imputed data from Table 8.16a.

$Y =$	1	0
$V = 1$		
$D = 1$	280	68
$D = 0$	190	849

TABLE 8.18: Bayesian analysis based on selected data of Table 8.16a.

Parameter	Mean	sd	Lower 2 1/2	Median	Upper 2 1/2
fpf	0.183	0.0127	0.1589	0.1827	0.2093
tpf	0.8091	0.0562	0.6909	0.8121	0.9101

TABLE 8.19: Imputed table for mammography study.

$Y =$	1	2	3	4	5
$V = 1$					
$D = 1$	$s_1 = 89$	$s_2 = 74$	$s_3 = 167$	$s_4 = 153$	$s_5 = 249$
$D = 0$	$r_1 = 383$	$r_2 = 173$	$r_3 = 108$	$r_4 = 35$	$r_5 = 78$
Total	$m_1 = 472$	$m_2 = 247$	$m_3 = 275$	$m_4 = 188$	$m_5 = 327$

If an improper prior distribution is used, the posterior analysis based on Table 8.18 and BUGS CODE 8.3, with 35,000 observations generated from the posterior distribution, with a burn in of 5,000 and a refresh of 100, the analysis gives essentially the same results for the ROC area as reported in Table 8.8. Note that BUGS CODE 8.3 can be used for the imputed Table 8.19 or the selected Table 8.7 with verification bias. Refer to Exercises 10 and 11 for additional information.

One would expect the IPW method would also apply for the case of two correlated ordinal tests. See Exercises 12 and 13 for additional information.

8.8 Without the Missing at Random Assumption

The MAR assumption assumes that the probability of referring a subject to the gold standard only depends on the results Y of the test. If the referral depends on the disease status of the patient, the MAR assumption is violated and another approach is taken. Consider the case of one binary test, as shown in Table 8.20.

The MAR assumption holds if

$$P[V = 1 \mid D, Y] = P[V = 1 \mid Y],$$

and if this is not possible, the probability of validation has to depend on the disease status. One way to do this is to base inferences on the probability model to be defined as

$$\theta_i = P[Y = i] \tag{8.46}$$

TABLE 8.20:   One binary test.

Y	1 (Positive)	0 (Negative)
V = 1		
$D = 1$	s_1	s_0
$D = 0$	r_1	r_0
V = 0	u_1	u_0
Total	m_1	m_0

and

$$\phi_i = P[D = 1 \mid Y = i], \tag{8.47}$$

for $i = 0, 1$.

The probability of validation is made to depend on the disease status by defining

$$\lambda_{ij} = P[V = 1 \mid D = i, Y = j], \tag{8.48}$$

for $i, j = 0, 1$.

Consequently,

$$
\begin{aligned}
P[V = 0 \mid Y = 0] &= P[V = 0 \mid D = 0, Y = 0]P[D = 0 \mid Y = 0] \\
&\quad + P[V = 0 \mid D = 1, Y = 0]P[D = 1 \mid Y = 0] \\
&= (1 - \lambda_{10})\phi_0 + (1 - \lambda_{00})(1 - \phi_0),
\end{aligned} \tag{8.49}
$$

and in a similar way,

$$P[V = 0 \mid Y = 1] = (1 - \lambda_{11})\phi_1 + (1 - \lambda_{01})(1 - \phi_1). \tag{8.50}$$

The likelihood function is based on the marginal distribution of Y, the conditional distribution of V given D and Y, and the conditional distribution of D given Y. Referring to Table 8.20 and the definition of the parameters above, gives

$$
\begin{aligned}
L(\theta, \phi, \lambda) &\propto \theta_0^{m_0}(1 - \theta_0)^{m_1}(\lambda_{11}\phi_1)^{s_1}(\lambda_{10}\phi_0)^{s_0}\lambda_{01}^{r_1}(1 - \phi_1)^{r_1}, \\
&\quad \lambda_{00}^{r_0}(1 - \phi_0)^{r_0}[(1 - \lambda_{10})\phi_0 + (1 - \lambda_{00})(1 - \phi_0)]^{u_0}, \\
&\quad [(1 - \lambda_{11})\phi_1 + (1 - \lambda_{01})(1 - \phi_1)]^{u_1},
\end{aligned} \tag{8.51}
$$

as the likelihood function, which depends on seven unknown parameters.

$$\propto \theta_0^{m_0}(1 - \theta_0)^{m_1},$$

$$\sum_{i=0}^{i=u_0} B(u_0, i)\lambda_{10}^{s_0}(1 - \lambda_{10})^i\lambda_{00}^{r_0}(1 - \lambda_{00})^{u_0-i}\phi_0^{s_0+i}(1 - \phi_0)^{r_0+u_0-i}, \tag{8.52}$$

$$\sum_{i=0}^{i=u_0} B(u_1, i)\lambda_{11}^{s_1}(1 - \lambda_{11})^i\lambda_{01}^{r_1}(1 - \lambda_{01})^{u_1-i}\phi_1^{s_1+i}(1 - \phi_1)^{r_1+u_1-i},$$

where $B(u_0, i)$ is the binomial coefficient "i from u_0." The Bayesian approach requires a prior distribution for the parameters, and if a uniform is used, the effect is to make the posterior density of all the parameters proportional to the likelihood function (Equation 8.52). Note that the posterior distribution has three components: (a) the marginal posterior distribution of θ_0; (b) the joint marginal posterior distribution of λ_{00}, λ_{10}, and ϕ_0; and (c) the joint marginal posterior distribution of λ_{01}, λ_{11}, and ϕ_1. Also note that λ_{00} and λ_{10} can be eliminated (using the properties of the beta distribution) from the second component, thus leaving the posterior distribution of ϕ_0 expressed as a mixture of beta distributions, where the ith component has a beta distribution with parameters $(s_0 - i + 1, r_0 + u_0 + 1 - i)$. In a similar fashion, eliminating λ_{01} from the third component leaves a mixture of beta distributions for the posterior distribution of ϕ_1, where the ith component has a beta distribution with parameters $(s_1 - i + 1, r_1 + u_1 + 1 - i)$.

What are the weights for the mixture of the posterior distribution of ϕ_0?

Consider the second component of Equation 8.52 and integrate with respect to λ_{00}, λ_{10}, and ϕ_0, then the result is proportional to the ith weight of the mixture, namely,

$$
\begin{aligned}
w_i' = BC(u_0, i)\Gamma(s_0 + 1)\Gamma(i + 1)\Gamma(r_0 + 1)\Gamma(u_0 + 1 - i)\Gamma(s_0 + 1 + i) \\
\times \Gamma(r_0 + u_0 + 1 - i),
\end{aligned} \tag{8.53}
$$

$$
\Gamma(s_0 + i + 2)\Gamma(r_0 + u_0 + 2 - i)\Gamma(m_0 + 2 - i).
$$

The ith component of the posterior distribution of ϕ_0 is

$$
w_i = w_i' \Big/ \sum_{i=0}^{i=u_0} w_i', \tag{8.54}
$$

for $i = 0, 1, \dots, u_0$, and a similar expression holds for the posterior distribution of ϕ_1, namely, in Equation 8.53 replace subscript 0 with subscript 1. The code below follows the notation given above.

BUGS CODE 8.5

```
model;
# w/o MAR assumption
# one binary test
# uniform prior for parameters
{
theta0~dbeta(23,18)
theta1<-1-theta0
# for lamda00
for( i in 1:9){x00[i]~dbeta(5,s00[i])}
# for ph1
for( i in 1:6){ y1[i]~dbeta(s1[i],r1[i])}
# for lamda10
```

```
for( i in 1:9){ x10[i]~dbeta(11,r10[i])}
# for ph0
for( i in 1:9){ y0[i]~dbeta(s0[i],r0[i])}
# mixture for lamda00
for ( i in 1:9){ lam00[i]<-w0[i]*x00[i]}
#for mixture for lamda10
for ( i in 1:9){ lam10[i]<-w0[i]*x10[i]}
# dist of lamda00
lamda00<-sum(lam00[])
# distribution for lamda10
lamda10<-sum(lam10[])
# mixture for ph1
for ( i in 1:6){ z1[i]<-w1[i]*y1[i]}
# dist of ph1
ph1<-sum(z1[])
# mixture for ph0
for ( i in 1:9){ z0[i]<-w0[i]*y0[i]}
# dist of ph0
ph0<-sum(z0[])
tpf<-ph1*theta1/(ph1*theta1+ph0*theta0)
fpf<-(1-ph1)*theta1/(1-ph1*theta1-ph0*theta0)
# d is the difference between lamda00 and lamda10
# d0010 =0 implies MAR assumption is valid
d0010<-lamda00-lamda10
}
list( s0=c(11,12,13,14,15,16,17,18,19),
      r0=c(13,12,11,10,9,8,7,6,5),
      s00=c(9,8,7,6,5,4,3,2,1),
      r10=c(1,2,3,4,5,6,7,8,9),
      r1=c(9,8,7,6,5,4),
      s1=c(10,11,12,13,14,15),
w0=c(.0000000000317598,
.0000000007254027,
.0000000160704596,
.0000003447113582,
.0000071495689118,
.0001432706582719,
.0027739125769611,
.0519594920665954,
.9451158135902790),
w1=c(.0000006087689438,
     .0000112068828285,
     .0001995892465651,
     .0034390762485069,
     .0574817030107593,
     .938867815))
```

TABLE 8.21: One binary test.

	Y	1 (Positive)	0 (Negative)
V = 1			
	$D = 1$	$s_1 = 9$	$s_0 = 10$
	$D = 0$	$r_1 = 3$	$r_0 = 4$
$V = 0$		$u_1 = 5$	$u_0 = 8$
Total		$m_1 = 17$	$m_0 = 22$

As an example of not assuming the MAR, consider the outcomes for a binary test (Table 8.21).

The posterior analysis was performed with a burn in of 5,000, a refresh of 100, and 50,000 observations are generated for the posterior distribution of the parameters (Table 8.22).

Recall that ϕ_1 is the probability of disease given the test is positive, while θ_0 is the probability the test is negative. The weights can be difficult to compute because of the large numbers of the gamma function. I used the log gamma function on w_i' in Equation 8.53, followed by exponentiation, then normalized w_i', giving the ith weight of the mixture as w_i in Equation 8.54. The results give the characteristics of the posterior distribution of the relevant parameters, including λ_{00} and λ_{10}. If these two are equal, there is a possibility that the MAR assumption holds. The parameter d0010 is the posterior distribution of the difference, and the 95% credible interval is $(-0.1073, 0.5604)$, which indicates the possibility that the MAR is valid; however, the other two lambda parameters, λ_{01} and λ_{11}, must also be equal for the MAR assumption to hold. The parameters λ_{00} and λ_{10} are part of the second component of the marginal joint distribution of λ_{00}, λ_{10}, and ϕ_0, while λ_{01} and λ_{11} are part of the third component, the marginal joint posterior distribution of λ_{01}, λ_{11}, and ϕ_1.

Plots of the posterior densities of the false and true positive fractions are given in Figures 8.3 and 8.4, respectively. Note the larger standard deviation for the false positive rate, which is also confirmed by Table 8.20. The densities

TABLE 8.22: The posterior analysis without MAR one binary test.

Parameter	Mean	sd	Lower 2 1/2	Median	Upper 2 1/2
fpf	0.4427	0.1489	0.1733	0.4381	0.7402
tpf	0.4381	0.0848	0.2756	0.4372	0.607
ϕ_0	0.7891	0.0772	0.6186	0.7969	0.916
ϕ_1	0.7863	0.08571	0.5946	0.796	0.9242
θ_0	0.5609	0.0766	0.4086	0.5617	0.708
λ_{00}	0.8266	0.1333	0.4908	0.8614	0.9815
λ_{10}	0.5517	0.103	0.3477	0.5531	0.7450
d0010	0.2749	0.1684	−0.1073	0.2919	0.5604

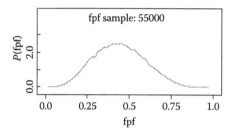

FIGURE 8.3: Posterior density of the false positive fractions.

of λ_{00} and λ_{10} appear as Figures 8.5 and 8.6, respectively, and the asymmetry of the former is evident.

How is the above analysis extended to two correlated binary tests? Consider Table 8.23. The Bayesian approach is based on the likelihood function, which is

$$L(\theta,\phi,\lambda) \propto \prod_{i=0}^{i=1}\prod_{j=0}^{j=1}\theta_{ij}^{m_{ij}} \prod_{i=0}^{i=1}\prod_{j=0}^{j=1}\phi_{ij}^{s_{ij}}(1-\phi_{ij})^{r_{ij}} \prod_{i=0}^{i=1}\prod_{j=0}^{j=1}\lambda_{1ij}^{s_{ij}} \prod_{i=0}^{i=1}\prod_{j=0}^{j=1}\lambda_{0ij}^{r_{ij}}, \quad (8.55)$$

$$[(1-\lambda_{111})\phi_{11}+(1-\lambda_{011})(1-\phi_{11})]^{u_{11}},$$
$$[(1-\lambda_{110})\phi_{10}+(1-\lambda_{010})(1-\phi_{10})]^{u_{10}},$$
$$[(1-\lambda_{101})\phi_{01}+(1-\lambda_{001})(1-\phi_{01})]^{u_{01}},$$
$$[(1-\lambda_{100})\phi_{00}+(1-\lambda_{000})(1-\phi_{00})]^{u_{00}},$$

which is of the same form as the likelihood function for one binary test, without the MAR assumption. Note, the likelihood function is for 16 unknown parameters, and with a uniform prior the posterior distribution is the likelihood function (Equation 8.55). Using the binomial theorem, the last four components of the likelihood function are expressed as a mixture of beta distributed random variables.

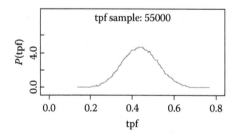

FIGURE 8.4: Posterior density of the true positive fraction.

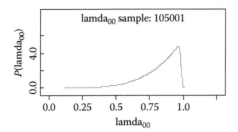

FIGURE 8.5: Posterior density of λ_{00}.

For example, the second component of the likelihood function is expressed as

$$\sum_{i=0}^{i=u_{11}} BC(u_{11}, i)(1 - \lambda_{111})^i \phi_{11}^i (1 - \lambda_{011})^{u_{11}-i}(1 - \phi_{11})^{u_{11}-i},$$

and when combined with λ_{111}, λ_{011}, and ϕ_{11} of the first component, gives

$$\sum_{i=0}^{i=u_{11}} BC(u_{11}, i)\lambda_{111}^{s_{11}}(1 - \lambda_{111})^i \lambda_{011}^{r_{11}}(1 - \lambda_{011})^{u_{11}-i}\phi_{11}^{s_{11}+i}(1 - \phi_{11})^{r_{11}+u_{11}-i}$$

as a mixture. Integrating this component with respect to λ_{111}, λ_{011}, and ϕ_{11}, expresses the marginal posterior distribution of ϕ_{11} as a mixture of random variables where the ith has a beta distribution with parameter $(s_{11}+1+i, r_{11}+1-i)$, where $i = 0, 1, \ldots, u_{11}$, namely,

$$\sum_{i=0}^{i=u_{11}} BC(u_{11}, i)\{\Gamma(s_{11}+1)\Gamma(i+1)\Gamma(r_{11}+1)\Gamma(u_{11}+1-i)\Gamma(s_{11}+i+1)$$
$$\times \Gamma(r_{11}+u_{11}-i+1)$$
$$\div \Gamma(s_{11}+2+i)\Gamma(r_{11}+u_{11}+2-i)\Gamma(m_{11}+2)\}$$
$$\times \phi_{11}^{s_{11}+i}(1 - \phi_{11})^{r_{11}+u_{11}-i}. \tag{8.56}$$

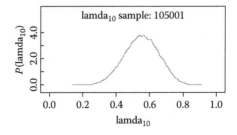

FIGURE 8.6: Posterior density of λ_{10}.

TABLE 8.23: Two binary scores with verification bias.

$Y_1 =$		1		0	
$V = 1$	$Y_2 =$	1	0	1	0
$D = 1$		s_{11}	s_{10}	s_{01}	s_{00}
$D = 0$		r_{11}	r_{10}	r_{01}	r_{00}
$V = 0$		u_{11}	u_{10}	u_{01}	u_{00}
Total		m_{11}	m_{10}	m_{01}	m_{00}

Of course, the last three components can be expressed in a similar fashion as mixtures of beta random variables for the posterior distribution of ϕ_{10}, ϕ_{01}, and ϕ_{00}, respectively. Our main focus is to estimate the true and false positive fractions for the two tests. For example, the tpf for the first test is

$$\text{tpf}_1 = \phi_{1.}\theta_{1.}/(\phi_{1.}\theta_{1.} + \phi_{0.}\theta_{0.}), \tag{8.57}$$

where

$$\varphi_{1.} \sim \text{beta}(s_{1.} + 1, r_{1.} + 1)$$

and

$$\theta_{1.} = \theta_{10} + \theta_{11}.$$
$$\text{fpf}_1 = (1 - \phi_{1.})\theta_{1.}/(1 - \phi_{1.}\theta_{1.} - \phi_{0.}\theta_{0.}). \tag{8.58}$$

8.9   One Ordinal Test and the Receiver Operating Characteristic Area

Consider one ordinal test with verification bias (Table 8.24). Not assuming MAR, what is the likelihood for making Bayesian inferences about the area under the ROC curve?

TABLE 8.24: Verification bias and one ordinal test.

$Y =$	1	2	3	k
$V = 1$				
$D = 1$	s_1	s_2		s_k
$D = 0$	r_1	r_2		r_k
$V = 0$	u_1	u_2		u_k
Total	m_1	m_2		m_k

The relevant parameters are:

$$\theta_i = P[Y = i],$$
$$\phi_i = P[D = 1 \mid Y = i],$$
$$\lambda_{ij} = P[V = 1 \mid D = i, Y = j],$$
$$P[V = 0 \mid Y = i] = (1 - \lambda_{1i})\phi_i + (1 - \lambda_{0i})(1 - \phi_i),$$
$$\alpha_i = P[Y = i \mid D = 1],$$
$$= \phi_i\theta_i \Big/ \sum_{i=1}^{i=k} \phi_i\theta_i,$$

and

$$\beta_i = (1 - \phi_i)\theta_i \Big/ \left(1 - \sum_{i=1}^{i=k} \phi_i\theta_i\right).$$

The area under the ROC curve is

$$A = A_1 + A_2, \tag{8.59}$$

where

$$A_1 = \alpha_2\beta_1 + \alpha_3(\beta_1 + \beta_2) + \cdots + \alpha_k(\beta_1 + \cdots + \beta_{k-1}),$$

and

$$2A_2 = \sum_{i=1}^{i=k} \alpha_i\beta_i..$$

Lastly, the likelihood function is

$$L(\theta, \phi, \lambda) \propto \prod_{i=1}^{i=k} \theta_i^{m_i} \prod_{i=1}^{i=k} \phi_i^{s_i}(1 - \phi_i)^{r_i} \prod_{i=1}^{i=k}[(1 - \lambda_{0i})(1 - \phi_i) + (1 - \lambda_{1i})\phi_i]^{u_i}, \tag{8.60}$$

$$\prod_{i=1}^{i=k} \lambda_{1i}^{s_i}\lambda_{0i}^{r_i},$$

and all inferences, including the area under the ROC curve, are based on it.

If a uniform prior is used for all parameters, the joint posterior distribution is proportional to the above likelihood. Recall that the marginal posterior distribution of ϕ_i is a mixture of beta distributed random variables. Consider then the ith component of the mixture, then $\phi_i \sim \text{beta}(s_i + i + 1, r_i + u_i + 1 - i)$, where $i = 1, 2, \ldots, i$.

8.10   Comments and Conclusions

This chapter presents a Bayesian approach to making inferences about medical test accuracy when verification bias is present, and this approach follows, to some extent, the presentation of Zhou, McClish, and Obuchowski [1], and Pepe [8]. The Zhou method is basically based on the likelihood function of the outcomes for various scenarios. The Bayesian approach uses the likelihood function, but uses a different method to compute the area under the ROC curve. The area under the ROC curve is based on a formula from Broemeling [7: 72], which is the form given also by Pepe [8].

The MAR assumption is used, as was done by Zhou, McClish, and Obuchowski, for one binary test, two binary tests, tests with ordinal outcomes, and two correlated tests with ordinal outcomes. Some variations of these basic designs include taking covariates into account and other aspects of the study design. Zhou, McClish, and Obuchowski [1] relaxed the MAR assumption for one binary test, using maximum likelihood for estimating test accuracy, while the Bayesian approach is extended to two binary tests, two correlated ordinal tests, where the posterior distribution of the ROC area is determined. I believe the latter Bayesian presentations without the MAR assumption have not been presented.

The following exercises review the basic concepts presented and extend those concepts to other experimental scenarios.

8.11   Exercises

1. Verify the analysis given by Table 8.3 for the posterior distribution of the tpf and fpf of the diabetes study of Table 8.2. Repeat the analysis using a uniform prior for the parameters, instead of an improper prior, with WinBUGS generating 55,000 observations from the posterior distribution, with a burn in of 5,000 observations and a refresh of 100. Plot the posterior densities of the tpf and the fpf. Report the posterior mean, standard deviation, median, and the lower and upper 2 1/2 percentiles, and display the trace of the various posterior distributions. Refer to BUGS CODE 8.1. In addition, report the MCMC error for all parameters.

2. For the two tests for detecting metastasis of colon cancer to the liver, reported in Table 8.5, perform a Bayesian analysis with an improper prior, as was done in Section 8.3, and compute the posterior distribution for the true and false positive fractions of both observers. Also, compare the true and false positive fractions of the two images. How do the two agree in their diagnosis of the disease? Refer to BUGS CODE 8.2, and

TABLE 8.25: CT study uterine cancer metastasis.

$Y =$	1	2	3	4	5
$V = 1$					
$D = 1$	$s_1 = 9$	$s_2 = 8$	$s_3 = 7$	$s_4 = 8$	$s_5 = 45$
$D = 0$	$r_1 = 12$	$r_2 = 2$	$r_3 = 4$	$r_4 = 5$	$r_5 = 6$
$V = 0$	$u_1 = 38$	$u_2 = 9$	$u_3 = 6$	$u_4 = 4$	$u_5 = 14$
Total	$m_1 = 59$	$m_2 = 19$	$m_3 = 17$	$m_4 = 17$	$m_5 = 65$

note that the code for all required computations for the second observer appears in the program statements.

3. Table 8.25 presents the results of a study using CT to study the metastasis of uterine cancer to the lymph nodes, where 1 indicates definitely no evidence of metastasis, 2 signifies very little evidence, 3 indicates very little evidence one way or the other, 4 is interpreted as some evidence of metastasis, while 5 indicates very strong evidence of metastasis. Note, a similar study was conducted by Gray, Begg, and Greenes [13], which is also reported and analyzed by Zhou, McClish, and Obuchowski [1].

Suppose the prior distributions are taken as a uniform over the parameters ϕ_i and θ_i for $i = 1, 2, 3, 4, 5$, where $\phi_i = P[D = 1 \mid Y = i]$ and $\theta_i = P[Y = i]$. Use the likelihood function as expressed by Equation 8.23 and combine with the uniform prior for the posterior analysis.

(a) What is the posterior distribution of ϕ_i for $i = 1, 2, 3, 4, 5$?

(b) What is the posterior distribution of θ_i for $i = 1, 2, 3, 4, 5$?

(c) What is the posterior distribution of $P[Y = i \mid D = 1]$ for $i = 1, 2, 3, 4, 5$?

(d) What is the posterior distribution of $P[Y = i \mid D = 0]$ for $i = 1, 2, 3, 4, 5$?

(e) What is the posterior distribution of the area under the ROC curve? Perform the analysis using WinBUGS with 45,000 observations, a burn in of 5,000 and a refresh of 100. Report the posterior mean, standard deviation, median, and the lower and upper 2 1/2 percentiles. Follow the analysis of Section 8.4, but with a uniform prior distribution. Refer to BUGS CODE 8.3. What are the MCMC errors for all parameters?

4. Consider Tables 8.9 and 8.11, which give the posterior analysis for estimating the ROC area for sites 1 and 2, respectively. How should the two estimates (using the posterior mean) of the two areas be combined? Consider weighting inversely proportional to the variance of the posterior distribution of the ROC area. What are the additional ways of combining the posterior means of the ROC areas?

5. Refer to Tables 8.13 and 8.14 and the melanoma staging study that compares a surgeon and a dermatologist with regard to the area under the ROC. Repeat the analysis, but use a uniform prior distribution for

the parameters. The prior is to be combined with the likelihood function (Equation 8.43). Hint: modify the list statement of BUGS CODE 8.4 using the information given by Equation 8.45!

6. The results in Table 8.26 report an experimental study for staging breast cancer, which involves MRI and mammography and involves staging 285 patients by two paired modalities, MRI and mammography. The first test, Y_1, is based on MRI, while the other, Y_2, corresponds to mammography, and the scores are 1, 2, and 3, which indicate the stage of the disease. Stage 1 indicates disease is confined to the primary tumor, while 2 indicates ambiguous information about metastasis, and 3 signals definite evidence of metastasis to lymph nodes. Megibow et al. [12] is a similar study.

I recommend using a uniform prior distribution for all the parameters and BUGS CODE 8.4 to perform a Bayesian analysis with a burn in of 5,000, a refresh of 100, and generate sufficient observations from the posterior distribution until the MCMC errors are <0.001 for all parameters.

   (a) Find the posterior characteristics of the ROC areas of the two modalities.

   (b) Test the hypothesis that the two modalities have the same area under the ROC curve.

   (c) Plot the densities of the posterior distributions of the two images.

   (d) Why do small cell frequencies cause problems for the analysis?

7. What is the MAR assumption for two ordinal tests with verification bias?

8. Refer to Table 8.13 for the melanoma staging study, where a dermatologist and a surgeon assign a stage to each patient. Tables 8.15a and b stratify the patients by gender, with Table 8.15a reporting the results for males. A Bayesian analysis, shown in Table 8.16, provides estimates of the ROC area for the two readers for the males.

   (a) Perform the analysis for the female results, reported in Table 8.15b, using BUGS CODE 8.4 with a burn in of 5,000 observations, a refresh of 100, and generating enough observations from the posterior distribution so that the MCMC error for all parameters is <0.0001. Assume a uniform prior distribution.

   (b) For females, compare the ROC area of the two readers. Do they differ?

   (c) Does gender have an effect on the ROC area for the dermatologist?

   (d) Does gender have an effect on the ROC area for the surgeon?

   (e) Combine the estimated ROC areas for males and females for the surgeon, and explain how they were combined and what principle is used.

   (f) Combine the estimated ROC areas for males and females for the dermatologist and explain how they were combined and what principle is used.

TABLE 8.26: Mammography and MRI for staging pancreatic cancer.

$Y_1 =$	1			2			3		
Y_2	1	2	3	1	2	3	1	2	3
$D=1$	$s_1 = 21$	$s_2 = 10$	$s_3 = 19$	$s_4 = 8$	$s_5 = 17$	$s_6 = 19$	$s_7 = 14$	$s_8 = 10$	$s_9 = 21$
$D=0$	$r_1 = 9$	$r_2 = 11$	$r_3 = 15$	$r_4 = 11$	$r_5 = 11$	$r_6 = 13$	$r_7 = 9$	$r_8 = 8$	$r_9 = 3$
$V=0$	$u_1 = 4$	$u_2 = 0$	$u_3 = 7$	$u_4 = 5$	$u_5 = 6$	$u_6 = 8$	$u_7 = 0$	$u_8 = 4$	$u_9 = 22$
Total	$m_1 = 34$	$m_2 = 21$	$m_3 = 41$	$m_4 = 24$	$m_5 = 34$	$m_6 = 40$	$m_7 = 23$	$m_8 = 22$	$m_9 = 46$

9. Refer to the IPW method for correcting for verification bias.

    (a) Verify Table 8.18, the Bayesian analysis from the selected Table 8.17a. Use an improper prior and generate 25,000 observations from the posterior distribution, with a burn in of 5,000 and a refresh of 100. Use BUGS CODE 8.1 and note that it is valid for data with or without verification bias.

    (b) Do the same analysis as part (a), but use information from the imputed data given by Table 8.17b. Show, computationally, that the results are essentially the same for the posterior distribution of the true and false positive fractions.

    (c) Prove theoretically that using either table—the imputed data table or the selected data (showing verification bias) table—and using the same prior, the posterior distribution is the same for the true and false positive fractions.

    (d) Pepe [8] reports that the maximum likelihood estimation of Begg and Greenes [6] are the same as those given by the inverse probability method. Show this is true.

    (e) What are the MCMC errors for the parameters?

10. Refer to the imputed Table 8.19 and show that the Bayesian analysis is essentially the same based on either the imputed table or the original selected table with verification bias. Use an improper prior distribution and BUGS CODE 8.3 with 35,000 observations generated from the posterior distribution, with a refresh of 100 and a burn in of 5,000.

11. When estimating the ROC area, prove theoretically that the Bayesian analysis gives the same result when based on either the imputed table or the selected table with verification bias. Assume the same prior is used in both cases. Refer to Section 8.4.

12. Prove that the Bayesian approach gives the same result for estimating two ROC areas for two correlated ordinal tests using either the imputed table or the original selected table with verification bias. Assume the same prior distribution for both.

13. Refer to Table 8.21, the results for one binary test. BUGS CODE 8.5 is implemented to perform the analysis without the MAR assumption. Perform an analysis assuming MAR using BUGS CODE 8.1 and compare your results to those of Table 8.22, the posterior analysis without the MAR condition.

    (a) Why do the results differ?

    (b) The analysis without the MAR assumption is more complex. Describe why.

14. In the case of two correlated binary tests, do not assume MAR and find the posterior distribution of ϕ_{10} based on the likelihood function, Equation 8.55. Refer to Equation 8.56, which expresses the marginal posterior

TABLE 8.27: CT and lung cancer risk.

$Y_1 =$		1		0	
$V = 1$ $Y_2 =$	1	0	1	0
$D = 1$	$s_{11} = 14$	$s_{10} = 12$	$s_{01} = 9$	$s_{00} = 3$
$D = 0$	$r_{11} = 4$	$r_{10} = 9$	$r_{01} = 13$	$r_{00} = 18$
$V = 0$	$u_{11} = 7$	$u_{10} = 8$	$u_{01} = 9$	$u_{00} = 10$
Total	$m_{11} = 25$	$m_{10} = 29$	$m_{01} = 31$	$m_{00} = 31$

distribution of ϕ_{11} as a mixture of beta random variables. The marginal posterior distribution of ϕ_{10} should be similar. Assume a joint uniform prior distribution for the parameters.

15. Refer to two correlated binary tests, without the MAR assumption determine the marginal posterior distribution of the true and false positive fractions for test 2, assuming a uniform prior for the parameters. Begin with the likelihood function Equation 8.55. Your result should be similar to Equations 8.57 and 8.58.

16. Consider the results of two correlated binary tests given in Table 8.27. The first test, Y_1, gives the results for a CT determination of lung cancer risk, where 0 indicates a small risk and 1 a high risk of lung cancer, while the second test, Y_2, is a determination of lung cancer risk using MRI.

    Write your own WinBUGS code and execute the analysis with 45,000 observations, with a burn in of 5,000 and a refresh of 100.
    (a) Assume a uniform prior distribution for the parameters and determine the posterior distribution of the true and false positive fractions for CT.
    (b) Find the posterior distribution of the true and false positive rates for MRI.
    (c) Plot the posterior densities of the true and false positive rates for CT and MRI.
    (d) Are CT and MRI correlated? How much?
    (e) Which test is more accurate?

17. Consider the case of one ordinal test, where mammography is used to screen for breast cancer. The results of the study are given in Table 8.28.

    Based on the likelihood function (Equation 8.60) and not assuming MAR and using a uniform prior for the parameters:
    (a) Write a WinBUGS program to find the area under the ROC curve.
    (b) Execute the program with 55,000 observations generated from the posterior distribution, with a burn in of 5,000 and refresh of 100. Determine the posterior distribution of the area under the ROC curve. What is the posterior mean, median, and standard deviation? What is a 95% credible interval for the area?

TABLE 8.28:   Verification bias and mammography.

$Y =$	1	2	3	4	5
$V = 1$					
$D = 1$	$s_1 = 7$	$s_2 = 23$	$s_3 = 21$	$s_4 = 44$	$s_5 = 78$
$D = 0$	$r_1 = 15$	$r_2 = 42$	$r_3 = 38$	$r_4 = 58$	$r_5 = 145$
$V = 0$	$u_1 = 67$	$u_2 = 18$	$u_3 = 43$	$u_4 = 21$	$u_5 = 10$
Total	$m_1 = 89$	$m_2 = 83$	$m_3 = 102$	$m_4 = 123$	$m_5 = 233$

(c) Repeat the above, but assuming MAR, using BUGS CODE 8.3. Use the same number of observations generated from the posterior distribution, and the same burn in and refresh as in (b).

(d) Compare the two estimated areas with the mean of the posterior distribution. Is there a difference? Why is there a difference?

18. Refer to the likelihood function (Equation 8.60) for one ordinal test without the MAR assumption. Extend the Bayesian analysis to two correlated ordinal tests.

(a) What is the likelihood function? Give the formula.

(b) What is the area under the ROC curve for the two tests? Refer to Equation 8.59.

19. Refer to the posterior density (Equation 8.52) for the parameters of one binary test without the MAR assumption, the results of one binary test (Table 8.20), and BUGS CODE 8.5. Find the posterior density of λ_{01} and λ_{11} from the third component of the posterior density. That is, eliminate ϕ_1 by integration using properties of the beta distribution and express the posterior distribution of ϕ_1 as a mixture. Referring to BUGS CODE 8.5, it will be obvious how to determine the posterior distribution of ϕ_1 as a mixture. Note, the same weights of the mixture are given by w1 in the list statement of BUGS CODE 8.5. A uniform prior distribution is assumed. Generate 50,000 observations from the joint posterior distribution, with a burn in of 5,000 and a refresh of 100.

(a) Report the characteristics of the posterior distribution of λ_{01}.

(b) Report the characteristics of the posterior distribution of λ_{11}.

(c) Determine the posterior distribution of the difference in the two parameters.

(d) Are they different? If so, why?

(e) Using the results of Table 8.22 and your previous results, is the MAR assumption valid?

References

[1] Zhou, X.H., McClish, D.K., and Obuchowski, N.A. *Statistical Methods for Diagnostic Medicine*. John Wiley, New York, 2002.

[2] Greenes, R. and Begg, C. Assessment of diagnostic technologies: Methodology for unbiased estimation from samples of selected verified patients. *Investigative Radiology*, 20:751, 1985.

[3] Bates, A.S., Margolis, P.A., and Evans, A.T. Verification bias in pediatric studies evaluating diagnostic tests. *Journal of Pediatrics*, 122:585, 1993.

[4] Philbrick, J.T., Horwitz, R.I., and Feinstein, A.R. Methodologic problems of exercise testing for coronary artery disease. *American Journal of Cardiology*, 46:807, 1980.

[5] Reid, M.C., Lachs, M.S., and Feinstein, A.R. Use of methodological standards in diagnostic test research. Getting better but still not good. *Journal of the American Medical Association*, 274: 645, 1995.

[6] Begg, C. and Greenes, R. Assessment of diagnostic tests when disease is subject to verification bias. *Biometrics*, 39:207, 1983.

[7] Broemeling, L.D. *Bayesian Biostatistics and Diagnostic Medicine*. Chapman & Hall/CRC, Boca Raton, 2007.

[8] Pepe, M.S. *The Statistical Evaluation of Medical Tests for Classification and Prediction*. Oxford University Press, Oxford, UK, 2003.

[9] Drum, D. and Christacopoulos, J. Hepatic scintigraphy in clinical decision making. *Journal of Nuclear Medicine*, 13:908, 1969.

[10] Broemeling, L.D. *Bayesian Methods for Measures of Agreement*. Chapman & Hall/CRC, Boca Raton, 2009.

[11] *AJCC Cancer Staging Handbook*. American Cancer Society, American College of Surgeons, Lippincott, Williams & Wilkins Healthcare, 1998.

[12] Megibow, A.J., Zhou, X.H., Rotterdam, H., Francis, I., Zerhouni, E.A., Balfe, D.M., Weinreb, J.C., et al. Pancreatic adenocarcinoma: CT versus MR imaging in the evaluation of resectability report of the Radiology Diagnostic Oncology Group. *Radiology*. 195:327, 1995.

[13] Gray, R., Begg, C., and Greenes, R. Correction of receiver operating characteristic curves when disease verification is subject to selection bias, *Medical Decision Making*, 4:151, 1984.

Chapter 9

Test Accuracy and Medical Practice

9.1 Introduction

The role of test accuracy in various forms of clinical practice is explored in this chapter. Of course, clinical practice involves the use of medical tests to assess the condition of the patients. Medical tests are used in many ways, as the patient soon finds out on entering the health care system, from diagnosing the underlying disease, to keeping track of the disease as the patient undergoes treatment.

One issue facing the practitioner is the choice of a threshold or cutoff value in order to declare that the patient has disease. For example, when undergoing a fasting blood glucose test, how high does the value of the blood glucose have to be in order for the doctor to treat the condition as type 2 diabetes? Another example is testing for coronary artery disease, when the patient complains of chest pain and seeks help from a physician. The doctor might send the subject for an exercise stress test, which involves injecting the patient with a radioactive nucleotide that is designed to target the heart and emit radiation that is detected by gamma cameras positron emission tomography or single photon emission tomography (PET or SPECT). At what point is the patient said to have heart disease? This is a case where the choice of a threshold is crucial. Of course, in this situation, the patient might be referred for further testing involving the gold standard, namely, a heart catheterization to examine the coronary arteries.

Another example is testing for breast cancer via a mammography where, as we have seen in previous chapters, on the basis of a five-point scale, the radiologist scores the likelihood that the lesion is present in the image. At what point does the clinician say that the patient has a malignant lesion? Again, we are faced with the choice of a threshold value to declare disease. Still another example that has been stressed repeatedly in the book is the diagnosis of prostate cancer using prostate-specific antigen (PSA) levels; at what point does the clinician declare that the disease is present? These examples will again illustrate the main topic of this chapter, namely, the role of test accuracy in clinical practice.

Once the patient is diagnosed with disease and therapy is initiated, the accuracy of various tests to monitor disease progress is crucial. The example considered in this chapter is the case of a Phase II clinical trial for cancer, where the progress of a patient is monitored by measuring the size of the tumor. At baseline, before treatment begins, the size of the tumor is measured by various imaging devices, such as computed tomography (CT) or magnetic resonance imaging (MRI) or both. The size is measured at various times during treatment, and at the termination of the trial, a final determination of the size is made. Next, a team of radiologists must decide for each patient the progress of the disease as measured by tumor size and other characteristics. In many cases the effect of treatment is categorized as: (a) a complete response, (b) a partial response, (c) no change, or (d) disease progression. Each category is defined in terms of the change in the size of the tumor, from the initiation of treatment to the termination of treatment. Such assessments are made for each patient and the total accumulated evidence, in turn, determines the overall success or failure of the trial. Of course, at issue is the accuracy of the imaging device and the agreement between the team of radiologists responsible for declaring the success or failure of the trial.

The last sections in the chapter are devoted to a new way to measure the accuracy of a medical test and are based on so-called decision curves, an approach whose foundation is the clinical benefit of a medical test. In turn, the clinical benefit is defined in terms of the true positive rate (TPR) and false positive rate (FPR) and a threshold value (that probability of disease, above which a patient would choose a biopsy). For example, in using PSA to diagnose prostate cancer, should the decision to biopsy a subject be based on the value of PSA and a threshold value, or should every patient be biopsied? A decision curve can be used to make this decision. The prostate cancer example and some others will illustrate the decision curve as a way to measure the accuracy of a medical test. The Bayesian approach to decision curves is novel and adds some features not provided with standard methods.

9.2   Choice of Optimal Threshold

Suppose that the scores of a medical test are available and that the scores are continuous, and that higher values of the score imply disease and that lower values imply non disease. Also suppose that sufficient studies have been performed so that the area under the receiver operating characteristic (ROC) curve is a reliable estimate of the accuracy of the medical test. Remember that the area under the ROC curve represents test accuracy for a given patient population. For the individual patient, the clinician is faced with the problem of choosing a cutoff value or threshold value that declares that the patient has the disease and the approach taken here considers two scenarios: (a) the score corresponding to the point on the ROC curve that is closest to the point (0,1), and (b) the score corresponding to the point on the ROC curve that minimizes the expected cost of conducting the test. Both scenarios will be illustrated with three examples: (1) the PSA test for prostate cancer, (2) the blood glucose test for type 2 diabetes, and (3) a biomarker test for the diagnosis of complications to head trauma. These three examples have been studied in earlier chapters for various methods of measuring test accuracy.

9.2.1 Optimal threshold for the prostate-specific antigen test for prostate cancer

Recall the prostate specific antigen (PSA) study of 12,000 men aged 50–65, which was a randomized study with a beta-carotene group as the treatment group vs. a placebo group. The data values for our use are from the study by Etzioni et al. [1], who used a subset of 683 subjects, on which the total PSA values were reported. See Pepe [2: 10] for additional details about the study, which was analyzed from a non-Bayesian approach. First to be conducted is a Bayesian analysis that estimates the ROC area under the assumption of binormality, where the ROC area is estimated as

$$\text{AUC} = \Phi\left[a/\sqrt{1+b^2}\right], \tag{4.14}$$

where X is normally distributed,

$$a = (\mu_d - \mu_{nd})/\sigma_d$$

and

$$b = \sigma_d/\sigma_{nd}.$$

The mean and standard deviation of X for the diseased population are μ_d and σ_D, respectively, while μ_{nd} and σ_{nd} are the mean and standard deviation, respectively, of X for the non diseased. Φ is the cumulative distribution function (CDF) of the standard normal distribution. Recall these developments from Chapter 4, where the ROC area is based on Equation 4.16 and the analysis is executed with BUGS CODE 4.3b, which is based on the Bayesian approach of O'Malley et al. [3]. The following BUGS CODE 9.1 contains the PSA data in the first list statement.

BUGS CODE 9.1

```
model;
        # Binormal model for area under ROC
        # Calculates posterior distribution of model parameters and the area
            under curve.
        # calculates the coordinates of the optimal psa threshold
        # yt= log transformed outcome, d=disease status, # var y[N], yt[N],
        d[N], mu[N], beta[P], vary[K], precy[K], auc, la1, la2;
        {
        # likelihood function
                for(i in 1:N) {
                        yt[i]~dnorm(mu[i],precy[d[i]+1]);
```

```
        yt[i] <- log(y[i]);
        mu[i] <- beta[1] + beta[2]*d[i];
                          }
# prior distributions - non-informative prior; similarly for informative
# priors
        for(i in 1:P) {
                beta[i] ~ dnorm(0, 0.000001);
        }

        for(i in 1:K) {
                precy[i]~dgamma(0.001, 0.001);
                vary[i] <- 1.0/precy[i];
                          }

# calculates area under the curve
        a <- beta[2]/sqrt(vary[2]);
        # ROC curve parameters
        la2 <- vary[1]/vary[2];
        # ROC area
        auc <- phi(a/sqrt(1+la2));
        b<- sqrt(la2) ;
   ka<-R*(1-p)/p
   nfpf<- a*b-sqrt(a*a+2*(1-b*b)*log(ka/b));
dfpf<-1-b*b;
fpf <- phi(nfpf/dfpf) ;
ntpf<-a - b*sqrt(a*a+(1-b*b)*log(ka/b));
tpf<- phi(ntpf/dfpf);
   }
   list(K=2, P=2, N=683, p=.5, R=1,
   y=c(.03,
   .09,.23,.27,.27,.29,.29,.29,.30,.31,.33,.35,.37,.37,.42,.43,.44,.45,
   .45,.46,.46,.47,.47,.48,.49,.49,.50,.50,.50,.51,.51,.55,.55,.56,.57,
   .57,.58,.58,.58,.58,.59,.59,.59,.61,.61,.62,.62,.63,.63,.64,.64,.64,
   .64,.65,.65,.65,.66,.66,.66,.66,.66,.66,.67,.67,.67,.67,.67,.68,.68,
   .69,.69,.69,.69,.69,.70,.71,.72,.72,.73,.74,.74,.75,.75,.75,.75,.75,
   .76,.76,.77,.77,.77,.77,.77,.77,.78,.78,.78,.78,.78,.78,.79,.79,.79,
   .79,.80,.80,.80,.81,.81,.81,.81,.82,.83,.83,.84,.85,.86,.87,.87,.87,
   .87,.87,.88,.89,.89,.89,.89,.89,.92,.92,.92,.93,.93,.93,.93,.93,.93,
   .94,.94,.95,.95,.95,.95,.96,.96,.97,.97,.98,.98,.98,.98,.98,.99,1.00,
   1.00,1.00,1.01,1.01,1.02,1.03,1.03,1.03,1.03,1.03,1.03,1.04,1.04,
   1.04,1.04,1.04,1.05,1.05,1.05,1.05,1.06,1.06,1.06,1.06,1.07,1.07,
   1.07,1.08,1.08,1.08,1.11,1.11,1.12,1.12,1.13,1.13,1.13,1.14,1.15,
   1.15,1.15,1.15,1.15,1.15,1.15,1.15,1.15,1.16,1.16,1.16,1.17,1.17,1.17,
   1.17,1.18,1.18,1.18,1.18,1.18,1.19,1.19,1.19,1.20,1.20,1.21,1.22,
   1.22,1.22,1.23,1.23,1.24,1.24,1.24,1.25,1.25,1.25,1.25,1.25,1.25,
   1.26,1.26,1.26,1.27,1.27,1.27,1.27,1.27,1.27,1.28,1.28,1.29,1.30,
```

1.30,1.31,1.31,1.32,1.32,1.33,1.34,1.35,1.35,1.35,1.35,1.35,1.35,
1.35,1.36,1.37,1.37,1.37,1.38,1.39,1.39,1.40,1.40,1.40,1.40,1.41,
1.41,1.41,1.41,1.41,1.41,1.43,1.43,1.43,1.43,1.44,1.44,1.45,1.46,
1.46,1.47,1.47,1.47,1.48,1.48,1.49,1.49,1.50,1.50,1.50,1.50,1.51,
1.51,1.51,1.51,1.53,1.54,1.54,1.55,1.55,1.56,1.57,1.57,1.58,1.58,
1.58,1.61,1.62,1.62,1.62,1.62,1.64,1.67,1.67,1.67,1.67,1.67,1.68,
1.69,1.69,1.70,1.70,1.70,1.71,1.71,1.71,1.71,1.71,1.71,1.71,1.71,
1.73,1.73,1.73,1.74,1.79,1.80,1.80,1.83,1.85,1.85,1.88,1.88,1.88,
1.89,1.89,1.89,1.91,1.91,1.91,1.92,1.93,1.93,1.94,1.95,1.96,2.01,
2.01,2.03,2.03,2.03,2.04,2.04,2.05,2.05,2.06,2.07,2.08,2.08,2.10,
2.11,2.13,2.13,2.14,2.16,2.17,2.19,2.19,2.19,2.22,2.22,2.23,2.24,
2.27,2.27,2.27,2.28,2.28,2.29,2.29,2.30,2.30,2.33,2.34,2.34,2.35,
2.36,2.36,2.37,2.40,2.41,2.42,2.43,2.43,2.43,2.43,2.46,2.50,2.50,
2.51,2.51,2.52,2.53,2.55,2.55,2.56,2.56,2.57,2.58,2.61,2.62,2.62,
2.63,2.63,2.63,2.66,2.69,2.70,2.71,2.73,2.77,2.79,2.82,2.82,2.82,
2.83,2.84,2.84,2.85,2.86,2.86,2.87,2.88,2.88,2.90,2.92,2.92,2.93,
2.95,2.96,2.96,2.96,2.97,2.98,3.03,3.03,3.04,3.05,3.05,3.08,3.10,
3.11,3.13,3.17,3.17,3.18,3.20,3.21,3.24,3.25,3.25,3.29,3.30,3.30,
3.32,3.32,3.33,3.34,3.35,3.38,3.41,3.42,3.43,3.45,3.51,3.55,3.57,
3.57,3.58,3.58,3.61,3.65,3.65,3.66,3.68,3.69,3.70,3.73,3.77,3.78,
3.78,3.78,3.80,3.84,3.88,3.89,3.95,3.97,3.97,4.00,4.03,4.03,4.04,
4.05,4.08,4.12,4.15,4.19,4.20,4.20,4.20,4.30,4.33,4.34,4.38,4.39,
4.40,4.41,4.44,4.47,4.47,4.48,4.52,4.54,4.60,4.62,4.64,4.70,4.75,
4.75,4.76,4.78,4.90,4.90,4.93,4.94,4.98,5.02,5.09,5.10,5.11,5.12,
5.13,5.13,5.25,5.28,5.37,5.39,5.44,5.44,5.53,5.54,5.64,5.65,5.67,
5.73,5.75,5.81,5.85,6.07,6.07,6.16,6.18,6.27,6.29,6.31,6.41,6.48,6.48,
6.50,6.52,6.52,6.54,6.54,6.56,6.56,6.77,6.92,6.93,7.09,7.19,7.21,
7.23,7.24,7.28,7.29,7.42,7.43,7.53,7.59,7.64,7.78,7.90,8.04,8.15,
8.31,8.37,8.57,8.62,8.69,9.07,9.11,9.15,9.15,9.17,9.24,9.30,9.33,
9.76,9.94,9.96,9.97,10.11,10.60,10.71,10.92,11.33,11.40,11.54,
11.62,11.65,12.69,12.69,13.61,13.94,14.82,15.41,15.84,15.84,15.89,
16.18,16.48,16.70,16.81,17.10,17.17,17.57,19.35,20.10,20.24,20.47,
20.53,21.48,22.50,23.81,24.63,25.06,26.67,27.68,29.31,31.46,33.02,
35.93,37.63,37.66,38.39,43.30,48.80,49.16,51.72,61.16,72.07,79.21,
90.66,99.97,99.98,99.98,99.98),
d=c(.00,1.00,.00,.00,.00,.00,.00,.00,.00,.00,.00,.00,.00,.00,.00,.00,
1.00,.00,.00,.00,.00,.00,.00,.00,.00,.00,.00,.00,.00,.00,.00,.00,.00,
.00,.00,.00,.00,.00,.00,1.00,.00,.00,1.00,.00,.00,.00,.00,.00,.00,.00,
.00,.00,.00,.00,.00,.00,.00,.00,.00,.00,.00,.00,.00,.00,.00,1.00,
.00,1.00,.00,.00,.00,.00,.00,.00,1.00,.00,.00,.00,.00,.00,.00,.00,
.00,.00,.00,.00,.00,.00,.00,.00,.00,.00,.00,.00,.00,.00,.00,.00,.00,
.00,.00,.00,.00,.00,.00,.00,.00,.00,.00,.00,.00,.00,.00,.00,.00,.00,
.00,.00,.00,.00,.00,.00,.00,.00,.00,.00,.00,.00,.00,.00,.00,.00,.00,
.00,.00,1.00,.00,.00,.00,.00,.00,.00,.00,.00,.00,.00,.00,.00,.00,.00,
.00,.00,.00,.00,.00,.00,.00,.00,.00,.00,.00,.00,.00,.00,.00,.00,.00,
.00,.00,.00,.00,.00,.00,.00,.00,1.00,1.00,.00,.00,1.00,.00,.00,.00,

.00,1.00,.00,.00,.00,.00,.00,.00,.00,.00,.00,.00,.00,.00,.00,.00,
.00,.00,1.00,.00,.00,.00,.00,.00,.00,.00,1.00,1.00,.00,.00,1.00,.00,
1.00,1.00,.00,.00,.00,.00,.00,.00,.00,1.00,.00,.00,.00,.00,.00,.00,
.00,.00,1.00,.00,.00,.00,.00,.00,1.00,.00,.00,.00,.00,.00,.00,.00,
.00,.00,.00,.00,.00,.00,.00,.00,.00,.00,1.00,.00,.00,.00,.00,.00,.00,
.00,.00,.00,.00,1.00,.00,.00,.00,.00,1.00,1.00,.00,.00,1.00,1.00,.00,
.00,.00,.00,.00,.00,.00,.00,.00,1.00,.00,.00,.00,.00,1.00,1.00,.00,.00,
.00,.00,.00,.00,.00,.00,.00,.00,.00,1.00,.00,.00,1.00,.00,.00,.00,.00,
.00,.00,.00,.00,.00,.00,1.00,.00,.00,.00,.00,.00,1.00,.00,.00,.00,.00,
.00,.00,1.00,1.00,.00,.00,1.00,.00,.00,.00,.00,1.00,.00,1.00,.00,.00,
.00,.00,1.00,1.00,.00,.00,.00,.00,.00,1.00,.00,1.00,.00,.00,.00,.00,.00,
.00,.00,.00,.00,1.00,1.00,1.00,.00,1.00,.00,.00,.00,1.00,1.00,.00,.00,.00,
.00,1.00,.00,.00,.00,.00,.00,1.00,1.00,.00,1.00,.00,1.00,.00,1.00,.00,
.00,.00,1.00,.00,.00,.00,.00,.00,.00,.00,.00,.00,1.00,1.00,.00,1.00,.00,
1.00,.00,1.00,.00,1.00,.00,.00,1.00,.00,1.00,.00,.00,.00,1.00,1.00,.00,.00,
.00,.00,.00,1.00,1.00,.00,.00,.00,1.00,.00,1.00,1.00,.00,1.00,1.00,.00,
1.00,.00,.00,.00,1.00,.00,.00,.00,.00,1.00,1.00,.00,1.00,.00,.00,.00,
1.00,.00,1.00,.00,.00,1.00,.00,1.00,.00,.00,.00,1.00,.00,1.00,1.00,.00,
1.00,.00,.00,1.00,1.00,.00,.00,1.00,.00,1.00,.00,.00,.00,.00,.00,.00,
.00,1.00,1.00,.00,1.00,1.00,.00,1.00,.00,1.00,1.00,.00,.00,.00,1.00,1.00,
.00,1.00,1.00,1.00,1.00,1.00,1.00,1.00,1.00,1.00,.00,1.00,1.00,1.00,
1.00,.00,1.00,1.00,1.00,1.00,.00,.00,.00,.00,1.00,1.00,1.00,1.00,.00,
1.00,1.00,.00,1.00,.00,1.00,1.00,1.00,.00,1.00,.00,.00,.00,1.00,1.00,
.00,1.00,1.00,1.00,1.00,.00,1.00,.00,1.00,1.00,1.00,.00,1.00,1.00,.00,
.00,.00,1.00,.00,1.00,1.00,1.00,.00,1.00,.00,1.00,.00,1.00,1.00,.00,
1.00,1.00,1.00,.00,1.00,.00,1.00,.00,1.00,.00,.00,1.00,1.00,1.00,.00,.00,
1.00,1.00,1.00,.00,1.00,.00,1.00,1.00,1.00,1.00,.00,1.00,1.00,.00,1.00,
1.00,1.00,1.00,1.00,1.00,1.00,1.00,1.00,1.00,1.00,1.00,1.00,1.00,1.00,
1.00,1.00,1.00,1.00,.00,1.00,1.00,1.00,1.00,1.00,1.00,1.00,1.00,1.00,
1.00,1.00,1.00,1.00,1.00,.00,1.00,1.00,.00,1.00,.00,1.00,1.00,1.00,
1.00,1.00,1.00,1.00,1.00,1.00,1.00,1.00,1.00,1.00,.00,1.00,1.00,
1.00,1.00,1.00,1.00,1.00,1.00,1.00,1.00,1.00,1.00,1.00,1.00))
list(beta=c(0,0),precy=c(1,1))

Also recall from Chapter 4 that under binormality, the ROC curve has the representation

$$\mathrm{ROC}(t) = \Phi(a + b\Phi^{-1}(t)), \quad 0 \le t \le 1, \tag{9.1}$$

where Φ is the CDF of the standard normal distribution. It can be shown that with 45,000 observations generated from the Markov Chain Monte Carlo (MCMC) simulation, the ROC area has a posterior mean of 0.8216(0.0179) and a 95% credible interval (0.7848,0.855), a has a posterior mean of 1.12(0.0908), and b has a posterior mean of 0.684(0.0394). The MCMC simulation has a burn in of 5,000 and a refresh of 100, and the MCMC error is <0.0001 for all parameters.

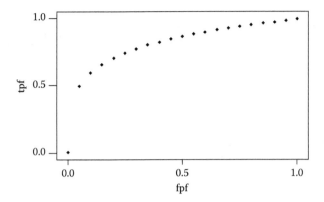

FIGURE 9.1: ROC curve for PSA test.

Our objective is to estimate the ROC curve via Equation 9.1 and to find the point on the ROC curve that is closest to the point (0,1) and declare the corresponding test score (PSA value) the optimal threshold for diagnosing prostate cancer.

The graph of the ROC curve shown in Figure 9.1 is based on Equation 9.1 and is calculated with the worksheet shown in Table 9.1. The first column is the false positive fraction (FPF), the second is the true positive fraction (TPF) based on Equation 9.1, where the posterior means of a and b (which are 1.12 and 0.684, respectively), are used in the formula, and the third column is the distance from a point on the ROC curve to the point (0,1).

The Euclidean distance is given by

$$d = \sqrt{\mathrm{fpf}^2 + (\mathrm{tpf} - 1)^2},\tag{9.2}$$

which is used to calculate the values in the third column, revealing that the point (0.25,0.744) is closest to the point (0,1); thus, the optimal point is estimated as (0.25,0.744), that is, the one with an fpf $-$ 0.25 and a true positive fraction of 0.744. What is the corresponding threshold PSA value on the log scale? The basic PSA information is given in natural log units, because the log transformation is needed to induce normality so that Equation 9.1 is the appropriate formula. Note, according to Pepe [2: 82], that the corresponding threshold is given by

$$c = \mu_{nd} - \sigma_{nd}\Phi^{-1}(t),\tag{9.3}$$

and the fourth column of Table 9.1 reports the test scores corresponding to the fpf values of the first column.

This approach to choosing the optimal threshold is not based on cost considerations, but is instead based on the point that is closest to the point where the fpf $= 0$ and the true positive fraction is 1. Cost considerations are often used to select a threshold value for a diagnostic test. Recall from Chapter 4

TABLE 9.1:   Worksheet for PSA study.

$t = $ fpf	$\Phi(a + b\Phi^{-1}(t)) = $ tpf	Distance	Log PSA values
0	0		
0.05	0.4979	0.5045	2.88302
0.10	0.5961	0.416037	2.46924
0.15	0.65949	0.372082	2.19006
0.20	0.70689	0.35484	1.96817*
0.25	0.74494	0.35715	1.77782
0.3	0.77676	0.373944	1.60687
0.35	0.80412	0.401083	1.44846
0.4	0.82811	0.435370	1.29815
0.45	0.84944	0.474518	1.15272
0.5	0.86864	0.516967	1.00960
0.55	0.88608	0.561674	0.86648
0.5	0.90204	0.607943	0.72105
0.65	0.91675	0.655309	0.57074
0.7	0.93901	0.703453	0.41233
0.75	0.94310	0.752155	0.24138
0.8	0.95503	0.801263	0.05103
0.85	0.99629	0.850668	−0.17086
0.9	0.87706	0.900292	−0.45004
0.95	0.98762	0.950081	−0.86832
1	1		

*Log PSA values corresponding to the smallest distance from (0,1).

that Zhou, Obuchowski, and McClish [4: Ch. 2] base the choice of an optimal cutoff value on minimizing the total cost:

$$C = \text{TPF}p(C_{tp} - C_{fn}) + \text{FPF}(1 - p)(C_{fp} - C_{tn}) + C_0 + pC_{fn} + (1 - p)C_{tn}, \tag{4.21}$$

where p is the disease incidence, C_0 is the cost of performing the test, while C_{tp}, C_{fn}, C_{fp}, and C_{tn} are the costs of a true positive (TP), false negative (FN), false positive (FP), and true negative (TN), respectively. When this expression is differentiated with respect to FPF, the slope of the curve at the optimal point is

$$\kappa = (1 - p)R/p, \tag{4.22}$$

where

$$R = (C_{tn} - C_{fp})/(C_{tp} - C_{fn}). \tag{4.23}$$

Assuming binormality, Somoza and Mossman [5] have shown that the optimal point is [FPF, TPF], where

$$\text{FPF}(a, b) = \Phi\left\{ \left[ab - \sqrt{a^2 + 2(1 - b^2)\ln(\kappa/b)}\right] \Big/ (1 - b^2)\right\}$$

and

$$\text{TPF}(a, b) = \Phi\left\{\left[a - b\sqrt{a^2 + (1 - b^2)\ln(\kappa/b)}\right] \Big/ (1 - b^2)\right\}. \qquad (4.24)$$

Treating κ as a constant, the coordinates of the optimal point are functions of the parameters a and b (see Equations 4.14 and 4.15) and have posterior distributions. In the approach to be presented, a and b will be considered random variables with a posterior distribution, but κ, and hence p and R (and hence the costs), will be considered fixed known constants, which is somewhat unrealistic (see Hans et al. [6]).

Basing the optimal point on cost considerations presents additional complexity in choosing a threshold score, because the costs or expected costs for the TP, FP, TN, and FN alternatives are not easy to determine and depend on one's point of view. By the latter, I mean taking cost into account by the provider and/or insurer will differ substantially from the patient's perspective. Ethical issues are also involved in the choice of an optimal threshold. Should costs be involved in the choice?

Returning to the PSA example, $R = 1$, which gives the Bayesian analysis shown in Table 9.2 for the coordinates of the optimal threshold based on the Somoza and Mossman formula 4.24.

Note that based on the posterior mean, the optimal point on the ROC curve has coordinates (0.1631,0.7097), which corresponds to the threshold log PSA value of between 1.968 and 2.190, or to a threshold between 7.156 and 8.944 on the original PSA scale. One needs to interpolate Table 9.1 to arrive at these values; thus, the optimal value based on cost considerations is quite close to that based on the distance criterion. Interpolation is required because the coordinates of the optimal point are based on a and b, which are random variables. Note also that the optimal point depends on κ, which in turn depends on R and disease incidence p. R given by Equation 4.23 is a ratio, whose numerator is the benefit of a TN and whose denominator is the benefit of a TP. It is difficult to know the value of the exact costs for the four alternatives, thus for purposes of illustration, consider taking multiples of the numerator vs. the denominator. As for the disease incidence, one must be careful of how the study is designed. If, for example, one is taking a true random sample from a well-defined population, then estimating p with the observed incidence is plausible, but in many studies this is not the case. For example, in the PSA study the observed incidence is 33.5% (of 683 cases), which must

TABLE 9.2: Posterior distribution of optimal point for $\kappa = 1$.

Parameter	Mean	sd	Error	2 1/2	Median	97 1/2
a	1.12	0.0908	<0.0001	0.9431	1.119	1.299
b	0.684	0.0394	<0.0001	0.6088	0.6832	0.7638
fpf	0.1631	0.0127	<0.0001	0.1394	0.1626	0.1894
tpf	0.7097	0.0205	<0.0001	0.6675	0.7101	0.748

TABLE 9.3:   Posterior analysis for optimal point PSA study.

Parameter	Mean	sd	Error	2 1/2	Median	97 1/2
fpf	0.0288	0.0020	<0.0001	0.0251	0.0288	0.0329
tpf	0.5777	0.0284	<0.0001	0.5194	0.5784	0.6319

Note: $p = 33\%$ and $R = 2$.

be taken with a grain of salt since the study is actually longitudinal with repeated values for some subjects.

Another problem with using Equation 4.24 is that certain restrictions are necessary because the expression under the square root must be positive, thus for the FPF point,

$$a^2 > 2(1 - b^2) \ln(\kappa/b), \tag{9.4}$$

and a similar constraint is necessary for the TPF of Equation 4.24. Recall that choices of p and R can result in Kappa values that give values that violate the constraint, which in turn makes the expression under the square root negative. Note also, that one cannot choose κ as negative, which would give an undefined value for the ln (log) function.

Suppose the benefit of a TN is double that of a TP, and the disease incidence is put at the observed value of 33%, then the posterior analysis is reported in Table 9.3. Based on cost considerations where the disease incidence is estimated at 33% and the benefit of a TN is twice that of a TP, the optimal point is $(0.0288, 0.5777)$, which corresponds to a threshold of approximately 2.90 on the log scale, or approximately 18 on the original PSA scale. The disease estimate of 33% is most likely much too high. Note that for screening studies, the disease incidence is usually quite small.

For example, suppose the disease incidence is 33% and the benefit of a TN is one half that of a TP, that is $R = 0.5$, then it can be verified that the coordinates of the optimal point are fpf $= 0.16023$ and tpf $= 0.7081$. This is left as an exercise, where 45,000 observations are generated from the simulation, with a burn in of 5,000 and a refresh of 500. All the MCMC errors are <0.0001, and a 95% credible interval for the tpf is $(0.6658, 0.7467)$. Compared to when $R = 2$ (the benefit of a TN is twice that of a TP) and $p = 33\%$, the change in the optimal point is dramatic. Note that the optimal threshold on the log scale is between 1.96 and 2.19, using interpolation of Table 9.1.

9.2.2   Test accuracy of blood glucose for diabetes

Attention is now given to a hypothetical example of diabetes in Chapter 4, which involves 59 subjects with diabetes and 19 without, where those with diabetes have a mean blood glucose value of 123.34 mg/dL, and those without have a mean value of 107.54 mg/dL. The corresponding standard deviations are 6.76 mg/dL for those with diabetes and 9.09 mg/dL for those without the

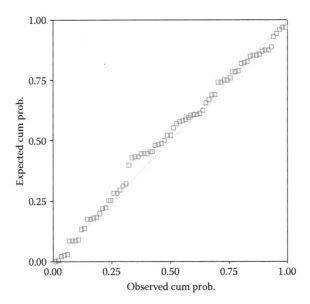

FIGURE 9.2: Normal P–P plot of glucose.

disease, and the actual values from the study are given in Table 4.11. From Figure 9.2, which gives the P–P plot of the blood glucose values, they appear to be normally distributed.

For the diabetes data of Table 4.11, the Bayesian analysis is executed with BUGS CODE 9.1 using 45,000 observations for the simulation, with a burn in of 5,000 and a refresh of 100. The blood glucose values discriminate well between those with and without the disease with a posterior mean(sd) of 0.9062(0.0433) for the ROC area, and based on the posterior mean(sd), the optimal point on the ROC has coordinates fpf = 0.2047(0.0665) and tpf − 0.8576(0.0328), see the asterisk of Table 9.4. Also based on the posterior mean, the values of a and b used for finding the point in the curve closest to (0,1) are $a = 2.33$ and $b = 1.405$.

Based on Equations 9.1 and 9.2, the second and third columns of Table 9.4 report the TP values and the distance from (0,1) for various values of the FPF given in the first column. It is seen that based on the distance criterion (Equation 9.2), the optimal point is (0.25,0.88), which corresponds to a threshold value of approximately 113 mg/dL. The table also gives the coordinates (fpf,tpf) of the ROC curve for the diabetes study.

The Bayesian analysis for the diabetes study is given in Table 9.5. The blood glucose test has good accuracy with a ROC area of 0.90 and the values a and b are used to compute the worksheet and the coordinates of the ROC curve of Figure 9.3. The TP and FP values reported above correspond to $R = 2$ and $p = 0.5$, where the latter is the disease incidence and the former assumes that the benefit of a TN is twice the benefit of a TP.

TABLE 9.4:   ROC curve diabetes study.

$t = \text{fpf}$	$\Phi(a + b\Phi^{-1}(t)) = \text{tpf}$	Distance
0	0	
0.05	0.307	0.694
0.10	0.557	0.454
0.15	0.718	0.318
0.20	0.822	0.267
0.25	0.889	0.273*
0.3	0.931	0.307
0.35	0.958	0.352
0.4	0.976	0.400
0.45	0.986	0.450
0.5	0.992	0.500
0.55	0.995	0.550
0.6	0.997	0.600
0.65	0.999	0.650
0.7	0.999	0.7004
0.75	0.999	0.7502
0.8	0.999	0.8000
0.85	1	0.8500
0.9	1	0.9000
0.95	1	0.9500
1	1	1.000

For additional information about choosing the optimal point on the ROC curve, refer to Exercises 1–12. Another way to study the accuracy of a medical test is to determine its clinical value as defined by decision curves.

9.3   Test Accuracy with Bayesian Decision Curves

9.3.1   Introduction

Vickers and Elkin [7] introduced decision curves to evaluate the clinical benefit of a diagnostic test. Using PSA as a diagnostic marker for prostate

TABLE 9.5:   Posterior analysis for diabetes study.

Parameter	Mean	sd	Error	2 1/2	Median	97 1/2
auc	0.9062	0.0433	<0.0001	0.802	0.9137	0.9683
a	2.33	0.4173	0.0060	1.559	2.316	3.19
b	1.405	0.2842	0.0013	0.9606	1.368	2.075
tpf	0.8576	0.0328	<0.0001	0.7889	0.8595	0.9164
fpf	0.2047	0.0665	<0.0001	0.1021	0.1955	0.3608

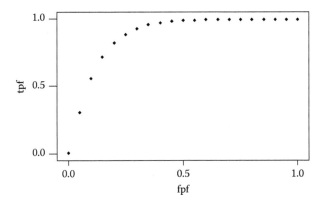

FIGURE 9.3: ROC curve for diabetes study.

cancer, three scenarios were considered. The first scenario bases a decision to biopsy on the value of the diagnostic marker, while the second calls for a biopsy of all persons, and no person has a biopsy for the third scenario. In order to compare the three scenarios, a decision curve is defined. A decision curve is a plot of the clinical benefit vs. a threshold probability. Also, Vickers et al. [8] extended the basic idea with confidence intervals that estimated the clinical benefit at given threshold probabilities and calculated the frequency properties of the estimated clinical benefits. Finally, Vickers [9] and, later, Steyerberg and Vickers [10] gave a general review of the topic. The approach taken here is to utilize a Bayesian method to make statistical inferences about decision curves.

To stress again the main point of this section, in order to assess the accuracy of a diagnostic test, Vickers and Elkin [7] introduced clinical benefits as the foundation for decision curves. They describe a prostate cancer study, where three scenarios are considered: (a) on the basis of the diagnostic test, perform a biopsy if the probability of cancer is greater than a threshold probability; (b) biopsy all persons; or (c) do not biopsy any person. Corresponding to the three alternatives, three decision curves are plotted on the same graph. A decision curve is a plot of the clinical benefit vs. a threshold probability.

With a traditional decision theory approach, a utility is assigned for each alternative and the alternative with the largest expected utility is selected as optimal. Using the prostate cancer example, Vickers [9] assigns a benefit to each of the four possible outcomes, TP, FP, TN, and FN on the familiar 2×2 table, where TP is a true positive (the person has the disease and the diagnostic test is positive), FP is a false positive, TN is a true negative, and FN is a false negative. Vickers continues by showing that the first alternative has the largest average benefit; however, he notes that it is often difficult to assign a benefit to the various outcomes and defines instead a net clinical benefit that depends only on the probabilities of the TPR and FPR and the incidence of disease. Then, using this definition of clinical benefit, a decision curve can be constructed and plotted for the three alternatives. The decision curve corresponding to alternative (a) dominates the other two curves, which implies that the biopsy decision of an individual should depend on the value of a diagnostic variable. This shows the clinical value, thus the accuracy, of the PSA test for prostate cancer.

The decision curve is independent of the assignment of utilities to the four outcomes of the 2×2 table; however, the clinical benefit does depend on sensitivity, specificity, and the prevalence of disease.

In a later paper, Vickers et al. [8] developed confidence intervals to estimate the clinical benefit at a given set of threshold probabilities and described the sampling properties of confidence intervals via a bootstrap.

The main purpose of this presentation is to utilize a Bayesian approach to making inferences about decision curves. This includes calculating the posterior distribution of the clinical benefit. In the forthcoming sections, decision curves are defined, Bayesian inferences are described, and the methodology is illustrated with a study based on a diagnostic test for prostate cancer. Determining the posterior distribution of a clinical benefit at a selected set of threshold probabilities is the foundation for computing decision curves for scenarios (a) and (b).

9.3.2   Decision curves

After reviewing a formal cost approach for choosing a therapy, and citing Baker, Kramer, and Srivastava [11], Vickers defines the net clinical benefit at threshold probability, p_t, as

$$p_{11} - p_{10}(p_t/(1 - p_t)), \tag{9.5}$$

where p_{11} is the probability of a TP, and p_{10} is the probability of an FP. Also, this can be expressed as

$$\text{Sens}\pi - (1 - \text{Spec})(1 - \pi)[p_t/(1 - p_t)], \tag{9.6}$$

where Sens is the sensitivity, Spec is the specificity, and π is the prevalence of disease. The definition of clinical benefit depends on the threshold probability p_t.

For scenario (a), if a person's estimate of disease is greater than the threshold probability p_t, the person opts for a biopsy, otherwise, they do not. Plotting the clinical benefit (1) vs. a reasonable range of p_t values, defines a decision curve for scenario (a). Note that the sensitivity and specificity depend on the threshold probability.

Assuming there is a diagnostic variable (e.g., PSA for prostate cancer), Vickers [9], following Vergouwe et al. [12], adopts the following algorithm in order to determine a positive test result for scenario (a):

(1) Select a threshold probability p_t.

(2) Using logistic regression, estimate the probability of disease.

(3) Define a test as positive if the probability of disease is greater than p_t.

(4) Calculate the net benefit using (1).

(5) Repeat (1) through (4) for a reasonable range of p_t.

Once one knows what patients test positive for disease, the sensitivity and specificity may be calculated for a given threshold probability.

On the other hand, for scenario (b), where all patients have a biopsy, the clinical benefit is calculated as

$$\pi - (1 - \pi)[p_t/(1 - p_t)], \tag{9.7}$$

for a range of p_t values. Note, for scenario (b) the sensitivity is 1, the specificity is 0, and the disease prevalence does not depend on p_t.

For scenario (c), when no one has a biopsy, the clinical benefit is zero, and the three decision curves can be plotted on the same graph and the clinical benefit of each scenario assessed and compared to the others.

For the range from 0.1 to 0.4 of p_t, Vickers [9] plots the three decision curves for a population of high-risk persons for prostate cancer and demonstrates that the first scenario is optimal because the decision curve for scenario (a) dominates the decision curve for the other two. Thus, for this population of high-risk patients, a biopsy is chosen depending on the probability of disease, which in turn depends on the value of the diagnostic variable PSA.

It can be shown that the assessment of clinical benefit is, in a sense, equivalent to choosing the optimal point on the ROC curve with a slope of 1, where the threshold probability is the same as the disease prevalence. Recall that the optimal point is chosen to minimize the total cost of performing a diagnostic test. Additional information is available from Broemeling [13] and Zhou, Obuchowski, and McClish [4: 150].

9.3.3 Bayesian inferences for decision curves

Consider Table 9.6 showing disease status vs. the results of a diagnostic test. A person at high risk for disease has a probability, p_{xy}, of disease status x and test result y, where $x, y = 0$ or 1. Note that p_{00} is the probability of

TABLE 9.6: Disease status vs. test result—cell probabilities and frequencies.

Diagnostic test	Disease	
	0	1
0	p_{00}, n_{00}	p_{01}, n_{01}
1	p_{10}, n_{10}	p_{11}, n_{11}

a TN, and p_{11} the probability of a TP, while p_{01} is the probability of an FN. The usual measures of test accuracy are given by the sensitivity:

$$\text{Sens} = p_{11}/(p_{01} + p_{11}), \tag{9.8}$$

and specificity

$$\text{Spec} = p_{00}/(p_{00} + p_{10}). \tag{9.9}$$

With regard to scenario (a), the diagnostic test is positive if the probability of disease is greater than the threshold probability p_t, where p_t is chosen over some reasonable range. Thus, the probability that the diagnostic test is positive depends on the threshold probability. For clinical benefit (Equation 9.6), the TPR and FPR depend on the threshold probability, and a decision curve is generated.

For scenario (b), the clinical benefit (Equation 9.7) depends on the disease prevalence, π, and the threshold probability, p_t, but the disease prevalence does not depend on the threshold probability!

A random sample of size n of high-risk individuals for disease is selected, where n_{ij} is the number with disease status j and test result i of Table 9.6. The conditional distribution of the cell frequencies, given the cell probabilities, is multinomial.

Which scenario, (a) or (b), is optimal, that is, does the decision curve of one scenario dominate the other? This will be answered by determining the posterior distribution of the clinical benefit for a given threshold probability. Scenario (a) is considered first, where the clinical benefit is given by Equation 9.6. Note the dependence of the cell frequencies and estimated cell probabilities on the threshold probability.

Assume a prior density for the probabilities of Table 9.6 and select a threshold probability from some range, then p_{11} and p_{10} have posterior distributions that induce a posterior distribution for the clinical benefit at that threshold probability. If the prior distribution of the cell probabilities is chosen as Dirichlet, the posterior distribution of the cell frequencies is also Dirichlet, and in particular if the prior is chosen as uniform, the posterior distribution is Dirichlet with parameter $(n_{00} + 1, n_{01} + 1, n_{10} + 1, n_{11} + 1)$.

Once the threshold probability is selected, a logistic regression determines the number of subjects who test positive and negative, which in turn affects the cell frequencies of Table 9.6, and thus the clinical benefit.

In a similar fashion, the clinical benefit of scenario (b) is given by Equation 9.7, with a posterior distribution that is induced by the posterior distribution of the disease prevalence:

$$\pi = p_{01} + p_{11}. \tag{9.10}$$

Recall for scenario (a) that a threshold probability is selected and a logistic regression determines the number of subjects that test positive, then the posterior distribution of a clinical benefit is generated from the joint posterior distribution of the four cell probabilities of Table 9.6. The posterior distribution of the clinical benefits is easily generated by an MCMC method of WinBUGS, and Bayesian inferences are implemented. The methodology is demonstrated with an example using PSA as a diagnostic marker to detect prostate cancer.

9.3.4   Bayesian decision curves for prostate cancer

Decision curves for scenarios (a) and (b) will be determined with the PSA study by Etzioni et al. [1]. The data can be found at http://labs.fhcrc.org/ pepe/book/index/html. This is an interesting dataset that is described by Pepe [2: 10], who assessed the accuracy of the test for prostate cancer, and it was used as an example for selecting the optimal point on the ROC curve in an earlier section. It should be noted that these cases are screen detected, which affects the over-diagnosis bias. For additional information, see Baker, Kramer, and Pierce [14].

Our purpose is to assess the clinical benefit of a diagnostic test by deriving the decision curve for scenarios (a) and (b) and calculating a credible interval for the estimated clinical benefits of both scenarios. In the Etzioni et al. [1] study, the free and total PSA levels of 683 men are available.

Some idea of the overall accuracy of total PSA is provided by the area under the ROC curve. See Broemeling [13], Pepe [2], and Zhou, Obuchowski, and McClish [4] for descriptions of measures of accuracy for medical tests. The area under the curve is estimated as $0.837(0.016)$ with a 95% confidence interval of $(0.805, 0.870)$, where a non-parametric technique of SPSS is employed to calculate the area. It shows that total PSA had fairly good accuracy. A Bayesian calculation using the regression method of O'Malley et al. [3] for the ROC area gives a posterior mean of 0.8215, a posterior standard deviation of 0.0178, and a 95% credible interval of $(0.7848, 0.8548)$.

A total of 454 men with no disease had a mean total PSA of $2.023(2.684)$ compared to 229 men with prostate cancer, with a mean total PSA of $10.313(17.451)$. This also indicates that total PSA is a good prognostic marker for the disease. The PSA for diseased and non-diseased individuals was highly skewed to the right. By analogy with breast cancer, interest in the ROC curve should be focused on "small" FPR values in the neighborhood of 0.01 and for moderate sensitivity values in a range around 0.80. See Baker, Kramer, and Pierce [14] for more information about the role of the ROC curve in evaluating diagnostic test accuracy. For the PSA example, in the region where $\text{FPR} = 0.01$, $\text{TPR} = 0.188$.

The test accuracy of total PSA appears to be good, but does it have a clinical benefit? In order to answer this question, the Vickers approach is taken and decision curves for scenarios (a) and (b) will be determined. For threshold probabilities, $p_t = 0.01, 0.05, 0.1, 0.15, 0.2, 0.25, 0.3, 0.35, 0.4$, the posterior distribution of clinical benefits (Equations 9.6 and 9.7) is computed as follows. The threshold probability range begins at 0.01, which is based on an analogy to breast cancer screening, where the ratio of "benefit" to "cost" is

TABLE 9.7: Posterior distribution of clinical benefits—scenarios (a) and (b) posterior mean(sd) and 95% credible intervals.

Threshold probability	Clinical benefit scenario (a)	Clinical benefit scenario (b)	95% credible interval for scenario (a)	95% credible interval for scenario (b)
0.01	0.3281(0.0181)	0.3295(0.0182)	0.2929,0.3639	0.2942,0.3656
0.05	0.3000(0.0189)	0.3014(0.0189)	0.2633,0.3375	0.2646,0.3390
0.1	0.2613(0.0200)	0.2626(0.0200)	0.2228,0.3011	0.2239,0.3024
0.15	0.2202(0.0211)	0.2190(0.0213)	0.1793,0.2621	0.1778,0.2613
0.2	0.2241(0.0196)	0.1703(0.0224)	0.1862,0.2631	0.1269,0.2147
0.25	0.2048(0.0186)	0.1151(0.0240)	0.1688,0.2419	0.0684,0.1625
0.3	0.1775(0.0177)	0.0517(0.0259)	0.1434,0.2128	0.0015,0.1033
0.35	0.1645(0.0169)	−0.0212(0.0276)	0.1320,0.1984	−0.0748,0.0340
0.4	0.1387(0.0163)	−0.1065(0.0299)	0.1072.0.1713	−0.1643,−0.0472

$1/0.0054 = 185$, and the corresponding risk threshold is $1/(185+1) = 0.011$. See Baker, Kramer, and Pierce [14] and Pauker and Kassirer [15] for interesting information on choosing a probability threshold.

Assume a uniform prior density for the cell probabilities of Table 9.1 and select a p_t and perform a logistic regression with a threshold probability p_t. The dependent variable is the occurrence of prostate cancer, and the independent variable is PSA. This determines the number of people who test positive and negative in Table 9.6. In all cases, PSA is a significant predictor of the occurrence of cancer. Lastly, determine the parameters of the joint Dirichlet posterior distribution of the cell probabilities of Table 9.6. Once this is computed, the posterior distributions of p_{11}, p_{10}, and π are determined, and so are those of the two clinical benefits (1) and (2). Repeat the above procedure for the other threshold probabilities. Using a refresh of 100 and a burn in of 5,000 observations, 45,000 values are generated from the Dirichlet distribution of the cell probabilities of Table 9.6, and the results are reported in Table 9.7. The MCMC error for estimating all model parameters is <0.0001, and the code is given below.

BUGS CODE 9.2

```
model;
# Pepe Caret PSA Etzioni et al. [1]
# Uniform prior
   {
# This is the calculation for the critical benefit
# method of Vickers
# uniform prior
# pt is threshold probability
# the parameters for the gamma variables are determined separately
   by logistic regression
```

```
# pt=.01
    g01[1,1]~dgamma(1,2)
    g01[1,2]~dgamma(1,2)
    g01[2,1]~dgamma(455,2)
    g01[2,2]~dgamma(230,2)
    h01<- sum(g01[,])
    for( i in 1 :2 ) {for( j in 1 :2 ){ theta01[i,j]<-g01[i,j]/h01}}

    cb01<-theta01[2,2]-theta01[2,1]*(.01/.99)
#pt=.05
    g05[1,1]~dgamma(1,2)
    g05[1,2]~dgamma(1,2)
    g05[2,1]~dgamma(455,2)
    g05[2,2]~dgamma(230,2)
    h05<- sum(g05[,])
    for( i in 1 :2 ) {for( j in 1 :2 ){ theta05[i,j]<-g05[i,j]/h05}}
    cb05<-theta05[2,2]-theta05[2,1]*(.05/.95)
#pt=.1
    g1[1,1]~dgamma(1,2)
    g1[1,2]~dgamma(1,2)
    g1[2,1]~dgamma(455,2)
    g1[2,2]~dgamma(230,2)

    h1<- sum(g1[,])
    for( i in 1 :2 ) {for( j in 1 :2 ){ theta1[i,j]<-g1[i,j]/h1}}
    cb1<-theta1[2,2]-theta1[2,1]*(1/9)
#pt=.15
    g15[1,1]~dgamma(16,2)
    g15[1,2]~dgamma(2,2)
    g15[2,1]~dgamma(440,2)
    g15[2,2]~dgamma(229,2)
     h15<- sum(g15[,])
    for( i in 1 :2 ) {for( j in 1 :2 ){ theta15[i,j]<-g15[i,j]/h15}}

    cb15<-theta15[2,2]-theta15[2,1]*(15/85)
# pt=.2
    g2[1,1]~dgamma(252,2)
    g2[1,2]~dgamma(26,2)
    g2[2,1]~dgamma(204,2)
    g2[2,2]~dgamma(205,2)

    h2<- sum(g2[,])
    for( i in 1 :2 ) {for( j in 1 :2 ){ theta2[i,j]<-g2[i,j]/h2}}

    cb2<-theta2[2,2]-theta2[2,1]/4
```

```
# pt=.25
    g25[1,1]~dgamma(338,2)
    g25[1,2]~dgamma(51,2)
    g25[2,1]~dgamma(118,2)
    g25[2,2]~dgamma(180,2)

    h25<- sum(g25[,])
    for( i in 1 :2 ) {for( j in 1 :2 ){ theta25[i,j]<-g25[i,j]/h25 }}
    cb25<-theta25[2,2]-theta25[2,1]/3
# pt=.3
    g3[1,1]~dgamma(379,2)
    g3[1,2]~dgamma(76,2)
    g3[2,1]~dgamma(77,2)
    g3[2,2]~dgamma(155,2)

    h3<- sum(g3[,])
    for( i in 1 :2 ) {for( j in 1 :2 ){ theta3[i,j]<-g3[i,j]/h3 }}
    cb3<-theta3[2,2]-theta3[2,1]*3/7
# pt=.35
    g35[1,1]~dgamma(406,2)
    g35[1,2]~dgamma(91,2)
    g35[2,1]~dgamma(50,2)
    g35[2,2]~dgamma(140,2)
    h35<- sum(g35[,])
    for( i in 1 :2 ) {for( j in 1 :2 ){ theta35[i,j]<-g35[i,j]/h35 }}
    cb35<-theta35[2,2]-theta35[2,1]*35/65
# pt=.4
    g4[1,1]~dgamma(416,2)
    g4[1,2]~dgamma(109,2)
    g4[2,1]~dgamma(40,2)
    g4[2,2]~dgamma(122,2)
    h4<- sum(g4[,])
    for( i in 1 :2 ) {for( j in 1 :2 ){ theta4[i,j]<-g4[i,j]/h4 }}
    cb4<-theta4[2,2]-theta4[2,1]*2/3
# differences in clinical benefit for a minus that for scenario b
d01<- cb01-clinb01
d05<-cb05-clinb05
d1 <- cb1-clinb1
d15<-cb15-clinb15
d2 <-cb2-clinb2
d3 <-cb3-clinb3
d25<- cb25-clinb25
d35<-cb35-clinb35
d4 <- cb4-clinb4
```

```
# probability that clinical benefit of a is greater than that for scenario b
p01<-step(d01)
p05<-step(d05)
p1<- step(d1)
p15<-step(d15)
p2<- step(d2)
p25<-step(d25)
p3<- step(d3)
p35<-step(d35)
p4<-step(d4)
p<- p01*p05*p1*p15*p2*p25*p3*p35*p4
# The following is the clinical benefit where all patients are biopsed and based on the
# Pepe total PSA data and the method of Vickers
# scenario b
# pt=.01
clinb01<- (theta01[1,2]+theta01[2,2]) -(1-(theta01[1,2]+theta01[2,2]))*(.01/.99)
# pt=.-5
clinb05<- (theta05[1,2]+theta05[2,2]) -(1-(theta05[1,2]+theta05[2,2]))*(.05/.95)
# pt=.1
clinb1<- (theta1[1,2]+theta1[2,2]) -(1-(theta1[1,2]+theta1[2,2]))*(1/9)
# pt=.15
clinb15<- (theta15[1,2]+theta15[2,2]) -(1-(theta15[1,2]+theta15[2,2]))*(15/85)
# pt =.2
clinb2<- (theta2[1,2]+theta2[2,2]) -(1-(theta2[1,2]+theta2[2,2]))*(1/4)
# pt =.25
clinb25<- (theta25[1,2]+theta25[2,2]) -(1-(theta25[1,2]+theta25[2,2]))*(1/3)
# pt=.3
clinb3<- (theta3[1,2]+theta3[2,2]) -(1-(theta3[1,2]+theta3[2,2]))*(3/7)
# pt=.35
clinb35<- (theta35[1,2]+theta35[2,2]) -(1-(theta35[1,2]+theta35[2,2]))*(35/65)
#pt=.4
clinb4<- (theta4[1,2] | theta4[2,2]) -(1-(theta4[1,2]+theta4[2,2]))*(2/3)
}
```

Figures 9.4a and b are the densities of the posterior distribution of the clinical benefit at threshold probabilities 0.05 and 0.30 for scenario (a), and demonstrate the dramatic effect of the threshold probability on the clinical benefit. The mean of the posterior distribution of the clinical benefit at threshold 0.05 is 0.30, for the threshold 0.30, the posterior mean is 0.178, and the density plots are based on BUGS CODE 9.2. The posterior means corresponding to the selected threshold probabilities for the two decision curves are shown in Figure 9.5.

It is seen that the decision curve for scenario (a) dominates (b), beginning with threshold probability 0.15. On inspection of Figure 9.5, one would

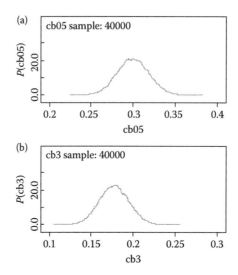

FIGURE 9.4: (a) Posterior density of clinical benefit at threshold 0.05, (b) Posterior density of clinical benefit at threshold 0.3.

conclude that the optimal strategy is to choose a biopsy on the basis of the total PSA level compared to every person having a biopsy. For threshold probabilities less than 0.15, the clinical benefits of the two scenarios are essentially the same. As the threshold probability increases, the posterior mean of the difference also increases, and can be observed from Figure 9.5.

The entries of Table 9.8 are computed with BUGS CODE 9.2, and present convincing evidence that the decision curve for scenario (a) dominates that for scenario (b), beginning with the threshold 0.15. The posterior probability that the clinical benefit for scenario (a) is greater than that for scenario (b) is 0.7478 at a threshold of 0.15 and 1 at a threshold of 0.2 and the larger

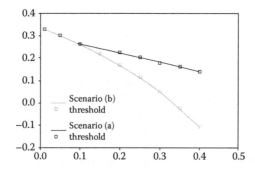

FIGURE 9.5: Clinical benefit for scenarios (a) and (b).

TABLE 9.8: The posterior probability that the clinical benefit of scenario (a) is greater than that for scenario (b).

Threshold	Mean	sd	2 1/2	Median	97 1/2
0.01	0.0098	0.0988	0	0	0
0.05	0.0493	0.2167	0	0	0
0.1	0.0969	0.2959	0	0	1
0.15	0.7478	0.4343	0	1	1
0.2	1	0	1	1	1
0.25	1	0	1	1	1
0.3	1	0	1	1	1
0.35	1	0	1	1	1
0.4	1	0	1	1	1

thresholds. Thus, if one's threshold probability is at least 0.15, one would benefit by using scenario (a), where the decision to biopsy depends on the value of PSA. In summary, this is an example where one scenario dominates the other, and becomes dominant in an obvious way when the threshold probability exceeds 0.15.

One advantage of the Bayesian approach is the way it uses prior information. For example, consider a threshold value of $p_t = 0.20$, then the 2×2 table (Table 9.6) would appear with cell frequencies: 251 for a TN, 203 for an FP, 204 for a TP, and 25 for an FN. If a uniform prior distribution is used, the resulting posterior distribution of the cell frequencies (TN, FP, TP, FN) is Dirichlet with parameter (252, 204, 205, and 26). Suppose a prior related study is available with cell frequencies (37, 20, 23, 11) for (TN, FP, TP, FN), then the prior information would be expressed as Dirichlet with parameter (38, 21, 24, 12) and would be combined with the experimental results of the present study to give a Dirichlet with parameter (289, 224, 228, 37). Of course, the effect of prior information is to lower the standard deviations of the posterior distribution of the clinical benefits.

9.3.5 Maximum likelihood and Bayes estimators

It is interesting to compare the maximum likelihood estimator (MLE) of the clinical benefit of scenario (a) with the Bayesian estimators of the clinical benefit. Vickers and Elkin [7] and others use MLE to estimate the clinical benefit of the two scenarios. Recall that the clinical benefit for scenario (a) is given by Equation 9.6, namely,

$$p_{11} - p_{10}(p_t/(1 - p_t)).$$

The MLE of this parameter is

$$n_{11}/n - n_{10}(p_t/(1 - p_t))/n, \tag{9.11}$$

TABLE 9.9: MLE(sd) and posterior mean(sd) of the clinical benefits for scenario (a).

Threshold probability	MLE	Posterior mean
0.01	0.3353(0.0180)	0.3281(0.0181)
0.05	0.3353(0.0180)	0.3000(0.0189)
0.1	0.3353(0.0180)	0.2613(0.0200)
0.15	0.3336(0.0180)	0.2202(0.0211)
0.2	0.2895(0.0176)	0.2241(0.0196)
0.25	0.2377(0.0176)	0.2048(0.0186)
0.3	0.1784(0.0176)	0.1775(0.0177)
0.35	0.1326(0.0181)	0.1645(0.0169)
0.4	0.0717(0.0188)	0.1387(0.0163)

where $n = n_{00} + n_{01} + n_{10} + n_{11}$. In Table 9.6, n_{ij} are the cell frequencies, where in particular n_{11} and n_{10} are the number of TP and FP counts for a particular threshold probability p_t. It should be emphasized that the cell frequencies of Table 9.6 vary with the threshold probability. The variance of the MLE is

$$p_{11}(1-p_{11})/n + p_{10}(1-p_{10})(p_t/(1-p_t))^2/n - 2(p_t/(1-p_t))p_{11}p_{10}/n, \tag{9.12}$$

which is estimated by the MLE

$$n_{11}(n-n_{11})/n^3 + (p_t/(1-p_t))^2 n_{10}(n-n_{10})/n^3 + 2(p_t/(1-p_t))n_{11}n_{10}/n^3. \tag{9.13}$$

It follows from the above that the MLEs of the clinical benefit and the MLE of the corresponding standard deviations are given by Table 9.9.

Recall that the Bayesian approach used a uniform prior for the cell probabilities, which accounts for the differences in the MLE and Bayesian estimators of the clinical benefits. The pattern of the MLEs and the posterior means is the same, but the former are somewhat larger than the latter, however, the standard deviations for each threshold probability are almost identical. I suppose that if an improper prior is used, the two estimators would be more alike, however, an improper prior was not used because it would result in an improper posterior density when the cell frequencies are zero. For "small" p_t (=0.01 and 0.05), some of the cell frequencies are in fact zero, thus, a uniform prior was employed instead. It would be interesting to compare the frequency properties of the posterior means with the corresponding MLEs.

9.3.6 Another example of decision curves

Suppose the head trauma study is revisited, where interest centers on investigating the accuracy of the biomarker for predicting complications of

TABLE 9.10: Posterior analysis for clinical benefits— scenarios (a) and (b) head trauma study.

Parameter	Mean	sd	2 1/2	Median	97 1/2
cb2	0.5777	0.0725	0.4308	0.5799	0.7138
cb3	0.5333	0.0789	0.3724	0.5358	0.6798
cb4	0.4687	0.0839	0.298	0.4708	0.626
cb5	0.4063	0.0898	0.2224	0.4093	0.5743
cb6	0.3156	0.0969	0.1132	0.3196	0.4944
cb7	0.2555	0.1108	0.0174	0.263	0.4512
cb8	0.1882	0.1294	−0.1045	0.203	0.4007
cb9	0.1253	0.1514	−0.2619	0.1588	0.331
clinb2	0.5895	0.0730	0.441	0.592	0.7259
clinb3	0.5624	0.0778	0.4039	0.5648	0.7073
clinb4	0.4531	0.0972	0.2535	0.4562	0.6337
clinb5	0.3436	0.1168	0.1058	0.3473	0.561
clinb6	0.1543	0.1454	−0.141	0.1581	0.4252
clinb7	−0.0938	0.1943	−0.4923	−0.0876	0.2677
clinb8	−0.6409	0.2917	−1.23	−0.6325	−0.0949
clinb9	−2.281	0.5809	−3.467	−2.263	−1.202

Note: cb values correspond to scenario (a) and clinb values to scenario (b).

head trauma by determining the clinical benefits for scenarios (a) and (b), defined by Equations 9.6 and 9.7, respectively. BUGS CODE 9.2 is revised by executing logistic regressions for threshold values ranging from 0.2 to 0.9 in steps of 0.1. Refer to the code where the cell counts for the 2 × 2 tables are entered as parameters of the gamma distribution, and note that by adding a 1 to the each of the four cell counts, a uniform prior is assumed. There is a 2 × 2 table for each threshold value and the cell counts are determined by logistic regression for the TP, TN, FP, and FN cells.

Based on 45,000 observations generated from the posterior distribution, with a burn in of 5,000 and a refresh of 100, the posterior analysis is reported in Table 9.10.

Beginning with threshold probability 0.4, that the clinical benefit of scenario (a) is larger than that of scenario (b), and it is even more obvious from Figure 9.6. This implies that beginning with threshold probability 0.4, if the probability of a complication exceeds the threshold value, then the subject should be further examined for complications due to head injury.

9.3.7 Comments and conclusions

Decision curves were introduced as a way to assess the clinical utility of a diagnostic test. The optimal strategy is to base the decision to biopsy on the value of a diagnostic variable compared to each person having a biopsy.

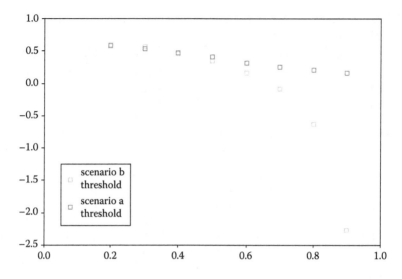

FIGURE 9.6: Clinical benefit scenario of (a) vs. (b) for head trauma study.

A Bayesian approach allows one to assess the uncertainty of a clinical benefit and to compare decision curves between various scenarios. It would be interesting to develop simultaneous credible bands about the decision curves.

In addition, it would be interesting to study the frequency properties of the Bayes estimators of the clinical benefits. For example, for a particular threshold probability, what are the frequency properties of the mode of the posterior distribution of the clinical benefit? With a uniform prior for the cell probabilities, one would expect the frequency properties of the posterior mode to be almost the same as the MLE of the clinical benefit. The 95% credible intervals for a clinical benefit should be very similar to the 95% confidence intervals of Vickers et al. [8]. With an improper prior, one would expect similar results with the Bayesian approach.

9.4   Test Accuracy and Clinical Trials

9.4.1   Introduction

This section describes the interplay between medical testing accuracy and clinical trials. Indeed, medical tests are a crucial element in the design and conduct of most clinical trials in oncology. For example, medical tests involving diagnostic imaging are present in Phase I trials for safety studies of new therapies, be they chemotherapy, radiotherapy, or biological therapy. In order to monitor the safety and efficacy endpoints of such trials, imaging procedures determine the advance and extent of the disease and produce the primary and secondary endpoints. This section introduces the three phases of clinical trials and how imaging plays a role in the conduct of the trial. This is followed by a description of the protocol for clinical trials in oncology and a brief description of the protocol review process at the University of Texas MD Anderson Cancer Center (MDACC). The response evaluation criteria in solid tumors (RECIST) criteria for response to therapy are introduced. This is a set of guidelines for the radiologist in their determination of the patient's response to therapy and hence on the conclusions for the success or failure of the trial.

Bayesian sequential stopping rules for the design and conduct of clinical trials are outlined and developed in the latter parts of the chapter, followed by a description of the software developed at MDACC for the design of such trials. The focus will be on Bayesian stopping rules for safety and efficacy in Phase I, II, and III trials.

Lastly, several examples are presented. The first example is a Phase I trial in renal cell carcinoma (RCC) that illustrates a Bayesian dose finding, based on logistic regression, while the second is a hypothetical Phase II study, developed by the author, but based on an actual study for inter observer agreement in lung cancer. The third example illustrates a statistical stopping rule for a Phase II trial in melanoma. For the three examples, the role of medical test accuracy is emphasized.

9.4.2 Clinical trials

Bayesian methods in clinical trials are best explained by Thall [16], who emphasizes the role that ethics and science play in the design and analysis of clinical trials. He stresses the complexity of such studies, because they involve decisions when selecting therapies, the choice of dose levels of a particular regimen, and above all, concern for the patient's safety. Primarily for patient safety, clinical trials should be conducted in a sequential fashion, which calls for interim monitoring of patient outcomes.

This section will review the three phases of a clinical trial and focus on the role of diagnostic imaging in such studies. Later, Bayesian sequential stopping rules for interim analysis of clinical trials will be given in more detail.

9.4.3 Phase I designs

Phase I trials evaluate how a treatment is to be administered and how that treatment affects the human body. First, consider a Phase I study that evaluates safety among a set of doses of a new treatment. The study will be designed to determine the maximum tolerable dose (MTD), which is the dose, whereby at higher doses, the safety of the patient would be compromised. We are assuming that as the dose level increases, the probability of toxicity increases and the probability of efficacy also increases. The main endpoint in a Phase I study is some measure of toxicity experienced by the patient as a result of the treatment, while the secondary endpoint is some measure of efficacy. To define the toxicity endpoint, the investigator characterizes the dose limiting toxicity (DLT), which is a set of toxicities that are severe enough to prevent giving more of the treatment at higher doses. The investigator bases the DLT on knowledge of the disease, treatment, and the patients who are eligible for the trial. Investigators are guided by the National Cancer Institute (NCI) list of common toxicities or in some other manner that is appropriate for the particular study.

Prior to implementing a Phase I trial, the investigator must have decided on the treatment route of administration and schedule. Also required for estimating the MTD, is the patient population defined via the eligibility and ineligibility criteria, a starting dose and a set of dose levels to test (DLT), and the dose escalation. The dose escalation includes decisions on selecting the MTD among a set of doses. The chosen starting dose is based on other similar Phase I studies and/or information from animal experiments. Once the investigator has chosen the dose levels to be tested, the dose escalation can be described.

There are many dose escalation rules, including the commonly used $3+3$ design and the continual reassessment method (CRM). Since the early days of the NCI, investigators have used traditional escalating rules, such as the $3+3$ design for determining the MTD in oncology trials, while the CRM (see Crowley [17]) is a newer development that is becoming more popular. The $3+3$ design is based on cohorts of size three or six, and there are several versions.

What is the role of medical test accuracy in Phase I clinical trials? The assessment of safety can include imaging of damage due to treatment and can also be an integral part of the assessment of efficacy. If a solid tumor is involved, imaging will measure the growth of that tumor during the course of the trial and this information will be used in planning the Phase II trial, and can serve as a source of prior information for a Bayesian sequential design of a trial. Thus, the accuracy of the imaging device to measure the size of the tumor is essential for the successful implementation of a Phase I trial, and as will be seen in the next section, for Phase II trials.

9.4.4   Phase II trials

Once a particular treatment or intervention has been studied with a Phase I trial and the MTD has been selected and we are quite satisfied that the treatment will be safe, studies of the treatment may progress to Phase II trials to determine if the treatment holds sufficient promise. Typically, the target population is patients with a specific disease, disease site, histology, or stage, or patients undergoing some surgical or anesthetic procedure. Often, the treatment dose is the MTD determined from previous Phase I trials. Although limited dose finding is sometimes allowed to accommodate different patient populations, the primary endpoints are measures of efficacy, while safety would be a secondary issue.

It is for Phase II trials that medical test accuracy, via diagnostic imaging, plays a crucial role. Often, the primary endpoint is the fraction of patients who experience a response to therapy, and frequently the response is based on the change in tumor size as measured from baseline to some future point at the end of the treatment cycle. The response categories can be classified as a complete response, a partial response, or no response, depending on the percent relative change from baseline. The World Health Organization (WHO) and RECIST criteria, as described by Padhani and Ollivier [18], define the actual response categories that must be carefully specified in the protocol. It is important to understand the uncertainty introduced into such trials by the disagreement between the radiologists who are responsible for assigning the response to therapy to each patient. This uncertainty is often unknown and unaccounted for by others, including statisticians, who are designing and analyzing trial information.

The efficacy information from a Phase I trial is also important and largely determines the type of Phase II trial to be designed. If little is known about the efficacy, a Phase IIa trial can be performed, with the goal of determining a certain minimum efficacy. On the other hand, if the efficacy information from Phase I trials indicates that the intervention does indeed have some benefit, a Phase IIb trial may be implemented to determine if the treatment has sufficient benefit compared to some standard treatment, either historical (from past patient data) or from an ongoing trial.

We will see how prior information from the relevant Phase I trials will be employed in the design of Bayesian sequential stopping rules and sample size information of the planning of the Phase II trial. Designs for Phase IIa trials include Gehan's two-stage and Simon's two-stage design, and also relevant are multi-stage designs, which are explained in Crowley [17]. Simon's [19] two-stage design is discussed here, because it is the most popular for a Phase IIa trial, however, Bayesian alternatives are becoming more widely used because they are more flexible and can easily incorporate information from prior related Phase I and II trials.

Phase II designs are based on statistical testing principals. Suppose p is the probability of a treatment response, then one tests the null hypothesis vs. the alternative hypothesis:

$$H : p < p_0 \text{ (e.g., } = 0.05) \text{ vs. } A : p > p_1 \text{ (e.g., } = 0.25). \tag{9.14}$$

The null hypothesis states that the proportion of responses is less than or equal to some specified proportion, p_0, that would not exhibit sufficient interest for further development. The alternative hypothesis states that the proportion of responses is greater than or equal to a proportion p_1 that the investigator considers clinically meaningful. If the alternative hypothesis is true, then further testing could be deemed reasonable. Of course, this decision is based on other considerations as well, such as any new information on safety.

The values for p_0 and p_1 are specified in advance and depend on the results of previous trials. Typical values for p_0 are from 0.1 to 0.4, and typical values for p_1 are from $p_0 + 0.15$ to $p_0 + 0.2$. To use a Simon two-stage design, investigators must also specify the probability of a type I error α, the probability of rejecting the null hypothesis when it is true (declaring that the new treatment has an effect above p_0 when it actually does not), and β, the probability of accepting the null hypothesis when it is false (declaring that the new treatment has no effect above p_0 when it actually does). Note that $(1 - \beta)$ is the power of the test.

Given these values, the Simon method will give the maximum sample size n, the stage 1 sample size n_1, and the rejection rule at each stage. DeVita, Hellman, and Rosenberg [20] provide tables for Simon's [19] two-stage design, and for example, when $\alpha = 0.05$, $\beta = 0.20$, $p_0 = 0.05$, and $p_1 = 0.25$, then $n_1 = 9$, $n = 17$, and the trial would be stopped early if there were zero out of nine responses. If there are one or more responses with 9 patients, the trial is continued, and if there are two or less responses among 17 patients, the null hypothesis is accepted, that is the intervention or treatment would not be of sufficient interest for further testing. Note with this design that the trial is stopped early for lack of efficacy. The Simon design can be used to justify the sample size and for stopping early. Stopping early protects future patients from receiving inefficacious treatments.

Some Bayesian designs allow more flexibility. For example, suppose the maximum sample size of N patients is accrued in k cohorts of size n, and that after observing the response of patients at the end of each cohort the investigator computes the probability that the observed proportion of responses p is greater than p_1, given the responses of the observed patients as $\Pr[p > p_1$ responses of patients]. If this probability is small, say 0.10 or 0.20, the trial is stopped for lack of efficacy. This is very much like the Simon design; however, a decision on lack of efficacy can be made after each cohort of patients. See Thall, Simon, and Estey [21,22] for additional information on Phase II trials that use Bayesian stopping rules.

If the intervention under investigation has shown some activity, a Phase IIb trial can be used to determine the extent of efficacy. This type of trial is usually comparative, since it has demonstrated prior efficacy, and the study intervention will be compared to some historical control, or to some standard current treatment via a randomized design. The advantage of using historical controls over concurrent controls is the smaller number of patients required, but the disadvantages of historical controls are that the patient populations may not be comparable to those used in the current clinical trial.

9.4.5   Phase III trials

We are now at the point where an intervention (drug or procedure) has been studied in a series of Phase I and Phase II trials and has demonstrated sufficient promise to be compared to the standard clinical treatment in a large randomized study.

Phase III trials are confirmatory, where the study procedure is to be compared to the standard therapy with the goal of providing evidence that the study drug will provide substantial improvement in survival time or in disease-free survival or some other time-to-event endpoint, such as time to response or time to hospitalization. Phase III trials should be designed to have a sufficient sample size to detect clinically relevant differences and are usually done in a multicenter setting. Provisions are made for interim looks by an independent Data Safety Monitoring Board, where the trial may be stopped early for reasons of safety and/or efficacy. The response to therapy may serve as a secondary endpoint in Phase III trials, thus diagnostic imaging plays a crucial role in the conduct of all clinical trials.

9.4.6 Protocol

What is a protocol? It states in detail how the medical study is to be organized and executed. There are generally two types: those submitted by a pharmaceutical or medical device company, and those that are initiated by a principal investigator at the institution. The protocol should include the following components: (1) an explanation of the scientific basis for the study; (2) a summary of the results of all previous related studies and experiments of the study intervention; (3) the patient eligibility and ineligibility criteria; (4) a list of the major and minor endpoints, including their definitions and how and when they will be measured; (5) the definitions of evaluable and intent-to-treat populations; (6) the estimated patient accrual rates by site; (7) a statistical section that outlines a detailed power analysis for sample size, a description of rules for stopping early, methods for randomizing patients, and the proposed statistical analysis; and (8) non-statistical stopping rules for safety considerations. Additional documentation that must accompany the protocol is a list of all NIH toxicities and the patient informed consent form. For protocols initiated by private companies, a biostatistician is assigned to review it, but for protocols initiated at MDACC, the study has one biostatistician assigned as a collaborator (the one who assisted the principal investigator [PI] in the statistical design) and a different statistician who reviews it and presents it to the department for approval.

Every protocol at MDACC is reviewed in three stages, first by the Department of Biostatistics and Applied Mathematics, next by the Clinical Research Committee (CRC), and lastly by the Institutional Review Board (IRB). During the first review, a biostatistician presents the protocol in written and oral form to the department, and there is a set procedure for this presentation. The presentation is concluded with a list of major and minor concerns regarding the revision of the protocol. Then, the department discusses the above-mentioned recommended revisions and votes to approve or disapprove, after that a directive is sent to the PI. If need be, the PI then revises the protocol accordingly, often with the help of the biostatistical collaborator and/or reviewer.

9.4.7   Guidelines for tumor response

The RECIST criteria provide the radiologist with guidelines for determining the change in tumor size in such a way that the response to therapy can be judged and the success or failure of the trial evaluated. The following outline will be useful for understanding the guidelines: eligibility, methods of measurement, baseline identification of target and non-target lesions, response criteria, evaluation of best overall response, confirmation and duration of response, and reporting of results. What follows is a very brief description.

Only patients with measurable lesions are eligible, namely, those that can be accurately measured with CT or MRI. Both targeted and non-target lesions are to be identified. A maximum of 10 lesions representative of all involved organs are identified as target lesions, which must be accurately and repeatedly measured by the longest diameter of the lesion. The primary endpoint is the sum of the longest diameters (SL) of the target lesions. All other lesions are identified as non-target lesions.

Based on the SL of the target lesions, each patient is classified into the following categories: complete response (CR), where all target lesions disappear; partial response (PR), where there is at least a 30% decrease in the SL of all target lesions, using the baseline SL as a reference; progressive disease (PD), where there is at least a 20% increase in the SL, relative to the smallest value of SL recorded since the treatment started; stable disease (SD), where there is neither sufficient shrinkage to qualify as PR or sufficient increase to qualify as PD.

There is also an evaluation of the non-target lesions, where the patient is classified as CR, incomplete response/SD, and PD. The patient is then given an overall best response, based on the response of the target and non-target lesions, and finally, the patient is put into one of four overall categories: CR, PR, SD, or PD. See Therasse et al. [23] for more detailed information on the RECIST guidelines and Padhani and Ollivier [18] for the implications of those guidelines for diagnostic radiologists.

Note, the guidelines are only that and do not include procedures for just how they are to be implemented. For example, there is no mention of the number of readers to be included or a procedure for the resolution of disagreement between radiologists in their determination of the patient's response to therapy. All these elements create an element of uncertainty, which is unknown by others involved in the design and conduct of a clinical trial. This creates uncertainty in the classification of a patient's response to therapy, and, consequently, is not accounted for by the statisticians in their design of Phase II trials. Now it is seen how accuracy affects the clinical trial, not only is there intrinsic inaccuracies with the medical devices but other inaccuracies introduced by the disagreement between radiologists who are classifying patients into the various response categories.

The study by Thiesse et al. [24] gives one some idea of the uncertainty in the process of assigning a patient's response to treatment. The study evaluated the impact of a review committee on the overall response status of a

TABLE 9.11: Agreement between the review committee and the original report—response by Review Committee.

Original report	CR	PR	MR	SD	PD	Total
CR	14	2	1	0	2	19
PR	4	38	4	6	10	62
MR	0	7	9	3	5	20
SD	0	0	1	4	15	20
PD	0	1	0	3	1	5
Total	18	48	11	16	33	126

Source: From Thiesse et al., *Journal of Clinical Oncology*, 15, 3507, 1997, with permission of American Society of Clinical Oncology.

patient for a large multicenter trial with 489 patients with renal cancer given cytokine therapy, see Negrier et al. [25]. There were five response categories: CR, PR, MR, SD, and PD, where MR stands for marginal response. A blinded peer review of all responders and all litigious cases was done by the review committee. The results for 126 reviewed files are given by Table 3 in Thiesse et al. [24] and Table 9.11.

Using the generalization of the G coefficient (see Chapter 4), its posterior distribution is easily found and provides a posterior mean of 0.019, a median of 0.018, and a standard deviation of 0.089. This implies that the agreement between the review committee and the original readers was very poor. Indeed, the Thiesse study itself gives 0.32 as the Kappa coefficient, which also confirms poor agreement. This shows that disagreement among radiologists is quite common in the conduct of a clinical trial, and in particular in the assignment of a patient's response to therapy. This fact usually remains unknown to others, including statisticians, in the design and analysis of such studies.

Many studies have demonstrated such lack of agreement between radiologists. For example, a recent investigation by Erasmus et al. [26] shows the lack of consistency in measuring tumor size and poor intra and inter observer agreement. In fact, for some lesions there was as much as a 50% difference in measuring the lesion size for two looks at the same image by the same reader! Again, it is important to recognize the importance of the role that test accuracy plays in assessing the success of a clinical trial.

9.4.8 Bayesian sequential stopping rules

Because of the complexity of clinical trials and the incorporation of prior information from other previous studies, the Bayesian approach to interim analysis is quite appropriate. What is to be presented here is for Phase II trials, where response to therapy is the primary endpoint, while toxicity is a secondary endpoint. Prior information on response and toxicity will be taken from previous Phase I and II trials that are relevant to the "new" therapy. The response to therapy is the main endpoint generated by radiologists using the RECIST criteria. The software to implement the design of the Bayesian stopping rule will be discussed and demonstrated in the next section.

Denote the following four probabilities of mutually exclusive and exhaustive events for a Phase II trial of an experimental therapy as: $\theta_1 =$ probability of response and toxicity, $\theta_2 =$ response and no toxicity, $\theta_3 =$ no response and toxicity, and $\theta_4 =$ no response and no toxicity. Suppose the corresponding probabilities of a previous standard relevant study are ϕ_1, ϕ_2, ϕ_3 and ϕ_4, respectively. Thus, the probability of a response with the experimental therapy is $\theta_r = \theta_1 + \theta_2$ and that for the standard is $\phi_r = \phi_1 + \phi_2$, the probability of toxicity with the experimental therapy is $\theta_t = \theta_1 + \theta_3$, while that for the standard is $\phi_t = \phi_1 + \phi_3$. It is known that $\theta = (\theta_1, \theta_2, \theta_3, \theta_4)$ and $\phi = (\phi_1, \phi_2, \phi_3, \phi_4)$ have Dirichlet distributions, thus so does (θ_r, θ_t) and (ϕ_r, ϕ_t).

Now suppose, based on historical information, that among n patients on the standard therapy, there are a responses, and among m patients, there are b toxicities, while for the "new" experimental therapy, a priori, there will be c responses and d toxicities.

Therefore, a priori,

$$\phi_r \sim \text{beta}(a, n - a) \tag{9.15}$$

and

$$\phi_t \sim \text{beta}(b, m - b). \tag{9.16}$$

Frequently, the prior information about experimental therapy is taken to be vague or non informative, and one lets θ_r and θ_t have uniform distributions.

The alternative hypothesis is

$$A : \theta_r < \phi_r \quad \text{or} \quad \theta_t > \phi_t$$

vs. the null

$$H : \theta_r \geq \phi_r \quad \text{or} \quad \theta_t \leq \phi_t.$$

The rule to stop the trial after observing the number of responses and toxicities is when

$$\Pr[\theta_r < \phi_r \mid \text{data}] > \eta$$

or

$$\Pr[\theta_t > \phi_t \mid \text{data}] > \varepsilon, \tag{9.17}$$

where η and ε are usually selected "large," say 0.90 or 0.95.

Thus, the trial is stopped if the posterior probability is high when the rate of responses with the experimental therapy is less than that of the standard, or if the posterior probability is large when the rate of toxicities with the experimental therapy exceeds that of the standard. Since ϕ_r and ϕ_t are correlated, the events $\theta_r < \phi_r$ and $\theta_t > \phi_t$ are not independent!

Medical test accuracy, via diagnostic imaging, plays an important role in this type of trial. It is important to know how the trial parameters are based on prior information. Such information about efficacy most likely is the result of imaging the tumor size in Phase I trials. For the trial at hand, the number of responses and the number of toxicities are based on imaging the size of the primary tumor and tumors at the sites of metastases. Note that θ_r, the probability of a response, is based on the RECIST criteria for categorizing patients into the various responses categories: CR, PR, SD, and PD. The protocol must specify in detail the definition of response that is used in the Bayesian stopping rule. Usually, response means the event CR or PR, which in turn, as has been explained above, depends on the change in tumor size from some reference time, defined in the protocol. The protocol will not mention the number of readers or how disagreements between readers are resolved. Of course, such information is not known to the statisticians who design the trial.

The following example is taken from Cook [27]. Suppose a previous related trial had 200 patients, among which $a = 60$ responded and 140 did not. Among 160 of these patients, $b = 40$ experienced toxicities, but 120 did not experience any serious side effects. Let the prior distribution of

$$\phi_r \sim \text{beta}(60,140)$$

and

$$\phi_t \sim \text{beta}(40,120),$$

while the prior distributions for the corresponding parameters of the melanoma group for response and toxicity are assumed to be uniform.

The trial is stopped when the null hypothesis is rejected:

$$\Pr[\theta_r < \phi_r \mid \text{data}] > 0.95$$

or

$$\Pr[\theta_t > \phi_t \mid \text{data}] > 0.95. \qquad (9.18)$$

The stopping rule for response is presented in Table 9.12.

TABLE 9.12: Stopping rule for response.

Response	Boundary
0	6
1	12
2	17
3	22
4	27
5	30

TABLE 9.13:   Stopping rule
for toxicity.

Toxicity	Boundary
3	3
3	4
4	6
5	8
6	10
6	11
7	13
7	14
8	16
8	17
9	19
10	21
10	22
11	24
11	25
12	27
12	28
13	30

Thus, if there are no responses among six patients, the trial is stopped. Therefore, one must know the response among at least six patients before the stopping rule for response takes effect. On the other hand, see Table 9.13 for the stopping rule for toxicity.

If the first three patients experience toxicity, the trial is stopped. What are the frequency properties of this test? Suppose the null hypothesis is "true" and that, hypothetically, $\theta_r = 0.4$, $\theta_t = 0.2$, $\phi_r = 0.3$, and $\phi_t = 0.25$, then using the above stopping rule, the probability of stopping the trial with various sample sizes is given in Table 9.14.

The probability of stopping is equivalent to the probability of a type I error. Note, with only three patients, the probability is 0.008 of stopping the trial, and as the sample size increases, the probability of stopping slowly increases up to 28 patients, then it has to increase to 1 at the maximum sample size of 30. Also for this scenario, the average number of patients is 26.7, experiencing an average of 5.34 toxicities and an average of 1.68 responses. The average number of patients treated is relatively large, because this scenario is when the null hypothesis is true.

Now suppose that the alternative hypothesis is "true" with $\theta_r = 0.2$, $\theta_t = 0.35$, $\phi_r = 0.3$, and $\phi_t = 0.25$, then the probability of stopping is equivalent to the "power" of the test and is shown in Table 9.15.

With this particular scenario of the alternative hypothesis, the probability of stopping or "power" gradually increases from 0.0429 with 3 patients to 1

TABLE 9.14: Probability of stopping when $\theta_r = 0.4$, $\theta_t = 0.2$, $\phi_r = 0.3$, and $\phi_t = 0.25$.

n	Probability of stopping
3	0.0080
4	0.0272
5	0.1037
6	0.1095
8	0.1120
9	0.1128
11	0.1198
12	0.1268
13	0.1277
14	0.1298
16	0.1305
17	0.1355
19	0.1361
20	0.1375
22	0.1395
23	0.1406
25	0.1409
26	0.1430
28	0.1433
29	0.439
30	1.000

with 30. The average number of responses is 2.8 with an average of 4.92 toxicities among an average of 14 treated patients. The probability of stopping or power is approximately 0.8 with 29 patients.

Thus, for any scenario of the probabilities for response and toxicity of the experimental and standard therapies, the probability of stopping the trial can be computed. This allows one to estimate the sampling properties of the Bayesian test for stopping the trial.

9.4.9 Software for clinical trials

The Department of Biostatistics and Applied Mathematics at MDACC has developed many programs for the analysis and design of clinical and scientific studies in medicine and biology. These can be accessed at http://Biostatistics/mdanderson.org/SoftwareDownload/.

This library includes many programs, and is easily accessible to the student. Only two of the most relevant for clinical trials will be described. The first is appropriate for Phase I dose-finding trials, while the second is used for Phase II trials, when the major endpoints are for response and toxicity. The latter program is called Multc Lean, and the former CRM Simulator.

TABLE 9.15: Probability of stopping with $\theta_r = 0.2$, $\theta_t = 0.35$, $\phi_r = 0.3$, and $\phi_t = 0.25$.

n	Probability of stopping
3	0.0429
4	0.1265
5	0.4235
6	0.4517
8	0.4680
9	0.4980
11	0.5132
12	0.5798
13	0.5881
14	0.6044
16	0.6125
17	0.6713
19	0.6780
20	0.6902
22	0.7261
23	0.7361
25	0.7409
26	0.7763
28	0.7802
29	0.7874
30	1.000

9.4.10   CRM simulator for Phase I trials

Recall that Phase I trials are the beginning of studying a new agent or therapy and the first concern is for the safety of the patient. The study is designed in order to determine the MTD, which is the dose whereby at higher doses the safety of the patient would be compromised. We are assuming that as the dose level increases, the probability of toxicity increases and the probability of efficacy also increases. The main endpoint in a Phase I study is some measure of toxicity experienced by the patient as a result of the treatment, while the secondary endpoint is some measure of efficacy. To define the toxicity endpoint, the investigator characterizes the DLT, which is a set of toxicities that are severe enough to prevent giving more of the treatment at higher doses. The investigator bases the DLT on knowledge of the disease, treatment, and the patients who are eligible for the trial. Investigators are guided by the NCI's list of toxicities, or in some other manner that is appropriate for the particular study.

Also required for estimating the MTD is the patient population defined via the eligibility and ineligibility criteria, a starting dose and a set of DLT, and the dose escalation. The dose escalation includes decisions on how to select the MTD among a set of doses. The chosen starting dose is based on other similar Phase I studies and/or information from animal experiments. Once the investigator has chosen the dose levels to be tested, the dose escalation can be described. The CRM simulator uses only one endpoint, namely toxicity, and is easily executed. The student should refer to the CRM Simulator Guide and the Methods of Description. Both technical reports can be accessed from the above internet address.

9.4.11 Multc Lean for Phase II trials

The example of a Phase II trial appearing earlier was implemented using Multc Lean. Recall that there are two major endpoints, one for the number of responses and one for the number of toxicities among the maximum number of patients to be accrued for the trial. The methods are best explained by the Multc Lean Statistical Tutorial by Cook [27], which together with the program, can be downloaded from the above address. The user must supply the maximum number of patients to be accrued, and information about prior related studies. Prior therapy is referred to as the standard therapy, while the therapy to be tested is referred to as the experimental therapy. Prior information about the new treatment is usually given as non informative or vague, while that for the standard is more informative and usually provided with the number of responses and the number of toxicities experienced by a given number of patients in earlier Phase I studies.

Multc Lean consists of four parts: model input, stopping criteria, scenario input, and scenario output. The model input statement specifies the prior information for the standard and experimental treatments. With regard to the stopping criteria, recall that

$$\Pr[\theta_r < \phi_r \mid \text{data}] > 0.95$$

or

$$\Pr[\theta_t > \phi_t \mid \text{data}] > 0.95. \qquad (9.19)$$

The first probability is for stopping the trial when the probability of a response for the experimental therapy is less than that for the standard therapy. If this probability exceeds 95%, the trial is stopped for lack of efficacy, relative to the standard treatment. On the other hand, the trial is stopped early if the probability of toxicity with the experimental treatment exceeds that of the standard with a high probability, in this case 0.95. All this information is specified in the stopping criteria section of Multc Lean. As a result of this information, the program provides stopping boundaries for response (Table 9.12) and toxicity (Table 9.13).

In order to know the frequency properties of the Bayesian stopping rule, Multc Lean computes the probability of stopping the trial for all sample sizes, given a particular scenario of values for $\theta = (\theta_1, \theta_2, \theta_3, \theta_4)$ and $\phi = (\phi_1, \phi_2, \phi_3, \phi_4)$ and thus for (θ_r, θ_t) and (ϕ_r, ϕ_t).

The stopping criteria (Equation 9.18) are given in terms of the response and toxicity parameters for the experimental (θ_r, θ_t) and standard therapies

(ϕ_r, ϕ_t), which must be kept in mind when running a particular scenario. For example, $\theta_r = 0.4$, $\theta_t = 0.2$, $\phi_r = 0.3$, and $\phi_t = 0.25$ was used when assuming the null hypothesis was true, while $\theta_r = 0.2$, $\theta_t = 0.35$, $\phi_r = 0.3$ and $\phi_t = 0.25$, was employed for an alternative hypothesis scenario. The program computes the probability of stopping, the average number of patients treated, the number of responses to be expected, and the average number of toxicities to be experienced by this average number of patients. See Tables 9.14 and 9.15 for the outcome of the two scenarios for the null and alternative hypotheses of this Phase II study.

9.4.12   A Phase I trial for renal cell carcinoma

Thall and Lee [28] give a nice description of three designs for Phase I trials. The $3 + 3$, CRM, and Bayesian logistic regression are compared with regard to the percentage of times the correct dose is selected, and they use prior information to design a Phase I trial for RCC. Patients were previously treated with interferon and are to be treated with a fixed dose of 5-FU and six dose levels of gemcitabine (GEM). These designs were briefly mentioned, but only the latter design will be described and illustrated with WinBUGS.

The logistic model for this design is

$$\text{Log}[\theta_i/(1 - \theta_i)] = \alpha + \beta x_i, \tag{9.20}$$

where there are d dose levels $x_1 < x_2 < \cdots < x_d$, θ_i is the probability of a DLT at dose x_i, and α and β are unknown parameters. Recall that the objective of a Phase I trial is to estimate the MTD, and to do this the investigator must specify the number of dose levels, d, the dose level, x_i, and a target toxicity level, T. The MTD is that dose where the probability of toxicity is as close as possible to T. That is, doses greater than the MTD have probabilities of toxicity that are at least as large as T, while for doses that are less than the MTD, the corresponding probabilities of a DLT are less than or equal to T.

Also to be specified is a rule for stopping the trial early. This can be problematic, and there is no unique way to do it. One can choose n patients and test all n to estimate the MTD, or one can have a rule that stops the trial early if a given number of patients has been treated at the next recommended dose. Usually, patients enter the trial in cohorts of size three or six, and after each cohort is treated, the next recommended dose level is selected.

The logit model assumes that as the dose level increases, so does the probability of toxicity. However, it is usually true that as the dose level increases, so does the probability of a favorable response, which creates somewhat of a dilemma, in that the two events, "toxicity" and "response," are competing with one another.

With the Bayesian approach, a prior probability for the parameters α and β must be specified, and this is usually done by selecting two probabilities of toxicity corresponding to two of the dose levels, d, and "solving" the resulting two equations for α and β. This information is given by the study investigator and is based on previous related human and animal studies. This way of

TABLE 9.16: Average probability (%) of toxicity by dose of GEM (mg/m^2).

Cohort	Assigned dose	Observed # toxicities	100	200	300	400	500	600
Prior			5.9	*25*	46.8	63.7	*75*	82.3
1	200	0	3.3	11.3	*26.1*	41.1	52.4	60.5
2	300	1	3.7	12.2	*29.1*	47.1	59.9	68.2
3	300	0	2.6	7.8	*19.2*	34.6	47.9	57.4
4	300	0	2.0	5.8	14.5	*27.7*	40.6	50.5
5	400	1	2.1	6.2	15.2	*29.4*	43.5	54.4
6	400	0	2.1	5.4	11.7	*21.7*	33.0	43.0
7	400	1	2.3	5.8	12.8	*23.9*	36.5	47.2
8	400	0	2.1	5.3	10.9	*19.9*	30.6	40.5
9	400	0	2.1	4.9	9.6	17.1	*26.4*	35.5
10	500	2	2.2	4.9	11.2	*22.4*	36.0	48.4
11	400	0	1.6	4.3	10.1	*20.1*	32.7	44.7
12	400	1	1.6	4.6	10.8	*21.3*	34.5	46.7

Source: From Thall, P.F. and Lee, S.J., *International Journal of Gynecological Cancer*, 13, 251, 2003, with permission of International Journal of Gynecological Cancer.

estimating the MTD is taken from Table 1 of Thall and Lee [28] and is presented as Table 9.16. They assume the target toxicity level is $T = 0.25$ and do not use an early stopping rule, but use the information from all 36 patients.

The prior distribution for α and β is chosen so that the probabilities of toxicity at doses 200 and 500 are 0.25 and 0.75, respectively, giving $\alpha = -1.1133$ and $\beta = 0.0031808$. This begins the process of selecting the MTD. With these initial values for the parameters as prior information, three patients enter the trial resulting in zero toxicities, then the logistic model estimates the parameters and the six probabilities of toxicity corresponding to the six dose levels. The dose level 300 is the dose that has a probability of toxicity of 26.1, which is closest to the target toxicity level, $T = 0.25$, thus 300 is selected as the next recommended dose, see the italicized entries of Table 9.16. The process is repeated for the remaining 33 patients in cohorts of size three. At the twelfth cohort, the next recommended dose level is 400, which is the estimated MTD.

It is important to remember that the primary aim of the Phase I trial is to provide information about the safety of the therapy, but an important secondary objective is to gather information on the efficacy of the treatment. Both the estimated MTD and the information on efficacy will be used in any following Phase II trials. Diagnostic imaging will determine the response of the primary kidney tumor size and the response of the size of any metastatic lesions to treatment.

9.4.13 An ideal Phase II trial

As the first case of a Phase II trial that employs diagnostic imaging, a hypothetical example based on real data from Erasmus et al. [26] is described. The study involves 5 readers, 40 lung cancer lesions, and 2 replications, that

TABLE 9.17: Tumor size mean(sd) by time and response category—averaged over 10 lesions.

Time	Response			
	CR	PR	SD	PD
0	3.77(1.57)	4.79(0.958)	4.20(1.69)	3.95(1.70)
1	2.16(1.51)	4.26(1.11)	4.14(1.93)	4.34(1.81)
2	1.37(1.31)	3.81(0.947)	4.19(1.68)	4.94(1.70)
% increase from baseline	−63.66	−20.45	0.0023	25.06

is, each reader views the same image twice. All readers read all 40 lesions, and the major endpoint is the size of the lesion as determined by CT. The main focus of this study is to estimate the inter and intra observer error, and the main conclusion is that tumor size measurements are often inconsistent and can lead to incorrect interpretations of response to therapy based on the WHO and RECIST criteria.

Using the first replication of the five readers, the study results are used as baseline measurements for a hypothetical Phase II study. The "first" 10 lesions are used for patients with an intended complete response, and repeat measurements are assigned at random for times 1 and 2. The "second" set of 10 lesions are used for an intended partial response category of patients, where the average lesion size decreased from time 0 to time 2 by 25%. The basic descriptive statistics for the trial are given in Table 9.17.

A normal random number generator is used to generate hypothetical tumor size measurements by category of response and by the repeated measurement times 0, 1, and 2. Thus, for the complete response category of 10 patients, there was a 64% decrease in the average lesion size, relative to baseline. On the other hand, for the progressive disease category, the average lesion size increased from 3.95 to 4.94 cm, an increase of 25.06%.

There were five readers in this study and the mean(sd) readings for the three times are given in Table 9.18.

Note that each reader has 10 lesions for each of the four response categories. Assuming no disagreement between readers, how should their readings be used

TABLE 9.18: Average (sd) lesion size for five readers by time—averaged over 40 lesions.

Time	Reader				
	1	2	3	4	5
0	3.92(2.59)	3.70(1.51)	4.42(1.55)	4.36(1.61)	4.14(1.55)
1	3.48(1.97)	3.09(1.68)	3.83(1.94)	3.92(2.00)	3.56(1.88)
2	3.24(2.04)	3.02(1.90)	3.72(2.13)	3.64(2.15)	3.37(2.02)

to assign lesions to response categories CR, PR, SD, and PD? Suppose that the lesions are assigned to two categories, response (including complete and partial response) if the percent decrease in lesion size is less than 30%, otherwise a lesion is assigned to the no response category.

Differences between readers will be tested with a logistic regression using the occurrence of response or no response as the dependent variable and using two factors for the independent variables: the patient label $(1, 2, \ldots, 40)$ and the reader number (1, 2, 3, 4, and 5). The logistic regression was performed using BUGS CODE 9.3.

BUGS CODE 9.3

```
model {
for( i in 1 : N ) {
y[i] ~ dbern(p[i])
logit( p[i]) <- beta[1] + beta[2]*n[i]+beta[3]*r[i]
}
phat <- mean(p[])
for (i in 1:3 ){
beta[i] ~ dnorm(0.0,0.0001)}
}
```

The list statement for the data includes the column y[ ] for the 200 occurrences or non occurrences of the overall response, while the n[ ] column contains the lesion id $(1, 2, \ldots, 40)$. The 200×1 reader id column is the coefficient of beta[3] in the logistic regression model. Zeros are given as the initial values of the three beta coefficients in the list statement for initial values of the WinBUGS program. The posterior analysis is given by Table 9.19. The lesion factor is included because the readers were paired with lesions, thus the effect for readers, given by beta[3], is adjusted for the lesion effect.

The lesion effect beta[2] is "significant." Its 95% credible interval excludes zero, however, the interval for the reader effect does include zero, implying that reader differences have a minimal effect in estimating the tumor response phat, see BUGS CODE 9.3. The posterior mean of the overall response is 0.32 with a standard deviation of 0.016 and a 95% credible interval of (0.287,0.352). Thus, the estimate of the overall response to therapy is 32%.

TABLE 9.19: Posterior distribution of tumor response.

Parameter	Mean	sd	Median	95% credible interval
beta[1]	6.240	1.268	6.174	3.968,8.906
beta[2]	−0.459	0.076	−0.453	−0.626,−0.325
beta[3]	−0.045	0.214	−0.045	−0.471,0.372
Phat: overall response	0.320	0.016	0.320	0.287,0.352

Of course, the fact that there was good reader agreement to begin with was by design for the hypothetical outcomes of lesion size.

9.4.14   A Phase II trial for advanced melanoma

Melanoma is a cancer of the skin, and about 55,000 new cases are diagnosed annually with approximately 8,000 deaths. If not successfully treated early intervention, it metastasizes to the brain, lungs, and liver, and in this advanced stage, there are few promising therapies. The protocol to be explained is for stage IV melanoma with a therapy that has shown some promise in other forms of cancer.

The therapy to be tested is an agent that is designed to be anti angiogenic, i.e., designed to destroy the blood supply to the tumor, and several Phase I and II trials have utilized this agent. In an early European Phase I trial with 37 patients with solid tumors, no serious toxicities were reported. In a Phase II study with 35 patients, this agent in combination with another produced no toxicities. In an NCI study with six patients, there were no objective responses but three patients experienced stable disease. In an ongoing Phase II trial, there have been some minor toxicities and reports of one confirmed CR. Thus, prior information leads us to use the following: with 72 patients, there have been no reported serious toxicities and, at the same time, little evidence of a favorable response to therapy.

Patients to be entered into this study must have a confirmed stage IV disease, must have measurable disease with at least one lesion that can be accurately measured over the course of the study, be at least 18 years of age, and have a performance status that shows they are well enough to complete the therapy. This is a randomized study, with patients randomly assigned to two dose levels of chemotherapy, where the endpoint is response to therapy.

In order to assign patients to a response category, the RECIST criteria will be followed. A patient's overall response is based on dynamic CT scans of the target lesions. The final category is based on the imaging results for the target lesions, the status of the non-target lesions, and the appearance of new lesions (Table 9.20).

TABLE 9.20:   Overall response to therapy.

Target lesions	Non-target lesions	New lesions	Overall response
CR	CR	No	CR
CR	SD/incomplete response	No	PR
PR	Non-PD	No	PR
SD	Non-PD	No	SSD
PD	Any	Yes or no	PD
Any	PD	Yes or no	PD
Any	Any	Yes	PD

In addition, the classification of response to the target lesions is based on the change in lesion size for the target lesions, relative to some reference time, either at baseline or at some earlier time when the size of the lesion was a minimum. One cycle of therapy is 4 weeks and the protocol must designate the times during this period when CT imaging of the target lesions will take place. Several treatment cycles of therapy must be experienced by patients in order for the patient to be assigned to an overall response category and for the category to be confirmed.

A statistics section of the protocol contains the power analysis, a justification for the sample size, and a description of the statistical analysis for the study results. It was decided that 57 patients can be accrued at the rate of 3 to 4 per month for this single center trial. A Bayesian stopping rule must be given that utilizes the information from prior Phase I and II studies. We have seen that with a total of 72 patients, no toxicities were reported and there was very little evidence of response to therapy (there were three of six who experienced SD in a European trial). Thus, there is good evidence of no toxicity, but very little evidence of response to therapy. There is very little evidence for treatment response because these trials were designed primarily to evaluate safety, not efficacy, thus the prior information for response is designated as vague or uninformative.

Multc Lean is used to design the stopping rule for this trial. The prior distribution, shown in Table 9.21, is for the probabilities of response and toxicity of the standard and the melanoma trial.

Thus, one is quite confident that there was very little toxicity among the 72 patients of previous trials. A uniform prior is given to the probabilities of a response for the standard and experimental therapies. Using Multc Lean, the stopping rule is

$$\Pr[\theta_r < \phi_r \mid \text{data}] > 0.85,$$
$$\Pr[\theta_t > \phi_t \mid \text{data}] > 0.85. \tag{9.21}$$

The stopping boundaries for response are shown in Table 9.22. The trial is stopped if there are three or less responses among the first 25 patients. On the other hand, the stopping boundaries for toxicity are shown in Table 9.23, and the trial is stopped early at one patient if there is at least one toxicity.

What are the sampling properties of this stopping rule? The third section of Multc Lean provides a way to estimate the probability of stopping for

TABLE 9.21: Prior beta distributions for the standard and experimental therapies.

Category	Therapy	Beta parameters
Response	Standard	(1,1)
Response	Experimental	(1,1)
Toxicity	Standard	(1,71)
Toxicity	Experimental	(0.0278,1.9722)

TABLE 9.22: Stopping
boundaries for response.

Responses	Boundary
0	5
1	12
2	19
3	25
4	32
5	39
6	45
7	52
8	57

TABLE 9.23: Stopping
boundaries for toxicity.

Toxicities	Boundary
1	1–12
2	14–45
3	47–57

various scenarios involving the probabilities of response and toxicity of the experimental therapy relative to the corresponding probabilities of $\phi_r = 0.5$ (beta(1,1)) and $\phi_t = 0.0135$ (beta(1,71)) for the standard therapy. The five scenarios shown in Table 9.24 were assumed for the experimental therapy.

For each scenario, the probability of stopping for a given number of patients can be computed with Multc Lean (Table 9.25). See scenario input and output sections of the program. In addition, Multc Lean gives the average number of patients, the average number of responses, and the average number of toxicities for each scenario (Table 9.26).

9.5   Summary and Conclusions

Chapter 9 explains the role that test accuracy plays in clinical practice. The choice of an optimal threshold for a medical test affects the accuracy of the test and two approaches are taken for choosing the cutoff point. The first approach is to base the choice on distance, where the optimal point on the ROC curve is the one that is closest to (0,1), and once this point is estimated, the corresponding test score threshold is determined. The choice of the optimal point, using both criteria, is demonstrated with two examples: (a) the PSA test

TABLE 9.24: Scenarios of the melanoma study.

Probability of response θ_r	Probability of toxicity θ_t	Scenario
0.5	0.5	1
0.01	0.01	2
0.2	0.2	3
0.21	0.02	4
0.60	0.011	5

for prostate cancer, and (b) the biomarker CK-BB test for complications from head trauma. BUGS CODE 9.1 executes the Bayesian analysis and method of selecting the optimal threshold of the test score.

The decision curve is a novel approach to measure the accuracy of a medical test and is demonstrated with the PSA and CK-BB medical tests for prostate cancer and head trauma complications. The decision curve is a plot of the clinical benefit vs. a range of threshold probabilities, and the clinical benefit is defined in terms of the TPF and FPF and the threshold probability. The decision curve is used to choose between two strategies to treat a patient: (1) biopsy a patient if the probability of disease is greater than a threshold probability, where the probability of disease is estimated with logistic regression; or (2) biopsy all patients, regardless of the score of the medical test. In both examples, the best approach is to biopsy the subject based on the value of the medical test.

Of course, clinical trials are an important part of clinical research and the accuracy of a medical test is crucial for the successful implementation of the trial. The emphasis here is on Phase I and II cancer trials, where the medical tests are the imaging devices that measure the size of the tumor during treatment. Based on the change in tumor size (measured by CT or MRI), each patient in the trial is classified into several categories (CR, PR, PD, etc.), thus, if the imaging devices are not accurate, patient progress can be misclassified and can lead to misleading conclusions about the efficacy of the treatment.

TABLE 9.25: Probability of stopping early for melanoma trial.

			Scenario		
n	1	2	3	4	5
1	0.5000	0.10000	0.0200	0.0200	0.0110
5	0.9697	0.8532	0.9548	0.3848	0.0639
10	0.9999	0.9133	0.9592	0.4439	0.1143
20	1	0.9867	0.9994	0.5654	0.1366
40	1	0.9994	1	0.6442	0.1666
56	1	0.9998	1	0.6707	0.1795
57	1	1	1	1	1

TABLE 9.26:   Average number of patients, responses, and toxicities.

Scenario	Average number of patients	Average number of responses	Average number of toxicities
1	1.99	0.999	0.999
2	5.17	0.517	0.517
3	5.13	0.102	0.1076
4	26.29	5.52	0.525
5	49.18	29.51	0.541

The role accuracy plays in clinical trials is illustrated with a Phase I trial for RCC and a Phase II trial for advanced melanoma. WinBUGS is not used for these examples, instead software available at the MDACC is employed to describe the Bayesian stopping rules for evaluating efficacy of treatment for Phase I and II trials.

9.6   Exercises

1. Verify Figure 9.1, the ROC curve for the PSA data. Use the information in Table 9.1, the TP and FP values of the first two columns. What point on the ROC curve gives a minimum distance between that point and the point (0,1)?

2. Construct the worksheet of Table 9.1, which gives the TP and FP values for the PSA study. How is the column for the distance from a point on the ROC curve to the point (0,1) calculated?

3. Show that Equation 9.2 gives the distance between (fpf,tpf) and (0,1).

4. Verify Equation 9.3.

5. Verify Table 9.2 using 45,000 observations generated from the joint posterior distribution, with a burn in of 5,000 and a refresh of 100. Note that the optimal point has coordinates given by the posterior mean of the fpf and tpf of the table and is for the value Kappa $= 1$. What is the optimal threshold on the log and the original scales corresponding to the optimal point on the ROC curve given by the table?

6. Verify Table 9.3, which is the posterior analysis for the optimal point on the ROC curve for the PSA study for $R = 2$ and $p = 0.33$. Use 45,000 observations generated from the posterior distribution, with a burn in of 5,000 and a refresh of 100. The posterior mean for the fpf and tpf is 0.0288 and 0.577, respectively. What is the corresponding threshold for PSA on the log and original scales?

7. Verify Table 9.4, the worksheet for the diabetes study. How is the second column of tpf values calculated? Be specific in your answer. How is third column of distances to the point (0,1) from points on the ROC calculated?

8. Verify Table 9.5, the posterior analysis for the diabetes study, using 45,000 observations, with a burn in of 5,000 and a refresh of 100. Note, the coordinates of the optimal point correspond to $R = 2$ and $p = 0.5$, where p is the incidence of diabetes. Thus, one is assuming that the benefit of a TN is twice that of TP. What is the corresponding threshold for the diabetes study?

9. Reproduce Figure 9.3, the ROC curve of the diabetes study. Use the first and second columns of Table 9.4 to plot the ROC curve.

10. Equation 9.5 defines the clinical benefit of scenario (a) and depends on the true and false positive fractions. Explain how the clinical benefit depends on the disease rate π and explain its effect on the clinical benefit. As the disease rate decreases to zero, what happens to the clinical benefit?

11. Using BUGS CODE 9.2, verify Table 9.7 with 45,000 observations generated from the posterior distribution, with a burn in of 5,000 and a refresh of 100. Plot the posterior densities of the clinical benefit for scenarios (a) and (b) at threshold probability 0.35.

12. Verify Table 9.8 using BUGS CODE 9.2.

13. Verify Table 9.10 by revising BUGS CODE 9.2 and generating 45,000 observations, with a burn in of 5,000 and a refresh of 100. Also, plot the posterior density of the clinical benefit for both scenarios (a) and (b) at threshold probability 0.75. What are the MCMC errors for the parameters?

14. Refer to Exercise 10 and BUGS CODE 9.2. Perform a Bayesian analysis for the clinical benefit of scenarios (a) and (b) using the blood glucose values given by the first list statement of the code. Also note that the disease indicator vector d is given in the first list statement of the code.

    Revise BUGS CODE 9.2 to perform the analysis. Determine the clinical benefit for a plausible range of threshold probabilities. Note that for each threshold probability, a logistic regression must be performed that computes the four counts in the 2×2 table. For each cell count, add 1 to the cell count and use those as the gamma parameters in the code. Execute the analysis with 55,000 observations, with a burn in of 10,000 and a refresh of 500.

15. Perform the posterior analysis for the G coefficient based on the information in Table 9.11. What is the posterior distribution of the Kappa parameter? See relevant formulas in Chapter 4.

16. Read Section 9.4.8 and using Multc Lean (download the program from MDACC: http://Biostatistics/mdanderson.org/SoftwareDownload/), verify the results of Tables 9.12 and 9.13 using the stopping rule (Equation 9.18).

17. Read Section 9.4.14 and using Multc Lean with the stopping rule (Equation 9.20), verify the results of the melanoma trial in Tables 9.20 through 9.25. Explain the difference in the probability of stopping between scenarios 1 and 5.

18. By choosing different beta prior distributions for the parameters of the standard and melanoma therapies and using Multc Lean, describe the effect of the prior distribution on the probability of stopping the trial.

19. Refer to Table 9.25 and explain why the average number of patients for scenario 5 is much greater than that for scenario 1. What is the effect of the stopping probabilities on the average number of patients?

20. Refer to Equation 9.20 where the probability of stopping for response and toxicity are both 0.85. Change these to 0.80 and 0.80, respectively, and determine the effect on the average number of patients, the average number of responses, and the average number of toxicities.

References

[1] Etzioni, R., Pepe, M., Longton, G., Hu, C., and Goodman, G. Incorporating the time dimension in receiver operating characteristic curves: A case study of prostate cancer. *Medical Decision Making*, 19:242, 1999.

[2] Pepe, M.S. *The Statistical Evaluation of Medical Tests for Classification and Prediction.* Oxford University Press, New York, 2003.

[3] O'Malley, J.A., Zou, K.H., Fielding, J.R., and Tampany, C.M.C. Bayesian regression methodology for estimating a receiver operating characteristic curve with two radiologic applications: Prostate biopsy and spiral CT for urethral stones. *Academic Radiology*, 8:713, 2001.

[4] Zhou, X., Obuchowski, N.A., and McClish, D.K. *Statistical Methods in Diagnostic Medicine.* John Wiley, New York, 2002.

[5] Somoza, E. and Mossman, D. Biological markers and psychiatric diagnosis: Risk benefit balancing using ROC analysis. *Biological Psychiatry*, 29:811, 1991.

[6] Hans, P., Albert, R., Born, D., and Chapelle, J.P. Derivation of a biochemical prognostic index in severe head injury. *Intensive Care Medicine*, 11:186, 1985.

[7] Vickers, A.J. and Elkin, B. Decision curve analysis: A novel method for evaluating prediction models. *Medical Decision Making*, 26:565, 2006.

[8] Vickers, A.J., Cronin, A.M., Elkin, E.B., and Gonen, M. Extensions to decision curves analysis, a novel method for evaluating diagnostic tests, prediction models and molecular markers. *BMC Medical Informatics and Decision Making*, 8:1, 2008.

[9] Vickers, A.J. Decision analysis for the evaluation of diagnostic tests, prediction models, and molecular markers. *The American Statistician*, 62:314, 2008.

[10] Steyerberg, E.W. and Vickers, A.J. Decision curve analysis: A discussion. *Medical Decision Making*, 28:146, 2008.

[11] Baker, S.G., Kramer, B.S., and Srivastava, S. Markers for early detection of cancer: Statistical issues for nested case-control studies. *BMC Medical Research Methodology*, 2:4, 2002

[12] Vergouwe, Y., Steyerberg, E.E., Eijkemans, M.J., and Habbema, J.D. Validity of prognostic models: When is a model clinically useful. *Seminars in Urologic Oncology*, 20:96, 2002.

[13] Broemeling, L.D. Letter to the editor. *The American Statistician*, 63:198, 2009.

[14] Baker, S.G. and Kramer, B.S. Peirce, Youden, and receiver operating characteristic curves. *The American Statistician*, 61:343, 2007.

[15] Pauker, S.G. and Kassirer, J.P. Therapeutic decision making: A cost-benefit analysis. *New England Journal of Medicine*, 293:229, 1975.

[16] Thall, P.F. Bayesian methods in early phase oncology trials. Proceedings of the section on Bayesian statistical science. *The American Statistical Association*, 30–39, 2000.

[17] Crowley, J. *Handbook of Statistics in Clinical Oncology*. Marcel-Dekker, New York, 2001.

[18] Padhani, A.R. and Ollivier, L. The RECIST criteria: Implications for diagnostic radiologists. *British Journal of Radiology*, 74:983, 2001.

[19] Simon, R. Optimal two-stage designs for Phase II clinical trials. *Controlled Clinical Trials*, 10:1, 1985.

[20] DeVita, V.T., Hellman, S., and Rosenberg, S.A. *Cancer, Principles and Practice of Oncology* (5th ed.). Lippincott & Raven, New York, 1997.

[21] Thall, P.F., Simon, R., and Estey, E.H. Bayesian sequential monitoring designs for single arm clinical trials with multiple outcomes. *Statistics in Medicine*, 14:357, 1995.

[22] Thall, P.F., Simon, R.M., and Estey, E.H. New statistical strategy for monitoring safety and efficacy in single arm clinical trials. *Journal of Clinical Oncology*, 14:296, 1996.

[23] Therasse, P., Arbuck, S.G., Eisenhauer, E.A. et al. New Guidelines to evaluate the response to treatment in solid tumors. *Journal of the National Cancer Institute*, 92:205, 2002.

[24] Thiesse et al. Response rate accuracy in oncology trials: Reasons for inter-observer variability. *Journal of Clinical Oncology*, 15:3507, 1997.

[25] Negrier, S., Escudier, B., Lasset, C. et al. The FNLCC CRECY trial: Interleukin 2 (IL2) + interferon (IFN) in the optimal treatment to induce responses in metastatic renal cell carcinoma. *Proceedings of the American Society of Clinical Oncology*, 15:39, 1996.

[26] Erasmus, J.J., Gladish, G.W., Broemeling, L.D., Sabaloff, B.S., Truong, M.T., Herbst, R.S., and Munden, R.F. Interobserver and intraobserver variability in measurement of non-small-cell carcinoma lung lesions: Implications for assessment of tumor response. *Journal of Clinical Oncology*, 21:2574, 2004.

[27] Cook, J.D. Multc Lean Statistical Tutorial. Department of Biostatistics and Applied Mathematics, University of Texas MD Anderson Cancer Center, March 28, 2005.

[28] Thall, P.F. and Lee, S.J. Practical model-based dose-finding in Phase I clinical trials: Methods based on toxicity. *International Journal of Gynecological Cancer*, 13:251, 2003.

Chapter 10

Accuracy of Combined Tests

10.1 Introduction

Chapter 10 introduces the reader to the methodology of measuring the accuracy of several medical tests that may be administered to a patient. Our main focus is on measuring the accuracy of a combination of two or more tests. For example, to diagnose type 2 diabetes, the patient is given a fasting blood glucose test, which is followed by an oral glucose tolerance test. What is the accuracy (true positive fraction [TPF] and false positive fraction [FPF]) of this combination of two tests? Or, in order to diagnose coronary artery disease, the subject's history of chest pain (CPH) is followed by an exercise stress test (EST). Still another example is for the diagnosis of prostate cancer, where a digital rectal exam is followed by measuring prostate-specific antigen (PSA). The reader is referred to Johnson and Sandmire [1] for a description of additional examples of multiple tests to diagnose a large number of diseases, including heart disease, diabetes, lung cancer, breast cancer, etc.

In certain situations, it is common practice to administer one or more tests to diagnose a given condition and we will explore two avenues. One is where it is common to administer several tests as standard medical practice, and the other is an experimental situation, where one test is compared to a standard medical test. As an example of the latter, magnetic resonance imaging (MRI) is now being studied as an alternative to standard mammography, as a method to diagnose breast cancer.

There are many studies that assess the accuracy of the combination of two or more tests. Two tests for the diagnosis of a disease measure different aspects or characteristics of the same disease. In the case of diagnostic imaging, two modalities have different qualities (resolution, contrast, and noise), thus, although they are imaging the same scene, the information is not the same from the two sources. When this is the case, the accuracy of the combination of two modalities is of paramount importance. For example, Buscombe et al. [2] have reported the accuracy of the combination of mammography and scintimammography for suspected breast cancer. Another study for diagnosing breast cancer was performed by Berg et al. [3], who measured the accuracy of mammography, clinical examination, ultrasound, and MRI in a preoperative assessment of the disease. The accuracy of each modality and various combinations of the modalities were measured. When investigating metastasis to the lymph nodes in lung cancer, Van Overhagen et al. [4] measured the accuracy of ultrasound and computed tomography (CT) and the combination of the two. Ultrasound conveys different information about metastasis compared to CT, but the combination of the two might provide a more accurate diagnosis than each separately. For an example of the diagnosis of head and neck cancer, Pauleit et al. [5] used two nuclear medicine modalities, ^{18}F-FET PET and ^{18}F-FDG PET, to assess the extent of the disease and estimate the accuracy of each, and the combination of the two. On the other hand, Schaffler et al. [6] evaluated pleural abnormalities with CT and ^{18}F-FDG PET and the combination of the two.

Now, switching from cancer to heart disease, Gerber et al. [7] used four-section multidetector CT and 3D Navigator MR for detecting stenosis of the coronary arteries, where the accuracy of each and the combination of the two was estimated. The above examples involve binary test scores where accuracy is measured by TPF, FPF, positive predictive value (PPV), and negative predictive value (NPV), but when the test scores are ordinal and involve more than two possible values, or when the test scores are continuous, the accuracy is measured by the area under the receiver operating characteristic (ROC) curve.

What is the optimal way to measure the accuracy for the combination of two binary tests? Pepe [8: 268] presents two approaches: (1) believe the positive rule (BP), where a positive test score on a subject means one or the other of the two tests is scored positive; and (2) believe the negative rule (BN), where a subject is scored positive if both tests are scored positive. Pepe [8: 268] also presents some properties about these rules.

Statement 10.1

a. The BP rule increases sensitivity relative to the two binary tests, but increases the FPF, but by no more than the sum of the two FPFs, namely,

$$\text{FPF}_1 + \text{FPF}_2. \tag{10.1}$$

b. The BN rule decreases the false positive rate relative to the false positive rates of the two tests, but at the same time, decreases the sensitivity, however, the sensitivity remains above $\text{TPF}_1 + \text{TPF}_2 - 1$.

This result is left as an exercise and is illustrated with many examples to be presented in the chapter.

The chapter is divided into four parts: (1) for two binary test scores, (2) for two ordinal test scores, (3) for continuous test scores, and (4) choosing an optimal test among several component tests.

For the first part on two binary tests, several examples are provided; then the idea is generalized to two binary tests with several readers, and to two binary tests when verification bias is present. For the section on two ordinal tests, the accuracy of the combination of the two tests is provided by the ROC curve, which in turn depends on the risk score of the component tests.

Again, the idea is generalized to several readers and when verification bias is present. For two tests with continuous scores, the ROC curve based on the risk score is the optimal way to measure the accuracy of the combination of the two tests, and the basic approach is generalized to several readers and to when verification bias is present.

10.2 Two Binary Tests

This section will employ a Bayesian approach to estimate the accuracy of two binary tests and the accuracy of the combination of the two using the BP rule and the BN rule given by Statement 10.1. Label the two tests Y_1 and Y_2, where both take on the values 0 or 1, where 0 indicates a negative test and 1 is a positive score for the medical test. A subject either has the disease or does not, as determined by the gold standard; thus, when $D = 1$, let

$$\theta_{ij} = P[Y_1 = i, Y_2 = j], \tag{10.2}$$

for $i, j = 0, 1$, and when $D = 0$, let

$$\phi_{ij} = P[Y_1 = i, Y_2 = j]. \tag{10.3}$$

Thus, the thetas are the four cell probabilities for the diseased subjects and the corresponding phis are the cell probabilities for the non-diseased subjects. The corresponding cell frequencies are denoted by n_{ij} and m_{ij} for the diseased and non-diseased subjects, respectively. Thus, assuming a uniform prior for the cell probabilities, the posterior distribution of the cell probabilities is Dirichlet for $\theta = (\theta_{00}, \theta_{01}, \theta_{10}, \theta_{11})$ with parameter $(n_{00} + 1, n_{01} + 1, n_{10} + 1, n_{11} + 1)$, and $\phi = (\phi_{00}, \phi_{01}, \phi_{10}, \phi_{11})$ is also Dirichlet with parameter vector $(m_{00} + 1, m_{01} + 1, m_{10} + 1, m_{11} + 1)$.

Once the posterior distribution of the cell probabilities is determined, the posterior distribution of the truncated cell probabilities is easily found. The truncated cell probabilities for the diseased subjects are given by

$$\theta_{ij}^* = \theta_{ij} \bigg/ \sum_{i=0}^{i=1} \theta_{ij}, \tag{10.4}$$

and for the non-diseased subjects the truncated cell probabilities are

$$\phi_{ij}^* = \phi_{ij} \bigg/ \sum_{i=0}^{i=1} \phi_{ij}, \tag{10.5}$$

for i and $j = 0$ or 1.

The TPF and FPF for the first test Y_1 are

$$\text{tpf1} = \theta_{1.} \tag{10.6}$$

and

$$\text{fpf1} = \phi_{1.}, \tag{10.7}$$

respectively, and the TPF and FPF for the second test are

$$\text{tpf2} = \theta_{.1} \tag{10.8}$$

and

$$\text{fpf2} = \phi_{.1}, \tag{10.9}$$

respectively.

The Equations 10.6 through 10.9 give the accuracy of the individual tests, but what about the combination of the two? Recall that there are two ways to measure the accuracy of combined tests, either by the BP rule or by the BN rule. With the former rule, the TPF is

$$\text{tpfbp} = \theta_{01} + \theta_{11} + \theta_{10} \tag{10.10}$$

and the FPF is

$$\text{fpfbp} = \phi_{01} + \phi_{11} + \phi_{10}. \tag{10.11}$$

On the other hand, using the BN rule the TPF is

$$\text{tpfbn} = \theta_{11}, \tag{10.12}$$

while the FPF is

$$\text{fpfbn} = \phi_{11}. \tag{10.13}$$

In what follows, the accuracies of the individual tests and the combined test will be estimated for several examples, but first, the computations necessary to achieve this goal will be executed with BUGS CODE 10.1.

BUGS CODE 10.1

```
Model;
# Two Binary Tests
{
# Dirichlet distribution generated
  g00~dgamma(a00,2)
  g01~dgamma(a01,2)
  g10~dgamma(a10,2)
```

```
  g11~dgamma(a11,2)
sumall<-g00+g01+g10+g11+
h00+h01+h10+h11
  h00~dgamma(b00,2)
  h01~dgamma(b01,2)
  h10~dgamma(b10,2)
  h11~dgamma(b11,2)
# cell probabilities for diseased
theta00<-g00/sumall
theta01<-g01/sumall
theta10<-g10/sumall
theta11<-g11/sumall
# cell probabilities for non diseased
ph00<-h00/sumall
ph01<-h01/sumall
ph10<-h10/sumall
ph11<-h11/sumall
# truncated distributions
stheta<-theta00+theta01+theta10+theta11
sph<-ph00+ph01+ph10+ph11
# truncated thetas
th00<-theta00/stheta
th01<-theta01/stheta
th10<-theta10/stheta
th11<-theta11/stheta
# truncated phi's
p00<-ph00/sph
p01<-ph01/sph
p10<-ph10/sph
p11<-ph11/sph
# sensitivities
# a designates row
# b designates column
# tpf a
tpfa<-th10+th11
#tpf b
tpfb<-th01+th11
# believe the positive tpf
bptpf<-th01+th11+th10
# believe the negative tpf
bntpf<-th11
# false positives
#fpf a
fpfa<-p10+p11
# fpf b
```

```
fpfb<-p01+p11
# believe the positive false positive fraction
bpfpf<-p01+p11+p10
# believe the negative fpf
bnfpf<-p11
}
# CASS data set
#Comparing est and cph
# a is est
# b is cph
list(a00=26, a01=184,a10=30,a11=787,
    b00=152,b01=177,b10=47,b11=70)
# Gerber et al. data set
# a is CT
# b is MRI
list(a00=13, a01=1,a10=11,a11=37,
    b00=169,b01=1,b10=31,b11=39)
# Berger et al. study for breast cancer
# a is mammography
# b is ultrasound
list(a00=11, a01=48,a10=21,a11=101,
    b00=22,b01=40,b10=8,b14=39)
# initial values
list(g00=1,g01=1,g10=1,g11=1,
    h00=1,h01=1,h10=1,h11=1)
```

10.2.1   Two binary tests for heart disease

The first example to be run is the coronary artery disease study dataset taken from Pepe [8: 47], consisting of two tests for coronary artery disease, where the first test is the EST and the second is the CPH. Tables 10.1a and b give the results for this paired study with 1465 subjects, where 1023 have the disease and 442 do not.

From the information portrayed in Tables 10.1a and b, and assuming a uniform prior, it can be shown that the posterior distribution of the cell probabilities, θ_{ij}, for the diseased subjects is Dirichlet (26, 184, 30, 787), while the cell probabilities, ϕ_{ij}, for the non-diseased subjects is Dirichlet with parameter vector (152, 177, 47, 70).

Using BUGS CODE 10.1 to execute the analysis with 55,000 observations generated from the posterior distribution, with a burn in of 5,000 and a refresh of 100, the posterior analysis is given in Table 10.2.

Thus, with the BN rule, the FPF and TPF are 0.1568 and 0.7662, respectively, with corresponding standard deviations 0.0171 and 0.0131. On the other hand, with the BP rule, the FPF and TPF are 0.6592 and 0.9747, respectively, with corresponding standard deviations 0.0224 and 0.0049. The simulations

TABLE 10.1a: Coronary artery disease for diseased patients.

EST	CPH 0	1	Total
0	25	183	208
1	29	786	815
Total	54	969	1023

TABLE 10.1b: Coronary artery disease for non-diseased patients.

EST	CPH 0	1	Total
0	151	176	327
1	46	69	115
Total	197	245	442

errors are quite small for all parameters, implying that 55,000 observations are sufficient to estimate the accuracy parameters.

Our main focus is on the combined test: for the TPF, the BP rule gives the largest value at 0.9745, while for the false positive rate, the BN rule gives the smallest value at 0.1568, compared to 0.6592 with the BP rule. Note that the TPF of 0.974 with the BP rule is greater than the TPF for the two modalities. On the other hand, the FPF with the BN rule is smaller than the FPF for the two modalities (EST and CPH). Figure 10.1 portrays the density of the posterior distribution of the TPF for the BP rule.

The main question remains, namely, which is the best rule, BP or BN, for measuring the accuracy of the combined test, where the first test is the EST and the second is the CPH? For the BN rule, the TPF of 0.7662 is reasonable, and the FPF of 0.1568 is outstanding, thus, I prefer the BN rule for this

TABLE 10.2: Posterior distribution of EST, CPH, and the combined test.

Parameter	Mean	sd	Error	2 1/2	Median	97 1/2
bnfpf	0.1568	0.0171	<0.00001	0.1247	0.1564	0.1919
bntpf	0.7662	0.0131	<0.00001	0.7399	0.7663	0.7915
bpfpf	0.6592	0.0224	<0.00001	0.6142	0.6594	0.7029
bptpf	0.9747	0.0049	<0.00001	0.9641	0.975	0.9833
fpfest	0.2622	0.0207	<0.00001	0.2228	0.2619	0.3035
fpfcph	0.5539	0.0234	<0.00001	0.5075	0.554	0.5996
tpfest	0.7954	0.0125	<0.00001	0.7704	0.7955	0.8195
tpfcph	0.9455	0.0071	<0.00001	0.9307	0.9457	0.9586

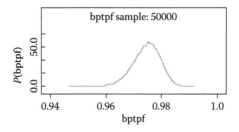

FIGURE 10.1: Posterior density of the TPF of the BP rule.

combined test, because the FPF of 0.6592 for the BP rule is much too high, and as a compromise, I would opt for the BN rule.

As a general rule of thumb for combined tests, for two tests that have high sensitivity, the accuracy should be measured by the BP rule, but on the other hand, if the two tests have relatively small false positive rates, the BN rule is to be preferred. Of course, this also implies that if the above is not true, then choosing the optimal rule to measure accuracy is difficult, and one must compromise. Note, for the prostate cancer study example, the TPF for both tests is "high," which implies accuracy should be measured by the BP rule, which is the choice stated above, but using a compromise approach.

10.2.2 Computed tomography and magnetic resonance imaging and coronary stenosis

The next example is based on the study by Gerber et al. [7], who investigated the use of both CT and MRI to determine the degree of stenosis in the coronary arteries, where 26 patients were suspected of having coronary artery disease.

The gold standard is coronary catheterization, which found 58 diseased segments (stenosis greater than 50%) and 236 non-diseased segments. This was an experimental study to determine the value of the two non-invasive imaging modalities to diagnose coronary artery disease. The study found that the sensitivity of CT and MRI was 79% and 62%, respectively, and that the specificity of CT and MRI was 71% and 84%, respectively, thus CT had higher sensitivity but smaller specificity compared to MRI. This is a very interesting study and only a brief synopsis is given here, thus the reader is invited to read the paper for more detail in order to know the value of the investigation. The information for the study is given in Table 10.3a.

Our goal is to determine the accuracy of the combined test using the BP and BN rules, thus BUGS CODE 10.1 is used to perform the Bayesian analysis. The simulation consists of generating 25,000 observations from the joint posterior distribution, with a burn in of 5,000 and a refresh of 100, and the results are shown in Table 10.4.

TABLE 10.3a: Study results of the CT-MRI study.

	MRI		
CT	**0**	**1**	**Total**
0	12	0	12
1	10	36	46
Total	22	36	58

Source: From Gerber, B.L. et al. *Radiology*, 234, 98, 2005, with permission of Radiological Society of North America.

TABLE 10.3b: Study results of the CT-MRI study.

	MRI		
CT	**0**	**1**	**Total**
0	168	0	168
1	30	38	68
Total	198	38	236

Source: From Gerber, B.L. et al. *Radiology*, 234, 98, 2005, with permission of Radiological Society of North America.

Which rule, the BP or BN rule, should be used to measure the accuracy of the combined test? Note, the TPF with the BP rule is higher than that with the BN rule, but on the other hand, the false positive rate is lower with the BN rule compared to the BP rule. This is a true quandary and it is not obvious which rule should be used to measure the accuracy of the combined test (see Exercise 3). Figure 10.2 displays the posterior density of the false positive rate for the BN rule of the combined test. Note that WinBUGS displays the posterior density of all the parameters listed in Table 10.4.

TABLE 10.4: Bayesian analysis for combined test of CT and MRI.

Parameter	Mean	sd	Error	2 1/2	Median	97 1/2
bnfpf	0.1627	0.0236	<0.0001	0.1191	0.1617	0.2116
bntpf	0.5965	0.0623	<0.0001	0.4731	0.5977	0.7148
bpfpf	0.2959	0.0293	<0.0001	0.2404	0.2953	0.3553
bptpf	0.7901	0.0515	<0.0001	0.68	0.7933	0.8817
fpfct	0.2918	0.0292	<0.0001	0.2361	0.2911	0.3511
fpfmri	0.1668	0.0238	<0.0001	0.1228	0.1658	0.2163
tpfct	0.7739	0.0527	<0.0001	0.662	0.766	0.8675
tpfmri	0.6127	0.0619	<0.0001	0.4883	0.614	0.7299

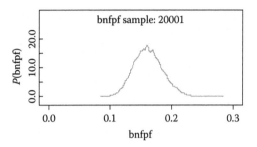

FIGURE 10.2: FPF for the BN rule of the combined test.

10.3   Two Binary Tests and Several Readers

When more than one reader is involved, the level of complexity increases. For now, only two readers will be considered, however, it will be obvious how to extend the results to follow more than two readers. Our approach is to assume a double paired design for two medical tests, where both tests are applied to all subjects and both readers look at exactly the same evidence, that is, for each subject, both readers interpret both tests.

Suppose, given $D = 1$,

$$\theta_{ijk} = P[Y_1 = i, Y_2 = j, K = k], \qquad (10.14)$$

where Y_1 is the first test, Y_2 is the second, with possible values $i, j = 0, 1$, and reader k, with possible values k, where $k = 1, 2$.

In a similar way, given $D = 0$,

$$\phi_{ijk} = P[Y_1 = i, Y_2 = j, K = k]. \qquad (10.15)$$

Therefore, θ_{ijk} are the cell probabilities for the diseased subjects and ϕ_{ijk} are the corresponding quantities for the non-diseased subjects for the kth reader. For each reader there will be a TPF and FPF for each of the two tests. In addition, for each rule, BP and BN, there will be a TPF and FPF for each reader. Suppose the cell frequencies are denoted by n_{ijk} for the diseased subjects and m_{ijk} for the non diseased, then assuming a uniform prior for all parameters, resulting in a posterior Dirichlet distribution for

$$\theta = (\theta_{001}, \theta_{011}, \theta_{101}, \theta_{111}; \theta_{002}, \theta_{012}, \theta_{102}, \theta_{112}), \qquad (10.16)$$

with parameter vector

$$n = (n_{001} + 1, n_{011} + 1, n_{101} + 1, n_{111} + 1; n_{002} + 1, n_{012} + 1, n_{102} + 1, n_{112} + 1),$$

and for

$$\phi = (\phi_{001}, \phi_{011}, \phi_{101}, \phi_{111}; \phi_{002}, \phi_{012}, \phi_{102}, \phi_{112}), \qquad (10.17)$$

a Dirichlet with parameter vector

$$m = (m_{001} + 1, m_{011} + 1, m_{101} + 1, m_{111} + 1; m_{002} + 1, m_{012} + 1,$$
$$m_{102} + 1, m_{112} + 1).$$

The parameters will be truncated as follows:
For reader 1 and the diseased patients, let

$$\theta_{ij1}^* = \theta_{ij1} \bigg/ \sum_{i,j=0}^{i,j=1} \theta_{ij1}, \tag{10.18}$$

and for reader 1 of the non-diseased subjects, let

$$\phi_{ij1}^* = \phi_{ij1} \bigg/ \sum_{i,j=0}^{i,j=1} \phi_{ij1}. \tag{10.19}$$

In all, there will be four truncated distributions for the cell frequencies, the two given by Equations 10.18 and 10.19, and the corresponding truncations for reader 2, with probabilities θ_{ij2}^* and ϕ_{ij2}^*, respectively.

The relevant accuracy parameters are the TPF and FPF for the two tests for reader 1, and the TPF and FPF for reader 2. Assuming the BP rule, there will be the TPF and FPF for readers 1 and 2, and assuming the BN rule the TPF and FPF for readers 1 and 2. This is a total of 16 accuracy parameters, thus the level of complexity has increased twofold from previous considerations, when only one reader is present.

In order to make inferences about the accuracy, consider the following.
The TPF of reader 1 for the first test is

$$\text{tpft1r1} = \theta_{1.1}, \tag{10.20}$$

where the dot denotes summation over the missing subscript, and the TPF for reader 1 of test 2 is

$$\text{tpft2r1} = \theta_{.12}. \tag{10.21}$$

As for the FPF of test 1 and reader 1,

$$\text{fpfr1t1} = \phi_{1.1}. \tag{10.22}$$

Assuming the BP rule, the TPF for reader 1 is

$$\text{tpfbpr1} = \theta_{011} + \theta_{111} + \theta_{101}, \tag{10.23}$$

and assuming the BN rule, the FPF for reader 2 is

$$\text{fpfbnr2} = \phi_{112}. \tag{10.24}$$

The remaining accuracy indices are defined in the obvious way, following the pattern provided by Equations 10.20 through 10.24.

The example considered for this section is a variation of the second example of the previous section, where the study information is given by Tables 10.3a and b. Recall that the two tests involved were imaging modalities CT and MRI for a coronary artery trial, where among all the patients, 58 coronary artery segments were determined to be diseased (stenosis >50%) and 236 were determined to be non diseased (stenosis <50%) and the reference standard was coronary angiography. Gerber et al. [7] is the basis for this study and the information provided in Tables 10.3a and b is for one reader, who was an expert in both CT and MRI and read the two images for all patients. I should point out that the information from the paper is portrayed in the margins of the tables (to give the actual TPF and FPF for each modality), however, the four cell frequencies are provided by me. This information was not provided by the Gerber et al. study.

I will change the four cell frequencies to give the information for a second reader (radiologist) and assume that the information in Tables 10.3a and b is that provided by the first radiologist, and that by the second reader is contained in the second list statement of BUGS CODE 10.2 provided by the second reader.

Note that I made a slight change in the cell frequencies so that the marginal totals are the same as those in Tables 10.3a and b for radiologist 1, however, this is not necessary, and will be explored further in Exercise 6. In order to execute the analysis, use BUGS CODE 10.2 below and note that the program statements closely follow Equations 10.14 through 10.24.

BUGS CODE 10.2

```
model;
# BUGS CODE 10.1
# Two Binary Tests with Two Readers
# Accuracy of Combined Tests
# BP or BN
# Dirichlet Distribution generated
{
g001~dgamma(a001,2)
g011~dgamma(a011,2)
g101~dgamma(a101,2)
g111~dgamma(a111,2)
g002~dgamma(a002,2)
g012~dgamma(a012,2)
g102~dgamma(a102,2)
g112~dgamma(a112,2)
h001~dgamma(b001,2)
h011~dgamma(b011,2)
```

```
h101~dgamma(b101,2)
h111~dgamma(b111,2)
h002~dgamma(b002,2)
h012~dgamma(b012,2)
h102~dgamma(b102,2)
h112~dgamma(b112,2)
sumall<-g001+g011+g101+g111+g002+g012+g102+g112+
        h001+h011+h101+h111+h002+h012+h102+h112
# cell probabilities for diseased
th001<-g001/sumall
th011<-g011/sumall
th101<-g101/sumall
th111<-g111/sumall
th002<-g002/sumall
th012<-g012/sumall
th102<-g102/sumall
th112<-g112/sumall
# cell probabilities for non diseased
ph001<-h001/sumall
ph011<-h011/sumall
ph101<-h101/sumall
ph111<-h111/sumall
ph002<-h002/sumall
ph012<-h012/sumall
ph102<-h102/sumall
ph112<-h112/sumall
# truncated cell probabilities
# reader 1 diseased
sth1<-th001+th011+th101+th111
theta001<- th001/sth1
theta011<- th011/sth1
theta101<- th101/sth1
theta111<- th111/sth1
# reader 2 diseaded
sth2<-th002+th012+th102+th112
theta002<- th002/sth2
theta012<- th012/sth2
theta102<- th102/sth2
theta112<- th112/sth2
# reader 1 non diseased
sph1<- ph001+ph011+ph101+ph111
phi001<-ph001/sph1
phi011<-ph011/sph1
phi101<-ph101/sph1
```

```
phi111<-ph111/sph1
# reader 2 non diseased
sph2<- ph002+ph012+ph102+ph112
phi002<-ph002/sph2
phi012<-ph012/sph2
phi102<-ph102/sph2
phi112<-ph112/sph2
# accuracy parameters
# TPF test 1 reader 1
tpft1r1<- theta101+theta111
# TPF test 2 reader 1
tpft2r1<- theta011+theta111
# TPF test 1 reader 2
tpft1r2<- theta102+theta112
# TPF test 2 reader 2
tpft2r2<- theta012+theta112
# FPF test 1 reader 1
fpft1r1<-phi101+phi111
#FPF test 1 reader 2
fpft1r2<- phi102+phi112
# FPF test 2 reader 1
fpft2r1<-phi011+phi111
# FPF test 2 reader 2
fpft2r2<- phi012+phi112
# accuracy for combined test
# accuracy parameters for BP
# TPF reader 1 BP
tpfbpr1<-theta101+theta111+theta011
# FPF reader 1 BP
fpfbpr1<-phi101+phi111+phi011
# TPF reader 2 BP
tpfbpr2<-theta102+theta112+theta012
# FPF reader 2 BP
fpfbpr2<- phi102+phi112+phi012
# accuracy parameters for BN
# TPF reader 1 BN
tpfbnr1<-theta111
# TPF reader 2 BN
tpfbnr2<- theta112
# FPF reader 1 BN
fpfbnr1<- phi111
# FPF reader 2 BN
fpfbnr2<- phi112
}
# two readers CT and MRI
```

list(a001=13 , a011=1,a101=11,a111=37,
a002=10, a012=4,a102=14,a112=34,
b001=169, b011=1,b101=31,b111=39,b002=167, b012=3,b102=33,b112=36)

two readers CT and MRI second version
list(a001=13 , a011=1,a101=11,a111=37,
a002=7, a012=7,a102=17,a112=31,
b001=169 , b011=1,b101=31,b111=39,
b002=151 , b012=19,b102=49,b112=21)

initial values
list(h001= 1, h011=1,h101=1,h111=1,h002= 1, h012=1,h102=1,h112=1,
g001=1 ,g011=1,g101=1,g111=1,g002=1 , g012=1,g102=1,g112= 1)

The parameter labels are as follows: fpft2r1 means FPF test 2 reader 1 and tpfbnr1 signifies TPF believe the negative reader 1, etc. It is not surprising that for a given measure of accuracy there is very little difference in the posterior characteristics of the two distributions for the two readers (see Table 10.5).

Simulations errors were <0.0001 for all parameters, thus we would expect the accuracy of the combined test (CT and MRI) to be the same for both radiologists, for both the BP and BN rules. Note that the analysis is generated using 45,000 observations, with a burn in of 5,000 and a refresh of 100. The input for this problem appears in the first list statement of BUGS CODE 10.2 and assumes that a uniform prior is appropriate for all 16 cell probabilities.

TABLE 10.5:   Posterior accuracy for two readers—coronary artery stenosis for CT and MRI: combined test accuracy.

Parameter	Mean	sd	2 1/2	Median	97 1/2
fpfbnr1	0.1639	0.0237	0.1187	0.1617	0.2119
fpfbnr2	0.1506	0.023	0.1082	0.1494	0.1984
fpfbpr1	0.2958	0.0293	0.2399	0.2954	0.3549
fpfbpr2	0.3011	0.0295	0.245	0.3006	0.3605
fpft1r1	0.2941	0.0293	0.2354	0.2913	0.3508
fpft1r2	0.2885	0.0291	0.2333	0.2881	0.3474
fpft2r1	0.1667	0.0240	0.1224	0.1658	0.2167
fpft2r2	0.1631	0.0237	0.1194	0.1662	0.2122
tpfbnr1	0.5965	0.0617	0.4734	0.5979	0.7135
tpfbnr2	0.5483	0.0625	0.4243	0.5487	0.6692
tpfbpr1	0.79	0.0514	0.681	0.7932	0.8816
tpfbpr2	0.8388	0.0464	0.7386	0.8424	0.9187
tpft1r1	0.7738	0.0527	0.6633	0.7767	0.8685
tpft2r2	0.774	0.0525	0.6635	0.7768	0.8683
tpft2r1	0.6127	0.0612	0.47	0.6141	0.729
tpft2r2	0.6131	0.0612	0.4896	0.6143	0.73

10.4 Accuracy of Combined Binary Tests with Verification Bias

Upcoming is a venture into an area that has not been studied, the case of two binary tests with verification bias. What is the accuracy of the combined test when verification bias is present?

When assessing the accuracy of two tests, the design in many cases is paired. For example, two imaging devices (e.g., CT and MRI) are procuring information from the same patients and both images would be expected to be quite similar. Another case of a paired design is for two readers who are imaging the same set of patients with the same imaging device. One expects the information gained from the two paired sources to be highly correlated, and in the case of two paired readers, agreement between the two is also of interest. Recall from Chapter 8, the experimental layout for a paired study when verification bias is present (Table 10.6).

Note that the number of subjects verified under the gold standard, when both tests are positive, is $s_{11} + r_{11}$, among which s_{11} had the disease and r_{11} did not have the disease, and the number who were not verified under the gold standard when both tests are positive is u_{11} etc. Also note that the total number of subjects is

$$\sum_{i,j=0}^{i,j=1} m_{ij} = m_{..}.$$

The following derivation is based on the missing at random (MAR) assumption, namely,

$$P[V = 1 \mid Y_1, Y_2, D] = P[V = 1 \mid Y_1, Y_2].$$

Thus, the probability that a subject's disease status is verified depends only on the outcomes of the two tests.

Suppose the unknown parameters are defined as follows:

$$\phi_{ij} = P[D = 1 \mid Y_1 = i, Y_2 = j] \tag{10.25}$$

TABLE 10.6: Two binary scores with verification bias.

$V = 1$	$Y_1 =$ $Y_2 =$	1		0	
		1	0	1	0
$D = 1$		s_{11}	s_{10}	s_{01}	s_{00}
$D = 0$		r_{11}	r_{10}	r_{01}	r_{00}
$V = 0$		u_{11}	u_{10}	u_{01}	u_{00}
Total		m_{11}	m_{10}	m_{01}	m_{00}

and

$$\theta_{ij} = P[Y_1 = i, Y_2 = j], \tag{10.26}$$

for $i, j = 0, 1$.

Also let

$$\phi_{i\cdot} = P[D = 1 \mid Y_1 = i] \tag{10.27}$$

and

$$\phi_{\cdot j} = P[D = 1 \mid Y_2 = j], \tag{10.28}$$

where $i, j = 0, 1$.

The likelihood for the parameters is

$$L(\theta, \phi) \propto \prod_{i=0}^{i=1} \prod_{j=0}^{j=1} \phi_{ij}^{s_{ij}} (1 - \phi_{ij})^{r_{ij}} \prod_{i=0}^{i=1} \prod_{j=0}^{j=1} \theta_{ij}^{m_{ij}}. \tag{10.29}$$

Assuming an improper prior distribution for the parameters, the posterior distributions are

$$\phi_{ij} \sim \text{beta}(s_{ij}, r_{ij}), \tag{10.30}$$

for $i, j = 0, 1$, and θ_{ij} are distributed Dirichlet with parameter vector $(m_{00}, m_{01}, m_{10}, m_{11})$.

Note that

$$\phi_{1\cdot} \sim \text{beta}(s_{1\cdot}, r_{1\cdot})$$

and

$$\phi_{\cdot 1} \sim \text{beta}(s_{\cdot 1}, r_{\cdot 1}), \tag{10.31}$$

where

$$s_{1\cdot} = s_{11} + s_{10}$$

and

$$r_{1\cdot} = r_{11} + r_{10}.$$

The main parameters of interest are the TPF and the FPF for the two tests, thus for the first test

$$\text{tpf}_1 = P[Y_1 = 1 \mid D = 1]$$

and is given by Bayes theorem as

$$\text{tpf}_1 = \phi_{1\cdot}\theta_{1\cdot}/(\phi_{1\cdot}\theta_{1\cdot} + \phi_{0\cdot}\theta_{0\cdot}), \tag{10.32}$$

where $\phi_{i.}$ are given by Equation 8.19 and

$$\theta_{1.} = \theta_{11} + \theta_{10}.$$

As for test 1, the FPF is given by

$$fpf_1 = (1 - \phi_{1.})\theta_{1.}/(1 - \phi_{1.}\theta_{1.} - \phi_{0.}\theta_{0.}). \tag{10.33}$$

With regard to test 2, the TPF is

$$tpf2 = \phi_{.1}\theta_{.1}/(\phi_{.1}\theta_{.1} + \phi_{.0}\theta_{.0})$$

and the FPF is

$$fpf2 = (1 - \phi_{.1})\theta_{.1}/(1 - \phi_{.1}\theta_{.1} - \phi_{.0}\theta_{.0}) \tag{10.34}$$

The main focus of this section is on measuring the accuracy of the combined test in the presence of verification bias of both tests using the BN and BP principles.

Assume the BP principle is in effect, then the TPF for the combined test is

$$tpfbp = P[Y_1 = 1 \text{ or } Y_2 = 1 \mid D = 1], \tag{10.35}$$

while

$$fpfbp = P[Y_1 = 1 \text{ or } Y_2 = 1 \mid D = 0]. \tag{10.36}$$

Now assume that the BN assumption is in effect, then the TPF is

$$tpfbn = P[Y_1 = 1, Y_2 = 1 \mid D = 1], \tag{10.37}$$

while

$$fpfbn = P[Y_1 = 1, Y_2 = 1 \mid D = 0]. \tag{10.38}$$

The above four accuracy measures can be expressed as follows: for the BP assumption,

$$tpfbp = (\phi_{11}\theta_{11} + \phi_{01}\theta_{01} + \phi_{10}\theta_{10})/P[D = 1] \tag{10.39}$$

and

$$fpfbp = ((1 - \phi_{11})\theta_{11} + (1 - \phi_{01})\theta_{01} + (1 - \phi_{10})\theta_{10})/P[D = 0], \tag{10.40}$$

where

$$P[D = 1] = (\phi_{11}\theta_{11} + \phi_{01}\theta_{01} + \phi_{10}\theta_{10} + \phi_{00}\theta_{00}). \tag{10.41}$$

TABLE 10.7: Screening test for Alzheimer's disease.

$Y_1 =$	1		0	
$V = 1$ $\quad Y_2 =$	1	0	1	0
$D = 1$	$s_{11} = 191$	$s_{10} = 25$	$s_{01} = 85$	$s_{00} = 5$
$D = 0$	$r_{11} = 10$	$r_{10} = 25$	$r_{01} = 95$	$r_{00} = 150$
$V = 0$	$u_{11} = 10$	$u_{10} = 10$	$u_{01} = 100$	$u_{00} = 700$
Total	$m_{11} = 211$	$m_{10} = 60$	$m_{01} = 280$	$m_{00} = 855$

For the BN assumption

$$\text{tpfbn} = \phi_{11}\theta_{11}/P[D = 1] \tag{10.42}$$

and

$$\text{fpfbn} = (1 - \phi_{11})\theta_{11}/P[D = 0]. \tag{10.43}$$

As a first example of two binary tests, consider a screening test for Alzheimer's disease, where the two tests correspond to two observers (psychiatrists) screening the same patients (Table 10.7).

Only 155 of 855 are referred to the gold standard when both readers give a negative score, but when both observers give a positive score, 201 of 211 are referred to the gold standard. What are the TPFs for both observers? What are the FPFs? Of the 1406 subjects, 855 received a negative assessment by both psychiatrists, while 211 received a positive assessment (judged as having the disease) from both. Recall that this study is analyzed in Chapter 8 where the accuracy is assessed for both readers, but for the present, the main emphasis will be on the accuracy of the combined readers (psychiatrists). BUGS CODE 10.3 closely follows the notation of the above derivation for the accuracy measures of the two readers.

BUGS CODE 10.3

```
model;
{
# two binary tests verification bias
# accuracy of combined tests
g00~dgamma(m00,2)
g01~dgamma(m01,2)
g10~dgamma(m10,2)
g11~dgamma(m11,2)
h<-g00+g01+g10+g11
th00<-g00/h
th01<-g01/h
th10<-g10/h
th11<-g11/h
```

```
ph00~dbeta(s00,r00)
ph01~dbeta(s01,r01)
ph10~dbeta(s10,r10)
ph11~dbeta(s11,r11)
s1.<-s11+s10
r1.<-r11+r10
s.1<-s01+s11
r.1<- r01+r11
r0.<- r00+r01
s0.<-s00+s01
s.0<-s00+s10
r.0<-r00+r10
ph1.~dbeta(s1.,r1.)
ph.1~dbeta(s.1,r.1)
ph0.~dbeta(s0.,r0.)
ph.0~dbeta(s.0,r.0)
th1.<-th11+th10
th.1<-th01+th11
th0.<-th01+th00
th.0<-th00+th10
# accuracy for test 1
tpf1<-ph1.*th1./pd1
fpf1<-(1-ph1.)*th1./(1-pd1)
# p[D-1]
pd1<-ph1.*th1.+ph0.*th0.
# accuracy for test 2
tpf2<-ph.1*th.1/pd2
fpf2<-(1-ph.1)*th.1/(1-pd2)
pd2<-ph.1*th.1+ph.0*th.0
# accuracy combined tests
# believe the positive, BP
tpfbp<-(ph11*th11+ph01*th01+ph10*th10)/pd
# P[d=1]
pd<- ph11*th11+ph10*th10+ph01*th01+ph00*th00
fpfbp<-((1-ph11)*th11+(1-ph10)*th10+(1-ph01)*th01)/(1-pd)
# believe the negative , BN
tpfbn<-ph11*th11/pd
fpfbn<- (1-ph11)*th11/(1-pd)
}
# Alzheimers two readers improper prior
list(s00=5,r00=150,s01=85,r01=95,s10=131,r10=25,s11=191,r11=10,
m00=211,m01=60,m10=280,m11=855)
# CT and MRI for lung cancer risk improper prior
list(s00=3,r00=18,s01=9,r01=13,s10=12,r10=9,s11=14,r11=4,
m00=31,m01=31,m10=29,m11=25)
```

TABLE 10.8: Posterior analysis for the accuracy of two readers—
Alzheimer's study with verification bias.

Parameter	Mean	sd	Error	2 1/2	Median	97 1/2
fpf1	0.358	0.0406	<0.0001	0.2788	0.3582	0.4372
fpf2	0.4768	0.0279	<0.0001	0.4218	0.4769	0.5311
fpfbn	0.1306	0.0354	<0.0001	0.0681	0.1282	0.2055
fpfbp	0.3674	0.0358	<0.0001	0.2986	0.3667	0.4394
tpf1	0.9336	0.0070	<0.0001	0.9189	0.9339	0.9468
tfp2	0.7554	0.0168	<0.0001	0.7214	0.7557	0.7876
tpfbn	0.7504	0.0138	<0.00001	0.723	0.7504	0.7771
tpfbp	0.9937	0.0027	<0.00001	0.9872	0.9941	0.9979

The list statement of BUGS CODE 10.3 contains the information from the Alzheimer's study in Table 10.7, and the analysis is executed with 45,000 observations, with a burn in of 5,000 and a refresh of 100, assuming an improper prior density for the parameters (Table 10.8). See the likelihood function (Equation 10.29) and put an improper prior density on θ_{ij} and ϕ_{ij} that appear in the formula.

According to the BP rule, the TPF and FPF are 0.9937 and 0.3374, respectively, while the corresponding entities for the BN rule are 0.7504 and 0.1306. How should accuracy be reported for the combined test? As usual, the BP rule gives higher true and false positive rates compared to the BN rule. I like the BN rule, which gives a fairly high TPF but an extremely small false positive rate. On the other hand, the BP rule provides an extremely large TPF. What rule would you choose to represent the accuracy of the combined test? See Exercise 7 for additional information about choosing the accuracy of the combined test.

Figure 10.3 presents the posterior density of the TPF (tpfbp) for the BP rule, with a 95% credible interval of (0.9872,0.997). This topic will be continued in Exercise 8 with an example based on Exercise 16 of Chapter 8, which uses two imaging modalities to assign risk of lung cancer in a paired design.

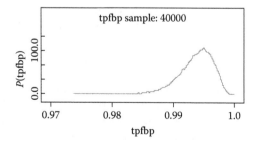

FIGURE 10.3: Posterior density of the TPF for the BP rule.

10.5   Likelihood Ratio, the Risk Score, the Neyman–Pearson Lemma, and the Accuracy of Multiple Ordinal Tests

A change of emphasis from binary to ordinal and continuous test scores brings us to some "new" ideas for measuring the accuracy by combining two tests. For ordinal and continuous scores, the area under the ROC curve measures the intrinsic accuracy of a medical test; but how should the area be computed when two tests are combined? The ROC curve of the risk score is the foundation for measuring the accuracy for the combined test, but in turn, the risk score is a monotone increasing function of the likelihood ratio, which is the optimal way to measure accuracy for the combined test.

The optimality of the risk function is a consequence of the Neyman–Pearson lemma, which is a familiar result from classical statistics for testing hypotheses.

In what follows, the likelihood ratio will be defined and the optimality of the ROC curve of the likelihood ratio will be demonstrated by referring to the Neyman–Pearson lemma, then the risk function will be defined and shown to be a monotone increasing function of the likelihood ratio, thus the ROC curve of the risk function is the same as the ROC curve of the likelihood ratio. Pepe's [8: 269–274] development of the subject is closely followed but is given a Bayesian emphasis, and the end result will be that the optimal way to measure the accuracy of the combined test is to estimate the area under the ROC curve of the risk function. Determining the risk function is equivalent to performing a logistic regression using the test scores of the two tests as predictors, then the ROC curve of the predicted probabilities (from the logistic regression) is computed, from which the area is then estimated. Such an area is the accuracy of the combined test, and the methodology is illustrated with various examples using ordinal test scores. The first example is from an imaging trial using MRI and CT to detect lung cancer, where one radiologist uses a five-point confidence score, and the ROC curve of the risk function of the combined test is computed and compared to the ROC curve of the individual tests.

This section continues with the definition of the likelihood ratio and concludes with the definition of the risk score.

Suppose $Y = (Y_1, Y_2, \ldots, Y_p)$ is the vector of scores of p ordinal tests, then the likelihood ratio is

$$\mathrm{LR}(Y) = P[Y \mid D = 1]/P[Y \mid D = 0], \tag{10.44}$$

where D is the indicator of disease. The numerator is the probability of the observed test scores when the disease is present, and the denominator is the probability of the observed scores, given the disease is not present.

Recall that the likelihood ratio is used as a test statistic for the null hypothesis

$$\mathrm{H} \colon D = 1$$

versus the alternative hypothesis

$$A: D = 0,$$

where larger values of $LR(Y)$ are evidence of the null hypothesis, and smaller values are evidence that the alternative is true.

It can be shown that the likelihood ratio has certain optimal properties, summarized by the following result.

Statement 10.2

Suppose a decision about the accuracy of a medical test is based on the criterion

$$LR(Y) > c, \qquad\qquad (10.45)$$

then the likelihood ratio:

a. Maximizes the TPF among all rules with FPF $= t$, for all $t \in (0,1)$.

b. Minimizes the FPF among all rules with TPF $= r$, for all $r \in (0,1)$.

c. Minimizes the overall misclassification probability

$$\rho(1 - \text{tpf}) + (1 - \rho)]\text{fpf},$$

  where ρ is the disease rate.

d. Minimizes the expected cost, regardless of the costs associated with false negative and false positive errors.

The threshold c above, appearing in Statement 10.1, depends on the objective at hand, but for our purposes the above result implies that the ROC curve based on the likelihood ratio is optimal, in the sense that its area is the largest.

Pepe [8: 269–275] states that the above results are a consequence of the Neyman–Pearson lemma, and also presents the correspondence between concepts involving medical test accuracy and the analogous classical approach to testing hypothesis (Table 10.9).

The likelihood function (Equation 10.45) is difficult to work with because of the complexity of determining its distribution, but, fortunately, the risk score

$$RS(Y) = P[D = 1 \mid Y] \qquad\qquad (10.46)$$

does not have this disadvantage and has the property that it is a monotone function of the likelihood ratio. Simply stated, the risk score assigns a probability of disease to each study subject.

TABLE 10.9: Correspondence between testing hypotheses and medical test accuracy.

	Hypothesis testing	Accuracy of test
Type I error	Significance level	FPF
	$\alpha = P[\text{reject null} \mid \text{null}]$	$\text{FPF} = P[D = 1 \mid D = 0]$
Type II error	Power	TPF
	$1 - \beta = P[\text{reject null} \mid \text{alt.}]$	$\text{TPF} = P[\text{choose } D = 1]D = 1]$
Possible states	H_0 vs. H_1	$D = 0$ vs. $D = 1$
Decision	Either H_0 or H_1	Either $D = 0$ or $D = 1$
Information	Sample results Y	Test results Y
Test statistic	$P[Y \mid H_1]/P[Y \mid H_0]$	$P[Y \mid D = 1]/P[Y \mid D = 0]$

Statement 10.3

The risk score (Equation 10.46) has the same ROC curve as the likelihood ratio (Equation 10.45). Also, it has the same optimal properties (see Statement 10.1) as the likelihood ratio. Observe that

$$
\begin{aligned}
\text{RS}(Y) &= P[D = 1 \mid Y] \\
&= P[Y \mid D = 1]P[D = 1]/P[Y] \\
&= P[Y \mid D = 1]P[D = 1]/\{P[Y \mid D = 1]P[D = 1] \\
&\quad + P[Y \mid D = 0]P[D = 0]\} \\
&= \text{LR}(Y)P[D = 1]/\{\text{LR}(Y)P[D = 1] + P[D = 1]\} \quad (10.47)
\end{aligned}
$$

which shows that the risk score is a monotone increasing function of the likelihood ratio, which implies that the ROC curve of the risk score is the same as that of the likelihood ratio. For our purposes, the risk score will be used to measure the accuracy of combined tests, namely, using the area of the ROC curve of the risk score. Pepe [8: 274–275] shows the utility of logistic regression for finding the ROC curve of the risk score. Note that the following statement shows why.

Statement 10.4

Suppose the risk score is expressed as

$$
\text{logit } P[D = 1 \mid Y] = \gamma + g(\lambda, Y), \quad (10.48)
$$

where g is a known function, then:

a. The parameter λ can be estimated, even for retrospective designs in which the sampling depends on D.

b. The function g is optimal for determining the ROC curve of the risk function.

From a practical point of view, logistic regression can be used to determine the ROC curve of the risk function, but it should be noted that finding a suitable function g can be a challenge. After all, g can be a complicated non-linear function of λ and/or Y, but it would be convenient if g is linear in the test scores Y. Of course, a Bayesian approach is taken in order to estimate the logistic regression function (Equation 10.48).

The approach taken here is based on the risk score and Pepe [8: 274–275] gives a good account, which is outlined in more detail in McIntosh and Pepe [9].

Suppose that there are two medical tests with ordinal scores, then for diseased subjects the layout is as shown in Tables 10.10a and b. Thus, there are n_{ij} diseased subjects with a score of i for test 1 and score j for test 2 and the cell probabilities for the diseased are

$$\theta_{ij} = P[T_1 = i, T_2 = j \mid D = 1], \tag{10.49}$$

for the first test, where $i, j = 1, 2, \ldots, k$.

For the non diseased, the cell probabilities are

$$\phi_{ij} = P[T_1 = i, T_2 = j \mid D = 0]. \tag{10.50}$$

Define the ROC area for test 1, the usual way, as

$$\text{Area1} = A_{11} + A_{12}/2, \tag{10.51}$$

TABLE 10.10a: Two medical tests with ordinal scores—diseased patients: frequencies and probabilities.

Test 1	Test 2 scores			
	1	2	...	k
1	n_{11}, θ_{11}	n_{12}, θ_{12}		n_{1k}, θ_{1k}
2	n_{21}, θ_{21}	n_{22}, θ_{22}		n_{2k}, θ_{2k}
	.			
	.			
k	n_{k1}, θ_{k1}	n_{k2}, θ_{k2}		n_{kk}, θ_{kk}

TABLE 10.10b: Two medical tests with ordinal scores—non-diseased patients: frequencies and probabilities.

Test	1	2	...	k
1	m_{11}, ϕ_{11}	m_{12}, ϕ_{12}		m_{1k}, ϕ_{1k}
2	m_{21}, ϕ_{21}	m_{22}, ϕ_{22}		m_{2k}, ϕ_{2k}
.				
.				
k	m_{k1}, ϕ_{k1}	m_{k2}, ϕ_{k2}		m_{kk}, ϕ_{kk}

where

$$A_{11} = \sum_{i=2}^{i=k} \theta_{i.} \left(\sum_{j=1}^{j=i-1} \phi_{j.} \right), \tag{10.52}$$

and $\theta_{i.}, i = 1, 2, \ldots, k$, are the sum of θ_{ij} over the missing subscript and

$$A_{12} = \sum_{i=1}^{i=k} \theta_{i.} \phi_{i.}. \tag{10.53}$$

The ROC area for the second test is defined in a similar fashion as

$$\text{Area2} = A_{21} + A_{22}/2, \tag{10.54}$$

where

$$A_{21} = \sum_{i=2}^{i=k} \theta_{.i} \left(\sum_{j=1}^{j=i-1} \phi_{.j} \right) \tag{10.55}$$

and

$$A_{22} = \sum_{i=1}^{i=k} \theta_{.i} \phi_{.i}.$$

Our goal is to use the area under the ROC of the risk score as a measure of the accuracy of the combined tests T_1 and T_2, where the risk scores are determined by logistic regression (if appropriate)

$$\text{logit}(\theta_{ij}) = \gamma + g(\lambda, T_1, T_2), \tag{10.56}$$

and the unknown parameters γ and λ (possibly a vector) are estimated by Bayesian techniques. From the logistic regression, the estimated (e.g., posterior means) cell probabilities are employed to estimate the area under the ROC curve of the risk score.

Note that the area under the ROC curve of the risk score is based on the posterior distribution of the $2k^2$ parameters, θ_{ij} and ϕ_{ij}, for $i, j = 1, 2, \ldots, k$, and this scenario is illustrated with the following example, where the area under the ROC curve is given by the usual formulas (Equations 4.47 and 4.48) employed earlier in Chapter 4. Of course, in addition, the area under the ROC curves for the individual tests will also be portrayed and compared to the area under the ROC curve of the risk score. It will be a challenge to develop a good logistic regression (Equation 10.51); however, in some cases it will turn out that the logit is a linear function of the two tests, T_1 and T_2. The risk score is assigned to each experimental unit and is the probability of disease, which is estimated from the raw scores of the two component tests!

Note, using the risk score is a statistical procedure and will ideally be utilized by the clinician working with a statistician.

10.5.1 Magnetic resonance imaging and computed tomography determination of lung cancer risk

When considering the accuracy of two ordinal tests, a paired study is envisioned, where each test is applied to each patient and one reader examines the results of both tests. It is important to remember that the reader uses the results of both tests for each patient in order to decide what score to assign to the patient.

Our first example involves the MRI and CT determination of lung cancer risk, where one radiologist interprets both images and gives a score from 1 to 5 for the presence of a malignant lesion with the following definition: A score of 1 indicates no evidence of malignancy, while a score of 2 indicates very little evidence of a lesion. A score of 3 designates a benign lesion, while a score of 4 indicates that there is some evidence of a malignancy, and finally a score of 5 signals that the lesion is definitely malignant. This is obviously a paired design in that both images are taken on each patient and one would expect a "large" correlation between the scores of MRI and CT images. A total of 261 patients have lung cancer and 674 do not, and the gold standard is lung biopsy (Tables 10.11a and b).

The above study is hypothetical, but many studies have investigated CT and MRI as alternatives to detecting lung cancer, and it should be noted that CT has shown good promise (in comparison to x-ray) in a recent national lung cancer screening trial, see Gierada, Pilgrim, and Ford [10] for additional information.

With regard to the accuracy of the combined test, the approach is to find the area under the ROC curve of the risk score, which is determined by logistic regression, namely,

$$\text{logit}(\text{theta}[i]) = b[1] + b[2]T_1 + b[3]T_2, \tag{10.57}$$

TABLE 10.11a: MRI and CT scores for detecting lung cancer—diseased patients.

CT scores	MRI scores					Total
	1	2	3	4	5	
1	15	10	6	2	1	34
2	9	21	10	3	2	45
3	5	6	32	6	3	52
4	2	0	6	47	2	57
5	0	1	2	5	65	73
Total	31	38	56	63	73	261

TABLE 10.11b: MRI and CT scores for detecting lung cancer—non-diseased patients.

CT scores	MRI scores					Total
	1	2	3	4	5	
1	92	62	41	8	5	208
2	58	81	10	8	4	161
3	38	30	65	31	18	182
4	16	2	21	35	12	86
5	5	1	3	11	17	37
Total	209	176	140	93	56	674

where theta[i] is the probability that the ith patient has disease, and $i = 1$, $2, \ldots, N$.

N is the number of patients in the study with 261 with disease (lung cancer) and 674 with no disease, and $b[i]$ are unknown regression coefficients. From a Bayesian viewpoint, the regression coefficients are given vague prior distributions of the form

$$b[i] \sim \text{dnorm}(0.000, 0.0001), \tag{10.58}$$

namely, a normal distribution with mean 0 and precision 0.0001.

A Bayesian analysis for the lung cancer study is executed in two stages: (1) using BUGS CODE 10.4, estimate the ROC areas of MRI and CT; and (2) using BUGS CODE 10.5, estimate the ROC area of the combined test, based on the risk score (Equation 10.57).

Consider first, estimating the area of the two tests, BUGS CODE 10.4.

BUGS CODE 10.4

```
model;
# BUGS CODE 10.4
{
# Scores for tests 1 and 2 diseased
for(i in 1:5){for(j in 1:5){ g[i,j]~dgamma(y[i,j],2)}}
sg<-sum(g[,])
# Scores for tests 1 and 2 non diseased
for(i in 1:5){for(j in 1:5){ h[i,j]~dgamma(z[i,j],2)}}
sh<-sum(h[,])
# cell probabilities for diseased
for( i in 1:5){for( j in 1:5){ theta[i,j]<-g[i,j]/sg }}
# cell probabilities for non diseased
for( i in 1:5){for( j in 1:5){ phi[i,j]<-h[i,j]/sh }}
# cell probabilities have a Dirichlet distribution
th1.<-sum(theta[1,])
th2.<-sum(theta[2,])
```

```
th3.<-sum(theta[3,])
th4.<-sum(theta[4,])
th5.<-sum(theta[5,])
th.1<-sum(theta[,1])
th.2<-sum(theta[,2])
th.3<-sum(theta[,3])
th.4<-sum(theta[,4])
th.5<-sum(theta[,5])
ph1.<- sum(phi[1,])
ph2.<-sum(phi[2,])
ph3.<- sum(phi[3,])
ph4.<- sum(phi[4,])
ph5.<- sum(phi[5,])
ph.1<- sum(phi[,1])
ph.2<- sum(phi[,2])
ph.3<-sum(phi[,3])
ph.4<-sum(phi[,4])
ph.5<-sum(phi[,5])
# ROC area test 1
area1<- area11+area12/2
area11<-th2.*ph1.+th3.*(ph1.+ph2.)+th4.*(ph1.+ph2.+ph3.)+
th5.*(ph1.+ph2.+ph3.+ph4.)
area12<-th1.*ph1.+th2.*ph2.+th3.*ph3.+th4.*ph4.+th5.*ph5.
# ROC area for test 2
area2<- area21+area22/2
area21<-th.2*ph.1+th.3*(ph.1+ph.2)+th.4*(ph.1+ph.2+ph.3)+
th.5*(ph.1+ph.2+ph.3+ph.4)
area22<-th.1*ph.1+th.2*ph.2+th.3*ph.3+th.4*ph.4+th.5*ph.5
}
# example of mri (test 2)and ct (test 1) for lung cancer
# assumes uniform prior
list(y=structure(.Data=c(16,11,7,3,2,
                         10,22,11,4,3,
                         6,7,33,7,4,
                         3,1,7,48,3,
                         1,2,3,6,66),.Dim = c(5,5)),
     z=structure(.Data=c(93,63,42,9,6,
                         59,82,11,9,5,
                         39,31,66,32,19,
                         17,3,22,36,13,
                         6,2,4,12,18),.Dim = c(5,5))))
# Two MRI versions for prostate cancer
# Assumes a uniform prior
list(y=structure(.Data=c(4,8,7,4,1,
                         2,6,9,8,7,
```

TABLE 10.12: Posterior analysis for MRI and CT study of lung cancer—individual ROC areas.

Parameter	Mean	sd	Error	2 1/2	Median	97 1/2
area ct	0.6836	0.0188	<0.0001	0.6459	0.6837	0.7198
area11	0.5931	0.0213	<0.0001	0.5509	0.5931	0.6346
area12	0.181	0.0057	<0.0001	0.1696	0.1811	0.1921
area mri	0.6886	0.0183	<0.0001	0.652	0.6889	0.7239
area21	0.5992	0.0207	<0.0001	0.5581	0.5994	0.6392
area22	0.1788	0.0049	<0.0001	0.1689	0.1789	0.1883

$$
\begin{array}{l}
3,3,5,12,8,\\
1,2,2,9,14,\\
1,1,2,5,24),.\text{Dim} = c(5,5)),\\
\text{z=structure(.Data=c(104,10,4,3,4,}\\
32,63,12,7,6,\\
26,23,32,14,10,\\
12,12,21,31,16,\\
14,16,29,22,14),.\text{Dim} = c(5,5))))
\end{array}
$$

Based on generating 45,000 observations from the posterior distribution, with a burn in of 5,000 and a refresh of 100, the Bayesian analysis is presented in Table 10.12.

The Markov Chain Monte Carlo (MCMC) errors are quite small and show that the presented estimated ROC areas are very "close" to the actual posterior areas, and the analysis also shows that the two areas are about the same, that is the accuracy of the two modalities are essentially the same. The probability of a tie with CT is estimated with a posterior mean of 0.181 and 0.1788 with MRI. Thus, one would expect the accuracy of the combined test, as measured by the ROC area of the risk score, to be about the same value in the area of 0.70.

The statements of BUGS CODE 10.5 closely follow the formulas given above.

BUGS CODE 10.5

```
model:
# logistic regression
{
for( i in 1:935){d[i]~dbern(theta[i])}
for( i in 1:935){logit(theta[i])<- b[1]+b[2]*T1[i]+b[3]*T2[i]}
# prior distributions
for( i in 1:3){ b[i]~dnorm(0.000,.0001)}
# area under the curve
for( i in 1 : 935){ p[i]~dnorm(mu[i], precy[d[i]+1])
mu[i]<-beta[1]+beta[2]*d[i]}
```

```
# prior distributions
for( i in 1:2){beta[i]~dnorm(0.000,.0001)
precy[i]~dgamma(0.0001,.0001)
vary[i]<-1/precy[i]}
la1y<-beta[2]/sqrt(vary[1])
la2y<-vary[2]/vary[1]
auc<-phi(la1y/sqrt(1+la2y))
}
list(T1=c(1,1,1,1,1,1,1,1,1,1,1,1,1,1,1,1,1,1,1,1,1,1,1,1,1,1,1,1,1,1,1,1,2,2,2,2,
2,2,2,2,2,2,2,2,2,2,2,2,2,2,2,2,2,2,2,2,2,2,2,2,2,2,2,2,2,2,2,2,2,2,2,2,2,2,2,2,3,3,
3,3,3,3,3,3,3,3,3,3,3,3,3,3,3,3,3,3,3,3,3,3,3,3,3,3,3,3,3,3,3,3,3,3,3,3,3,3,3,3,3,3,
3,3,3,3,3,3,3,4,4,4,4,4,4,4,4,4,4,4,4,4,4,4,4,4,4,4,4,4,4,4,4,4,4,4,4,4,4,4,4,4,4,4,
4,4,4,4,4,4,4,4,4,4,4,4,4,4,4,4,4,4,5,5,5,5,5,5,5,5,5,5,5,5,5,5,5,5,5,5,5,5,5,5,5,5,
5,5,5,5,5,5,5,5,5,5,5,5,5,5,5,5,5,5,5,5,5,5,5,5,5,5,5,5,5,5,5,5,5,5,5,5,5,5,5,5,5,5,
5,5,5,5,5,5,5,5,1,1,1,1,1,1,1,1,1,1,1,1,1,1,1,1,1,1,1,1,1,1,1,1,1,1,1,1,1,1,1,1,1,1,
1,1,1,1,1,1,1,1,1,1,1,1,1,1,1,1,1,1,1,1,1,1,1,1,1,1,1,1,1,1,1,1,1,1,1,1,1,1,1,1,1,1,
1,1,1,1,1,1,1,1,1,1,1,1,1,1,1,1,1,1,1,1,1,1,1,1,1,1,1,1,1,1,1,1,1,1,1,1,1,1,1,1,1,1,
1,1,1,1,1,1,1,1,1,1,1,1,1,1,1,1,1,1,1,1,1,1,1,1,1,1,1,1,1,1,1,1,1,1,1,1,1,1,1,1,1,1,
1,1,1,1,1,1,1,1,1,1,1,1,1,1,1,1,1,1,1,1,1,1,1,1,1,1,1,1,1,1,1,1,1,1,1,1,1,1,1,1,1,1,
1,2,2,2,2,2,2,2,2,2,2,2,2,2,2,2,2,2,2,2,2,2,2,2,2,2,2,2,2,2,2,2,2,2,2,2,2,2,2,2,2,
2,2,2,2,2,2,2,2,2,2,2,2,2,2,2,2,2,2,2,2,2,2,2,2,2,2,2,2,2,2,2,2,2,2,2,2,2,2,2,2,2,2,
2,2,2,2,2,2,2,2,2,2,2,2,2,2,2,2,2,2,2,2,2,2,2,2,2,2,2,2,2,2,2,2,2,2,2,2,2,2,2,2,2,2,
2,2,2,2,2,2,2,2,2,2,2,2,2,2,2,2,2,2,2,2,2,2,2,2,2,2,2,2,2,2,2,2,3,3,3,3,3,3,3,3,3,3,
3,3,3,3,3,3,3,3,3,3,3,3,3,3,3,3,3,3,3,3,3,3,3,3,3,3,3,3,3,3,3,3,3,3,3,3,3,3,3,3,3,3,
3,3,3,3,3,3,3,3,3,3,3,3,3,3,3,3,3,3,3,3,3,3,3,3,3,3,3,3,3,3,3,3,3,3,3,3,3,3,3,3,3,3,
3,3,3,3,3,3,3,3,3,3,3,3,3,3,3,3,3,3,3,3,3,3,3,3,3,3,3,3,3,3,3,3,3,3,3,3,3,3,3,3,3,3,
3,3,3,3,3,3,3,3,3,3,3,3,3,3,3,3,3,3,3,3,3,3,3,3,3,3,3,3,3,3,3,3,3,3,3,3,3,3,3,3,3,3,
4,4,4,4,4,4,4,4,4,4,4,4,4,4,4,4,4,4,4,4,4,4,4,4,4,4,4,4,4,4,4,4,4,4,4,4,4,4,4,4,4,4,
4,4,4,4,4,4,4,4,4,4,4,4,4,4,4,4,4,4,4,4,4,4,4,4,4,4,4,4,4,4,4,4,4,4,4,4,4,4,4,4,4,4,
5,5,5,5,5,5,5,5,5,5,5,5,5,5,5,5,5,5,5,5,5,5,5,5,5,5,5,5,5,5,5,5,5,5,5,5,5),
T2=c(1,1,1,1,1,1,1,1,1,1,1,1,1,1,1,2,2,2,2,2,2,2,2,2,2,3,3,3,3,3,3,4,4,5,1,1,1,1,1,1,
1,1,1,2,2,2,2,2,2,2,2,2,2,2,2,2,2,2,2,2,2,2,3,3,3,3,3,3,3,3,3,4,4,4,5,5,1,1,1,1,
1,2,2,2,2,2,2,3,3,3,3,3,3,3,3,3,3,3,3,3,3,3,3,3,3,3,3,3,3,3,3,3,3,3,3,3,3,3,4,4,4,4,
4,4,5,5,5,1,1,3,3,3,3,3,3,4,4,4,4,4,4,4,4,4,4,4,4,4,4,4,4,4,4,4,4,4,4,4,4,4,4,4,4,4,
4,4,4,4,4,4,4,4,4,4,4,4,4,4,5,5,2,3,3,4,4,4,4,4,5,5,5,5,5,5,5,5,5,5,5,5,5,5,5,
5,5,5,5,5,5,5,5,5,5,5,5,5,5,5,5,5,5,5,5,5,5,5,5,5,5,5,5,5,5,5,5,5,5,5,5,5,5,5,5,5,5,
5,5,5,5,5,5,1,1,1,1,1,1,1,1,1,1,1,1,1,1,1,1,1,1,1,1,1,1,1,1,1,1,1,1,1,1,1,1,1,1,1,1,
1,1,1,1,1,1,1,1,1,1,1,1,1,1,1,1,1,1,1,1,1,1,1,1,1,1,1,1,1,1,1,1,1,1,1,1,1,1,1,1,1,1,
1,1,1,1,1,1,1,1,1,1,1,1,2,2,2,2,2,2,2,2,2,2,2,2,2,2,2,2,2,2,2,2,2,2,2,2,2,2,2,2,2,2,
2,2,2,2,2,2,2,2,2,2,2,2,2,2,2,2,2,2,2,2,2,2,2,2,2,2,2,2,2,2,2,2,3,3,3,3,3,3,3,3,3,3,
3,3,3,3,3,3,3,3,3,3,3,3,3,3,3,3,3,3,3,3,3,4,4,4,4,4,4,4,5,5,5,5,5,1,
1,1,1,1,1,1,1,1,1,1,1,1,1,1,1,1,1,1,1,1,1,1,1,1,1,1,1,1,1,1,1,1,1,1,1,1,1,1,1,1,1,1,
1,1,1,1,1,1,1,1,1,1,2,2,2,2,2,2,2,2,2,2,2,2,2,2,2,2,2,2,2,2,2,2,2,2,2,2,2,2,2,2,2,2,
2,2,2,2,2,2,2,2,2,2,2,2,2,2,2,2,2,2,2,2,2,2,2,2,2,2,2,2,2,2,2,2,2,2,2,2,2,2,2,2,2,2,
2,2,2,2,2,2,2,2,2,3,3,3,3,3,3,3,3,3,4,4,4,4,4,4,4,5,5,5,5,1,1,1,1,1,1,1,1,1,1,1,1,
```

1,1,1,1,1,1,1,1,1,1,1,1,1,1,1,1,1,1,1,1,1,1,1,1,1,1,2,2,2,2,2,2,2,2,2,2,2,2,2,2,2,2,2,
2,2,2,2,2,2,2,2,2,2,2,2,2,3,3,3,3,3,3,3,3,3,3,3,3,3,3,3,3,3,3,3,3,3,3,3,3,3,3,3,3,3,3,
3,3,3,3,3,3,3,3,3,3,3,3,3,3,3,3,3,3,3,3,3,3,3,3,3,3,3,3,3,3,3,3,3,4,4,4,4,4,4,4,4,
4,4,4,4,4,4,4,4,4,4,4,4,4,4,4,4,4,4,4,4,4,4,5,5,5,5,5,5,5,5,5,5,5,5,5,5,5,5,5,5,5,5,1,1,
1,1,1,1,1,1,1,1,1,1,1,1,1,2,2,3,3,3,3,3,3,3,3,3,3,3,3,3,3,3,3,3,3,3,4,4,4,4,4,4,
4,4,4,4,4,4,4,4,4,4,4,4,4,4,4,4,4,4,4,4,4,4,4,4,5,5,5,5,5,5,5,5,5,5,5,1,1,
1,1,1,2,3,3,3,4,4,4,4,4,4,4,4,4,4,5,5,5,5,5,5,5,5,5,5,5,5,5,5,5,5),
d=c(1,1,1,1,1,1,1,1,1,1,1,1,1,1,1,1,1,1,1,1,1,1,1,1,1,1,1,1,1,1,1,1,1,1,1,1,1,1,1,
1,1,1,1,1,1,1,1,1,1,1,1,1,1,1,1,1,1,1,1,1,1,1,1,1,1,1,1,1,1,1,1,1,1,1,1,1,1,1,1,1,
1,1,1,1,1,1,1,1,1,1,1,1,1,1,1,1,1,1,1,1,1,1,1,1,1,1,1,1,1,1,1,1,1,1,1,1,1,1,1,1,1,
1,1,1,1,1,1,1,1,1,1,1,1,1,1,1,1,1,1,1,1,1,1,1,1,1,1,1,1,1,1,1,1,1,1,1,1,1,1,1,1,1,
1,1,1,1,1,1,1,1,1,1,1,1,1,1,1,1,1,1,1,1,1,1,1,1,1,1,1,1,1,1,1,1,1,1,1,1,1,1,1,1,1,
1,1,1,1,1,1,1,1,1,1,1,1,1,1,1,1,1,1,1,1,1,1,1,1,1,1,1,1,1,1,1,1,1,1,1,1,1,1,1,1,1,
1,1,1,1,1,0,0,0,0,0,0,0,0,0,0,0,0,0,0,0,0,0,0,0,0,0,0,0,0,0,0,0,0,0,0,0,0,0,0,0,0,0,
0,0,0,0,0,0,0,0,0,0,0,0,0,0,0,0,0,0,0,0,0,0,0,0,0,0,0,0,0,0,0,0,0,0,0,0,0,0,0,0,0,0,
0,0,0,0,0,0,0,0,0,0,0,0,0,0,0,0,0,0,0,0,0,0,0,0,0,0,0,0,0,0,0,0,0,0,0,0,0,0,0,0,0,0,
0,0,0,0,0,0,0,0,0,0,0,0,0,0,0,0,0,0,0,0,0,0,0,0,0,0,0,0,0,0,0,0,0,0,0,0,0,0,0,0,0,0,
0,0,0,0,0,0,0,0,0,0,0,0,0,0,0,0,0,0,0,0,0,0,0,0,0,0,0,0,0,0,0,0,0,0,0,0,0,0,0,0,0,0,
0,0,0,0,0,0,0,0,0,0,0,0,0,0,0,0,0,0,0,0,0,0,0,0,0,0,0,0,0,0,0,0,0,0,0,0,0,0,0,0,0,0,
0,0,0,0,0,0,0,0,0,0,0,0,0,0,0,0,0,0,0,0,0,0,0,0,0,0,0,0,0,0,0,0,0,0,0,0,0,0,0,0,0,0,
0,0,0,0,0,0,0,0,0,0,0,0,0,0,0,0,0,0,0,0,0,0,0,0,0,0,0,0,0,0,0,0,0,0,0,0,0,0,0,0,0,0,
0,0,0,0,0,0,0,0,0,0,0,0,0,0,0,0,0,0,0,0,0,0,0,0,0,0,0,0,0,0,0,0,0,0,0,0,0,0,0,0,0,0,
0,0,0,0,0,0,0,0,0,0,0,0,0,0,0,0,0,0,0,0,0,0,0,0,0,0,0,0,0,0,0,0,0,0,0,0,0,0,0,0,0,0,
0,0,0,0,0,0,0,0,0,0,0,0,0,0,0,0,0,0,0,0,0,0,0,0,0,0,0,0,0,0,0,0,0,0,0,0,0,0,0,0,0),
p=c(.10,.10,.10,.10,.10,.10,.10,.10,.10,.10,.10,.10,.10,.10,.10,.13,.13,.13,.13,.13,
.13,.13,.13,.13,.13,.17,.17,.17,.17,.17,.17,.23,.23,.29,.13,.13,.13,.13,.13,.13,.13,
.13,.13,.17,.17,.17,.17,.17,.17,.17,.17,.17,.17,.17,.17,.17,.17,.17,.17,.17,.17,
.17,.17,.23,.23,.23,.23,.23,.23,.23,.23,.23,.29,.29,.29,.37,.37,.18,.18,.18,.18,
.18,.23,.23,.23,.23,.23,.23,.30,.30,.30,.30,.30,.30,.30,.30,.30,.30,.30,.30,.30,.30,
.30,.30,.30,.30,.30,.30,.30,.30,.30,.30,.30,.30,.30,.30,.30,.30,.30,.37,.37,.37,
.37,.37,.37,.45,.45,.45,.23,.23,.38,.38,.38,.38,.38,.38,.46,.46,.46,.46,.46,.46,.46,
.46,.46,.46,.46,.46,.46,.46,.46,.46,.46,.46,.46,.46,.46,.46,.46,.46,.46,.46,.46,.46,
.46,.46,.46,.46,.46,.46,.46,.46,.46,.46,.46,.46,.46,.46,.46,.46,.46,.54,.54,
.38,.46,.46,.55,.55,.55,.55,.55,.63,.63,.63,.63,.63,.63,.63,.63,.63,.63,.63,.63,.63,
.63,.63,.63,.63,.63,.63,.63,.63,.63,.63,.63,.63,.63,.63,.63,.63,.63,.63,.63,.63,.63,
.63,.63,.63,.63,.63,.63,.63,.63,.63,.63,.63,.63,.63,.63,.63,.63,.63,.63,.63,.63,
.63,.63,.63,.63,.63,.63,.63,.63,.63,.10,.10,.10,.10,.10,.10,.10,.10,.10,.10,.10,
.10,.10,.10,.10,.10,.10,.10,.10,.10,.10,.10,.10,.10,.10,.10,.10,.10,.10,.10,.10,.10,
.10,.10,.10,.10,.10,.10,.10,.10,.10,.10,.10,.10,.10,.10,.10,.10,.10,.10,.10,.10,
.10,.10,.10,.10,.10,.10,.10,.10,.10,.10,.10,.10,.10,.10,.10,.10,.10,.10,.10,.10,.10,
.10,.10,.10,.10,.10,.10,.10,.10,.10,.10,.10,.10,.10,.10,.10,.10,.10,.13,.13,.13,

.13,.13,.13,.13,.13,.13,.13,.13,.13,.13,.13,.13,.13,.13,.13,.13,.13,.13,.13,.13,
.13,.13,.13,.13,.13,.13,.13,.13,.13,.13,.13,.13,.13,.13,.13,.13,.13,.13,.13,.13,
.13,.13,.13,.13,.13,.13,.13,.13,.13,.13,.13,.13,.13,.13,.13,.13,.17,.17,.17,.17,
.17,.17,.17,.17,.17,.17,.17,.17,.17,.17,.17,.17,.17,.17,.17,.17,.17,.17,.17,.17,
.17,.17,.17,.17,.17,.17,.17,.17,.17,.17,.17,.17,.17,.17,.17,.23,.23,.23,.23,.23,
.23,.23,.23,.29,.29,.29,.29,.29,.13,.13,.13,.13,.13,.13,.13,.13,.13,.13,.13,.13,
.13,.13,.13,.13,.13,.13,.13,.13,.13,.13,.13,.13,.13,.13,.13,.13,.13,.13,.13,.13,
.13,.13,.13,.13,.13,.13,.13,.13,.13,.13,.13,.13,.13,.13,.13,.13,.13,.13,.13,.13,
.13,.13,.13,.17,.17,.17,.17,.17,.17,.17,.17,.17,.17,.17,.17,.17,.17,.17,.17,.17,
.17,.17,.17,.17,.17,.17,.17,.17,.17,.17,.17,.17,.17,.17,.17,.17,.17,.17,.17,.17,
.17,.17,.17,.17,.17,.17,.17,.17,.17,.17,.17,.17,.17,.17,.17,.17,.17,.17,.17,.17,
.17,.17,.17,.17,.17,.17,.17,.17,.17,.17,.17,.17,.17,.17,.17,.17,.17,.17,.17,.17,
.23,.23,.23,.23,.23,.23,.23,.23,.23,.23,.29,.29,.29,.29,.29,.29,.29,.29,.37,.37,.37,
.37,.18,.18,.18,.18,.18,.18,.18,.18,.18,.18,.18,.18,.18,.18,.18,.18,.18,.18,.18,
.18,.18,.18,.18,.18,.18,.18,.18,.18,.18,.18,.18,.18,.18,.18,.18,.18,.23,.23,.23,
.23,.23,.23,.23,.23,.23,.23,.23,.23,.23,.23,.23,.23,.23,.23,.23,.23,.23,.23,.23,
.23,.23,.23,.23,.23,.23,.30,.30,.30,.30,.30,.30,.30,.30,.30,.30,.30,.30,.30,.30,.30,
.30,.30,.30,.30,.30,.30,.30,.30,.30,.30,.30,.30,.30,.30,.30,.30,.30,.30,.30,.30,.30,
.30,.30,.30,.30,.30,.30,.30,.30,.30,.30,.30,.30,.30,.30,.30,.30,.30,.30,.30,.30,.30,
.30,.30,.30,.30,.30,.30,.30,.30,.37,.37,.37,.37,.37,.37,.37,.37,.37,.37,.37,.37,.37,
.37,.37,.37,.37,.37,.37,.37,.37,.37,.37,.37,.37,.37,.37,.37,.37,.37,.45,.45,.45,
.45,.45,.45,.45,.45,.45,.45,.45,.45,.45,.45,.45,.45,.23,.23,.23,.23,.23,.23,
.23,.23,.23,.23,.23,.23,.23,.23,.23,.23,.30,.30,.38,.38,.38,.38,.38,.38,.38,.38,.38,
.38,.38,.38,.38,.38,.38,.38,.38,.38,.38,.38,.46,.46,.46,.46,.46,.46,.46,.46,.46,
.46,.46,.46,.46,.46,.46,.46,.46,.46,.46,.46,.46,.46,.46,.46,.46,.46,.46,.46,.46,
.46,.46,.46,.46,.46,.54,.54,.54,.54,.54,.54,.54,.54,.54,.54,.54,.54,.30,.30,.30,.30,
.30,.38,.46,.46,.46,.55,.55,.55,.55,.55,.55,.55,.55,.55,.55,.55,.63,.63,.63,.63,.63,
.63,.63,.63,.63,.63,.63,.63,.63,.63,.63,.63))
list(precy=c(1,1), beta=c(0,0), b=c(0,0,0))

As before, when estimating the ROC area of the risk score, 45,000 observations are generated for the MCMC simulation, with a burn in of 5,000 and
a refresh of 100. The first list statement of BUGS CODE 10.5 gives the data for
executing the analysis. There are 935 observations in each vector, where the
first 261 correspond to the diseased subjects and the remaining 674 correspond
to the non-diseased patients. Vector T_1 is for the first test and T_2 gives the
values (1, 2, 3, 4, 5) for the second test. After running the logistic regression,
I put the estimated probability of lung cancer (the risk score) into a vector
labeled p. This vector is the input to the algorithm for estimating the ROC
area, which is clearly specified in the code. This part of the code has been used
before in Chapter 4 for continuous normally distributed observations, and the
posterior analysis is presented as Table 10.13.

The ROC area of the risk score is the area under the curve (AUC) and is
estimated as 0.7246(0.0192) with the posterior mean, and the median is about
the same value, indicating very little skewness in the posterior distribution.
The implication is that the combined test has an accuracy that is about the

TABLE 10.13: Posterior accuracy of the combined test—ROC area of the risk score.

Parameter	Mean	sd	Error	2 1/2	Median	97 1/2
auc	0.7246	0.0192	<0.0001	0.6858	0.725	0.7616
b[1]	−2.952	0.2162	<0.0001	−3.381	−2.949	−2.533
b[2]	0.3562	0.0752	<0.0001	0.2089	0.3562	0.504
b[3]	0.3392	0.0723	<0.0001	0.1965	0.3391	0.481
beta[1]	0.2412	0.0053	<0.0001	0.2309	0.2412	0.2516
beta[2]	0.1385	0.0127	<0.0001	0.1136	0.1385	0.1638
precy[1]	53.13	2.904	0.0143	47.59	53.1	58.92
precy[2]	28.79	2.528	0.0122	24.08	28.71	33.99

same as the accuracy of the individual tests (see Table 10.13), which portrays the individual area as approximately 0.68. Of course, this is not surprising because the individual ROC area for CT and MRI are essentially the same, thus, one would expect the accuracy of the combined test to be about the same as the individual values.

Note, that b are the regression coefficients for the logistic regression, and beta are the regression coefficients in the normal regression for the ROC area of the risk score. As shown in BUGS CODE 10.5, the logistic regression is linear in the two test variables, T_1 and T_2, but I did add the squares and cross product of the two and the ROC area remained the same, thus, the linear association appears to be adequate for estimating the risk score for the combined test. The risk scores are not normally distributed, but can be transformed to normality approximately via the log transformation, however, when this is done the ROC area remains at about 0.72.

There are many interesting examples involving two or more modalities for assessing risk of disease. For example, Utsunomiya et al. [11] used SPECT/CT scintigraphy and CT to study bone metastasis in cancer patients, while Mazaheri et al. [12] combined diffusion weighted MRI imaging with MR spectrographic imaging to identify malignant lesions of the prostate. In another prostate cancer study, Futterer et al. [13] studied the accuracy of prostate cancer localization with a combination of contrast enhanced MRI and proton MR spectroscopic imaging. In all these investigations, the accuracy of the tests was measured by the ROC area.

Prostate cancer is an active area of imaging studies as illustrated by Coakley et al. [14], who employed endorectal MR imaging with MR spectroscopic imaging with the objective of detecting recurrence of the disease.

For additional information about combination imaging modalities in prostate cancer see Heijmink et al. [15], who used body array vs. endorectal coil MR for a comparison of image quality for localization and staging of prostate cancer.

It should be emphasized that the combined test is the one that assigns a risk score (the probability of disease) to each experimental unit and is obtained by logistic regression, thus, it is envisioned that the clinician or the responsible person assigns the risk score. The ROC area of the risk score is the accuracy of the combined test and is optimal, in the sense that its ROC area is at least as large as the ROC areas of the component tests. In other words, the optimal combined test is a statistical procedure for assigning scores to the experimental unit (often a patient).

10.5.2   Body array and endorectal coil magnetic resonance imaging for localization of prostate cancer

Heijmink et al.'s [15] study is the basis of the next example and illustrates the use of the risk score to measure the accuracy of using two tests to identify those parts (segments) of the prostate gland that have malignant lesions. MRI is the basis of this important study, where the modality is applied in two ways to each patient, either with a body array or by the rectum. There are many objectives to the study, including measuring the quality (contrast, resolution, and noise) of two tests, but the emphasis for our purposes is on localization, that is, determining what part of the prostate contains malignant lesions.

The reader assigns a score from 1 to 5 to each of the 14 segments of the gland, where 1 means a lesion is definitely absent, 2 signifies a lesion is probably absent, 3 indicates an ambiguous result, 4 signifies a lesion is probably present, and 5 indicates a lesion is definitely present. After total prostatectomy, histopathology of the prostate gland serves as the gold standard for the 46 patients, where 124 segments among a total of 644 segments were identified as having separate cancer foci; the imaging results are shown in Tables 10.14a and b.

BUGS CODE 10.4 and 10.5 are used to execute the Bayesian analysis where the latter deals with the ROC area for the two tests, test 1 and test 2, where test 1 is the body array coil MRI and test 2 is the endorectal coil MRI. The second list statement of BUGS CODE 10.4 is the information from Tables 10.14a and b, and the analysis is executed with 45,000 observations

TABLE 10.14a: Body array and endorectal MRI for localization of prostate cancer for 124 diseased segments.

Body array scores	Endorectal scores					Total
	1	2	3	4	5	
1	3	7	6	3	1	20
2	1	5	8	7	6	27
3	2	2	4	11	7	26
4	0	1	1	8	13	23
5	0	0	1	4	23	28
Total	6	15	20	33	50	124

Source: From Heijmink, S.W.T.P., et al., *Radiology*, 244, 184, 2007, with permission of Radiological Society of North America.

TABLE 10.14b: Body array and endorectal MRI for localization of prostate cancer for 520 non-diseased segments.

Body array scores	Endorectal scores					Total
	1	**2**	**3**	**4**	**5**	
1	103	9	3	2	3	120
2	31	62	11	6	5	115
3	25	22	31	13	9	100
4	11	19	20	30	15	95
5	13	15	28	21	13	90
Total	183	127	93	72	45	520

Source: From Heijmink, S.W.T.P. et al., *Radiology*, 244, 184, 2007, with permission of Radiological Society of North America.

generated by MCMC, with a burn in of 5,000 and a refresh of 100, where the Bayesian analysis is given in Table 10.15.

It is apparent from Table 10.15 that the ROC area of the endorectal coil MRI images are more accurate than the body array MRI coil images, where the latter area has a posterior mean of 0.7519 compared to 0.5521 for the former. Assuming one radiologist is reading both images for all patients, does the combined test have higher accuracy? Remember, the radiologist is using the scores of both images to assign a score for the combined test.

In order to assess the accuracy of the combined test, BUGS CODE 10.5 is executed using the study information of Table 10.16, which should be appended to BUGS CODE 10.5 as a list statement. The information includes the d vector, the disease indicator, results of the first test T_1, results of the second test T_2, and the vector p of risk scores, which are the theta values produced by the logistic regression (Table 10.16).

Using 55,000 observations for the MCMC simulation, with a burn in of 5,000 and a refresh of 100, the posterior analysis for the accuracy of the combined test is given in Table 10.17.

This is a very encouraging result because the ROC area of the combined test has a posterior mean of 0.8038, which is larger than the ROC areas of the component tests, which are 0.5521 for test 1, the body array coil, and 0.7517 for the endorectal coil, and one sees the value of using two imaging methods to identify the segment of the prostate gland that has a malignant lesion.

TABLE 10.15:   Posterior analysis for the prostate cancer study with two MRI images—ROC areas of the two tests.

Parameter	Mean	sd	Error	2 1/2	Median	97 1/2
Body coil array MRI area	0.5521	0.0257	<0.0001	0.5011	0.5523	0.602
Endorectal coil MRI area	0.7519	0.0216	<0.00001	0.7078	0.7525	0.7927

TABLE 10.16: Information for the prostate cancer study—values for the first test, second test, disease indicator, and risk scores.

---

list(N=644,
T1=c(1,1,1,1,1,1,1,1,1,1,1,1,1,1,1,1,1,1,1,1,2,2,2,2,2,2,2,2,2,2,2,2,2,2,2,2,
2,2,2,2,2,2,2,2,3,3,3,3,3,3,3,3,3,3,3,3,3,3,3,3,3,3,3,3,3,3,3,3,3,3,4,4,4,4,4,4,
4,4,4,4,4,4,4,4,4,4,4,4,4,4,4,4,4,5,5,5,5,5,5,5,5,5,5,5,5,5,5,5,5,5,5,5,5,5,5,5,
5,5,5,5,1,1,1,1,1,1,1,1,1,1,1,1,1,1,1,1,1,1,1,1,1,1,1,1,1,1,1,1,1,1,1,1,1,1,1,1,
1,1,1,1,1,1,1,1,1,1,1,1,1,1,1,1,1,1,1,1,1,1,1,1,1,1,1,1,1,1,1,1,1,1,1,1,1,1,1,1,
1,1,1,1,1,1,1,1,1,1,1,1,1,1,1,1,1,1,1,1,1,1,1,1,1,1,1,1,1,1,1,1,1,1,1,1,1,1,1,1,
1,2,2,2,2,2,2,2,2,2,2,2,2,2,2,2,2,2,2,2,2,2,2,2,2,2,2,2,2,2,2,2,2,2,2,2,2,2,2,2,
2,2,2,2,2,2,2,2,2,2,2,2,2,2,2,2,2,2,2,2,2,2,2,2,2,2,2,2,2,2,2,2,2,2,2,2,2,2,2,2,
2,2,2,2,2,2,2,2,2,2,2,2,2,2,2,2,2,2,2,2,2,2,2,2,2,2,2,2,3,3,3,3,3,3,3,
3,3,3,3,3,3,3,3,3,3,3,3,3,3,3,3,3,3,3,3,3,3,3,3,3,3,3,3,3,3,3,3,3,3,3,3,3,3,3,3,
3,3,3,3,3,3,3,3,3,3,3,3,3,3,3,3,3,3,3,3,3,3,3,3,3,3,3,3,3,3,3,3,3,3,3,3,3,3,3,3,
3,3,3,3,3,3,3,3,3,3,4,4,4,4,4,4,4,4,4,4,4,4,4,4,4,4,4,4,4,4,4,4,4,4,4,4,4,4,4,4,
4,4,4,4,4,4,4,4,4,4,4,4,4,4,4,4,4,4,4,4,4,4,4,4,4,4,4,4,4,4,4,4,4,4,4,4,4,4,4,4,
4,4,4,4,4,4,4,4,4,4,4,4,4,4,4,4,4,4,4,4,5,5,5,5,5,5,5,5,5,5,5,5,5,5,
5,5,5,5,5,5,5,5,5,5,5,5,5,5,5,5,5,5,5,5,5,5,5,5,5,5,5,5,5,5,5,5,5,5,5,5,5,5,5,5,
5,5,5,5,5,5,5,5,5,5,5,5,5,5,5,5,5,5,5,5,5,5,5,5,5,5),
T2=c(1,1,1,2,2,2,2,2,2,2,3,3,3,3,3,3,4,4,4,5,1,2,2,2,2,2,3,3,3,3,3,3,3,3,4,4,4,4,
4,4,4,5,5,5,5,5,5,1,1,2,2,3,3,3,3,4,4,4,4,4,4,4,4,4,4,4,5,5,5,5,5,5,5,2,3,4,4,4,4,
4,4,4,4,5,5,5,5,5,5,5,5,5,3,4,4,4,4,4,5,5,5,5,5,5,5,5,5,5,5,5,5,5,5,5,5,5,5,5,5,
5,5,5,5,1,1,1,1,1,1,1,1,1,1,1,1,1,1,1,1,1,1,1,1,1,1,1,1,1,1,1,1,1,1,1,1,1,1,1,1,
1,1,1,1,1,1,1,1,1,1,1,1,1,1,1,1,1,1,1,1,1,1,1,1,1,1,1,1,1,1,1,1,1,1,1,1,1,1,1,1,
1,1,1,1,1,1,1,1,1,1,1,1,1,1,1,1,1,1,1,1,1,2,2,2,2,2,2,2,2,3,3,3,4,4,5,5,
5,1,1,1,1,1,1,1,1,1,1,1,1,1,1,1,1,1,1,1,1,1,1,1,1,1,1,1,2,2,2,2,2,2,2,2,2,
2,2,2,2,2,2,2,2,2,2,2,2,2,2,2,2,2,2,2,2,2,2,2,2,2,2,2,2,2,2,2,2,2,2,2,2,2,2,2,2,
2,2,2,2,2,2,2,2,2,2,3,3,3,3,3,3,3,3,3,3,3,4,4,4,4,4,4,5,5,5,5,5,1,1,1,1,1,1,1,
1,1,1,1,1,1,1,1,1,1,1,1,2,2,2,2,2,2,2,2,2,2,2,2,2,2,2,2,2,2,2,2,2,2,2,2,2,3,
3,3,3,3,3,3,3,3,3,3,3,3,3,3,3,3,3,3,3,3,3,3,3,3,3,3,3,3,4,4,4,4,4,4,4,4,4,4,
4,4,5,5,5,5,5,5,5,5,5,1,1,1,1,1,1,1,1,1,1,1,1,2,2,2,2,2,2,2,2,2,2,2,2,2,2,2,2,2,2,
3,3,3,3,3,3,3,3,3,3,3,3,3,3,3,3,3,3,3,4,4,4,4,4,4,4,4,4,4,4,4,4,4,4,4,4,4,4,
4,4,4,4,4,4,4,4,5,5,5,5,5,5,5,5,5,5,5,5,1,1,1,1,1,1,1,1,1,1,1,1,1,2,2,2,2,
2,2,2,2,2,2,2,3,3,3,3,3,3,3,3,3,3,3,3,3,3,3,3,3,3,4,4,
4,4,4,4,4,4,4,4,4,4,4,4,4,4,4,4,5,5,5,5,5,5,5,5,5,5,5,5),
d=c(1,1,1,1,1,1,1,1,1,1,1,1,1,1,1,1,1,1,1,1,1,1,1,1,1,1,1,1,1,1,1,1,1,1,1,1,1,1,
1,1,1,1,1,1,1,1,1,1,1,1,1,1,1,1,1,1,1,1,1,1,1,1,1,1,1,1,1,1,1,1,1,1,1,1,1,1,1,1,
1,1,1,1,1,1,1,1,1,1,1,1,1,1,1,1,1,1,1,1,1,1,1,1,1,1,1,1,1,1,1,1,1,1,1,1,1,1,1,1,
1,1,1,0,0,0,0,0,0,0,0,0,0,0,0,0,0,0,0,0,0,0,0,0,0,0,0,0,0,0,0,0,0,0,0,0,0,0,0,0,
0,0,0,0,0,0,0,0,0,0,0,0,0,0,0,0,0,0,0,0,0,0,0,0,0,0,0,0,0,0,0,0,0,0,0,0,0,0,0,0,
0,0,0,0,0,0,0,0,0,0,0,0,0,0,0,0,0,0,0,0,0,0,0,0,0,0,0,0,0,0,0,0,0,0,0,0,0,0,0,0,
0,0,0,0,0,0,0,0,0,0,0,0,0,0,0,0,0,0,0,0,0,0,0,0,0,0,0,0,0,0,0,0,0,0,0,0,0,0,0,0,
0,0,0,0,0,0,0,0,0,0,0,0,0,0,0,0,0,0,0,0,0,0,0,0,0,0,0,0,0,0,0,0,0,0,0,0,0,0,0,0,

(continued)

TABLE 10.16 (continued):   Information for the prostate cancer study—values for the first test, second test, disease indicator, and risk scores.

0,0,0,0,0,0,0,0,0,0,0,0,0,0,0,0,0,0,0,0,0,0,0,0,0,0,0,0,0,0,0,0,0,0,0,0,0,0,0,0,
0,0,0,0,0,0,0,0,0,0,0,0,0,0,0,0,0,0,0,0,0,0,0,0,0,0,0,0,0,0,0,0,0,0,0,0,0,0,0,0,
0,0,0,0,0,0,0,0,0,0,0,0,0,0,0,0,0,0,0,0,0,0,0,0,0,0,0,0,0,0,0,0,0,0,0,0,0,0,0,0,
0,0,0,0,0,0,0,0,0,0,0,0,0,0,0,0,0,0,0,0,0,0,0,0,0,0,0,0,0,0,0,0,0,0,0,0,0,0,0,0,
0,0,0,0,0,0,0,0,0,0,0,0,0,0,0,0,0,0,0,0,0,0,0,0,0,0,0,0,0,0,0,0,0,0,0,0,0,0,0,0,
0,0,0,0,0,0,0,0,0,0,0,0,0,0,0,0,0,0,0,0,0,0,0,0,0,0,0,0,0,0,0,0,0,0,0,0,0,0,0,0,
0,0,0,0,0,0,0,0,0,0,0,0,0,0,0,0,0,0,0,0,0,0,0,0,0,0,0,0,0,0,0,0,0,0,0,0,0,0,0,0,
0,0,0,0,0,0,0,0,0,0,0,0,0,0,0,0,0,0,0,0,0,0,0,0,0,0,0,0,0,0,0,0),
p=c(.06,.06,.06,.15,.15,.15,.15,.15,.15,.15,.33,.33,.33,.33,.33,.33,.59,.59,.59,
.80,.04,.10,.10,.10,.10,.10,.25,.25,.25,.25,.25,.25,.25,.25,.48,.48,.48,.48,.48,
.48,.48,.72,.72,.72,.72,.72,.72,.03,.03,.07,.07,.17,.17,.17,.17,.37,.37,.37,.37,
.37,.37,.37,.37,.37,.37,.37,.63,.63,.63,.63,.63,.63,.63,.05,.12,.28,.28,.28,.28,
.28,.28,.28,.28,.52,.52,.52,.52,.52,.52,.52,.52,.52,.52,.52,.52,.52,.08,.20,.20,
.20,.20,.41,.41,.41,.41,.41,.41,.41,.41,.41,.41,.41,.41,.41,.41,.41,.41,.41,.41,
.41,.41,.41,.41,.41,.06,.06,.06,.06,.06,.06,.06,.06,.06,.06,.06,.06,.06,.06,.06,
.06,.06,.06,.06,.06,.06,.06,.06,.06,.06,.06,.06,.06,.06,.06,.06,.06,.06,.06,.06,
.06,.06,.06,.06,.06,.06,.06,.06,.06,.06,.06,.06,.06,.06,.06,.06,.06,.06,.06,.06,
.06,.06,.06,.06,.06,.06,.06,.06,.06,.06,.06,.06,.06,.06,.06,.06,.06,.06,.06,.06,
.06,.06,.06,.06,.06,.06,.06,.06,.15,.15,.15,.15,.15,.15,.15,.15,.15,.33,.33,.33,
.59,.59,.80,.80,.80,.04,.04,.04,.04,.04,.04,.04,.04,.04,.04,.04,.04,.04,.04,.04,
.04,.04,.04,.04,.04,.04,.04,.04,.04,.04,.04,.04,.04,.04,.04,.04,.10,.10,.10,.10,
.10,.10,.10,.10,.10,.10,.10,.10,.10,.10,.10,.10,.10,.10,.10,.10,.10,.10,.10,.10,
.10,.10,.10,.10,.10,.10,.10,.10,.10,.10,.10,.10,.10,.10,.10,.10,.10,.10,.10,.10,
.10,.10,.10,.10,.10,.10,.10,.10,.10,.10,.10,.10,.10,.10,.10,.10,.10,.10,.25,.25,
.25,.25,.25,.25,.25,.25,.25,.25,.25,.48,.48,.48,.48,.48,.48,.72,.72,.72,.72,.72,
.03,.03,.03,.03,.03,.03,.03,.03,.03,.03,.03,.03,.03,.03,.03,.03,.03,.03,.03,.03,
.03,.03,.03,.03,.03,.07,.07,.07,.07,.07,.07,.07,.07,.07,.07,.07,.07,.07,.07,.07,
.07,.07,.07,.07,.07,.07,.07,.17,.17,.17,.17,.17,.17,.17,.17,.17,.17,.17,.17,.17,
.17,.17,.17,.17,.17,.17,.17,.17,.17,.17,.17,.17,.17,.17,.17,.17,.17,.37,.37,
.37,.37,.37,.37,.37,.37,.37,.37,.37,.63,.63,.63,.63,.63,.63,.63,.63,.63,
.02,.02,.02,.02,.02,.02,.02,.02,.02,.02,.02,.05,.05,.05,.05,.05,.05,.05,.05,.05,
.05,.05,.05,.05,.05,.05,.05,.05,.05,.05,.12,.12,.12,.12,.12,.12,.12,.12,.12,.12,
.12,.12,.12,.12,.12,.12,.12,.12,.12,.12,.28,.28,.28,.28,.28,.28,.28,.28,.28,.28,
.28,.28,.28,.28,.28,.28,.28,.28,.28,.28,.28,.28,.28,.28,.28,.28,.28,.28,.28,.28,
.52,.52,.52,.52,.52,.52,.52,.52,.52,.52,.52,.52,.52,.52,.52,.01,.01,.01,.01,.01,
.01,.01,.01,.01,.01,.01,.01,.01,.03,.03,.03,.03,.03,.03,.03,.03,.03,.03,.03,.03,
.03,.03,.03,.08,.08,.08,.08,.08,.08,.08,.08,.08,.08,.08,.08,.08,.08,.08,.08,.08,
.08,.08,.08,.08,.08,.08,.08,.08,.08,.08,.08,.20,.20,.20,.20,.20,.20,.20,.20,.20,
.20,.20,.20,.20,.20,.20,.20,.20,.20,.20,.20,.20,.41,.41,.41,.41,.41,.41,.41,.41,
.41,.41,.41,.41,.41))
list(precy=c(1,1), beta=c(0,0), b=c(0,0,0))

TABLE 10.17: Accuracy of the combined test for the prostate cancer study.

Parameter	Mean	sd	Error	2 1/2	Median	97 1/2
auc	0.8038	0.0223	<0.0001	0.7579	0.8047	0.8451
b[1]	−3.405	0.3392	0.0026	−4.088	−3.397	−2.758
b[2]	−0.441	0.1035	<0.0001	−0.6495	−0.439	−0.2419
b[3]	1.052	0.1081	<0.0001	0.846	1.05	1.269
beta[1]	0.1518	0.00702	<0.0001	0.1381	0.1519	0.1656
beta[2]	0.2098	0.0181	<0.0001	0.1742	0.2097	0.2456

Recall that the radiologist assigns a risk score (via logistic regression) to each segment based on the scores of the two MRI images, and the ROC curve of the combined test is shown in Figure 10.4.

10.5.3 Accuracy of combined test with a covariate for lung cancer study

An analysis of the above example is continued by taking into account the effect of a covariate on the risk score. Suppose the age of each patient in the study is known and is included in the logistic regression. What is the effect on the accuracy of the test? The age is listed in Table 10.18 and is used with BUGS CODE 10.5 to estimate the accuracy of the combined test.

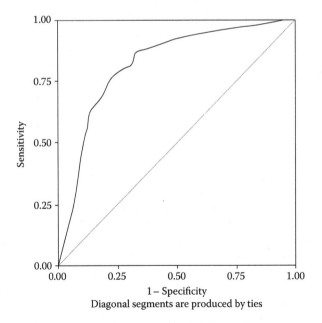

FIGURE 10.4: ROC curve of the combined test for prostate cancer study.

TABLE 10.18:   The covariable age for the prostate cancer study and risk scores.

age=c(68,70,67,75,75,72,73,84,81,91,61,81,85,75,75,83,74,80,80,57,89,100,
68,84,99,88,83,78,84,89,80,91,70,81,85,78,80,85,76,76,81,89,89,100,64,79,82,
82,69,92,75,95,65,75,81,83,84,89,80,67,72,94,85,73,84,77,79,70,69,77,75,76,
73,77,67,76,88,87,88,63,76,85,79,74,77,86,73,83,87,78,83,78,70,101,94,100,
78,82,69,85,78,85,84,87,81,85,88,87,82,76,90,73,69,77,89,75,85,69,76,83,71,
81,77,101,84,93,94,72,68,96,93,79,71,75,86,87,80,96,76,70,74,77,97,98,93,65,
79,65,71,82,78,63,87,74,73,84,90,70,81,85,76,72,75,88,48,77,92,85,74,87,90,
95,64,66,104,80,74,96,79,84,69,73,92,79,84,72,84,6,88,47,98,88,47,78,77,58,
47,95,48,175,72,80,68,99,58,99,98,82,73,88,80,65,82,97,83,78,81,96,76,80,86,
73,67,101,80,71,91,87,61,72,83,76,64,80,83,89,57,90,75,99,73,80,66,87,83,70,
75,77,64,52,84,71,81,64,90,92,76,74,75,91,80,76,68,55,69,56,70,71,67,81,69,
58,78,66,74,53,53,67,73,76,70,61,66,61,66,56,60,48,83,69,62,66,42,62,62,70,
57,69,68,83,75,87,65,62,65,59,67,54,71,81,68,64,47,55,51,62,63,65,52,80,48,
70,65,75,69,63,65,94,56,59,61,73,58,40,66,76,60,79,50,82,74,70,73,71,55,76,
67,73,76,90,68,67,49,58,66,57,56,77,64,61,65,60,33,53,70,69,83,65,81,74,63,
71,63,75,64,71,64,58,59,59,58,64,76,56,68,67,65,76,73,64,53,80,45,62,72,85,
58,81,58,52,61,63,64,58,64,70,62,60,78,69,79,60,45,62,73,80,88,73,63,73,63,
69,63,73,51,66,62,64,47,72,49,63,61,57,63,75,82,56,71,71,56,62,73,70,65,69,
59,71,56,70,61,70,86,36,59,77,66,66,60,70,71,76,73,61,73,63,59,50,60,61,82,
61,55,73,53,69,65,59,67,63,69,82,73,60,52,51,62,67,67,50,57,67,92,68,72,53,
46,60,54,73,49,65,58,74,57,72,74,84,62,82,70,56,69,63,66,67,59,57,66,60,56,
63,66,80,97,67,76,77,38,66,49,45,54,75,79,94,61,82,74,66,56,65,55,78,68,61,
68,54,67,89,81,61,36,77,66,76,66,53,49,72,85,83,76,70,62,81,58,76,55,63,57,
84,58,66,54,64,59,53,46,65,58,56,39,69,65,62,63,65,37,70,72,58,66,70,58,55,
85,53,72,74,70,49,56,58,76,63,65,74,61,54,56,65,54,66,68,69,65,64,76,65,52,
72,73,77,65,72,70,59,65,56,60,54,59,65,58,66,72,62,56,62,58,51,61,83,67,69,
50,70,75,55,48,72,71,52,57,68,79,54,58,47,53,62,56,49,68,62,62,58,80,62,77,
69,51,69,46,84,49,57,67,74,52,80,61,64,71,45,61,56,56,65,69,80,37,60,57,72,
59,55,82,62,66,77,82,47,87,77,71,66,60,72,58,47,55,78,57,69,76,61,77,83,53,
44,71,69,51,73,75,49,41,47,64,68,60,69,75,74,77,68,64,70,46,64,67,50,55,59,
63,63,68,62,69,56,88,71,82,49,68,42,62,54,57,71,74,65,51,82,57,72,59,62,73,
64,63,53,67,75,72,62,61,82,81,71,46,70,52,57,58,68,74,69,74,75,64,75,72,54,
75,54,58,61,72,51,75,81,61,63,66,64,62,86,52,59,61,59,76,59,50,54,57,73,56,
50,70,64,85,75,55,55,53,61,64,73,60,44,82,65,53,77,58,63,66,82,45,43,63,62,
72,73,71,76,64,50,57,69,76,78,63,61,77,65,66,65,75,59,70,73,64,61,67,60,50,
52,55,77,63,73,61,76,63,69,58,46,71,60,49,51,78,71,52,61,64,78,47,72,70,95,
72,62,49,65,73,54,53,70,58,63,73,80,63,75,72,41,59,42,66,72,65,59,64,64,54,
71,56,69,54,40,48,73,60,73,57,77,79,52,57)
p=c(.05,.07,.04,.13,.13,.09,.10,.36,.27,.62,.02,.27,.39,.14,.14,.39,.15,.32,.29,
.01,.61,.90,.07,.44,.88,.66,.48,.29,.51,.68,.35,.79,.14,.54,.51,.28,.35,.50,.21,
.21,.38,.65,.67,.93,.06,.38,.48,.47,.11,.80,.25,.86,.06,.23,.42,.51,.54,.72,.39,
.09,.16,.85,.57,.20,.61,.37,.44,.17,.15,.38,.29,.32,.25,.37,.14,.38,.81,.82,.83,

TABLE 10.18 (continued): The covariable age for the prostate cancer study and risk scores.

.06,.28,.60,.39,.23,.39,.72,.26,.61,.75,.44,.68,.50,.23,.97,.91,.96,.48,.63,.20,
.73,.50,.75,.72,.79,.60,.73,.82,.79,.65,.43,.86,.33,.20,.46,.84,.40,.74,.21,.43,
.69,.25,.61,.53,.98,.76,.93,.94,.34,.29,.97,.94,.52,.24,.48,.84,.85,.68,.96,.53,
.38,.54,.63,.97,.98,.95,.22,.69,.22,.41,.79,.67,.19,.89,.55,.50,.83,.93,.39,.77,
.85,.60,.47,.57,.90,.02,.65,.94,.85,.55,.89,.93,.96,.20,.25,.99,.73,.54,.97,.70,
.83,.35,.48,.94,.72,.83,.47,.88,.39,.81,.73,.91,.88,.72,.93,.66,.88,.82,.11,.87,
.72,.64,.86,.48,.99,.17,.99,.99,.88,.66,.95,.85,.39,.89,.99,.90,.81,.87,.98,.76,
.84,.93,.66,.45,.99,.84,.60,.97,.94,.25,.63,.90,.75,.33,.84,.91,.95,.15,.96,.73,
.99,.68,.85,.42,.94,.89,.56,.72,.77,.34,.08,.91,.60,.87,.33,.96,.97,.74,.69,.73,
.97,.85,.16,.05,.01,.06,.01,.06,.08,.04,.27,.06,.01,.20,.04,.11,.01,.01,.04,.10,
.15,.06,.02,.04,.02,.04,.01,.02,.00,.33,.06,.02,.04,.00,.02,.02,.07,.01,.06,.05,
.33,.13,.50,.03,.02,.03,.01,.05,.01,.07,.27,.05,.03,.00,.01,.00,.02,.03,.03,.00,
.24,.00,.07,.04,.14,.06,.02,.04,.73,.01,.01,.02,.10,.01,.00,.04,.15,.02,.21,.00,
.30,.11,.06,.10,.07,.01,.15,.05,.10,.14,.59,.05,.04,.00,.02,.05,.01,.01,.22,.04,
.03,.04,.02,.00,.01,.09,.07,.42,.04,.35,.14,.03,.10,.03,.17,.04,.10,.04,.02,.02,
.02,.02,.04,.19,.01,.07,.06,.04,.18,.14,.04,.01,.29,.00,.03,.12,.46,.02,.33,.02,
.01,.03,.03,.04,.02,.04,.08,.03,.02,.24,.07,.28,.02,.00,.03,.12,.37,.67,.17,.05,
.18,.04,.10,.04,.17,.01,.06,.04,.05,.00,.15,.01,.04,.03,.02,.04,.21,.42,.02,.12,
.13,.02,.04,.16,.12,.06,.10,.02,.14,.02,.11,.03,.12,.58,.00,.03,.27,.08,.09,.04,
.15,.16,.30,.20,.05,.26,.08,.04,.01,.05,.03,.40,.03,.01,.15,.01,.09,.05,.02,.07,
.04,.09,.39,.14,.02,.01,.01,.03,.06,.07,.01,.02,.06,.76,.08,.14,.01,.00,.03,.01,
.15,.00,.05,.02,.17,.02,.13,.16,.47,.03,.38,.10,.01,.09,.04,.06,.07,.02,.01,.06,
.02,.01,.04,.05,.32,.87,.07,.22,.30,.00,.08,.01,.00,.01,.25,.38,.83,.04,.48,.22,
.08,.02,.06,.02,.33,.09,.03,.11,.01,.09,.72,.44,.04,.00,.30,.07,.27,.08,.01,.01,
.17,.57,.50,.26,.13,.04,.43,.02,.27,.01,.05,.02,.55,.02,.08,.01,.05,.03,.01,.00,
.07,.03,.02,.00,.11,.06,.04,.05,.07,.00,.13,.17,.02,.08,.12,.02,.02,.58,.01,.17,
.22,.12,.01,.02,.02,.26,.05,.06,.22,.05,.02,.02,.09,.02,.09,.12,.14,.08,.07,.41,
.12,.02,.27,.28,.43,.11,.25,.26,.07,.14,.04,.04,.01,.03,.08,.03,.08,.20,.05,.02,
.05,.03,.01,.04,.54,.09,.13,.01,.14,.27,.02,.01,.19,.17,.01,.02,.11,.41,.02,.03,
.01,.01,.05,.02,.01,.11,.05,.05,.03,.52,.07,.39,.16,.01,.17,.01,.63,.01,.03,.13,
.30,.01,.49,.06,.08,.20,.01,.06,.03,.03,.10,.17,.51,.00,.05,.03,.25,.04,.02,.65,
.08,.14,.48,.64,.01,.79,.46,.26,.15,.07,.28,.05,.01,.03,.50,.04,.20,.42,.07,.47,
.67,.02,.01,.26,.20,.02,.31,.41,.01,.00,.01,.10,.19,.06,.20,.38,.35,.47,.19,.11,
.24,.01,.11,.17,.01,.03,.06,.10,.10,.19,.09,.21,.03,.83,.27,.65,.01,.19,.00,.08,
.03,.04,.27,.36,.16,.02,.69,.05,.35,.07,.11,.40,.15,.13,.03,.20,.48,.34,.10,.09,
.71,.68,.33,.01,.28,.03,.05,.06,.23,.44,.25,.44,.46,.14,.44,.42,.04,.54,.05,.08,
.12,.42,.03,.53,.75,.12,.16,.22,.19,.15,.86,.04,.10,.07,.05,.39,.05,.01,.02,.04,
.30,.03,.01,.20,.09,.72,.36,.03,.03,.03,.09,.16,.43,.09,.01,.73,.18,.04,.56,.07,
.14,.21,.73,.01,.01,.14,.13,.37,.43,.35,.53,.15,.03,.08,.35,.61,.66,.18,.13,.65,
.23,.26,.24,.55,.10,.38,.49,.21,.15,.30,.12,.03,.04,.06,.63,.17,.50,.14,.60,.18,
.36,.09,.02,.41,.13,.03,.04,.72,.50,.06,.17,.24,.74,.03,.54,.44,.97,.54,.20,.02,
.16,.40,.03,.03,.35,.11,.21,.51,.82,.24,.68,.58,.01,.15,.01,.36,.55,.33,.16,.33,
.33,.10,.59,.14,.52,.10,.01,.04,.67,.23,.66,.16,.78,.82,.08,.16)

TABLE 10.19: Posterior analysis for the accuracy of the combined test.

Parameter	Mean	sd	2 1/2	Median	97 1/2
auc	0.889	0.0128	0.8623	0.8896	0.9125
b[1]	−12.5	0.8524	−14.23	−12.48	−10.88
b[2]	0.4446	0.0936	0.2638	0.4443	0.63
b[3]	0.2784	0.0895	0.1031	0.2781	0.4552
b[4]	0.1319	0.0104	0.1119	0.1317	0.1529
beta[1]	0.1608	0.0074	0.1462	0.1609	0.1755
beta[2]	0.4238	0.0192	0.3859	0.4237	0.4615

Note: Age is a covariable: lung cancer test.

Add the age vector and the risk scores p to the second list statement of BUGS CODE 10.5 and revise the code accordingly, and remember that the output of the logistic regression is the theta vector of risk scores, which should be included in the second list statement as the vector p. I executed the analysis using 45,000 observations, with a burn in of 5,000 and a refresh of 100, and the results are listed in Table 10.19.

The effect of age on the ROC area of the combined test is dramatic, with a value of 0.889(0.0128) compared to a posterior mean of 0.7246 when the covariate is ignored. Compare Table 10.13 with Table 10.19 and remember that the b coefficients are the regression coefficients for the logistic regression, while the beta values are the values of the regression coefficient for the normal regression, which is necessary for calculating the ROC area. It appears that the regression coefficients have a substantial effect and are needed to calculate the risk score (via logistic regression) and the ROC area. Perhaps it is not surprising after all, because the average age of the patients with disease is 79.81(9.83) years compared to 64.63(10.44) years for the patients without disease.

The effect of age on the ROC for each modality (MRI and CT) is not calculated, but is left as an exercise. See Chapter 5 and the use of ordinal regression techniques for calculating the ROC area for ordinal test scores and including covariates in the analysis.

TABLE 10.20: Two ordinal tests.

$Y_1 =$	1			2			3		
Y_2	1	2	3	1	2	3	1	2	3
$D = 1$	s_1	s_2	s_3	s_4	s_5	s_6	s_7	s_8	s_9
$D = 0$	r_1	r_2	r_3	r_4	r_5	r_6	r_7	r_8	r_9
$V = 0$	u_1	u_2	u_3	u_4	u_5	u_6	u_7	u_8	u_9
Total	m_1	m_2	m_3	m_4	m_5	m_6	m_7	m_8	m_9

10.5.4 Accuracy of a combined test with two components and verification bias

Our objective is to estimate the accuracy of the combined test with two components, when the design is paired and each test is subject to verification bias. As far as I know, this is "new" material, and recall from Section 8.5 in Chapter 8, the Bayesian approach to estimating the accuracy of two paired test, where the general layout is given by Table 8.11, which is portrayed in Table 10.20.

The total number of observations is

$$m = \sum_{i=1}^{i=9} m_i,$$

for two ordinal tests with scores 1, 2, 3, where the total number of subjects who are not subject to verification (the gold standard) is

$$u = \sum_{i=1}^{i=9} u_i.$$

Also recall from Chapter 8, Equations 8.32 through 8.42, which provide the Bayesian analysis for the ROC areas of the two paired tests. The paired design can occur in basically two ways: (a) there are two medical tests, both administered to each subject; or (b) there are two readers who are assigning scores to each patient, based on the results of one medical test. What is the accuracy of the combined test?

It should be recalled that our approach assumes the MAR assumption and uses inverse probability weighting to create an imputed table corresponding to Table 10.20. That is, for each pair of tests scores (Y_1, Y_2) in Table 10.20, multiply the two cell frequencies by the inverse of the verification rate. For example, corresponding to $(1,1)$, the verification rate is $(s_1 + r_1)/m_1$, and s_1 is replaced by $s_1 m_1/(s_1 + r_1)$, an operation that is performed for all nine cells of the table. Once the imputed table is constructed, the risk score is found by using logistic regression to compute the risk scores, then the risk scores are employed to compute the ROC area of the combined test. For additional information on the inverse probability method, see Section 8.7 in Chapter 8 or Pepe [8: 171–2]. The methodology is introduced by an example taken from Chapter 8 (Table 8.13), which is depicted in Table 10.21.

Table 10.21 gives the results of a paired study, where a dermatologist and a surgeon score the stage of disease of a patient, where those patients who are diseased actually have melanoma, but where those that do not have the disease do not have melanoma, but have some other form of skin disease.

In Table 8.13 of Chapter 8, the ROC area for the surgeon is reported as 0.7867(0.01192) and 0.6351(0.0145) for the dermatologist. Using inverse probability weighting, Table 10.21 is converted to the imputed Tables 10.22a and b.

TABLE 10.21: Staging melanoma by a dermatologist and a surgeon.

$Y_1 =$	1			2			3		
Y_2	1	2	3	1	2	3	1	2	3
$D=1$	$s_1 = 8$	$s_2 = 26$	$s_3 = 51$	$s_4 = 43$	$s_5 = 81$	$s_6 = 94$	$s_7 = 117$	$s_8 = 140$	$s_9 = 208$
$D=0$	$r_1 = 101$	$r_2 = 105$	$r_3 = 83$	$r_4 = 67$	$r_5 = 72$	$r_6 = 40$	$r_7 = 41$	$r_8 = 30$	$r_9 = 4$
$V=0$	$u_1 = 2$	$u_2 = 18$	$u_3 = 62$	$u_4 = 14$	$u_5 = 83$	$u_6 = 67$	$u_7 = 63$	$u_8 = 40$	$u_9 = 108$
Total	$m_1 = 111$	$m_2 = 149$	$m_3 = 196$	$m_4 = 124$	$m_5 = 236$	$m_6 = 201$	$m_7 = 221$	$m_8 = 210$	$m_9 = 320$

TABLE 10.22a: Imputed values for 1078 diseased patients.

Y_1	1	2	3	
Y_2				Total
1	8	30	75	113
2	48	125	141	314
3	164	173	314	651
Total	220	328	530	1078

TABLE 10.22b: Imputed values for 689 non diseased.

Y_1	1	2	3	
Y_2				Total
1	103	119	121	343
2	76	110	60	246
3	57	37	6	100
Total	236	266	187	689

The posterior analysis for the accuracy of the combined test is executed with BUGS CODE 10.5 using 45,000 observations for the simulation, with a burn in of 5,000 and a refresh of 100, where the logistic regression produced a risk score vector with ROC area 0.825(0.010), with a 95% credible interval of (0.806,0.844), which compares with a ROC area of 0.7867 for the surgeon and 0.6351 for the dermatologist. The logistic regression had two strong predictors, where the percentage of correct predictions for the 1078 diseased patients is 85.1 compared to 60.8 for the 689 non-diseased patients.

To execute the analysis with BUGS CODE 10.5, the list statement must contain the vector of values d, the disease indicator, the values for T_1 and T_2 for each of the 1767 patients, and lastly, the vector of risk scores produced by the logistic regression. Note that each of the vectors in the list statement is of length 1767! See Exercise 12 for additional information about the dermatology study.

10.6 Accuracy of the Combined Test for Continuous Scores

When considering continuous scores, the accuracy of the combined test is based on the ROC area of the risk scores, thus, the approach for this section is to employ the O'Malley et al. [16] Bayesian determination of the ROC area, where the risk scores r are regressed on the disease indicator d, that is to say:

$$r[i] \sim \text{dnormal}(m(u)), \text{precision}(d[i]+1), \tag{10.59}$$

where $i = 1, 2, \ldots, N$,

$$m(u) = \text{beta}[1] + \text{beta}[2]^* d, \tag{10.60}$$

where d is the indicator variable, $\text{beta}[i] \sim \text{dnorm}(0, 0.0001)$,

$$\text{precision}[i] \sim \text{dgamma}(0.0001, 0.0001), \text{ and } i=1, 2.$$

The AUC is computed as

$$\text{AUC<-phi(la1/sqrt(1+la2))}, \tag{10.61}$$

where phi is the distribution function of the standard normal,

$$\text{la1<-beta}[2]/\text{sqrt(var}[1]), \tag{10.62}$$
$$\text{la2<-var}[2]/\text{var}[1], \tag{10.63}$$

Also

$$\text{var}[i]=1/\text{precision}[i], \tag{10.64}$$

and $i = 1, 2$.

The risk score, vector r, is obtained by logistic regression where the dependent variable is probability of disease (the risk score) and the independent variables are the continuous scores of the component tests, but additional covariate information can be included.

10.6.1   Two biomarkers for pancreatic cancer

A good example for illustrating the methodology is one previously considered in Chapter 4, the pancreatic cancer study by Wieand, Gail, and James [17] that investigated the effect of two biomarkers on the disease incidence. The data can be downloaded at http://www.fhcrc.org/labs/books/pepe and is referenced by Pepe [8: 9].

The first biomarker is CA19-9 and the second biomarker is CA125. The original values were transformed by logs to achieve approximate normality, because the original values of CA19-9 and CA125 were highly skewed to the right. On the original scale the mean(sd) of CA19-9 is 18.03(20.81) for the 51 control patients and 1715(3681) for the cancer patients, whereas for the CA125 marker, the mean(sd) for the control patients is 21.81(30.29) and 55.04(138.8) for the diseased. The median for the first biomarker is 10 for the control and 249 for the cancer patients, and for the second biomarker, the medians are 11.4 vs. 21.8 for the control and diseased patients, respectively. Note the large variability of both biomarkers, but based on the difference in

TABLE 10.23: Biomarkers, risk scores, and disease incidence of pancreatic study.

T1=c(28.00,15.50,8.20,3.40,17.30,15.20,32.90,11.10,87.50,16.20,107.90,5.70,
25.60,31.20,21.60,55.60,8.80,6.50,22.10,14.40,44.20,3.70,7.80,8.90,18.00,
6.50,4.90,10.40,5.00,5.30,6.50,6.90,8.20,21.80,6.60,7.60,15.40,59.20,5.10,
10.00,5.30,32.60,4.60,6.90,4.00,3.65,7.80,32.50,11.50,4.00,10.20,2.40,719.00,
2106.67,24000.00,1715.00,3.60,521.50,1600.00,454.00,109.70,23.70,464.00,
9810.00,255.00,58.70,225.00,90.10,50.00,5.60,4070.00,592.00,28.60,6160.00,
1090.00,10.40,27.30,162.00,3560.00,14.70,83.30,336.00,55.70,1520.00,3.90,
5.80,8.45,361.00,369.00,8230.00,39.30,43.50,361.00,12.80,18.00,9590.00,
555.00,60.20,21.80,900.00,6.60,239.00,3100.00,3275.00,682.00,85.40,
10290.00,770.00,247.60,12320.00,113.10,1079.00,45.60,1630.00,79.40,
508.00,3190.00,542.00,1021.00,235.00,251.00,3160.00,479.00,222.00,15.70,
2540.00,11630.00,1810.00,6.90,4.10,15.60,9820.00,1490.00,15.70,45.80,7.80,
12.80,100.53,227.00,70.90,2500.00),
T2=c(13.30,11.10,16.70,12.60,7.40,5.50,32.10,27.20,6.60,9.80,10.50,7.80,
9.10,12.30,12.00,42.10,5.90,9.20,7.30,6.80,10.70,15.70,8.00,6.80,47.35,17.90,
96.20,108.90,16.60,9.50,179.00,12.10,35.60,15.00,12.60,5.90,10.10,8.50,
11.40,54.65,9.70,11.20,35.70,22.50,21.20,5.60,9.40,12.00,9.80,17.20,10.60,
79.10,31.40,15.00,77.80,25.70,11.70,8.25,14.95,8.70,14.10,123.90,12.10,
99.10,18.60,10.50,6.60,74.00,43.90,45.70,13.00,7.30,8.60,17.20,15.40,14.30,
93.10,66.30,26.70,32.40,9.90,30.30,11.20,202.00,35.70,9.20,103.60,21.40,
8.10,29.90,17.50,30.80,57.30,6.50,33.80,53.60,17.20,94.20,33.50,3.70,11.70,
19.90,38.70,27.30,20.10,86.10,844.00,36.90,6.90,27.70,9.90,38.60,142.60,
12.50,11.60,21.20,13.20,19.20,1024.00,14.10,34.80,35.30,35.00,15.50,12.10,
31.60,184.80,24.80,10.40,34.50,19.40,22.20,53.90,15.40,17.30,36.80,49.80,
26.57,9.70,19.20,14.20),
d = c(0,0,0,0,0,0,0,0,0,0,0,0,0,0,0,0,0,0,0,0,0,0,0,0,0,0,0,0,0,0,0,0,0,0,0,0,
0,0,0,0,0,0,0,0,0,0,0,0,0,0,1,1,1,1,1,1,1,1,1,1,1,1,1,1,1,1,1,1,1,1,1,1,1,1,1,
1,1,1,1,1,1,1,1,1,1,1,1,1,1,1,1,1,1,1,1,1,1,1,1,1,1,1,1,1,1,1,1,1,1,1,1,1,1,1,
1,1,1,1,1,1,1,1,1,1,1,1,1,1,1,1,1,1,1,1,1,1,1,1),
p=c(.38,.30,.28,.24,.30,.28,.49,.33,.74,.30,.84,.23,.35,.40,.34,.68,.24,.24,.32,
.28,.48,.25,.25,.25,.45,.27,.56,.64,.26,.24,.84,.25,.34,.35,.25,.24,.29,.57,.24,
.43,.24,.40,.32,.29,.27,.22,.25,.41,.27,.25,.27,.47,.99,.99,.99,.99,.24,.99,.99,
.99,.85,.77,.99,.99,.99,.58,.99,.90,.65,.36,.99,.99,.37,.99,.99,.28,.69,.98,.99,
.37,.73,.99,.56,.99,.31,.24,.61,.99,.99,.99,.47,.56,.99,.27,.40,.99,.99,.85,.42,
.99,.25,.99,.99,.99,.99,.91,.99,.99,.99,.99,.86,.99,.89,.99,.71,.99,.99,.99,.99,
.99,.99,.99,.99,.99,.30,.99,.99,.99,.25,.31,.33,.99,.99,.31,.52,.34,.42,.85,.99,
.69,.99))

---

the means and medians between the diseased and non-diseased patients, one would expect a high value for the ROC area of CA19-9. The original values are presented in Table 10.23, where T_1 is the CA19-9 biomarker, T_2 is CA125, and d is the vector that indicates disease or non disease among the subjects.

TABLE 10.24: Bayesian analysis for the accuracy of the combined test—two biomarkers for pancreatic cancer.

Parameter	Mean	sd	Error	2 1/2	Median	97 1/2
auc	0.9127	0.0227	<0.0001	0.8625	0.915	0.951
beta[1]	0.3568	0.0223	<0.0001	0.3129	0.3567	0.4015
beta[2]	0.4361	0.0368	<0.0001	0.3633	0.4362	0.5084

BUGS CODE 10.4 is executed with 45,000 observations, with a burn in of 5,000 and a refresh of 100, in order to determine the effect of the biomarkers on the probability of disease with the following result. The logistic regression gave posterior means(sd) of $-1.464(0.388)$ for the constant, $0.027(0.009)$ for the coefficient of CA19-9, and $0.016(0.008)$ for the coefficient of CA125, and gave a very good fit to the data, where 86.3% of the patients without disease were correctly predicted, and 77.8% of the diseased patients were correctly predicted with the model. Note that the risk scores estimated by the logistic regression are the risk scores indicated by the vector r above.

In order to determine the accuracy of the combined test, BUGS CODE 10.5 is executed with 45,000 observations, with a burn of 5,000 and a refresh of 100, and the results are reported in Table 10.24.

A ROC area of 0.9127 implies very good accuracy for the combined test, which compares to a ROC area of 0.8733(0.0275), based on CA19-9, and 0.6786(0.0438) for CA125, where the Bayesian analysis of the individual biomarkers are reported in the discussion after Table 4.21 of Chapter 4. Simulation errors of <0.0001 for the parameters indicate 45,000 observations are sufficient, and imply that the reported values are close to the "true" posterior quantities (Figure 10.5).

Figure 10.5 shows three ROC curves: (a) the combined test in blue, (b) the CA125 test in green, and (c) the C19-9 test in red.

10.6.2 Blood glucose tests and type 2 diabetes

As a second example for finding the accuracy of a combined test for diabetes, a special study is designed involving 120 subjects, 60 of whom have the disease and 60 who do not, where the first test is the fasting blood glucose test and the second is the glucose tolerance test. The first test is performed after fasting for 8 hours and consists of taking a blood sample and measuring the amount of glucose in milligrams per deciliter. Normal levels are between 70 and 110 mg/dL, whereas a fasting level between 111 and 125 mg/dL indicates some problems with sugar metabolism, but levels in excess of 126 mg/dL are considered those of a true diabetic.

As for the oral glucose tolerance test, fasting for at least 10 hours is required and a first sample of blood is taken, then the subject is given a bottle of "glucola" with a high amount of sugar, and the glucose level is measured at 30 minutes, 1 hour, and 2 hours later. If the 2-hour measurement is between

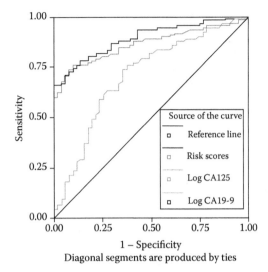

FIGURE 10.5: ROC curves for two biomarkers for the combined test of the pancreatic cancer study.

140 and 200 mg/dL the subject is judged to have prediabetes, whereas 2-hour levels in excess of 200 mg/dL indicate that the subject is definitely diabetic.

For more information about tests for type 2 diabetes, see Johnson and Sandmire [1: 146]. If it is assumed that one reader interprets both tests, what is the accuracy of the individual tests and the combined test?

The descriptive statistics for the diabetes study is shown in Table 10.25. It appears that there is good discrimination between the diseased and non-diseased groups by both tests. The Bayesian analysis shows that the posterior mean(sd) of the ROC area for the oral glucose test is 0.9196(0.0238) and 0.9945(0.0034) for the glucose tolerance test, while that for the combined test is 1(0.0000001611). Thus, the combined test is perfect, but not much of an improvement on the glucose tolerance test, therefore, one does not have to rely on the combined test, the glucose tolerance test is accurate enough! If cost considerations are taken into account, the fasting oral test is probably sufficient because the cost of the oral glucose test is much less than that of the glucose tolerance test, but the former is also quite accurate with a ROC area of

TABLE 10.25: Means and standard deviations for type 2 diabetes: the oral glucose test and the glucose tolerance test.

D test	0	1
T_1 (oral glucose)	108.24(14.37)	135.25(12.28)
T_2 (glucose tolerance)	139.80(20.11)	215.84(20.62)

TABLE 10.26:   Posterior ROC areas for the blood glucose test and the glucose tolerance test.

Parameter	Mean	sd	Error	2 1/2	Median	97 1/2
Combined	1	<0.0000001	<0.000000001	1	1	1
Oral	0.9196	0.0238	<0.0001	0.8658	0.9222	0.9585
Tolerance	0.9945	0.0034	<0.00001	0.9857	0.9954	0.9988

0.9196. The Bayesian analysis is executed with BUGS CODE 10.5 using 45,000 observations with a burn in of 5,000 and a refresh of 100 (see Table 10.26).

Figure 10.6 portrays the ROC areas of the combined tests, where the combined test is in blue, the glucose tolerance test is in green, and the fasting oral test is in red, and note that the combined test is perfect, but there is little improvement on the glucose tolerance test. Recall that the ROC area for the combined test is based on the risk scores, which in turn is determined by a logistic regression which regresses the risk score on the scores of the two tests. For the logistic regression use BUGS CODE 10.4, with the values shown in Table 10.27 for the two tests.

10.7   Observations and Conclusions

This chapter introduces the reader to Bayesian techniques for measuring the accuracy of multiple tests that diagnose the same disease, and it is

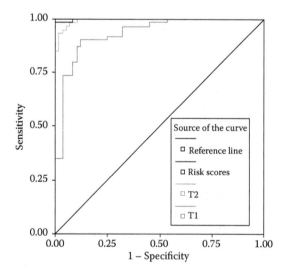

FIGURE 10.6: ROC area of the oral glucose test and the glucose tolerance test.

TABLE 10.27:   List statement for the type 2 diabetes study.

list(K=2, N=120,
d=c(0,0,0,0,0,0,0,0,0,0,0,0,0,0,0,0,0,0,0,0,0,0,0,0,0,0,0,0,0,0,0,0,0,0,0,0,0,0,0,0,
0,0,0,0,0,0,0,0,0,0,0,0,0,0,0,0,0,0,0,0,0,1,1,1,1,1,1,1,1,1,1,1,1,1,1,1,1,1,1,1,1,1,
1,1,1,1,1,1,1,1,1,1,1,1,1,1,1,1,1,1,1,1,1,1,1,1,1,1,1,1,1,1,1,1,1,1,1,1,1,1,1,1),
T1=c(117,139,101,94,94,108,124,88,122,111,112,83,118,111,90,92,107,116,
83,94,115,115,118,125,123,65,115,139,84,115,113,98,121,114,104,99,127,95,
109,111,118,128,118,107,103,94,117,118,103,99,93,93,97,108,119,120,100,
115,128,108,124,109,150,134,117,158,160,126,145,148,138,155,146,145,139,
147,126,130,141,126,125,117,130,137,156,152,124,146,113,134,130,146,138,
123,127,128,133,116,149,130,123,130,135,149,129,118,133,134,128,142,164,
125,128,148,135,129,143,134,137,134),
T2=c(136,125,115,134,159,196,144,130,179,141,111,99,105,146,164,122,
136,139,166,147,159,143,183,126,137,159,156,157,152,115,119,153,103,125,
131,110,138,126,129,127,174,170,146,134,129,119,144,138,140,152,147,132,
130,132,144,131,138,170,159,124,220,220,219,220,214,194,200,221,205,177,
171,207,225,226,224,217,215,237,232,167,180,212,214,201,219,230,207,235,
263,201,233,215,215,199,217,220,208,214,217,242,248,224,242,246,238,221,
231,198,218,211,203,233,193,217,188,277,229,184,203,192))

Note: d is the disease indicator, T_1 is the blood glucose test, and T_2 is the glucose tolerance test.

interesting to note that the subject is not dealt with in a substantial way by contemporary textbooks like Pepe [8] and Zhou, Obuchowski, and McClish [19], however, the Pepe account does introduce the idea of risk score, which is the approach (with a Bayesian flavor) taken in this book. An older account by Kraemer [18] does introduce the central ideas for dealing with multiple tests, but with an emphasis on taking costs into account.

The accuracy of the combined binary test is introduced by using the BP and BN rules, and the concepts are illustrated with two examples: (a) two tests for heart disease, and (b) the combined test using both CT and MRI to diagnose coronary stenosis.

The scenario of working with two binary tests is generalized to multiple readers and to the case when verification bias (assuming MAR) is present. I believe the latter section has not been studied before. Following Pepe [8], the risk score is defined and used to measure the overall accuracy of multiple binary tests. The risk score is the probability that the experimental unit has the disease, given the outcomes the ordinal tests, and the ROC area of the risk scores estimates the accuracy of the combined test, and the risk score is determined by a logistic regression, where the dependent variable is the risk score and the independent variables are the scores of the various tests. A nice feature of the risk score is that it is optimal and should be used by the practitioner to measure the accuracy of the combined test. It should be emphasized that the risk score is a statistical procedure that is available to the clinician in order to estimate accuracy via the ROC area. Such methodology is illustrated with an example of a combined test using MRI and CT to determine the risk of lung cancer, and another example with two versions (body array coil and endorectal coil) of MRI for the localization of prostate cancer. Patient covariate information is easily accommodated in the logistic regression to determine the risk scores of the combined test.

Lastly, the chapter deals with continuous test scores, where logistic regression determines the risk score for estimating the accuracy of the combined test, and two examples illustrate the methodology. The first example involves two biomarkers for prostate cancer while the second involves two blood glucose tests (fasting blood glucose and the glucose tolerance test).

The exercises at the end of the chapter give the student additional information about the important concepts involved in measuring the accuracy of the combined test, be they binary, ordinal, or continuous test scores, and the necessary code and data are provided to make it easier for the student to learn the fundamental ideas.

There are areas for generalizing the concepts of the chapter, namely: (a) not assuming MAR for two binary or ordinal tests with verification bias, see Section 10.4, and Section 10.5.4; and (b) measuring the accuracy of the combined test with three or more components. A good example of the latter is the Berg et al. [3] study that investigates the accuracy of mammography, clinical examination, ultrasound, and MRI for the preoperative assessment of breast cancer, thus, there are four diagnostic tests that can be combined into one, for which the accuracy can be measured by the risk score.

10.8 Exercises

1. Verify the posterior analysis for the dataset given by Table 10.2. Use BUGS CODE 10.1 and generate 55,000 observations from the joint posterior distribution, with a burn in of 5,000 and a refresh of 100. Which is the "best" rule, BP or BN, for estimating the accuracy of the combined test?

2. Prove Statement 10.1 about the TPF and FPF for the BP and BN rules.

3. Verify Table 10.4, the Bayesian analysis for the combined CT and MRI tests for coronary artery disease. Use the information in Tables 10.3a and b along with BUGS CODE 10.1. Note, the second list statement of the code is the data from Tables 10.3a and b, assuming a uniform prior, which is used because two of the cells have zero counts. I used 25,000 observations, with a burn in of 5,000 and a refresh of 100, which resulted in simulation errors of less than 0.0001 for all parameters. Based on your analysis, which rule, the BP or BN rule, should be employed to measure the accuracy of the combined test? Plot the posterior density of the TPF and FPF for the BP rule.

The student should refer to Gerber et al. [7] for additional information about using CT and MRI to diagnose coronary artery disease, which gives much more detail than provided here about the image quality of the two modalities. Based on the Bayesian analysis of Table 10.4, is it reasonable to replace coronary angiography (the gold standard) with MRI or CT?

4. Refer to the Berg et al. [3] study about the accuracy of multiple tests (mammography, clinical examination, ultrasound, and MRI) to detect malignant lesions in preoperative breast cancer. Two of the tests were the standard mammography and ultrasound. What is presented here is just a synopsis of a very detailed investigation into multiple tests to diagnose breast cancer. The data presented in Tables 10.28a and b are for two tests, mammography and ultrasound, and the marginal totals of the table are taken from the paper, but the cell frequencies are provided by me.

From Tables 10.28a and b, it is obvious that the sensitivity of mammography and ultrasound are $120/177 = 0.67$ and $147/177 = 0.83$, respectively, while the specificity for mammography and CT are $61/81 = 0.75$ and $28/81 = 0.34$, respectively. Note this is not a screening trial for breast cancer but a clinical study that identifies malignant lesions before an operation. What is the accuracy of the test that combines these two modalities?

TABLE 10.28a: Data for 177 malignant lesions and two tests.

Mammography	0	1	Total
0	10	47	57
1	20	100	120
Total	30	147	177

Source: From Berg, W.A. et al., *Radiology*, 233, 830, 2004, with permission of Radiological Society of North America.

TABLE 10.28b: Data for 82 non-malignant lesions and two tests.

Mammography	0	1	Total
0	21	40	61
1	7	13	20
Total	28	53	81

Source: From Berg, W.A. et al., *Radiology*, 233, 830, 2004, with permission of Radiological Society of North America.

TABLE 10.29:   Posterior analysis for mammography and ultrasound.

Parameter	Mean	sd	Error	2 1/2	Median	97 1/2
bnfpf	0.1647	0.0399	<0.0001	0.9045	0.1622	0.25
bntpf	0.5579	0.0366	<0.0001	0.4858	0.5579	0.6293
bpfpf	0.7411	0.0463	<0.0001	0.6443	0.7427	0.8275
bpfpf	0.9392	0.0177	<0.0001	0.9003	0.9409	0.9691
fpfmam	0.2586	0.0470	<0.0001	0.1723	0.2565	0.3554
fpfus	0.6472	0.0514	<0.0001	0.5435	0.6482	0.7447
tpfmam	0.6471	0.0345	<0.0001	0.6047	0.6749	0.74
tpfus	0.823	0.0283	<0.0001	0.7641	0.8242	0.8747

Note: Berg et al. [3] study.

Use BUGS CODE 10.1 and note that the data taken from the above tables appear in the third list statement of the program. Using 35,000 observations generated for the simulation, with a burn in of 5,000 and a refresh of 500, verify the Bayesian analysis that appears in Table 10.29.
   (a) Based on Table 10.29, what is the accuracy of the combined test with the BP rule?
   (b) Based on Table 10.29, what is the accuracy of the combined test with the BN rule?

5. Extend the BP and BN rules to more than two binary tests. For three binary tests, what is the best way to express the accuracy of the combined test? Read the Berg et al. [3] study that uses several modalities (mammography, clinical examination, ultrasound, and MRI) to detect malignant breast lesions. Use the information in Table 1 of the paper and determine the accuracy of the combined test for MRI, mammography, and ultrasound, by employing your generalization for the BP and BN rules.

6. Consider a second version of the Gerber et al. [7] study with two readers, but now the reader 2 scores have marginal totals that are not the same as those of the first reader, thus the accuracy measures for the two modalities are not the same as those of the first reader (Tables 10.30a and b).

TABLE 10.30a: Study results for CT-MRI study: radiologist 2 and 58 diseased segments.

CT	MRI		
	0	1	Total
0	6	6	12
1	16	30	46
Total	22	36	58

TABLE 10.30b: Study results for CT-MRI study: radiologist 2 and 236 non-diseased segments.

CT	0	1	Total
0	150	18	168
1	48	20	68
Total	198	36	236

(a) Perform a Bayesian analysis using Tables 10.30a and b for the second reader, and Tables 10.3a and b for the first reader. With BUGS CODE 10.2, note that the second list statement included the data for this version of the Gerber et al. study. Also, note that the author of this book provided the information for the cell frequencies. A uniform prior is assumed for all parameters! Generate 45,000 observations for the posterior distribution, with a burn in of 5,000 and a refresh of 100.

(b) You should get the results shown in Table 10.31 for the posterior distribution of accuracy parameters; thus, verify Table 10.31.

The differences in the posterior mean between reader 1 and reader 2 for the same accuracy parameter begins with fpfbn1 (the FPF for BP), which is quite different from the first scenario portrayed in Exercise 5, when the two readers had very high agreement.

(c) Verify that the simulations errors are <0.0001 for all parameters.

TABLE 10.31: Posterior distribution of accuracy for two readers: second version.

Parameter	Mean	sd	2 1/2	Median	97 1/2
fpfbn1	0.1625	0.0238	0.1187	0.1616	0.2117
fpfbnr2	0.0874	0.0181	0.0552	0.0863	0.126
fpfbpr1	0.2958	0.0249	0.2396	0.2954	0.3551
fpfbpr2	0.3709	0.0309	0.3111	0.3706	0.4327
fpft1r1	0.2917	0.0293	0.2356	0.2912	0.351
fpft1r2	0.2916	0.0291	0.2362	0.291	0.3506
fpft2r1	0.1666	0.0241	0.1221	0.1657	0.2164
fpft2r2	0.1667	0.0240	0.1223	0.1658	0.2162
tpfbnr1	0.5972	0.0618	0.4733	0.598	0.7159
tpfbnr2	0.5005	0.0632	0.377	0.5008	0.6249
tpfbpr1	0.7905	0.0513	0.6819	0.7935	0.8818
tpfbpr2	0.8875	0.0397	0.7985	0.8917	0.9529
tpft1r1	0.7744	0.0527	0.6632	0.7775	0.8682
tpft1r2	0.7745	0.0529	0.6635	0.7775	0.869
tpft2r1	0.6133	0.0612	0.4907	0.614	0.7303
tpft2r2	0.6135	0.0612	0.4901	0.6149	0.7302

(d) What is your conclusion about the accuracy of the combined test? Do you prefer the BN rule compared to the BP rule? Why?

7. Verify Table 10.8 using the information in Table 10.7, which gives the study results for the Alzheimer's study. Execute the analysis with 45,000 observations for the simulation, with a burn in of 5,000 and a refresh of 100.

   Which rule, BN or BP, gives the "best" estimates for the accuracy of the combined test? Explain your answer in detail!

8. Consider the results of two correlated binary tests when verification bias is present, as given in Table 10.32. The first test, Y_1, gives the results for a CT determination of lung cancer risk, where 0 indicates a small risk, and 1 is a high risk of lung cancer, while the second test, Y_2, is a determination of lung cancer risk using MRI.

   Using BUGS CODE 10.3, execute the analysis with 45,000 observations, with a burn in of 5,000 and a refresh of 100. The second list statement of the code gives the data for this example, assuming an improper prior distribution.

   (a) Assume an improper prior distribution for the parameters and determine the posterior distribution of the TPF and FPF for CT.
   (b) Find the posterior distribution of the true and false positive rates for MRI.
   (c) Find the posterior distribution of the TPF and FPF for the BP rule.
   (d) Find the posterior distribution of the TPF and FPF for the BN rule.
   (e) Plot the posterior densities of the true and false positive rates for the BP and BN rules.
   (f) Verify the posterior analysis reported in Table 10.33, and base your answers to (a) to (e) on the results.
   (g) What is your estimate of the accuracy of the combined test?
   (h) Repeat (a) to (e) but assume a uniform prior for the parameters appearing in the likelihood function (Equation 10.29).

9. Generalize Section 10.4 to two binary tests with verification bias and two readers. There will be 16 measures of test accuracy, including eight for the combined test based on the BP and BN rules. Add the relevant

TABLE 10.32: CT and lung cancer risk.

	Y_1	1		0	
$V = 1$	$Y_2 =$	1	0	1	0
$D = 1$		$s_{11} = 14$	$s_{10} = 12$	$s_{01} = 9$	$s_{00} = 3$
$D = 0$		$r_{11} = 4$	$r_{10} = 9$	$r_{01} = 13$	$r_{00} = 18$
$V = 0$		$u_{11} = 7$	$u_{10} = 8$	$u_{01} = 9$	$u_{00} = 10$
Total		$m_{11} = 25$	$m_{10} = 29$	$m_{01} = 31$	$m_{00} = 31$

TABLE 10.33: Posterior analysis for CT and MRI determination of lung cancer risk with verification bias.

Parameter	Mean	sd	Error	2 1/2	Median	97 1/2
fpf1	0.2864	0.06201	<0.0001	0.1725	0.284	0.4141
fpf2	0.3809	0.0668	<0.0001	0.253	0.3799	0.5144
fpfbn	0.0880	0.0697	<0.0001	0.0264	0.0829	0.1795
fpfbp	0.5766	0.0653	<0.0001	0.4559	0.5777	0.7021
tpf1	0.6755	0.0712	<0.0001	0.5305	0.6779	0.8074
tpf2	0.6007	0.0737	<0.0001	0.4538	0.6018	0.7414
tpfbn	0.3663	0.0694	<0.0001	0.2367	0.3645	0.5081
tpfbp	0.9165	0.0439	<0.0001	0.8131	0.9233	0.9812

statement to BUGS CODE 10.3 that will execute the Bayesian analysis. Illustrate your results with an example.

10. (a) Refer to Tables 10.11a and b, the information about the MRI-CT study of lung cancer and verify Table 10.12, which gives the posterior analysis for the ROC areas of the two modalities separately. In order to do this, execute the analysis using BUGS CODE 10.4 with 45,000 observations generated for the MCMC simulation, with a burn in of 5,000 and a refresh of 100.

(b) Using BUGS CODE 10.5, verify Table 10.13, the Bayesian analysis for the ROC area of the combined test. Again, use 45,000 observations, with a burn in of 5,000 and a refresh of 100.

11. For both parts below, generate 45,000 observations, with a burn in of 5,000 and a refresh of 100.

(a) Using Tables 10.14a and b and BUGS CODE 10.4, verify Table 10.15, the Bayesian analysis for the prostate cancer study using two forms of MRI, the body array coil and the endorectal coil. This table gives the ROC areas separately for the two tests.

(b) Using Tables 10.14a and b and 10.16, and using BUGS CODE 10.5, verify Table 10.17, the posterior analysis for the ROC area of the combined test.

12. From the information in Tables 10.22a and b, use BUGS CODE 10.5 with 45,000 observations generated from the joint posterior distribution, with a burn in of 5,000 and a refresh of 100, and execute a Bayesian analysis for the melanoma study. Show the following:

(a) The posterior mean(sd) of the coefficients for the logistic regression are: $-4.932(0.287)$ for $b[0]$, the constant, $1.683(0.085)$ for $b[1]$, the coefficient of Y_1, and $0.892(0.081)$ for $b[2]$, the coefficient of Y_2.

(b) The posterior mean(sd) for the ROC area of the combined test is $0.825(0.010)$.

(c) Comment on the accuracy of the combined test found in (b) relative to the ROC areas of 0.786(0.0119) for the surgeon and 0.6351(0.0145) for the dermatologist. How would you interpret the accuracy of the combined test? What does it mean?

13. (a) Verify Table 10.24, the Bayesian analysis for the pancreatic study, using BUGS CODE 10.5 and the data about the biomarkers in Table 10.24, with 45,000 observations generated from the posterior distribution, with a burn in of 5,000 and a refresh of 100.

(b) Via BUGS CODE 10.4, perform a logistic regression using the risk score theta as the dependent variable and the values of T_1 and T_2 as independent variables, and verify that the estimated risk scores are given by the vector p of Table 10.23. Note that the p vector is used as the dependent variable for the normal regression of BUGS CODE 10.5, which is used to estimate the ROC area of the combined test.

14. Verify Figure 10.5, the plot of the three ROC curves for the combined test, for the CA19-9 biomarker and for the CA125 biomarker of the pancreatic cancer study.

15. (a) Verify Table 10.26, the posterior analysis for the type 2 diabetes study using the oral glucose test, T_1, and the glucose tolerance test, T_2, by referring to Table 10.27, which contains the values of the two tests and the disease incidence vector d. Use BUGS CODE 10.5 with 45,000 observations for the simulation, with a burn in of 5,000 and a refresh of 100.

(b) Estimate the risk score vector by performing a logistic regression using the values of the two tests as independent variables and the risk score theta as the dependent variable. Refer to BUGS CODE 10.4 and use the estimated risk vector as the dependent variable in BUGS CODE 10.5.

16. Reproduce Figure 10.6, a plot of the ROC curves for the combined test, and for the two diagnostic tests T_1 and T_2, for type 2 diabetes.

17. Read the Berg et al. [3] study that investigates the accuracy of four binary tests for the preoperative assessment of breast cancer.

(a) Using Table 5 of that reference, develop a rule that combines the TPF and FPF of the four-into-one combined test, and measure the accuracy of the combined test. Write the code and generate sufficient observations for the simulation so that the MCMC errors are <0.0001 for all parameters. Report the TPF and the FPF for the combined test with a Bayesian analysis that gives the posterior mean, sd, median, and lower and upper 2 1/2% points of each posterior distribution.

(b) In order to combine the four tests into one combined test, use the risk scores.

(c) Plot the ROC for the combined test based on the risk scores of part (b).

References

[1] Johnson, D. and Sandmire, D. *Medical Tests that can Save Your Life.* Rodale and St. Martin's Press, New York, 2004.

[2] Buscombe, J.R., Cwikla, J.B., Holloway, B., and Hilson, A.J.W. Prediction of the usefulness of combined mammography and scintimammography in suspected primary breast cancer using ROC curves. *Journal of Nuclear Medicine*, 42:3, 2001.

[3] Berg, W.A., Gutierrez, L., NessAlver, M.S., Carter, W.B., Bhargavan, M., Lewis, R.S., and Loffe, O.B. Diagnostic accuracy of mammography, clinical examination, US and MR, imaging in preoperative assessment of breast cancer. *Radiology*, 233:830, 2004.

[4] Van Overhagen, H., Brakel, K., Heijenbrok, M.W., van Kasteren, J.H.L.M., van de Moosdijk, C.N.F., Roldaan, A.C., van Gils., A.P., and Hansen, B.E. Metastases in supraclavicular lymph nodes in lung cancer: Assessment with palpation, US, and CT. *Radiology*, 232:75, 2004.

[5] Pauleit, D., Zimmerman, A., Stoffels, G., Bauer, D., Risse, J., Fluss, M.O., Hamacher, K., Coenene, H.H., and Langen, K.J. ^{18}F-FET PET compared with ^{18}F-FDG PET and CT in patients with head and neck cancer. *Journal of Nuclear Medicine*, 47:256, 2006.

[6] Schaffler, G.J., Wolf, G.W., Schoellnast, H., Groell, R., Maier, A., Smolle-Juttner, F.M., Woltsche, M., Fasching, G., Nicolletti, R., and Aigner, R.M. Non-small cell lung cancer: Evaluation of pleural abnormalities on CT scans with ^{18}F-FET PET. *Radiology*, 231:858, 2004.

[7] Gerber, B.L., Coche, E., Pasquet, A., Ketelslegers, E., Vancraeynest, D., Grandin, C., van Beers, B.E., and Vanocerschelde, J.L.J. Coronary artery stenosis: Direct comparison of four-section multi-detector row CT and 3D navigator MR imaging for detection—Initial results. *Radiology*, 234:98, 2005.

[8] Pepe, M.S. *The Statistical Evaluation of Medical Tests for Classification and Prediction.* Oxford University Press, 2003.

[9] McIntosh, M.W. and Pepe, M.S. Combining screening tests: Optimality of the risk score. *Biometrics*, 58:657, 2002.

[10] Gierada, M., Pilgrim, T.K., and Ford, M. Lung cancer interobserver agreement on interpretation of pulmonary findings at low-dose CT screening. *Radiology*, 246(1):265, 2008.

[11] Utsunomiya, D., Shiraishi, S., Imuta, M., Tomiguchi, S., Kawanaka, K., Morishita, S., Awai, K., and Yamashita, Y. Added valued of SPECT/PET fusion in assessing suspected bone metastasis: Comparison with scintigraphy alone and nonfused scintigraphy and CT. *Radiology*, 238:264, 2005.

[12] Mazaheri, Y., Shukla-Dave, A., Hricak, H., Fine, S.W., Zhang, J., Inurrigarro, G., Moskowitz, C.S. et al. Prostate cancer: Identification with combined diffusion-weighted MR imaging and 3D H MR spectroscopic imaging correlation with pathologic findings. *Radiology*, 246:480, 2008.

[13] Futterer, J.J., Heijmink, S.W.T.P., Scheenen, T.W.J., Veltman, J., Huisman, H.J., Vos, P., Hulsbergen-Van de Kaa, C.A. et al. Prostate cancer localization with dynamic contrast-enhanced MR imaging and proton MR spectroscopic imaging. *Radiology*, 241:449, 2006.

[14] Coakley, F.V., Teh, H.S, Quayyum, A., Swanson, M.G., Lu, Y., Roach, M., Pickett, B., Shinohara, K., Vigneron, D.B., and Kurhanewicz, J. Endorectal MR imaging and MR spectroscopic imaging for locally recurrent prostate cancer after external beam radiation therapy: Preliminary experience. *Radiology*, 233:441, 2004.

[15] Heijmink, S.W.T.P., Futterer, J.J., Hambrock, T., Takahashi, S., Scheenen, T.W.J., Huisman, H.J., Hulsbergen-Van de Kaa, C.A., et al. Prostate cancer: Body array versus endorectal coil MR imaging at 3T-comparison of image quality localization and staging performance. *Radiology*, 244:184, 2007.

[16] O'Malley, A.J., Zou, K.H., Fielding, J.R., and Tempany, C.M.C. Bayesian methodology for estimating a receiver operating characteristic curve with two radiologic applications: Prostate biopsy and spriral CT of ureteral stones. Academic *Radiology*, 8:713, 2001.

[17] Wieand, S., Gail, M.H., and James, B.R. A family of nonparametric statistics for comparing diagnostic markers with paired or unpaired data. *Biometrics*, 76:585, 1989.

[18] Kraemer, H.C. *Evaluating Medical Tests, Objective and Quantitative Guidelines*. Sage, Newbury Park, 1992.

[19] Zhou, X.H., Obuchowski, N.A., and McClish, D.K. *Statistical Methods in Diagnostic Medicine*. John Wiley, New York, 2002.

Chapter 11

Bayesian Methods for Meta-Analysis

11.1 Introduction

Determining the accuracy of a medical test is quite difficult because accuracy is an elusive thing to estimate. It is well known that the accuracy of a particular test will vary because of intra and inter study variation. Even though a test can be replicated under identical conditions, nevertheless, the accuracy will vary between repetitions. The literature on the accuracy of a particular test continues to grow, with each study giving some estimate of its accuracy, but, of course, the estimates vary because of different study conditions. Accuracy varies because the study population of patients vary, the test itself changes somewhat, and the readers of the test scores also vary. Because of the various sources of variation, it is important to conduct studies that summarize the accuracy of a particular medical test.

For example, take the case of heart disease and the accuracy of the exercise stress test. Hundreds of studies estimate the accuracy (the true and false positive fractions or the receiver operating characteristic [ROC] area) of this test. Suppose the exercise stress test could be replicated under identical conditions with the same patients and the same readers to interpret the test scores, and the test accuracy estimated. Of course, the accuracy of the test will vary from test to test and such variation is called experimental error. It is difficult and, in fact, almost impossible to estimate the experimental error for most medical tests, and it is more common to summarize the accuracy of the exercise stress test of different studies that involve different patient populations and different sets of radiologists. There are studies where it is possible to estimate the replication variation within readers and this is presented in Chapter 6 on agreement.

The first scenario to be considered is estimating the common true positive rate (TPR) and the false positive rate (FPR) from different studies and arriving at a common value of the accuracy of the test. The accuracy is expressed with an estimate similar to the area under the ROC curve. When it is assumed that the various studies have different TPR and FPR, because the decision threshold varies, one is assuming that the studies have the same ROC curve, called the SROC curve. The estimate of the accuracy of the summary receiver operating characteristic (SROC) curve is the ordinate of the point of intersection, where the SROC curve intersects the line with equation TPR + FPR = 1.

The approach taken here is appropriate for ordinal and continuous data that employ a threshold to declare a positive test and follows the presentation of Kardaun and Kardaun [1], and Moses, Shapiro, and Littenberg [2], who do not work directly with the TPF and FPR, but with their logits and consequently with a regression model that allows one to work easily with the SROC curve. It should be noted that this approach is also adopted by Zhou, Obuchowski, and McClish [3], and will be followed to some extent here, except a Bayesian model is established. Much has been accomplished using the Bayesian approach and the reader is referred to Stangl and Berry [4], who edit a book consisting of many papers that address various issues in meta-analysis from a Bayesian viewpoint; however, the book focuses on summarizing various estimates of treatment efficacy in clinical trials and not on estimating various measures of test accuracy.

Chapter 11 begins with summarizing information about test accuracy for tests with ordinal and continuous scores, where it is assumed that the tests share a common ROC curve, thus, the tests may differ in the threshold used to declare a positive test. The FPR and TPR can be plotted to determine a common ROC curve, called the SROC curve. The TPR and FPR are transformed so that one may use bilogistic regression to determine the accuracy of the combined tests, where the posterior distribution of the parameters of the model is determined. The slope and intercept of the regression determine the SROC curve. The chapter is continued by allowing the inclusion of study covariates that allow for inter study variation between the various studies that comprise the meta-analysis. Two or more versions of the same tests are compared with the bilogistic regression methodology, using inter study covariates. Bayesian inference is illustrated with a well-known example of DeVries, Hunink, and Polak [5], which is based on two versions of ultrasonography (US) for the diagnosis of stenosis of the peripheral arteries, and the accuracy of the two versions is compared with a regression model and by comparing the Q-statistics, which measure the accuracy of the SROC. Additional examples include diagnostic studies for coronary artery disease, inflammatory bowel disease, osteomyelitis, breast cancer, and recurrent colorectal cancer. The conclusion of the chapter emphasizes the summarization of tests with a common ROC area, where the posterior distribution of the ROC area and its standard deviation allow one to compute the common area, which is estimated as a weighted average of the individual ROC areas weighted by the inverse of the posterior variance.

Several of the examples are based on recent studies of meta-analyses. For example, a meta-analysis by Vanhoenacker et al. [6] summarizes the accuracy of multidetector computed tomography (CT) angiography for the diagnosis of coronary artery disease, while the Horsthuis et al. [7] meta-analysis explores the detection of inflammatory bowel disease with ultrasound, MR scintigraphy, and CT. The latter study allows one to compare the accuracy of the three modalities. A third example by Pakos et al. [8] is a meta-analysis of a nuclear medicine procedure to diagnose osteomyelitis, and in all three examples the

SROC curve is inferred by Bayesian methods. Such papers have a standard way of presenting their results, including an introduction, a description of the methods, a report of the results, and a section for comments and conclusions. Such papers usually have enough detail so that others may check and replicate their results. Of paramount importance is describing just how the various studies of the meta-analysis are included in the study, and the description should include enough information to determine the heterogeneity between the various studies, including the number of readers used in each study, and the threshold value stated. If the SROC curve is to be determined, the homogeneity needs to be demonstrated so that one has confidence in the overall accuracy of the combined studies.

11.2 Summary Receiver Operating Characteristic Curve and Bilogistic Regression

The first scenario for meta-analysis is the least complicated, namely, one assumes that the various studies have a common ROC curve. If the test scores are continuous or ordinal, the threshold value may vary, giving different FPR and TPR values that can be plotted to give the common ROC curve, referred to as the SROC curve. Thus, there is enough homogeneity between studies to assume a common ROC curve and this assumption needs to be checked with the information given for the various studies. The reader should remember that the inclusion and exclusion criteria may vary between studies, that, of course, the readers interpreting the test scores will not be the same between studies, that the threshold (the value that declares a positive test) value can vary, and lastly that the various tests, although related, may not be the same. For example, using CT to diagnose lung cancer, the CT equipment will not be the same and will not be operated in the same manner from study to study. On examining the recent issues of the imaging literature, the summary accuracy is usually based on a SROC analysis.

The approach taken here is to base the regression analysis on the approach of Moses, Shapiro, and Littenberg [2], who do not work with the (FPR,TPR) points directly, but with a transformation

$$B = V - U \tag{11.1}$$

and

$$S = V + U, \tag{11.2}$$

where

$$U = \text{logit}(\text{FPR})$$

and

$$V = \text{logit}(\text{TPR}). \tag{11.3}$$

The B values are regressed on the S values

$$B = \text{beta}[1] + \text{beta}[2]S, \tag{11.4}$$

where beta[1] and beta[2] are unknown parameters.

It is assumed that U and V have logistic distributions, thus, it is reasonable to assume that B and S also have logistic distributions. The interpretation of B and S is very informative, because B is the log odds ratio where the numerator is the odds of a positive test given the disease is present, and the denominator is the odds that the test score is positive given the disease is not present. The variable S can be thought of as measuring the effect of the threshold value, in the sense that if S is zero, the TPR and FPR are equal, and if S is positive, the sensitivity (TPR) is greater than the specificity $(1 - \text{FPR})$. On the other hand, when the sensitivity is less than the specificity, the values of S are negative. Note, the interpretation of the intercept beta[1] is the average value of B (the log odds ratio) when $S = 0$, and that beta[2] measures the effect of S on B, that is for each unit increase in S, B increases on the average by beta[2] units. It is probably safe to say that if beta[2] is close to zero, then the test at hand has the same power to detect a difference in the two populations (diseased vs. non diseased) for all values of the threshold.

It can be shown that the SROC curve is defined as

$$\text{SROC}(\text{FPR}) = \left[1 + \exp{-\text{beta}[1]}/\right.$$
$$\left.(1 - \text{beta}[2])[(1 - \text{FPR})/\text{FPR}]^{(1+\text{beta}[2])/(1-\text{beta}[2])}\right]^{-1} \tag{11.5}$$

and the curve is determined by plotting (SROC,TPR) where the SROC and FPR values are given by Equation 11.5. Note that if beta[2] $= 0$, Equation 11.5 can be modified accordingly by letting beta[2] $= 0$.

There are many ways to estimate the parameters beta[1] and beta[2] of the regression of B on S, and the reader is referred to Moses, Shapiro, and Littenberg [2] for some non-Bayesian estimation techniques. The approach here is Bayesian, where first the posterior density of beta[1] and beta[2] is determined using Equation 11.4 with a non-informative prior for the regression coefficients and scale parameter tau. The prior distribution of beta[i] is normal (0.0001,0.0001) and the prior distribution of tau is gamma with parameters 0.0001 and 0.0001. Note, the logistic distribution has two parameters and is similar in shape to the normal distribution. The posterior distribution of the ordinates of the SROC curve (Equation 11.4) is induced by the posterior distribution of beta[1] and beta[2], and will be illustrated with several examples in a later section.

11.3   Bayesian Analysis for Summary Accuracy

The obvious measure of test accuracy for a meta-analysis is the area under the SROC curve, however, since the TPR values tend to be concentrated over a relatively small range of FPR values, the entire SROC curve must be estimated by extrapolation to the entire range of FPR values over (0,1). Because of this impediment, other measures of test accuracy have been devised, e.g., Moses, Shapiro, and Littenberg [2], who proposed the ordinate of the intersection between the SROC curve and the line with equation

$$(\text{TPR} + \text{FPR}) = 1, \tag{11.6}$$

which is the negative diagonal of the unit square.

It can be shown that the ordinate of the intersection is

$$Q = (1 + e^{-\text{beta}[1]/2})^{-1}, \tag{11.7}$$

where beta[1] is the intercept term of the bilogistic regression (Equation 11.4). Therefore, the posterior distribution of beta[1] induces the posterior distribution of Q. Note that "large" values of Q close to 1 indicate excellent accuracy because the intersection point is close to the (0,1) point of the unit square, and in a similar way, values of Q close to 0.5 imply that the SROC curve is close to the main diagonal of the unit square, indicating very poor accuracy.

11.3.1   A meta-analysis with one test

The first set of examples of a meta-analysis are for studies with a common SROC curve, where the sample information consists of true and false positive ratios or, equivalently, the true negative (TN), true positive (TP), false positive (FP), and false negative (FN) of the 2 × 2 square for each study. Consider the DeVries, Hunink, and Polak [5] meta-analysis, which summarizes nine studies where peripheral artery stenosis is determined with regular duplex US. The information from the study is in the form of a 2 × 2 table of TN, TP, FP, and FN values for each study and appears in the following list statement of BUGS CODE 11.1.

BUGS CODE 11.1

```
# one test
model;
{
for(i in 1:N){tpr[i]<-(tp[i]+.5)/(tp[i]+fn[i]+.05)}
for(i in 1:N){fpr[i]<-(fp[i]+.5)/(fp[i]+tn[i]+.05)}
for(i in 1:N){u[i]<-logit(fpr[i])}
for(i in 1:N){v[i]<-logit(tpr[i])}
for(i in 1:N){b[i]<-v[i]-u[i]}
```

```
for(i in 1:N){s[i]<-v[i]+u[i]}
# bilogistic regression of b on s
for(i in 1:N){b[i]~dlogis(mu[i],tau)
mu[i]<-beta[1]+beta[2]*s[i]}
for(i in 1:2){beta[i]~dnorm(.0000,.0001)}
tau~dgamma(.0001,.0001)
P<-1+exp(-beta[1]/2)
# accuracy of test
Q<-1/P
#sroc curve assumes slope is 0
r1<-exp(-beta[1])
for(i in 1:N){r2[i]<-(1-fpr[i])/fpr[i]}
for(i in 1:N){r3[i]<-1+r1*r2[i]}
for(i in 1:N){sroc[i]<-1/r3[i]}
}
# data from DeVries et al. [5] duplex mode
list(N=8, tn=c(516,89,235,262,488,48,156,376),
fn=c(28,8,23,20,14,7,2,31),
fp=c(20,12,5,22,9,3,14,12),
tp=c(78,59,75,89,118,48,39,121))
# data from Meijer et al.
# a 1 replaces a 0 in the data of Meijer et al.
list(N=20,
tn=c(23,35,10,9,23,30,60,48,12,38,23,60,9,42,27,50,35,5,37,20),
fn=c(2,1,1,1,1,1,1,1,1,1,1,1,1,1,1,2,1,1,1,1,1),
fp=c(1,2,1,1,1,6,7,4,4,1,6,7,1,4,3,5,1,1,3,3),
tp=c(36,28,29,29,16,19,20,18,88,12,35,14,25,53,38,25,13,26,44,76))
# initial values
list(beta=c(0,0), tau=1))
```

For example, the TN, FN, FP, and TP values for the first study are 516, 28, 20, and 78, respectively, and the program calculates the TPR and FPR. First, a bilogistic regression is performed, which produces the posterior characteristics of the regression coefficients beta[1] and beta[2], then the posterior distribution of the Q parameter is calculated, which expresses the summary accuracy of the eight studies. Lastly, the posterior characteristics of the ordinates of the SROC curve are determined corresponding to the eight FPR values of the eight studies. The analysis is executed with 65,000 observations generated from the joint posterior distribution, with a burn in of 5,000 and a refresh of 100 (Table 11.1).

Regression analysis of B on S reveals that the slope is "small" with a 95% credible interval $(-0.6842, 0.5995)$, indicating that it is not unreasonable to let beta[2] $= 0$, and implying that ultrasound duplex is discriminating between the diseased and non-diseased populations in the same way for all values of the test threshold. BUGS CODE 11.1 contains statements that calculate the SROC values assuming the slope is zero. The coefficient beta[1] has a posterior

TABLE 11.1: Posterior analysis for peripheral artery stenosis.

Parameter	Mean	sd	Error	2 1/2	Median	97 1/2
Q	0.9104	0.0270	<0.0001	0.861	0.9132	0.946
beta[1]	4.7	0.551	0.0098	3.647	4.707	5.728
beta[2]	−0.0275	0.3372	0.0060	−0.6842	−0.0231	0.5995
sroc[1]	0.802	0.0857	0.0015	0.604	0.8149	0.9244
sroc[2]	0.9319	0.0455	<0.0001	0.8411	0.9399	0.9775
sroc[3]	0.7104	0.1046	0.0018	0.4735	0.7219	0.8781
sroc[4]	0.8946	0.0587	0.0011	0.7674	0.905	0.9636
sroc[5]	0.673	0.1102	0.0019	0.4277	0.6832	0.8569
sroc[6]	0.8796	0.0636	0.00125	0.7384	0.8907	0.9577
sroc[7]	0.9017	0.0563	0.0011	0.7814	0.9116	0.9663
sroc[8]	0.7737	0.9023	0.0016	0.5607	0.7865	0.9109
tau	2.063	0.6882	0.0051	0.9117	1.995	3.602

mean of 4.7, which is the average value of B when $S = 0$, indicating that the odds ratio has a posterior mean of 4.7, that is to say, the odds of a positive test for the diseased population (those with stenosis in the peripheral arteries) is 4.7 times more than the odds of a positive test result for the non-diseased patients. Q has a posterior mean of 0.91, which implies good accuracy with ultrasound to detect stenosis.

The slope beta[2] of the regression of B on S is close to zero and the analysis calculates the posterior characteristics of the SROC values corresponding to the FPR values, which are calculated as the FPR vector. The posterior density of beta[2] is depicted in Figure 11.1 and shows that the posterior probability is large in the neighborhood of zero.

Simulation errors were quite small for all parameters, demonstrating that 60,000 observations are sufficient for a reliable posterior analysis. The above determination of the SROC values assumes beta[2] = 0, however, I performed the analysis assuming beta[2] is not zero and the results were quite similar to Table 11.1; the analysis was executed with BUGS CODE 11.2.

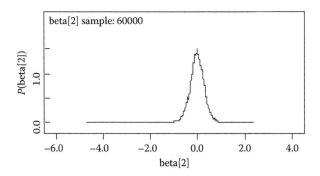

FIGURE 11.1: Posterior density of the slope beta[2].

BUGS CODE 11.2

```
# one test with beta[2] not zero
model;
{
for(i in 1:N){tpr[i]<-(tp[i]+.5)/(tp[i]+fn[i]+.05)}
for(i in 1:N){fpr[i]<-(fp[i]+.5)/(fp[i]+tn[i]+.05)}
for(i in 1:N){u[i]<-logit(fpr[i])}
for(i in 1:N){v[i]<-logit(tpr[i])}
for(i in 1:N){b[i]<-v[i]-u[i]}
for(i in 1:N){s[i]<-v[i]+u[i]}
# bilogistic regression of b on s
for(i in 1:N){b[i]~dlogis(mu[i],tau)
mu[i]<-beta[1]+beta[2]*s[i]}
for(i in 1:2){beta[i]~dnorm(.0000,.0001)}
tau~dgamma(.0001,.0001)
P<-1+exp(-beta[1]/2)
# accuracy of test
Q<-1/P
# sroc curve, does not assume slope is zero
r1<-exp(-beta[1]/(1-beta[2]))
for(i in 1:N){r2[i]<-(1-fpr[i])/fpr[i]}
r3<-(1+beta[2])/(1-beta[2])
for(i in 1:N){r4[i]<-pow(r2[i],r3)}
for(i in 1:N){r5[i]<-1+r1*r4[i]}
for(i in 1:N){sroc[i]<-1/r5[i]}
}
# data from DeVries et al. duplex mode
list(N=8, tn=c(516,89,235,262,488,48,156,376),
fn=c(28,8,23,20,14,7,2,31),
fp=c(20,12,5,22,9,3,14,12),
tp=c(78,59,75,89,118,48,39,121))
# initial values
list( beta=c(0,0), tau=1))
```

11.3.2 Summary accuracy for scintigraphy

Our second example of a meta-analysis involves the diagnosis of osteomyelitis using antigranulocyte scintigraphy with ^{99m} radiolabeled monoclonal antibodies, and the analysis consists of 19 non-overlapping studies with reference standards of cell culture, histologic examination, and clinical follow-up. The study was conducted by Pakos et al. [8] with MEDLINE and EMBASE searches, and a SROC curve was constructed; the reader is referred to the paper for additional important information as to the heterogeneity of the meta-analysis. Our approach is Bayesian and will consist of determining the overall accuracy as expressed by the Q parameter, performing the regression analysis of the B scores on the S scores, and, finally, computing the posterior characteristics of the SCROC values. Note, for this example, the input values are the TPR and FPR of the 19 studies contained in the list statement of BUGS CODE 11.3.

BUGS CODE 11.3

```
# one test
model;
{
for(i in 1:N){u[i]<-logit(fpr[i])}
for(i in 1:N){v[i]<-logit(tpr[i])}
for(i in 1:N){b[i]<-v[i]-u[i]}
for(i in 1:N){s[i]<-v[i]+u[i]}
# bilogistic regression of b on s
for(i in 1:N){b[i]~dlogis(mu[i],tau)
mu[i]<-beta[1]+beta[2]*s[i]}
for(i in 1:2){beta[i]~dnorm(.0000,.0001)}
tau~dgamma(.0001,.0001)
P<-1+exp(-beta[1]/2)
# accuracy of test
Q<-1/P
#sroc curve assumes slope is 0
r1<-exp(-beta[1])
for(i in 1:N){r2[i]<-(1-fpr[i])/fpr[i]}
for(i in 1:N){r3[i]<-1+r1*r2[i]}
for(i in 1:N){sroc[i]<-1/r3[i]}
}
# data from Pakos et al.
list(N=19,tpr=c(.67,.75,.85,.35,.99,.95,.90,.99,.80,.78,.95,.61,.90,.89,.88,.93,.93,
        .43,.38),
fpr=c(.15,.05,.11,.17,.22,.33,.33,.01,.33,.60,.43,.04,.17,.01,.10,.70,.01,.01,.57))
# initial values
list( beta=c(0,0), tau=1))
```

A Bayesian analysis is performed with 65,000 observations generated for the Markov Chain Monte Carlo (MCMC) simulation, with a burn in of 5,000 and a refresh of 100, and the results appear in Table 11.2.

It should be noted that the 19 SROC values correspond to the following (in that order) FPR values

FPR=(0.15,0.05,0.11,0.17,0.22,0.33,0.33,0.01,0.33,0.60,0.43,0.04,0.17,0.01,
        0.10,0.70,0.01,0.01,0.57),

where the SROC values are computed according to Equation 11.5, assuming beta[2]=0. The 95% credible interval for beta[2] is $(-0.7051, 0.2755)$ and

TABLE 11.2:   Bayesian analysis for the osteomyelitis meta-analysis.

Parameter	Mean	sd	Error	2 1/2	Median	97 1/2
Q	0.8524	0.0346	<0.0001	0.7756	0.8555	0.9113
beta[1]	3.559	0.5512	0.0048	2.481	3.558	4.658
beta[2]	−0.2063	0.2454	0.0020	−0.7051	−0.2032	0.2755
sroc[1]	0.8486	0.0702	<0.0001	0.6783	0.8609	0.949
sroc[2]	0.6899	0.1189	0.0010	0.3861	0.6487	0.8473
sroc[3]	0.7994	0.08594	<0.0001	0.5963	0.8126	0.9287
sroc[4]	0.8661	0.064	<0.0001	0.7099	0.8778	0.9557
sroc[5]	0.8981	0.0511	<0.0001	0.7712	0.9082	0.9675
sroc[6]	0.9382	0.0335	<0.0001	0.8548	0.9453	0.9811
sroc[7]	0.9382	0.0335	<0.0001	0.8548	0.9453	0.9811
sroc[8]	0.2745	0.1052	<0.0001	0.1077	0.2616	0.5158
sroc[9]	0.9382	0.0335	<0.0001	0.8548	0.9453	0.9811
sroc[10]	0.9785	0.0127	<0.0001	0.9472	0.9813	0.9937
sroc[11]	0.9585	0.0235	<0.0001	0.9001	0.9636	0.9876
sroc[12]	0.5881	0.1243	<0.0001	0.3324	0.5938	0.8146
sroc[13]	0.8661	0.064	<0.0001	0.7099	0.8778	0.9557
sroc[14]	0.2745	0.1052	<0.0001	0.1077	0.2616	0.5158
sroc[15]	0.7826	0.0906	<0.0001	0.5704	0.7958	0.9214
sroc[16]	0.9861	0.0084	<0.0001	0.9654	0.9879	0.996
sroc[17]	0.2745	0.1052	<0.0001	0.1077	0.2616	0.5185
sroc[18]	0.2745	0.1052	<0.0001	0.1077	0.2616	0.5185
sroc[19]	0.9758	0.0142	<0.0001	0.9406	0.9789	0.9929
tau	0.7552	0.1549	<0.0001	0.4822	0.7445	10.087

contains zero, thus, I let beta[2] = 0 in the code for sroc of BUGS CODE 11.3.
A regression of B on S values gives a posterior mean(sd) for the intercept of
3.559(0.5512), implying that the odds ratio for a positive test is 3.6 times
larger for those patients with osteomyelitis compared to those without the
disease (when $S = 0$). Figure 11.2 is a plot of the SROC curve for the Pakos
et al. [8] meta-analysis, whose ordinates are the posterior means portrayed in
Table 11.2.

The green line is the minor diagonal of the unit square. What is the area
under the curve? This is left as an exercise! It appears that the accuracy
implied by the SROC curve is quite good, and this is confirmed by the value
of the Q parameter, which has a posterior mean of 0.8524, a 95% credible
interval of (0.7756,0.9113), and a posterior density portrayed by Figure 11.3.

11.4   Meta-Analysis with Two Tests

When two tests are used to diagnose the same disease, the main emphasis
is on comparing the accuracy of the two tests. There are many examples of

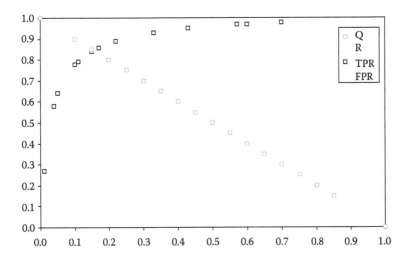

FIGURE 11.2: SROC curve of the Pakos et al. [8] study.

meta-analyses that summarize the accuracy of two sets of studies. The first set is a series of studies that use one test to diagnose disease, and the other set focuses on a series of tests with another test for diagnosing the same disease. For such a situation, two regressions of B on S are performed, two values of Q are computed, and the SROC curve for each test is determined.

Thus, let

$$B_1 = \text{beta1}[1] + \text{beta1}[2]S_1 \qquad (11.8)$$

be the first regression, and

$$B_2 = \text{beta2}[1] + \text{beta2}[2]S_2 \qquad (11.9)$$

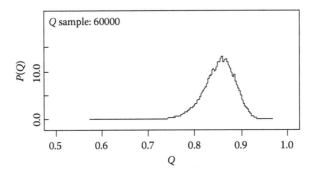

FIGURE 11.3: Posterior density of Q for Pakos et al. [8].

be the second, where

$$B_i = V_i - U_i,$$

and U_i and V_i are the logits of the TPR and FPR for the ith test $i = 1, 2$. See Equations 11.1 and 11.2 for the formal definitions.

In a similar way, the accuracy of the ith test is based on the Q parameters

$$Q_i = [1 + e^{-\mathrm{beta}i[2]/2}]^{-1} \tag{11.10}$$

for $i = 1, 2$.

Finally, the SROC curve is defined by its ordinates for the ith test as

$$\mathrm{SROC}i(\mathrm{FPR}) = [1 + \exp -\mathrm{beta}i[1]/$$
$$(1 - \mathrm{beta}i[2])[(1 - \mathrm{FPR})/\mathrm{FPR}]^{(1+\mathrm{beta}i[2])/(1-\mathrm{beta}i[2])}]^{-1} \tag{11.11}$$

corresponding to the FPR of the ith test, where $i = 1, 2$. Note that if the regression analysis implies beta[2] = 0, then one should modify Equation 11.11 accordingly.

For the first example, consider the meta-analysis of Horsthuis et al. [7], which is a study of the use of ultrasound and magnetic resonance imaging (MRI) to diagnose inflammatory bowel disease; nine studies were based on ultrasound and seven on MRI. On the basis of the meta-analysis, our purpose is to compare the two modalities. The authors found 1406 papers in a MEDLINE search and reduced the number to 16 for the present analysis. Much information was extracted concerning the number of readers, the threshold values for a positive test, and other study covariates. A later section will be devoted to including study covariates into the determination of test accuracy. The Bayesian analysis will consist of performing regression analyses for the two modalities and determining if the slope coefficient is zero. Depending on the value of beta[2], the appropriate formula for the SROC values for each test is used to determine the summary curve, and lastly, the Q values are computed to compare the two modalities for accuracy.

Consider BUGS CODE 11.4.

BUGS CODE 11.4

```
# two tests
model;
{
# for test 1
for(i in 1:N1){u1[i]<-logit(fpr1[i])}
for(i in 1:N1){v1[i]<-logit(tpr1[i])}
for(i in 1:N1){b1[i]<-v1[i]-u1[i]}
```

```
for(i in 1:N1){s1[i]<-v1[i]+u1[i]}
# bilogistic regression test 1
for(i in 1:N1){b1[i]~dlogis(mu1[i],tau1)
mu1[i]<-beta1[1]+beta1[2]*s1[i]}
for(i in 1:2){beta1[i]~dnorm(.0000,.0001)}
tau1~dgamma(.0001,.0001)
P1<-1+exp(-beta1[1]/2)
# accuracy of test 1
Q1<-1/P1
#sroc test 1, assumes slope is 0
r11<-exp(-beta1[1])
for(i in 1:N1){r12[i]<-(1-fpr1[i])/fpr1[i]}
for(i in 1:N1){r13[i]<-1+r11*r12[i]}
for(i in 1:N1){sroc1[i]<-1/r13[i]}
# for test 2
for(i in 1:N2){u2[i]<-logit(fpr2[i])}
for(i in 1:N2){v2[i]<-logit(tpr2[i])}
for(i in 1:N2){b2[i]<-v2[i]-u2[i]}
for(i in 1:N2){s2[i]<-v2[i]+u2[i]}
# bilogistic regression test 2
for(i in 1:N2){b2[i]~dlogis(mu2[i],tau2)
mu2[i]<-beta2[1]+beta2[2]*s2[i]}
for(i in 1:2){beta2[i]~dnorm(.0000,.0001)}
tau2~dgamma(.0001,.0001)
#sroc test 2, assumes slope is zero
r21<-exp(-beta2[1])
for(i in 1:N2){r22[i]<-(1-fpr2[i])/fpr2[i]}
for(i in 1:N2){r23[i]<-1+r21*r22[i]}
for(i in 1:N2){sroc2[i]<-1/r23[i]}
P2<-1+exp(-beta2[1]/2)
# accuracy of test 2
Q2<-1/P2
# difference in accuracy of two tests
d<-Q1-Q2
}
# below is data from DeVries et al. [5] for duplex and color
# duplex is test 1
# color is test 2
list(N1 = 8, fpr1=c(.04,.12,.02,.08,.02,.06,.08,.03),
    tpr1=c(.74,.88,.77,.82,.89,.87,.95,.80),
N2=6, fpr2=c(.01,.01,.02,.06,.05,.05),
    tpr2=c(.90,.99,.88,.89,.99,.96))
# horsthuis et al. [7] study
# test 1 is US and test 2 is MR
list(N1 =9, fpr1=c(.10,.01,.06,.12,.07,.07,.01,.33,.04),
```

tpr1=c(.78,.93,.90,.81,.93,.88,.87,.96,.92),
N2=7, fpr2=c(.01,.01,.39,.39,.01,.08,.15),
    tpr2=c(.99,.99,.91,.87,.82,.95,.89))
list(beta1=c(0,0), tau1=1, beta2=c(0,0),tau2=1)

The second list statement is the information for the Horsthuis et al. [7] meta-analysis, where the first test is ultrasound and the second is MRI. The third list statement is for the initial values of the parameters. A Bayesian analysis is executed with 65,000 observations for the simulations, with a burn in of 5,000 and a refresh of 100.

The two Q values are quite close, where for ultrasound the posterior mean is 0.9098 and for MRI the posterior mean is 0.9324, and the difference d between the two have a 95% credible interval of $(-0.0956, 0.0991)$, implying that the accuracy of US and MRI are about the same, which is confirmed to some extent by a plot of the two SROC curves portrayed by Figure 11.4. Note, the green curve corresponds to US and the red to MRI (Table 11.3).

The Bayesian regression analysis indicates that that the slope of each test is zero, thus, the SROC values for both tests are computed assuming beta1[2] and beta2[2] are zero. Also, the intercept term, beta1[1], has a posterior mean of 4.668, implying that the odds (when $S_1 = 0$) of a positive test result for those with inflammatory bowel disease is 4.6 times that of the odds of a positive result for those without the disease. Also, since the two slopes are close to zero, the implication is that both tests are differentiating equally between disease and non disease for all values of the threshold. See Horsthuis et al. [7] for additional information about the choice of threshold for the various studies for each modality.

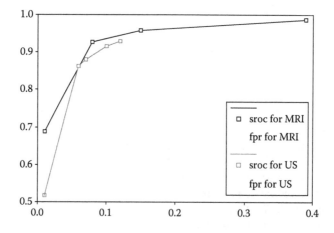

FIGURE 11.4: SROC curves for US and MRI.

TABLE 11.3: Posterior analysis for the inflammatory bowel disease meta-analysis.

Parameter	Mean	sd	Error	2 1/2	Median	97 1/2
$Q1$	0.9098	0.0203	<0.0001	0.8629	0.9123	0.942
$Q2$	0.9324	0.0493	<0.0001	0.8148	0.9426	0.9863
beta1[1]	4.668	0.4757	0.0048	3.679	4.683	5.575
beta1[2]	−0.5686	0.3238	0.0034	−1.242	−0.5631	0.0664
beta2[1]	5.647	1.418	0.0129	2.963	5.597	8.557
beta2[2]	−0.901	1.003	0.0112	−2.929	−0.8595	0.9871
d	−0.02263	0.05346	<0.0001	−0.0956	−0.03004	0.09919
srocl[1]	0.9151	0.0409	<0.0001	0.8148	0.9232	0.967
srocl[2]	0.5176	0.1107	0.0011	0.2857	0.522	0.7271
srocl[3]	0.8626	0.0594	<0.0001	0.7165	0.8734	0.9439
srocl[4]	0.9294	0.03521	<0.0001	0.8438	0.9365	0.9729
srocl[5]	0.8805	0.0535	<0.0001	0.7488	0.8906	0.952
srocl[6]	0.8805	0.0535	<0.0001	0.7488	0.8906	0.952
srocl[7]	0.5176	0.1107	0.0011	0.2857	0.522	0.7271
srocl[8]	0.9791	0.0120	<0.0001	0.9512	0.9816	0.9924
srocl[9]	0.806	0.0754	<0.0001	0.6227	0.8184	0.9166
sroc2[1]	0.6875	0.2199	0.0019	0.1636	0.7314	0.9813
sroc2[2]	0.6875	0.2199	0.0019	0.1636	0.7314	0.9813
sroc2[3]	0.986	0.0410	<0.0001	0.9253	0.9942	0.9997
sroc2[4]	0.986	0.0410	<0.0001	0.9253	0.9942	0.9997
sroc2[5]	0.6875	0.2199	0.0019	0.1636	0.7314	0.9813
sroc2[6]	0.926	0.1041	<0.0001	0.6274	0.9591	0.9978
sroc2[7]	0.9581	0.0748	<0.0001	0.7736	0.9794	0.9989
tau1	1.538	0.4736	0.0030	0.735	1.498	2.588
tau2	0.5946	0.2134	0.0015	0.2411	0.5732	1.072

11.5 Meta-Analysis with Study Covariates and One Test

Often, a meta-analysis contains information about the various studies and should be included in the meta-analysis. Of course, patient covariates and individual study information do indeed affect the accuracy of the medical test being assessed, and need to be taken into account when estimating test accuracy. This will be done using the following bilogistic regression model for the meta-analysis of one test:

$$B = \beta[1] + \beta[2]S + \sum_{i=1}^{i=k} \eta_i X_i, \qquad (11.12)$$

where $B = V - U$, $S = V + U$, and X_i are k study covariates, while the parameters are unknown. Thus, for a given study of the analysis, there are

k covariates such as age, the number of readers, the percentage of males to females, and other information. Of course, it depends on the meta-analysis just what information is available and often the information is not available for some studies. The information for a particular meta-analysis will consist of a column for the FPR, the TPR, and a separate column for each study covariate. Once the regression is performed, the SROC ordinates can be computed along with the accuracy parameter Q.

The meta-analysis of Horsthuis et al. [7] provides an excellent example of using covariates to estimate the summary accuracy of ultrasound in order to diagnose inflammatory bowel disease, with the fraction of patients with Crohn's disease and the fraction of males per study serving as covariates. Recall in the previous example with the ultrasound information of Horsthuis et al., the accuracy of ultrasound was 0.9098 as measured by the posterior mean of the Q parameter. Will the accuracy change when study age and fraction of males are included in the analysis?

In order to answer that question, the Bayesian regression (Equation 11.12) is executed with 65,000 observations for the simulation, with a burn in of 5,000 and a refresh of 100. Note, the first list statement of BUGS CODE 11.5 contains the necessary information to execute the Bayesian analysis, where the emphasis will be on assessing the effects of the covariates on B and using Q and the SROC values to assess the accuracy of ultrasound to detect inflammatory bowel disease.

BUGS CODE 11.5

```
# one test with covariates
model;
{
for(i in 1:N){u[i]<-logit(fpr[i])}
for(i in 1:N){v[i]<-logit(tpr[i])}
for(i in 1:N){b[i]<-v[i]-u[i]}
for(i in 1:N){s[i]<-v[i]+u[i]}
# bilogistic regression of b on s
for(i in 1:N){b[i]~dlogis(mu[i],tau)
mu[i]<-beta[1]+beta[2]*s[i]+neta[1]*x1[i]+neta[2]*x2[i]}
for(i in 1:2){beta[i]~dnorm(.0000,.0001)}
for(i in 1:2){neta[i]~dnorm(.0000,.0001)}
tau~dgamma(.0001,.0001)
P<-1+exp(-beta[1]/2)
# accuracy of test Q<-1/P
# sroc curve assumes slope is 0
r1<-exp(-beta[1])
for(i in 1:N){r2[i]<-(1-fpr[i])/fpr[i]}
for(i in 1:N){r3[i]<-1+r1*r2[i]}
for(i in 1:N){sroc[i]<-1/r3[i]}
}
```

data from Horsthuis et al.
x1 is fraction with Crohn disease
x2 is fraction of males
true and false positive ratios are on per patient basis
list(N=9,tpr=c(.78,.935,.90,.812,.931,.884,.87,.96,.92),
        fpr=c(.10,.01,.05,.12,.07,.07,.01,.33,.03),
# the first component of x2 is the average of the remaining 8
        x1=c(.322,.093,.322,1,.406,.643,1,1,.642),
        x2=c(.4088,.097,.457,.571,.423,.465,.366,.571,.321))
# Pakos et al.
list(N=19,
tpr=c(.67,.75,.85,.35,.99,.95,.90,.99,.80,.78,.95,.61,.90,.89,.88,.93,.93,.43,.38),
fpr=c(.15,.26,.11,.17,.22,.33,.33,.01,.33,.6,.43,.04,.17,.01,.10,.70,.01,.01,.57),
# x1 is the average age
# x2 is the percentage of males
# components 2,3,4,14, and 17 of x1were given the age 54, the avg of remaining
# components 2,3,4,14, and 17 of x2 were given the value.70, the avg of
    remaining
x1=c(59,54,54,54,48,57,61,56,66,45,58,48,47,54,60,58,54,48,46),
x2=c(83,70,70,70,81,53,68,67,86,72,82,76,71,70,58,68,70,59,67))
# initial values
list(beta=c(0,0),neta=c(0,0), tau=1))

The surprising result is that the adjusted accuracy of ultrasound is 0.9748 with covariates compared to 0.9098 without covariates, even though the effect of the covariates is somewhat negligible (Table 11.4). For example,

TABLE 11.4: Bayesian analysis for meta-analysis of ultrasound for diagnosis of inflammatory bowel disease.

Parameter	Mean	sd	Error	2 1/2	Median	97 1/2
Q	0.9748	0.0394	0.0015	0.9034	0.9826	0.996
beta[1]	7.978	1.591	0.0827	4.472	8.006	11.04
beta[2]	−0.0813	0.3484	0.0141	−0.8162	−0.0714	0.5889
neta[1]	0.9398	1.443	0.0598	−1.916	0.988	3.858
neta[2]	−8.634	4.293	0.2369	−16.65	−8.773	0.4042
sroc[1]	0.9868	0.0528	0.0021	0.9067	0.9972	0.9999
sroc[2]	0.9262	0.1329	0.0064	0.4691	0.9698	0.9984
sroc[3]	0.9773	0.0749	0.0030	0.8216	0.9941	0.9997
sroc[4]	0.9887	0.0546	0.0019	0.9227	0.9977	0.9999
sroc[5]	0.9825	0.0662	0.0026	0.8682	0.9958	0.9998
sroc[6]	0.9825	0.0662	0.0026	0.8682	0.9958	0.9998
sroc[7]	0.9262	0.1329	0.0064	0.4691	0.9698	0.9984
sroc[8]	0.9954	0.0384	0.0012	0.9773	0.9994	1.0
sroc[9]	0.9665	0.0903	0.0039	0.7302	0.9899	0.9995
tau	2.167	0.7886	0.0190	0.8561	2.089	3.919

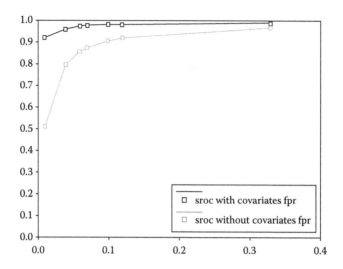

FIGURE 11.5: Effect of covariates on SROC.

the 95% credible interval for beta[1], which is the effect of the proportion of patients with ulcerated colitis, is $(-1.916, 3.858)$, which implies a small effect of that covariate on the average value of B. Also, it should be noted that the 95% credible interval for beta[2] contains zero, thus the ordinates of the points on the SROC curve are computed assuming beta[2]=0. See Equation 11.5.

The effect of the covariates on the SROC curve is evident from Figure 11.5, because the SROC curve corresponding to the inclusion of the two covariates dominates the SROC curve when the covariates are not included over the range where $0 < \text{FPR} < 0.35$.

11.6   Meta-Analysis with Covariates for Several Tests

A Bayesian analysis provides an excellent approach to comparing two medical tests with covariates when the information is provided by a meta-analysis, and Horsthuis et al. [7] paper presents such an example. It consists of two modalities, ultrasound and MRI, to diagnose inflammatory bowel disease, where the two covariates are the fraction of patients with Crohn's disease and the fraction of male patients. The disease presents as Crohn's disease or as ulcerative colitis, while the disease attacks both genders with about the same frequency. The analysis is based on BUGS CODE 11.6.

BUGS CODE 11.6

```
# two tests with covariates
model;
{
for(i in 1:N1){u1[i]<-logit(fpr1[i])}
for(i in 1:N1){v1[i]<-logit(tpr1[i])}
for(i in 1:N1){b1[i]<-v1[i]-u1[i]}
for(i in 1:N1){s1[i]<-v1[i]+u1[i]}
# bilogistic regression of b on s for test 1
for(i in 1:N1){b1[i]~dlogis(mu1[i],tau1)
mu1[i]<-beta1[1]+beta1[2]*s1[i]+neta1[1]*x11[i]+neta1[2]*x12[i]}
for(i in 1:2){beta1[i]~dnorm(.0000,.0001)}
for(i in 1:2){neta1[i]~dnorm(.0000,.0001)}
tau1~dgamma(.0001,.0001)
P1<-1+exp(-beta1[1]/2)
# accuracy of test 1
Q1<-1/P1
# sroc curve assumes slope is 0
# sroc values for test 1
r11<-exp(-beta1[1])
for(i in 1:N1){r12[i]<-(1-fpr1[i])/fpr1[i]}
for(i in 1:N1){r13[i]<-1+r11*r12[i]}
for(i in 1:N1){sroc1[i]<-1/r13[i]}
# for test 2
for(i in 1:N2){u2[i]<-logit(fpr2[i])}
for(i in 1:N2){v2[i]<-logit(tpr2[i])}
for(i in 1:N2){b2[i]<-v2[i]-u2[i]}
for(i in 1:N2){s2[i]<-v2[i]+u2[i]}
# bilogistic regression of b2 on s2 for test 2
for(i in 1:N2){b2[i]~dlogis(mu2[i],tau2)
mu2[i]<-beta2[1]+beta2[2]*s2[i]+neta2[1]*x21[i]+neta2[2]*x22[i]}
for(i in 1:2){beta2[i]~dnorm(.0000,.0001)}
for(i in 1:2){neta2[i]~dnorm(.0000,.0001)}
tau2~dgamma(.0001,.0001)
P2<-1+exp(-beta2[1]/2)
# accuracy of test 2
Q2<-1/P2
# sroc curve assumes slope is 0
r21<-exp(-beta2[1])
for(i in 1:N2){r22[i]<-(1-fpr2[i])/fpr2[i]}
for(i in 1:N2){r23[i]<-1+r21*r22[i]}
for(i in 1:N2){sroc2[i]<-1/r23[i]}
# difference of accuracy of two tests
d<-Q1-Q2
}
```

```
# data from Horsthuis et al. [7] for US and MRI
# US is test 1 and MRI is test 2
# x11 is fraction with Chron's disease
# x12 is fraction of males
# tpr and fpr are on a per patient basis
list(N1=9,tpr1=c(.78,.935,.90,.812,.931,.884,.87,.96,.92),
          fpr1=c(.10,.01,.05,.12,.07,.07,.01,.33,.03),
# the first component of x12 is the average of the remaining 8
          x11=c(.322,.093,.322,1,.406,.643,1,1,.642),
          x12=c(.4088,.097,.457,.571,.423,.465,.366,.571,.321),
# the following is for test 2 or MRI
          N2=7,
          x21=c(.6,.54,1,1,.34,.36,1),
# below.49 is the average of other six components of x22
          x22=c(.6,.42,.46,.36,.49,.56,.52),
# tpr and fpr are on a per patient basis
          tpr2=c(.99,.99,.913,.87,.818,.956,.889),
          fpr2=c(.18,.01,.29,.29,.01,.08,.143))
# initial values
list( beta1=c(0,0),neta1=c(0,0), tau1=1,beta2=c(0,0),neta2=c(0,0), tau2=1))
```

The analysis consists of two bilogistic regressions corresponding to the two tests.

$$B_1 = \beta_1[1] + \beta_1[2]S_1 + \sum_{i=1}^{i=k} \eta_{1i}X_{1i} \tag{11.13}$$

for the first test and

$$B_2 = \beta_2[1] + \beta_2[2]S_2 + \sum_{i=1}^{i=k} \eta_{2i}X_{2i} \tag{11.14}$$

for the second, where the same k covariates apply to both tests. B and S values are defined the usual way in terms of U and V values, which are the logits of the false and true positive fractions, respectively. See Equations 11.1 through 11.3 for definitions of U, V, B, and S values used in the above regression.

Once the regressions are performed, the posterior distribution of the regression parameters induce a posterior distribution for the accuracy parameters, namely,

$$Q_1 = (1 + e^{-\beta_1[1]/2})^{-1} \tag{11.15}$$

for the first test and

$$Q_2 = (1 + e^{-\beta_2[1]/2})^{-1} \tag{11.16}$$

for the second test.

For the last part of the analysis, the posterior distribution of the SROC values is generated, which allows one to compare the SROC curves, and hence to compare graphically the accuracy of the two tests. Equation 11.11 should be modified in order to compute the SROC values for the two tests (Table 11.5).

The analysis is executed with 325,000 observations, with a burn in of 5,000 and a refresh of 100. Comparing US and MRI, based on the posterior means of $Q1$ and $Q2$, indicates that US has more accuracy, but this comparison should be made with caution because the posterior distribution of $Q2$ is highly skewed with a median for $Q2$ of 0.998 and a posterior mean of 0.866, thus, if the comparison is made with the means there appears to be a difference, but if based on the medians there appears to be no difference. From Figure 11.6, the two SROC curves are plotted for the two tests, implying that the US test has more accuracy than MRI.

TABLE 11.5: Bayesian analysis for comparing US and MRI for inflammatory bowel disease.

Parameter	Mean	sd	Error	2 1/2	Median	97 1/2
$Q1$	0.9774	0.0335	<0.0001	0.9318	0.9828	0.9962
$Q2$	0.866	0.2919	0.0109	0.00021	0.998	1
beta1[1]	8.106	1.53	0.0442	5.231	8.093	11.13
beta1[2]	−0.058	0.3414	0.0075	−0.7215	−0.0595	0.6222
beta2[1]	12.15	12.58	0.4865	−16.93	12.45	35.1
beta2[2]	0.2252	0.8867	0.0199	−1.656	0.2409	1.931
d	0.1114	0.2933	0.0109	−0.05664	−0.01007	0.9798
neta1[1]	1.195	1.43	0.0313	−1.568	1.172	4.075
neta1[2]	−9.288	4.252	0.1304	−17.69	−9.193	−1.522
neta2[1]	−6.287	5.884	0.1805	−17.61	−6.324	6.22
neta2[2]	−5.574	20.13	0.7688	−42.63	−6.059	40.85
sroc1[1]	0.9903	0.0489	0.00125	0.9541	0.9973	0.9999
sroc1[2]	0.9392	0.1068	0.0028	0.6538	0.9706	0.9985
sroc1[3]	0.9828	0.0609	0.0015	0.9077	0.9942	0.9997
sroc1[4]	0.9916	0.0462	0.0011	0.9623	0.9978	0.9999
sroc1[5]	0.9869	0.0547	0.0014	0.9336	0.996	0.9998
sroc1[6]	0.9869	0.0547	0.0014	0.9336	0.996	0.9998
sroc1[7]	0.9392	0.1068	0.0028	0.6538	0.9706	0.9985
sroc1[8]	0.9966	0.0338	0.000821	0.9893	0.9994	1
sroc1[9]	0.9742	0.0723	0.0019	0.8525	0.9902	0.9995
sroc2[1]	0.852	0.3289	0.01212	0	1	1
sroc2[2]	0.7841	0.377	0.0140	0	0.9996	1
sroc2[3]	0.8627	0.3192	0.0117	0	1	1
sroc2[4]	0.8627	0.3192	0.0117	0	1	1
sroc2[5]	0.7841	0.377	0.0140	0	0.9996	1
sroc2[6]	0.8344	0.3436	0.0127	0	1	1
sroc2[7]	0.847	0.3332	0.0123	0	1	1
tau1	2.174	0.7841	0.0056	0.8769	2.094	3.93
tau2	0.8198	0.3824	0.0066	0.2321	0.7694	1.699

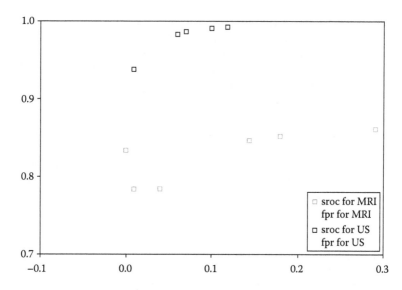

FIGURE 11.6: SROC curves for US and MRI.

Based on the 95% credible interval, the slope of the two regressions is zero, that is, the effect of the S values on the B value is negligible, but note that the intercept for the regression of US is 8.106, which implies that the odds of a positive test with US for the diseased population is 8 times the odds of a positive US for those without inflammatory disease, whereas for MRI, the intercept posterior mean is 12.15 implying the odds of a positive test with MRI for the diseased group is about 12 times the odds of a positive test for the non-diseased group. Also, based on the 95% credible interval, for all values of the threshold both tests discriminate equally between those with inflammatory bowel disease and those without.

Do the covariates have any effect on the accuracy of the two tests? The 95% credible interval for netal[2] is $(-17.69, -1.522)$ which implies that the percentage of male patients has an effect on the B score for US, however, additional analysis is needed in order to determine if the effect is non negligible by calculating the Q value for US when covariates are not used in the regression.

One note of caution in using regression for the meta-analysis to assess the accuracy is that the relatively small sample size (the number of studies) might not produce a model with a good fit to the data, which in turn implies that one may not have high confidence in one's assessment of the accuracy of the medical test.

It is interesting to observe the posterior density of $Q2$ (see Figure 11.7).

Note, the "small" jump in a small neighborhood of 0 and the "large" jump in a small neighborhood of 1!

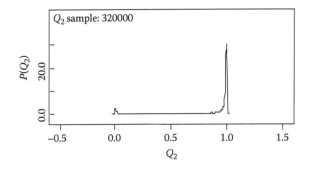

FIGURE 11.7: Posterior density of Q_2: the accuracy of MRI.

11.7   Other Meta-Analyses

Up to this point, the emphasis has been on summarizing the accuracy of various medical tests, but now the focus will be on estimating the complications of various medical tests. As is well known, many tests are accompanied by various complications: (a) coronary angiography can result in stroke and damage to the coronary arteries, and (b) certain contrast media in diagnostic imaging can result in damage to the kidneys. It is for the latter scenario that a Bayesian approach will be taken and is based on the meta-analysis of Heinrich et al. [9], who compare the nephrotoxity of iso-osmolar iodixanol with non-ionic, low-osmolar contrast media.

The meta-analysis included a thorough search of MEDLINE, EMBASE, and BIOSIS databases, trial registries, conference proceedings, and requests from companies. Also, randomized clinical trials assessing the serum creatinine levels before and after the intravascular administration of iodixanol (LOCM) were included. The main endpoint was the incidence of contrast media-induced nephropathy (kidney disease), which is measured by a change in serum creatinine values. Their main conclusion was based on 25 trials and on conventional statistical procedures, which showed that iodixanol is not associated with a significantly reduced risk of contrast-induced nephropathy.

Our approach is Bayesian, where the main endpoint is the incidence of kidney complications (defined as a certain percent increase in serum creatinine, which is measured before and after the administration of the procedure). There are two groups, one with contrast media, labeled LOCM, while the other group of patients did not have contrast medium administered, and is referred to as the iodixanol group. Some of the trials are randomized, and there are various types of contrast media, as well as many other study covariates, and are of interest in how they impact the complication rate.

Suppose the analysis is based on the logistic model, where for the control group (patients not receiving the contrast media)

$$X_i \sim \text{binomial}(n_i, p_i) \tag{11.17}$$

for $i = 1, 2, \ldots, n$, and for the patients receiving the contrast agent

$$Y_i \sim \text{binomial}(m_i, q_i) \tag{11.18}$$

with $i = 1, 2, \ldots, m$.

In addition, let

$$\log(p_i/(1 - p_i)) = \theta_i \tag{11.19}$$

for $i = 1, 2, \ldots, n$, and

$$\log(q_i/(1 - q_i)) = \phi_i + \theta_i \tag{11.20}$$

with $i = 1, 2, \ldots, m$.

Prior information for the parameters is specified as

$$\theta_i \sim \text{nid}(\mu_\theta, \tau_\theta) \tag{11.21}$$

and

$$\phi_i \sim \text{nid}(\mu_\phi, \tau_\phi). \tag{11.22}$$

At this stage there are several choices: one could let θ_i and ϕ_i have non-informative normal distributions (0.0000,0.00001), or let μ_θ and μ_ϕ have normal distributions (informative or non informative) and the precisions τ_θ and τ_ϕ have gamma distributions (informative or non informative). It depends on the amount of prior information available to the analyst. For additional information about the model, see Berry [10]. The first part of the analysis will be based on the model defined by Equations 11.17 through 11.22, then the analysis will be expanded to include the study covariates.

Our analysis is executed with BUGS CODE 11.7 with 65,000 observations for the simulation, a burn in of 5,000 and a refresh of 100.

BUGS CODE 11.7

```
model;
{
# binomial distributions for two tests
for(i in 1:17){x[i]~dbin(p[i],m[i])}
for(i in 1:17){y[i]~dbin(q[i],n[i])}
# logit models for the two complication rates
for(i in 1:17){logit(p[i])<-theta[i]}
```

```
for(i in 1:17){logit(q[i])<-phi[i]}
# prior distributions
for(i in 1:17){theta[i]~dnorm(muth,tauth)}
for(i in 1:17){phi[i]~dnorm(muphi,tauphi)}
muth~dnorm(0.000,.0001)
muphi~dnorm(0.000,.0001)
tauth~dgamma(.00001,.00001)
tauphi~dgamma(.00001,.00001)
# mean of complication rates
pee<-mean(p[])
qee<-mean(q[])
d<-pee-qee
}
# Heinrich et al. meta analysis
list( m=c(123,72,54,210,58,105,76,134,35,25,64,54,32,100,20,59,101),
n=c(125,76,48,204,56,116,77,125,35,25,65,49,32,50,19,60,99),
x=c(6,5,1,14,0,4,2,12,2,4,2,5,1,7,1,0,0),
y=c(7,0,5,9,1,1,0,17,9,4,17,14,0,3,0,0,2))
# initial values
# must initiate other chain with specification tool
list(muth=0,muphi=0,tauth=1,tauphi=1)
```

The above code is labeled with the appropriate identification for the Bayesian analysis, such as pee and qee are the means of the complication rates for the two groups of patients, and the data for the Heinrich et al. [9] meta-analysis, where the components of the x vector are the number of complications for the control or iodixanol patients, and the y vector contains the number of complications for the treatment or the patients receiving the contrast media. In addition, the m vector is the number of patients in the control group and the n vector contains the number of patients in the treatment group. Note that only 17 of the 22 studies are used for the analysis because they have complete information on the complication rates. Table 11.6 portrays the Bayesian analysis for the Heinrich et al. meta-analysis.

TABLE 11.6: Bayesian analysis for the safety of contrast media study.

Parameter	Mean	sd	Error	2 1/2	Median	97 1/2
d	−0.0337	0.01098	<0.0001	−0.0556	−0.0336	−0.0125
muphi μ_ϕ	−3.319	0.547	0.0033	−4.532	−3.274	−2.365
muth μ_θ	−3.085	0.2284	0.0050	−3.617	−3.057	−2.722
pee	0.0489	0.0063	<0.0001	0.0374	0.0486	0.0622
qee	0.0826	0.0089	<0.00001	0.0660	0.0823	0.1009
tauphi τ_ϕ	0.3825	0.2161	0.0017	0.1003	0.3378	0.9219
tauth τ_θ	2596	14400	429	0.6974	5.058	28040

Key parameters are d, the difference in the two averages of the complication rates, and muphi, which is the difference in the logits of the complication rates, and both imply there is a difference in the complication rates. The average complication rate for the group not receiving the contrast agent is 0.0489, compared to 0.0826 for the group receiving the contrast agent, and their difference d has a 95% credible interval of $(-0.0556, -0.0125)$, while the 95% credible interval for muphi is $(-4.532, -2.365)$. Note the large posterior mean for τ_θ, however, the median is 5.058, and should be used as the estimate of the precision parameter.

This example is generalized to include study covariates in the exercises.

11.8   Comments and Conclusions

This chapter has introduced the reader to the Bayesian approach to meta-analysis with emphasis on the SROC curve, that is where one assumes that the separate studies have a common ROC curve. The Bayesian approach is based on a regression analysis and consists of computing the posterior distribution of the SROC values and the accuracy Q of the combined studies. If covariates are involved, the posterior distribution of the relevant regression coefficients tells one if they contribute to the analysis.

The chapter begins by considering a meta-analysis involving only one test and, via an MCMC simulation, determines the posterior distribution of the relevant parameters. A regression of B on S is the basis of the analysis, where the intercept and slope give us valuable information about the worth of the test in differentiating between diseased and non-diseased patients; $B = V - U$ and $S = V + U$, where U and V are the logits of the FPF and TPF, respectively. It can be shown that the intercept is the odds ratio of a positive test result for those with and without the disease (when $S = 0$), and that the value of the slope is very informative about the accuracy. For example, if the slope is close to zero, the implication is that the test's ability to discriminate between the two groups of patients is independent of the threshold values used in the separate studies. Once the posterior distribution of the slope and intercept are determined, their joint posterior distribution induces a posterior distribution for the SROC values and the accuracy parameter Q.

Various interesting examples exemplify the Bayesian analysis and are taken from the medical research literature. For example, the Horsthuis et al. [7] study is especially interesting because it is well documented and involves several tests, namely, US, MR, scintigraphy, and CT, to diagnose inflammatory bowel disease, and is one of the few meta-analyses to study several modalities simultaneously. Another positive feature of the Horsthuis et al. analysis is the inclusion of many covariates to assess the inter study variation.

The example is examined in three stages: (a) US was only considered without any covariates; (b) US and MR were compared, without using any covariates; and (c) US was used along with two covariates, the fraction of patients with Crohn's disease and the fraction of male patients. Thus, this example allows one to display the full complement of Bayesian methods in order to conduct the analysis.

As mentioned earlier, the foundation of the Bayesian approach is the bilogistic regression model of B on S, however, Zhou, Obuchowski, and McClish [3: 403] describe another method based on a binary regression model. This was not considered, but instead the bilogistic regression model was employed. Also, very little has appeared in the literature concerning the meta-analysis of tests with continuous scores from a Bayesian viewpoint, however, Zhou, Obuchowski, and McClish [3: 409] do present such an approach, which is based on a weighted average of estimated ROC areas of the various studies. Note, it would be very difficult to do a Bayesian analysis, because most, if not all, such meta-analyses are not done in a Bayesian fashion, but instead are done with non-Bayesian methodology, that is, each individual ROC area is estimated by conventional methods along with an associated estimated standard error (standard deviation).

The approach presented here follows Miller et al. [13], but other approaches can be found in Harbord et al. [14], Leeflang et al. [15], Paul et al. [16]; and Rutter and Gatsonins [17]. Lastly, the chapter concludes with many exercises that expand on the methods introduced earlier in the chapter, which gives the student additional information that should prove advantageous in a consulting environment.

11.9 Exercises

1. Confirm Table 11.1, the Bayesian analysis for the meta-analysis of the DeVries, Hunink, and Polak [5] study for the diagnosis of peripheral artery stenosis using regular duplex ultrasound. Based on BUGS CODE 11.1, generate 65,000 observations for the MCMC simulation, with a burn in of 5,000 and a refresh of 100.
   (a) What is the posterior mean and 95% credible interval for the accuracy parameter Q? Does it indicate good accuracy for ultrasound?
   (b) Do you agree that the slope of the regression of B on S can be considered zero?
   (c) Explain how the SROC values are computed. What are the false positive ratios corresponding to the nine SROC values.
   (d) Display the posterior density of Q. Is the distribution skewed?
   (e) Using the MCMC errors as a guide, are 65,000 observations sufficient for the simulation?

2. Consider the meta-analysis of Meijer et al. [11], which consists of 22 studies for the diagnostic performance of thin slice CT coronary angiography to detect coronary stenosis of at least 50% occlusion of the artery.

The data appear as a list statement in BUGS CODE 11.1 and a Bayesian analysis is performed with 65,000 observations, a burn in of 5,000 and a refresh of 100 with the following results. Note that this is a very involved meta-analysis and the reader is encouraged to carefully read the paper. The purpose here is to focus on the Bayesian analysis and not critique the meta-analysis. The authors do a conventional analysis, which should be compared to the results of Table 11.7.

The plot of the posterior density of the slope parameter beta[2] appears in Figure 11.8 and shows that the posterior probability is concentrated in a "small" neighborhood of zero, which is confirmed by a 95% credible interval of $(-0.33, 0.19)$, thus the SROC values are computed assuming beta[2]=0.

(a) Do you agree that one should let beta[2] $= 0$ for computing the SROC values?

(b) Plot the posterior density of Q.

(c) Does the posterior density of Q imply that CT angiography has good accuracy?

TABLE 11.7:   Posterior analysis for the meta-analysis of Meijer et al. [11].

Parameter	Mean	sd	Error	2 1/2	Median	97 1/2
Q	0.9526	0.0059	<0.0001	0.9397	0.953	0.9628
beta[1]	6.016	0.2573	0.0059	5.492	6.021	6.51
beta[2]	−0.0676	0.1334	0.0030	−0.3304	−0.0675	0.1938
sroc[1]	0.9664	0.0087	<0.0001	0.9417	0.9648	0.9781
sroc[2]	0.9842	0.0042	<0.00001	0.9461	0.9675	0.9798
sroc[3]	0.9842	0.0042	<0.00001	0.9744	0.9848	0.9906
sroc[4]	0.9858	0.0038	<0.00001	0.9771	0.9863	0.9916
sroc[5]	0.9635	0.0094	<0.0001	0.9417	0.9648	0.9781
sroc[6]	0.9887	0.0030	<0.00001	0.9816	0.9891	0.9933
sroc[7]	0.9804	0.0052	<0.0001	0.9397	0.953	0.9628
sroc[8]	0.9741	0.0068	<0.0001	0.9583	0.975	0.9845
sroc[9]	0.9936	0.0017	<0.00001	0.9895	0.9938	0.9962
sroc[10]	0.9408	0.0149	<0.0001	0.9065	0.9427	0.9641
sroc[11]	0.9913	0.0023	<0.00001	0.9859	0.9916	0.9949
sroc[12]	0.9804	0.0052	<0.0001	0.9397	0.953	0.9628
sroc[13]	0.9858	0.0038	<0.00001	0.9771	0.9863	0.9916
sroc[14]	0.9773	0.0060	<0.0001	0.9634	0.9781	0.9864
sroc[15]	0.9812	0.0050	<0.0001	0.9697	0.9819	0.9888
sroc[16]	0.9778	0.0058	<0.0001	0.9642	0.9786	0.9868
sroc[17]	0.9453	0.0138	<0.0001	0.9134	0.947	0.9668
sroc[18]	0.9924	0.0020	<0.00001	0.9877	0.9927	0.9955
sroc[19]	0.9744	0.0067	<0.0001	0.9588	0.9753	0.9847
sroc[20]	0.9861	0.0037	<0.00001	0.9775	0.9866	0.9918
tau	3.011	0.5821	0.0037	1.977	2.98	4.238

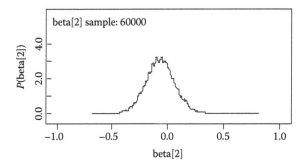

FIGURE 11.8: Posterior density of the slope parameter beta[2].

(d) Do the SROC values indicate good accuracy for CT angiography?

(e) Confirm that the plot of the SROC values vs. the corresponding FPR values appear as in Figure 11.9.

(f) Does the plot of the SROC values confirm a Q value with a posterior mean of 0.9526?

(g) Are any of the posterior distributions of the parameters skewed? If so, identify the parameter.

(h) Interpret the posterior mean of the intercept beta[1].

3. This is another example of a meta-analysis with no covariates where only the TPR and FPR are reported, and is based on the research of Huebner et al. [12], who determined the summary accuracy of whole-body FDG-PET for detecting recurrent colorectal cancer. It consists of 10 studies that vary from 18 to 130 patients. The FPR and TPR appear as a list statement in BUGS CODE 11.8 and the analysis is executed

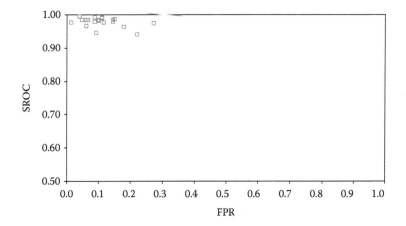

FIGURE 11.9: SROC curve of the Meijer et al. [11] meta-analysis.

TABLE 11.8:   Bayesian analysis of meta-analysis for FDG-PET of recurrent colorectal cancer.

Parameter	Mean	sd	Error	2 1/2	Median	97 1/2
Q	0.9581	0.0078	<0.00001	0.9396	0.959	0.971
beta[1]	6.291	0.3801	0.003794	5.489	6.30	7.024
beta[2]	−0.3434	0.1435	0.0014	−0.642	−0.3415	−0.0596
sroc[1]	0.94	0.023	<0.0001	0.8821	0.9441	0.972
sroc[2]	0.8386	0.0526	<0.0001	0.7097	0.8465	0.919
sroc[3]	0.996	0.0017	<0.0001	0.9917	0.9963	0.9982
sroc[4]	0.9904	0.0040	<0.0001	0.9802	0.9911	0.9957
sroc[5]	0.9921	0.0033	<0.0001	0.9837	0.9927	0.9965
sroc[6]	0.8386	0.0526	<0.0001	0.7097	0.8465	0.919
sroc[7]	0.9956	0.0018	<0.0001	0.9909	0.9959	0.998
sroc[8]	0.9951	0.0020	<0.0001	0.99	0.9955	0.997
sroc[9]	0.9926	0.0031	<0.0001	0.9847	0.9932	0.996
sroc[10]	0.8386	0.0526	<0.0001	0.7097	0.8465	0.919
tau	1.773	0.5353	<0.0001	0.8855	1.719	2.965

with 65,000 observations for the simulation, with a burn in of 5,000 and refresh of 100. See Table 11.8 for the results of the posterior distribution for the parameters. Note, this code uses an analysis that does not assume the slope beta[2] = 0 and the reader is referred to Equation 11.5.

BUGS CODE 11.8

```
# one test and assumes beta[2] is not zero
model;
{
for(i in 1:N){u[i]<-logit(fpr[i])}
for(i in 1:N){v[i]<-logit(tpr[i])}
for(i in 1:N){b[i]<-v[i]-u[i]}
for(i in 1:N){s[i]<-v[i]+u[i]}
# bilogistic regression of b on s
for(i in 1:N){b[i]~dlogis(mu[i],tau)
mu[i]<-beta[1]+beta[2]*s[i]}
for(i in 1:2){beta[i]~dnorm(.0000,.0001)}
tau~dgamma(.0001,.0001)
P<-1+exp(-beta[1]/2)
# accuracy of test
Q<-1/P
# sroc curve, does not assume slope is zero
r1<-exp(-beta[1]/(1-beta[2]))
for(i in 1:N){r2[i]<-(1-fpr[i])/fpr[i]}
r3<-(1+beta[2])/(1-beta[2])
```

```
for(i in 1:N){r4[i]<-pow(r2[i],r3)}
for(i in 1:N){r5[i]<-1+r1*r4[i]}
for(i in 1:N){sroc[i]<-1/r5[i]}
}
# data from Huebner et al.
list(N=10, tpr=c(.96,.90,.99,.98,.92,.96,.99,.99,.95,.94),
    fpr=c(.03,.01,.33,.17,.20,.01,.31,.29,.21,.01))
# initial values
list( beta=c(0,0), tau=1))
```

The 95% credible interval for beta[2] is $(-0.642, -0.0596)$, which does not contain zero, thus I used Equation 11.5 in the above code to compute the SROC curve.

(a) Is FDG-PET an accurate test for detecting recurrent colorectal cancer?

(b) Are 65,000 observations sufficient for the simulation? Explain your answer.

(c) Would you use Equation 11.5 to compute the SROC values? Explain your answer.

(d) Display the posterior density of beta[2].

(e) The intercept has a posterior mean of 6.291. What does this imply about the odds ratio of a positive test for those diseased vs. a positive test for those without recurrent colorectal cancer?

(f) Plot the posterior means of the SROC values vs. the FPR values appearing in the list statement in the above code.

4. Verify Table 11.3, the Bayesian analysis for the Horsthuis et al. [7] meta-analysis for the diagnosis of inflammatory bowel disease using ultrasound, MRI, scintigraphy, and CT. Use BUGS CODE 11.4 with 65,000 observations for the MCMC simulation, with a burn in of 5,000 and a refresh of 100.

(a) Is ultrasound a good test for the diagnosis of inflammatory bowel disease? Explain your answer.

(b) Is the slope beta[2] $= 0$? Explain your answer.

(c) Display the posterior density of beta[2].

(d) What is your interpretation of the intercept of the regression of B on S? What does the value imply about the detection of disease with US?

(e) If the simulation is increased to 120,000 observations, describe the effect on the MCMC error.

5. Verify Table 11.4, the posterior analysis for the Horsthuis et al. [7] meta-analysis with covariates for the diagnosis of inflammatory bowel disease. There are two covariates, namely, the percentage of patients with Crohn's disease and the percentage of male patients on a per study basis. Execute the analysis with BUGS CODE 11.5 with 65,000 observations for the simulation, with a burn in of 5,000 and a refresh of 100.

    (a) Are the covariates needed for the analysis?

    (b) Display the posterior densities of beta[1] and beta[2].

    (c) What is the 95% credible interval for beta[2]?

    (d) What is the overall accuracy of ultrasound based on Q for the meta-analysis?

    (e) Using Q, compare the accuracy of US with covariates to that of not using covariates. That is, compare the results of Table 11.4 to Table 11.3.

    (f) What is your overall conclusion about using covariates for the diagnosis of inflammatory bowel disease with ultrasound?

6. Using BUGS CODE 11.5, execute a Bayesian analysis for the Pakos et al. [8] meta-analysis, which summarizes the accuracy of antigranulocyte scintigraphy with ^{99m}Tc radiolabeled monoclonal antibodies for the diagnosis of osteomyelitis. Two covariates are used, namely, the average age and the percentage of males on a per study basis. Note, a list statement of BUGS CODE 11.5 contains the relevant information for the Pakos et al. [8] study. Using 65,000 observations for the simulation, with a refresh of 100 and a burn in of 5,000, I got the results shown in Table 11.9. Confirm the results of the table.

    (a) Note the highly skewed distribution for Q with a posterior mean of 0.6112 and posterior median of 0.799. Because of the skewness, I prefer the median and assign an accuracy of 0.799. Do you agree?

    (b) The posterior distributions of the SROC values are also highly skewed, thus, I prefer the medians. If you plot the posterior medians vs. the corresponding false positive ratios, you arrive at Figure 11.10. Verify Figure 11.10.

    (c) From the plot of the SROC curve, I believe that the accuracy corresponds to approximately the posterior median of Q. Do you agree?

    (d) An accuracy of 0.799 is respectable, thus I affirm that the nuclear medicine procedure of using scintigraphy to diagnose osteomyelitis is useful in medical practice. Do you agree?

    (e) Display the posterior density of Q.

    (f) Are the two covariates useful for estimating accuracy?

    (g) Compare the accuracy using covariates to that of not using covariates. Use BUGS CODE 11.1 with the Pakos et al. [8] data for the Bayesian analysis without covariates (Table 11.9) or revise BUGS CODE 11.5, either way it is your choice!

7. Verify Table 11.5 by executing the analysis with BUGS CODE 11.6 for comparing ultrasound with MRI for the diagnosis of inflammatory bowel disease. There are nine studies with US and seven with MRI and the list statement of the code gives the necessary information for the analysis, including the two covariates, which are the percentage of patients with Crohn's disease and the percentage of males in each study.

TABLE 11.9:   Bayesian analysis for the Pakos et al. [8] meta-analysis.

Parameter	Mean	sd	Error	2 1/2	Median	97 1/2
Q	0.6112	0.4034	0.0253	0.000167	0.799	0.999
beta[1]	2.559	9.528	0.6029	-17.39	2.76	18.17
beta[2]	-0.2215	0.3121	0.0083	-0.84	$-.2196$	0.4051
neta[1]	0.0270	0.1435	0.0090	-0.2604	0.0404	0.3229
neta[2]	-0.0078	0.0801	0.004987	-0.1666	-0.01264	0.1708
sroc[1]	0.5453	0.4546	0.0283	$<10^{-9}$	0.7359	1
sroc[2]	0.5738	0.4504	0.0821	10^{-9}	0.8473	1
sroc[3]	0.5306	0.4562	0.0284	10^{-9}	0.6612	1
sroc[4]	0.5514	0.4538	0.02833	10^{-9}	0.7639	1
sroc[5]	0.5647	0.4519	0.0282	10^{-9}	0.8167	1
sroc[6]	0.5879	0.448	0.0279	10^{-9}	0.8861	1
sroc[7]	0.5879	0.448	0.0279	10^{-9}	0.8861	1
sroc[8]	0.4301	0.4531	0.0279	10^{-9}	0.1376	1
sroc[9]	0.5879	0.448	0.0279	10^{-9}	0.8861	1
sroc[10]	0.6336	0.438	0.0273	10^{-9}	0.9595	1
sroc[11]	0.6056	0.4446	0.0277	10^{-9}	0.9226	1
sroc[12]	0.4867	0.458	0.0286	10^{-9}	0.3969	1
sroc[13]	0.5514	0.4538	0.0283	10^{-9}	0.7639	1
sroc[14]	0.4301	0.4531	0.0283	10^{-9}	0.1376	1
sroc[15]	0.5263	0.4566	0.0285	10^{-9}	0.637	1
sroc[16]	0.6512	0.4332	0.0270	10^{-9}	0.9736	1
sroc[17]	0.4301	0.4531	0.0283	10^{-9}	0.1376	1
sroc[18]	0.4301	0.4531	0.0283	10^{-9}	0.1376	1
sroc[19]	0.6286	0.4393	0.0274	10^{-9}	0.9544	1
tau	0.6768	0.1493	0.0037	0.4135	0.6672	0.9975

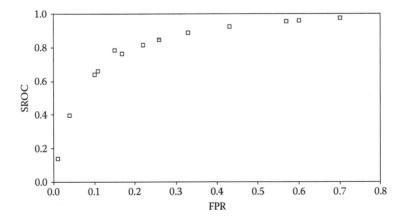

FIGURE 11.10: SROC curve of the Pakos et al. [8] meta-analysis.

I used 325,000 observations for the simulation, with a burn in of 5,000 and refresh of 100.

(a) Note the highly skewed distribution for $Q2$ for the accuracy of MRI and beta1[1], the intercept for the regression of the first test. I prefer the posterior median of 0.9982 for the accuracy of MRI. Do you agree? Explain your answer.

(b) Are the covariates necessary in order to estimate the accuracy of the two tests?

(c) Is there a difference in the accuracy of MRI and US for diagnosing inflammatory bowel disease?

(d) Plot the posterior density of $Q2$.

(e) Based on Figure 11.6, is there a difference in the two modalities? Explain your answer.

8. The meta-analysis of Heinrich et al. [9] is continued by including the covariates, average age per study, and percentage of women per study, and the student is referred to the paper for additional information about this important contribution to contrast safety. Recall that such contrast media as used in the Heinrich et al. analysis is common for coronary angiography, but its use runs the risk of kidney damage. The two logistic models (Equations 11.19 and 11.20) are easily generalized to

$$\log(p_i/(1-p_i)) = \theta_i + \eta_1[1]x_1 + \eta_1[2]x_2 \qquad (11.23)$$

and

$$\log(q_i/(1-q_i)) = \phi_i + \theta_i + \eta_2[1]y_1 + \eta_2[2]y_2, \qquad (11.24)$$

and the code to execute the analysis with covariates follows. One must specify the prior densities of the parameters in the above model in the obvious way and one should refer to BUGS CODE 11.9 for the definition of those priors.

BUGS CODE 11.9

```
model;
{
# binomial distributions for the two tests
# with covariates
for(i in 1:17){x[i]~dbin(p[i],m[i])}
for(i in 1:17){y[i]~dbin(q[i],n[i])}
# logit models for the two complication rates
for(i in 1:17){logit(p[i])<-theta[i]+neta1[1]*x1[i]+neta1[2]*x2[i]}
for(i in 1:17){logit(q[i])<-theta[i]+phi[i]+neta2[1]*y1[i]+neta2[2]*y2[i]}
# prior distributions
neta1[1]~dnorm(0.000,.0001)
```

```
neta1[2]~dnorm(0.000,.0001)
neta2[1]~dnorm(0.000,.0001)
neta2[2]~dnorm(0.000,.0001)
for(i in 1:17){theta[i]~dnorm(muth,tauth)}
for(i in 1:17){phi[i]~dnorm(muphi,tauphi)}
muth~dnorm(0.000,.0001)
muphi~dnorm(0.000,.0001)
tauth~dgamma(.00001,.00001)
tauphi~dgamma(.00001,.00001)
# mean of complication rates
pee<-mean(p[])
qee<-mean(q[])
d<-pee-qee
}
# heinrich et al. meta analysis
list(m=c(123,72,54,210,58,105,76,134,35,25,64,54,32,100,20,59,101),
n=c(125,76,48,204,56,116,77,125,35,25,65,49,32,50,19,60,99),
x=c(6,5,1,14,0,4,2,12,2,4,2,5,1,7,1,0,0),
y=c(7,0,5,9,1,1,0,17,9,4,17,14,0,3,0,0,2),
# x1 is the age for control group
x1=c(68.3,65.4,65,70.5,60.6,60.6,67,71,70,73.8,71.1,62,67,55,61.9,62.4,61),
# y1 is age for the treatment group
y1=c(69.5,67.1,66,72.4,61.1,62.1,67.3,72.8,72,72,70.6,65,69,52,60.2,63,59),
# x2 is the % women for control group
x2=c(50,36,52,40,31,25,33,31,31,44,36,28,12,18,23,19,20),
# y2 is % women for treatment group
y2=c(57,24,33,32,27,24,30,28.4,28.4,44,46,31,16,22,23,7,12))
# initial values
# must initiate other chain with specification tool
list(muth=0,muphi=0,tauth=1,tauphi=1, neta1=c(0,0), neta2=c(0,0))
```

A posterior analysis for the kidney complications of the contrast media is presented in Table 11.10 and is executed with 75,000

TABLE 11.10: Posterior analysis of complications of contrast media.

Parameter	Mean	sd	Error	2 1/2	Median	97 1/2
d	−0.032	0.0109	<0.00001	−0.0545	−0.0328	−0.0118
μ_ϕ	−1.28	3.611	0.2133	−8.428	−1.324	6.122
μ_θ	−6.466	2.26	0.1343	−10.51	−6.7	−1.57
$\eta_1[1]$	0.0456	0.0399	0.0022	−0.0408	0.0499	0.1157
$\eta_1[2]$	0.0105	0.0251	<0.0001	−0.0344	0.0087	0.0652
$\eta_2[1]$	0.0357	0.0644	0.0036	−0.0947	0.0397	0.1515
$\eta_2[2]$	0.0702	0.0471	0.0011	−0.0124	0.0666	0.1736
pee	0.0493	0.0064	<0.00001	0.0376	0.049	0.0628
qee	0.0822	0.0089	<0.00001	0.0656	0.0819	0.1006

observations generated for the MCMC simulation, with a burn in of 5,000 and a refresh of 100.

What does this imply about comparing the complication rates between the two groups?

(a) Are 75,000 observations sufficient for the simulation?

(b) The MCMC errors for μ_ϕ and μ_θ are relatively large. Does this affect your conclusions about comparing the two tests? Explain your answer!

(c) Which covariates impact the comparison of the two groups?

(d) Note the 95% credible interval for μ_ϕ. Why is μ_ϕ the key parameter to comparing the complication rates adjusted for the other variables?

(e) Plot the posterior density of μ_ϕ.

(f) After observing Tables 11.6 and 11.10, what is the implication for the impact of the covariates?

9. Refer to Miller et al. [13], which is a Bayesian adaptation of the SROC for meta-analysis, and refer to the following derivations:

(a) Derive the equation of the SROC curve (Equation 11.5).

(b) Derive the formula for Q (Equation 11.7).

Remember the definitions of U, V, B, and S given by Equations 11.1 through 11.4, and that Q is the ordinate of the intersection of the SROC curve and the line $\text{TPR} + \text{FPR} = 1$.

The Miller et al. [13] approach is similar to the one taken in this chapter and the paper illustrates the methodology with two interesting examples.

References

[1] Kardaun, L. and Kardaun, O. Comparative diagnostic performance of three radiological procedures for the detection of lumbar disk herniation. *Methods of Information in Medicine*, 29:12, 1990.

[2] Moses, L., Shapiro, D., and Littenberg, B. Combining independent studies of a diagnostic test into a summary ROC curve: Data-analytic approaches and some additional considerations. *Statistics in Medicine*, 13:1293, 1993.

[3] Zhou, X.-H., Obuchowski, N., and McClish, D.K. *Statistical Methods in Diagnostic Medicine*. John Wiley, New York, 2002.

[4] Stangl, D.K. and Berry, D.A. *Meta-Analysis in Medicine and Health Policy*. Marcel Dekker, New York, 2000.

[5] DeVries, S.O., Hunink, M.G.M., and Polak, J.F. Summary receiver operating characteristic curves as a technique for meta-analysis for the diagnostic performance of duplex ultrasonography in peripheral artery disease. *Academic Radiology*, 3:361, 1996.

[6] Vanhoenacker, P.K., Heijenbrok-Kal, M.H., Van Heste, R., Decramer, I., Van Hoe, L.R., Wijns, W., and Hunik, M.G.M. Diagnostic performance of multidetector CT angiography for assessment of coronary artery disease. *Radiology*, 244:419, 2007.

[7] Horsthuis, K., Bipat, S., Bennink, R.J., and Stoker, J.S. Inflammatory bowel disease diagnosed with US, MR, scintigraphy, and CT: Meta-analysis of prospective studies. *Radiology*, 247:64, 2008.

[8] Pakos, E.E., Koumoulis, H.D., Fotopoulos, A.D., and Ioannidis, J.P.A. Osteomyelitis: Antigranulocyte scintigraphy with ^{99m}Tc radiolabeled monoclonal antibodies for diagnosis-meta-analysis. *Radiology*, 245:732, 2007.

[9] Heinrich, M.C., Haberle, L., Muller, V., Bautz, W., and Uder, M. Nephrotoxicity of iso-osmolar iodixanol compared with nonionic low-osmolar contrast media: Meta-analysis of randomized controlled trials. *Radiology*, 250:68, 2009.

[10] Berry, S.M. Meta-analysis versus large trials: Resolving the controversy. In *Meta-Analysis in Medicine and Health Policy*, eds. D.K. Stangl and D.A. Berry, 65. Marcel-Dekker New York, NY, 2000.

[11] Meijer, A.B., Ying, L.O., Geleijns, J., and Kroft, L.J.M. Meta-analysis of 40- and 64-MDCT angiography for assessing coronary artery stenosis. *American Journal of Radiology*, 191:1667, 2008.

[12] Huebner, R.H., Park, K.C., Shepherd, J.E., Schwimmer, J., and Czernin, J. A meta-analysis of the literature for whole-body FDG PET detection of recurrent colorectal cancer. *Journal of Nuclear Medicine*, 41:1177, 2000.

[13] Miller, S.W., Sinha, D., Slate, E.H., Garrow, D., and Romagnuolo, J. Bayesian adoption of the summary ROC curve method for meta-analysis of diagnostic test performance. *Journal of Data Science*, 7:349, 2009.

[14] Harbord, R.M., Deeks, J.J., Egger, M., Whiting, P., and Sterne, J.A. A unification of models for meta-analysis of diagnostic accuracy studies. *Biostatistics*, 8:239–251, 2007.

[15] Leeflang, M.M.G., Deeks, J.J., Gatsonins, C., and Bossuyt, P.M.M. On behalf of the Cochrane diagnostic test accuracy working group. Systemataic reviews of diagnostic tests accuracy. *Annals of Internal Medicine*, 149:889–897, 2008.

[16] Paul, M., Riebler, A., Bachmann, L.M., Rue, H., and Held, L. Bayesian bivariate meta-analysis of diagnostic test studies using integrated nested Laplace approximations. *Statistics in Medicine*, 29:1325–1339, 2010.

[17] Rutter, C.M. and Gatsonins, C. A hierarchical regression approach to meta-analysis of diagnostic test accuracy evaluations. *Statistics in Medicine*, 20:2865–2884, 2001.

Appendix

Introduction to WinBUGS

A.1 Introduction

WinBUGS is the statistical package that is used for the book and it is important that the novice be introduced to the fundamentals of working in the language. This is a brief introduction to the package and for the first-time user it will be necessary to gain more knowledge and experience by practicing with the numerous examples provided in the download. WinBUGS is specifically designed for Bayesian analysis and is based on Markov Chain Monte Carlo (MCMC) techniques for simulating samples from the posterior distribution of the parameters of the statistical model. It is quite versatile and once the user has gained some experience, there are many rewards.

Once the package has been downloaded, the essential features of the program are described, first by explaining the layout of the BUGS document. The program itself is made up of two parts, one part for the program statements, and the other for the input of the sample data and the initial values for the simulation. Next to be described are the details of executing the program code and what information is needed for the execution. Information needed for the simulation are the sample sizes of the MCMC simulation for the posterior distribution and the number of such observations that will apply to the posterior distribution.

After execution of the program statements, certain characteristics of the posterior distribution of the parameters are computed, including the posterior mean, median, credible intervals, and plots of the posterior densities of the parameters. In addition, WinBUGS provides information about the posterior distribution of the correlation between any two parameters and information about the simulation. For example, one may view the record of simulated values of each parameter and the estimated error of estimation of the process. These and other activities involving the simulation and interpretation of the output will be explained.

Examples based on accuracy studies illustrate the use of WinBUGS and include estimation of the true positive fraction (TPF) and false positive fraction (FPF) for the exercise stress test, and modeling for the receiver operating characteristic (ROC) area. Of course, this is only a brief introduction, but it should be sufficient for the novice to begin the adventure of analyzing data. Because the book's examples provide the necessary code for all examples, the program can easily be executed by the user. After the book is completed by the dedicated student will have a good understanding of WinBUGS and the Bayesian approach to measuring test accuracy.

A.2 Download

I downloaded the latest version of WinBUGS at http://www.mrc-bsu .cam.ac.uk/bugs/winbugs/contents.shtml, which you can install in your program files, then you will be requested to download a decoder, which allows one to activate the full capabilities of the package.

A.3 Essentials

The essential feature of the package is a WinBUGS file or document that contains the program statements and space for input information.

A.3.1 Main body

The main body of the software is a WinBUGS document, which contains the program statements, the major part of the document, and a list statement or statements, which include the data values and some initial values for the MCMC simulation of the posterior distribution. The document is given a title and saved as a WinBUGS file, which can be accessed as needed.

A.3.2 List statements

List statements allow the program to incorporate certain necessary information that is required for successful implementation. For example, experimental or study information is usually inputted with a list statement, e.g., in a typical agreement study, the cell frequencies of the two raters are contained in the list statement. As another example, consider estimation of the two most basic measures of test accuracy, namely, the TPF and FPF, where the medical test scores are binary. The example is taken from Chapter 4 where the exercise stress test is binary and a positive test indicates coronary heart disease. The basic information is given by Table A.1 and the Bayesian analysis is executed with BUGS CODE 4.1. Thus, for the (0,0) cell, the number of non-diseased subjects that score 0 is 327, and the number of diseased subjects that score 1 is 818.

The execution of the code is explained in terms of the following statements that make up the WinBUGS file or document.

TABLE A.1: Exercise
stress test and heart disease.

EST	CAD	
	$D = 0$	$D = 1$
$X = 0$	327	208
$X = 1$	115	818

WinBugs Document
Measures of accuracy
Binary Scores
Model;
{
# Dirichlet distribution for cell probabilities
g00~dgamma(a00,2)
g01~dgamma(a01,2)
g10~dgamma(a10,2)
g11~dgamma(a11,2)
h<-g00+g01+g10+g11
# the theta have a Dirichlet distribution
theta00<-g00/h
theta01<-g01/h
theta10<-g10/h
theta11<-g11/h
# the basic test accuracies are below
tpf<-theta11/(theta11+theta01)
se<-tpf
sp<-1-fpf
fpf<-theta10/(theta10+theta00)
tnf<-theta00/(theta00+theta10)
fnf<-theta01/(theta01+theta11)
ppv<-theta11/(theta10+theta11)
npv<-theta00/(theta00+theta01)
pdlr<-tpf/fpf
ndlr<-fnf/tnf
}
# Exercise Stress Test Pepe [6]
# Uniform Prior (add one to each cell of the table frequencies !)
list(a00=328,a01=209,a10=116,a11=819)
# chest pain history
# Uniform prior
list(a00=198,a01=55,a10=246,a11=970)
# initial values
list(g00=1,g01=1,g10=1,g11=1)

The objective of this example is to determine the posterior distribution of the true and false positive rates for the exercise stress test. The program statements are included between the first and last curly brackets { }, and the code for the true positive fraction is

$$\text{tpf}<\text{-theta11/(theta11+theta01)}$$

where theta11 is the cell frequency for the (1,1) cell of the table, namely, that corresponding to the diseased subjects that score positive. The information from Table A.1 appears in the first list statement and the number 1 is added to each cell frequency from the table, which places a uniform prior distribution on the four cell parameters. The first nine statements of the code put a posterior Dirichlet distribution on the four cell parameters, theta00, theta01, thetra10, and theta11.

The last list statement contains the initial values for the MCMC simulation.

A.4   Execution of Analysis

A.4.1   Specification tool

The tool bar of WinBUGS is labeled as follows, from left to right: file, edit, attributes, tools, info, **model, inference**, doodle, maps, text, windows, examples, manuals, and help; I have highlighted the model and inferences labels. When the user clicks on one of the labels, a pop-up menu appears. In order to execute the program, the user clicks on **model**, then clicks on **specification**, and the specification tool appears (Figure A.1).

The specification tool is used together with the BUGS document as follows: (a) click on the word "model" of the document, (b) click on the check model box of the specification tool, (c) click on the compile box of the specification tool, (d) click on the word "list" of the list statement of the document,

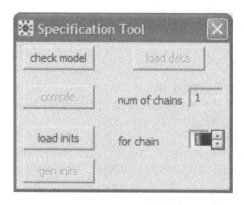

FIGURE A.1: Specification tool.

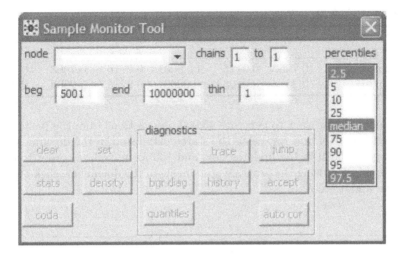

FIGURE A.2: Sample monitor tool.

and lastly, (e) click on load inits box of the tool. Now close the specification tool and go to the next step below.

A.4.2 Sample monitor tool

The sample tool is activated by first clicking on the inference menu of the tool bar, then clicking on sample, and the sample monitor tool appears as below. Type fpf then click on "set," then type tpf in the node box and click on "set," then type an $*$ in the node box. Type 5000 in the beg box, which means that the first 5001 observations generated for the posterior distribution of tpf and fpf will not be used when reporting the results of the simulation. The 5000 observations typed in beg are referred to as the "burn in" (Figure A.2).

A.4.3 Update tool

In order to activate the update tool, click on the **model** menu of the tool bar, then click on updates (see Figure A.3).

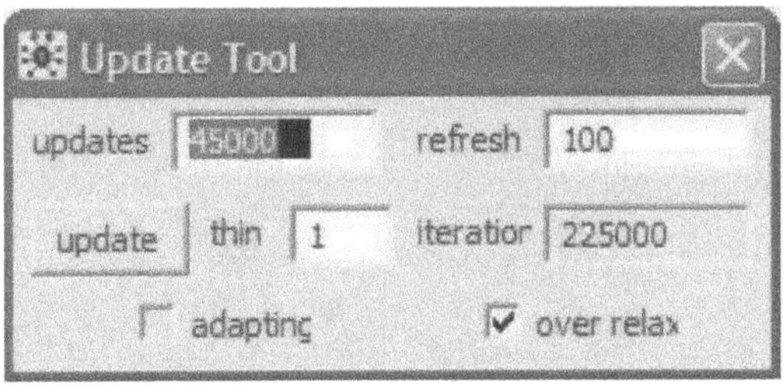

FIGURE A.3: Update tool.

TABLE A.2: Posterior analysis for tpf and fpf.

Parameter	Mean	sd	Error	Lower 2 1/2	Median	Upper 2 1/2
tpf	0.7967	0.0125	<0.00001	0.7716	0.7968	0.8208
fpf	0.2612	0.0208	<0.00001	0.2215	0.2608	0.3033

Suppose you want to generate 45,000 observations from the posterior distributions of tpf and fpf, using the statements listed in the document above, then type 45000 in the updates box, and 100 for refresh. In order to execute the simulation using the program statements in the document, click on update of the update tool.

A.5   Output

After the 40,000 observations have been generated from the joint posterior distribution of tpf and fpf, click on the stats box of the sample monitor tool (see Figure A.2). Certain characteristics of the joint distribution are displayed. Clicking on the history box, the values of 40,000 observations from the joint density of tpf and fpf are displayed, and the output for the posterior analysis is shown in Table A.2.

The first entry 0.7967 is the posterior mean of the true positive fraction with a standard deviation of 0.0125 for the posterior distribution of the TPF and a 95% credible interval for TPF is (0.7716,0.8208). MCMC simulations errors are <0.00001 for both parameters, indicating that 40,000 observations are sufficient for estimating the relevant posterior characteristics of the TPF and FPF.

Another feature of the output is that the posterior densities of the nodes can be displayed by clicking on the "density" box of the sample monitor tool. For example, Figure A.4 presents the posterior density of the TPF. The posterior density of the TPF appears to be symmetric about the mean of 0.79, which is confirmed by the posterior median of 0.79.

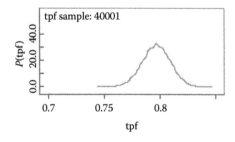

FIGURE A.4: Posterior density of the tpf.

TABLE A.3: Mammography results.

	Test result					
Status	**Normal** 1	**Benign** 2	**Probably** benign 3	**Suspicious** 4	**Malignant** 5	**Total**
Cancer	1	0	6	11	12	30
No cancer	9	2	11	8	0	30

A.6 Another Example

In order to illustrate the use of WinBUGS and to gain additional insight, an example of test accuracy will be presented, which has to do with estimating the area under the ROC curve for the accuracy of mammography for detecting breast cancer.

Consider the results of mammography administered to 60 women, of which 30 had the disease. This example appears in Zhou et al. [1: 21].

Mammography uses a score from 1 to 5, where 1 indicates a normal view, 2 indicates a benign tumor, 3 implies probably a benign tumor, 4 is a score for a suspicious lesion, and 5 indicates a malignant tumor, and for those with breast cancer it seems that the frequency of the scores increases with the increasing score for malignancy, while the opposite is true for those without the disease. A Bayesian analysis will put a uniform prior distribution on the 10 cell frequencies and will be executed with BUGS CODE 4.2. The formula for the ROC area is given by Broemeling [2: 72] and is coded as:

```
auc<-A1+A2/2
A1<-theta2*ph1+theta3*(ph1+ph2)+theta4*(ph1+ph2+ph3)
    +theta5*(ph1+ph2+ph3+ph4)

A2<-theta1*ph1+theta2*ph2+theta3*ph3+theta4*ph4+theta5*ph5
```

in the code below. There are three list statements: the first is for the mammography study, the second is from another study not considered here, and the third contains the initial values for the variables that generate the Dirichlet distribution of the cell probabilities for Table A.3.

```
WinBUGS Document
# Area under the curve
# Ordinal values
# Five values
Model;
{
# generate Dirichlet distribution
g11~dgamma(a11,2)
```

```
g12~dgamma(a12,2)
g13~dgamma(a13,2)
g14~dgamma(a14,2)
g15~dgamma(a15,2)
g01~dgamma(a01,2)
g02~dgamma(a02,2)
g03~dgamma(a03,2)
g04~dgamma(a04,2)
g05~dgamma(a05,2)
g1<-g11+g12+g13+g14+g15
g0<-g01+g02+g03+g04+g05
# posterior distribution of probabilities for response of diseased patients
theta1<-g11/g1
theta2<-g12/g1
theta3<-g13/g1
theta4<-g14/g1
theta5<-g15/g1
# posterior distribution for probabilities of response of non-diseased
   patients
ph1<-g01/g0
ph2<-g02/g0
ph3<-g03/g0
ph4<-g04/g0
ph5<-g05/g0
# auc is area under ROC curve
#A1 is the P[Y>X]
#A2 is the P[Y=X]
# from Broemeling [2: 72]
auc<-A1+A2/2
A1<-theta2*ph1+theta3*(ph1+ph2)+theta4*(ph1+ph2+ph3)+
theta5*(ph1+ph2+ph3+ph4)
A2<-theta1*ph1+theta2*ph2+theta3*ph3+theta4*ph4+theta5*ph5
}
# Mammography Example Zhou et al. [1]
# Uniform Prior
# see Table 4.7
list(a11=2,a12=1,a13=7,a14=12,a15=13,a01=10,a02=3,a03=12,
a04=9,a05=1)
# Gallium citrate Example Zhou et al. [1: 159]
# Uniform Prior
list(a11=13,a12=7,a13=4,a14=2,a15=19,a01=12,a02=3,a03=4,
a04=2,a05=4)
# initial values
list(g11=1,g12=1,g13=1,g14=1,g15=1,g01=1,g02=1,g03=1,
g04=1,g05=1)
```

TABLE A.4: Posterior distribution of area under the ROC curve—mammography example.

Parameter	Mean	sd	Error	Lower 2 1/2	Median	Upper 2 1/2
auc	0.7811	0.0514	<0.0001	0.6702	0.7848	0.8709
A1	0.688	0.0635	<0.0001	0.5564	0.6909	0.8036
A2	0.1861	0.0307	<0.0001	0.128	0.1854	0.2484

To begin the analysis, click on the model menu of the tool bar and pull down the specification tool: (a) click on the word "model" of the BUGS document; (b) click on the check model box of the specification tool (see Figure A.1); (c) activate the word "list" of the first list statement; (d) click on the compile box of the specification tool; and lastly (e) click on the word "list" of the third list statement of the document. If a mistake is made, the user will be notified, but you are now ready to execute the analysis.

In order to continue the process, from the inference menu, pull down the sample monitor tool (see Figure A.2) and type auc in the node box, followed by clicking the set box, then repeat the operation for nodes A1 and A2. For the final operation, put an * in the node box, and type 5000 in the beg box for the burn in.

Pull down the update tool (see Figure A.3) from the model menu and type 45,000 in the updates box, put 100 in the refresh box, and click on the update box. The simulation now begins with 45,000 observations generated from the posterior distribution of auc, A1, and A2 of the sample monitor tool, and the output shown in Table A.4 will appear.

The ROC area is estimated with a posterior mean of 0.7811 and a 95% credible interval of (0.6702,0.8709), while the other parameters have posterior means of 0.688 and 0.1861 for A1 and A2, respectively. The latter A2 is the posterior probability of a tie between the score of a diseased subject and the score of a non-diseased subject. Note the small MCMC errors of <0.001 for the three parameters, which imply that the value 0.7811 is within two decimal places of the "true" posterior mean for the auc parameter, the area under the ROC curve.

A.7 Summary

The appendix introduces the reader to WinBUGS and the novice should be able to begin Chapter 1 and learn the main topic, namely, how a Bayesian analyzes accuracy studies. To gain additional experience, refer to the manual and to the numerous examples that come with the downloaded version of the package. Practice, practice, and more practice is the key to understanding the importance of analyzing actual data with a Bayesian approach. There are many references about WinBUGS, including Broemeling [2], Woodworth [3], and the WinBUGS link [4], which in turn refer to many books and other resources about the package.

References

[1] Zhou, X.H., Obuchowski, N.A., and McClish, D.K. *Statistical Methods in Diagnostic Medicine*. John Wiley & Sons Inc., New York, 2002.

[2] Broemeling, L.D. *Bayesian Biostatistics and Diagnostic Medicine*. Chapman & Hall/CRC, Taylor & Francis Group, Boca Raton, 1996.

[3] Woodworth, G.G. *Biostatistics: A Bayesian Introduction*. John Wiley, Hoboken, NJ, 2004.

[4] The BUGS Project Resource at http://www.mrc-bsu.cam.ac.uk/bugs/winbugs/contents.shtml.

Index

A

Agreement with no gold standard, 151–152, 162
  combining reader information, 186, 200
  conditional Kappa and diagnosing depression, 166
  covariates, agreement, information, 156, 160, 161
  binary scores, 190, 202
  G-coefficient, 178–181, 187, 206
  intraclass Kappa, 151, 173–178, 186–195, 202–208
  J index, 178–179
  Kappa and association, 183–187, 195, 207, 208
  Kappa and stratification, 169
  ordinal categories, 165, 186, 187, 202
  other indices, 178–206
  two Kappa parameters, comparing, 206
  two or more readers, 151, 165, 186–195, 202
  weighted Kappa, 171–173, 186, 187, 202, 208
Agreement with a gold standard, 151–155, 180–182, 202–206
  blood glucose values and type II diabetes, 156, 157, 160–162
  melanoma example and four readers, 151–155
  with binary and ordinal scores, 151–155
  with continuous scores
  partial agreement, 186–195, 208
  Kappa and many raters, 190–195
  precursors of Kappa, 162–165
Audiology study, 99–107
  regression,
    log link function, 101, 107
    logistic link, 106, 107, 136

B

Bayes inference, 1–7
Bayesian methods for diagnostic accuracy, *See also Bayesian statistics*
  accuracy, 45–97
  binary scores and Bayesian methods,
    classification probabilities, 52, 53, 61, 79
    diagnostic likelihood ratios, 45, 52, 56
    predictive values, 52, 55, 56
    ROC curves, 45–98, 45–48, 57–74, 98
  clustered data, detection, localization, 46, 73–78, 95–97
  Bayesian receiver operating characteristic curve, 45–98
  mammography clustered information, 74
  optimal threshold value, choice of, 70

quantitative variables and test
    accuracy Bayesian
    methods
receiver operating characteristic
    curve
  defined, area of, 70
  Spokane Heart Study, 63,
      67, 93
Bayesian sequential stopping rules,
    327–329, 333, 352
Bayesian statistics, 1–8
  Bayes theorem, 3–5
    likelihood function, 5
    posterior distribution, 4–6
    prior distribution, 5, 7
  Monte Carlo Markov Chain
      techniques
    Gibbs sampling
Bayesian inference, 45–79
  credible intervals, 45, 72
  estimation, 45–48, 56,
      70, 79
  hypothesis testing, 94
Biopsy, non-small lung
    cancer, 75

C

Clinical trials, 326–352
  Phase I, 327–330, 335–344
  Phase II, 301, 327–348, 357
  Phase III, 330, 331, 347
  Protocols, 331, 327, 329, 331,
      335, 344, 345
    Bayesian sequential stopping
        rules, 327–329, 333, 352
    design, 46
    informed consent, 331
    introduction,
    NIC toxicities, 328, 338
    prior information, 333
    renal cell carcinoma example,
        327, 340, 352
    statistical considerations, 50
    tumor response guidelines,
        332, 343, 352

Combined tests, 353–412
  accuracy of
    believe the negative rule(BN),
        354, 355–361, 406–408
    believe the positive rule(BP),
        354, 355–361, 404–408
    false positive fractions, 353,
        357, 395, 403–408
    true positive fractions, 353,
        357, 395, 403–408
    with verification bias, 354, 355,
        370, 373, 395–409
  Bayesian inference for
    two binary tests, 354, 355, 358,
        368–371, 374, 403–408
    computed tomography
        (CT), 354, 360, 368,
        374–382, 386, 394, 403
    for heart disease, 353–358,
        403,
    magnetic resonance
        imaging(MRI), 406–409
    with several readers, 354,
        355, 362, 363, 367–373,
        403–408
  ordinal and continuous scores,
      354, 374, 377, 394–400,
      403, 404
    accuracy using ROC area of
        risk score, 377–382, 386,
        388, 394–403, 409, 410
    likelihood ratio and the risk
        score, 374–376
    localization of prostate cancer,
        386–388, 404, 412
Coronary artery calcium, 58, 61
  Spokane heart study, 63, 93
Coronary artery disease, 353, 358,
    360, 404, 405
  exercise stress test, 49, 54, 55

D

Decision curves, 302–326, 351
  Bayesian decision curves, 317
  clinical benefit, 314, 318

definition of, 314

prostate cancer example,
317–323

test accuracy in clinical practice,
312

three scenarios for, 313–325

threshold probability, 313–318,
321–325

Depth of tumor for melanoma
diagnosis, 14, 28

Discrete rating, 58, 61, 72

binary scores, 45, 46, 57

for multiple readers, 49, 72

ordinal scores, 45, 46, 73, 82, 94

agreement among multiple
readers, 83, 151, 155

Documentation for clinical
protocols, 47, 331

E

Extreme verification bias, 261

F

False positive fraction, 45, 52, 55,
56, 57, 63, 70, 71, 80,
82, 92

G

G coefficient, 178–181, 187, 206

comparing with Kappa and the
Jacquard coefficient, 179

definition of, 179

examples of, 179

measure of agreement, 179

with multiple readers, 200

I

Imaging

imaging modalities

computed tomography(CT),
9–11, 13, 15, 16, 19, 23, 24

fluoroscopy, 9, 10

gamma cameras, 12, 13

positron emission
tomography(PET), 12, 13

single photon emission
computed
tomography(SPECT), 12

magnetic resonance
imaging(MRI), 9, 11, 12,
13, 22, 23

mammography, 9, 11, 13, 19, 20

nuclear medicine, 11–14

X-ray, 9–11

Imperfect diagnostic test
procedure, 209–258

Bayesian inference for

lack of gold standard, 210–212,
244, 246

multiple readers, 209, 211, 236,
237, 241, 244, 246, 255

test accuracy for

area of ROC curve, 246–248,
256, 257

false positive
fraction(1-specificity),
210–216, 218, 221–252

true positive
fraction(sensitivity),
210–216, 218, 221–252

inflammatory bowel disease,
424–438, 443–416

informed consent, 331

inverse probability weighting, 261,
272, 279, 282, 293, 296

J

Jacquard coefficient, 178, 179

as a measure of agreement, 178

compared to Kappa and the G
coefficient, 179

K

Kappa, 151–208

and binary scores, 151–176

and melanoma example, 152, 154

and ordinal scores, 151, 152, 155,
162, 165, 171, 186, 187, 195

and partial agreement, 166, 170,
185–195

intraclass Kappa, 173–178, 187,
        195, 205, 208
precursors of Kappa, 162–166
stratified Kappa, 171, 187, 200,
        206, 207
weighted Kappa, 171–173, 186,
        187, 202, 208
with multiple readers(raters),
        151–166

L
Lack of gold standard in test
        accuracy, 210–212,
        244, 246
Latent variables
Bayesian inferences with latent
        variables, 216–218
infectious disease example,
        218–226
with imperfect reference
        standard, 209–258
Localization and detection with
        clustered data, 46, 72, 73,
        96, 97, 386–388
prostate cancer example,
        386, 387
Lung cancer, non-small cell, biopsy,
        74–78

M
Mammography, 46, 57, 58, 60, 61,
        68, 72, 93, 96
accuracy with ROC curve, 57–61
with verification bias, 260,
        270–273, 283, 294,
        297, 298
MCMC (Monte Carlo Markov
        Chain)
Bayesian inference and MCMC,
        451–460
MCMC and WinBUGS, 451–460
MCMC errors, 457
Melanoma, 152, 154, 327, 338,
        344–350
nuclear medicine procedures, 14

phase II trials for
sentinel lymph node
tumor depth and metastasis, 14
Meta-analysis and test accuracy,
        413–448
accuracy with Q-score, 414, 417
examples of
        comparing MRI and US for
                inflammatory bowel
                disease, 424–427
        scintigraphy for diagnosis of
                osteomyelitis, 420
        with covariates and one binary
                test, 427–430
        with covariates and several
                tests, 430–436
summary ROC curve (SROC),
        413–445
        bilogistic regression, 416
        definition of, 416

N
National Cancer Institute
list of toxicites for clinical trials,
        328

O
Optimal threshold, 303–315
Bayesian formula, 308
Blood glucose for Type II
        diabetes, 310–311
definition of, 307
for PSA (prostate specific
        antigen), 308
Somoza and Mossman
        coordinates, 308
Other meta analyses, 435–440
a contrast media study, 437
studies comparing side effects,
        435–438

P
Pharmaceutical companies
imaging protocols example,
        47–51

protocols for Phase II trials,
42, 331
Posterior information, 3–7
Bayesian inference, 4–6
posterior mean, 5, 456
posterior median, 5, 456
95% credible intervals, 5, 456
Protocol for clinical trials, 45–51,
327, 329, 331, 335, 344, 345

Q

Q-statistic, 414–417
measure of accuracy, 417
with SROC curves, 416
Quantitative variables,
coronary artery calcium, 62, 97
examples of
prostate specific antigen, 61,
92, 95
Type II diabetes, 64, 66, 67

R

Receiver operating characteristic
(ROC) curve
area of ROC curve, 45, 48,
61–69, 82, 91, 93
Bayesian inference and ROC
curves, 45, 61–70, 82, 84,
87, 91, 93
definition of, 45, 58, 64
for studies with verification bias,
272–292
measure of accuracy, 45, 58
with an imperfect reference
standard, 246–248,
256, 257
with clustered data, 74–78
RECIST criterion, 327, 329, 332,
334–347
Regression techniques for ROC
area, 111–135
examples
Audiolog, 133–136
coronary artery calcium for
heart disease, 62

mammography and clustered
data, 72–78
melanoma, 117–121
pancreatic cancer
prostate cancer, 127–133
Regression techniques
audiology example
Bayesian regression and
continuous scores, 87–91
examples
accuracy of prostate specific
antigen, 92, 95
audiology, 101–110,
133–135
blood glucose and type II
diabetes, 65
the O'Malley et al. approach
with regression, 97–148
WinBUGS with continuous
scores for ROC area, 88,
124–132
Bayesian regression for ordinal
data, 99, 100, 111, 112
Bayesian regression with binary
scores, 99, 100, 107,
110, 148
logistic link and false positive
fraction, 99, 106, 107, 136
log-link function and false
positive fraction, 101, 107
examples
staging metastasis for
melanoma, 117–123
tumor response study,
112–116
WinBUGS and ordinal
regression for ROC area,
113
the Congdon code, 148
Renal cell carcinoma, 327, 340,
341, 352

S

Sentinel lymph node biopsy, 14
melanoma metastasis, 13, 14

Sequential Bayesian analysis for
    Phase II trials, 327, 328,
    333, 352
Software for clinical trials, 327, 348
  CRM (continual reassessment
    method), 328, 337–340
  Mult-c for response and toxicity
    of Phase II trials, 337,
    339, 345, 346, 350
Software for Bayesian inference
  WinBUGS and R, 451–460
SROC (summary receiver
    operating) curve, 413–417
  Bayesian meta-analysis for
    binary and ordinal scores,
    415–439
  Q-statistic for accuracy, 417
Staging melanoma by a surgeon
    and dermatologist,
    395–397

T
Test accuracy *see medical test
    accuracy*
  area of ROC curve, *see area of
    ROC curve*
  classification probabilities, 51–53,
    61, 79
  diagnostic likelihood ratios, 45,
    52, 56
  positive and negative predictive
    values, 52–56
  Q-statistic for accuracy of
    meta-analysis, 414–417
  SROC curve, 415–417
Threshold, 57–63, 70
True positive fraction, 45, 52, 55, 56,
    57, 63, 70, 71, 80, 82, 92
Tumor response, 327–342
  guidelines for tumor response,
    332, 333
  use of in Phase II clinical trials,
    333
Tumor depth and metastasis in
    melanoma, 14

V
Verification bias, 259–300
  accuracy of studies with missing
    at random(MAR)
    assumption, 260, 283, 287,
    288, 290, 292, 294
  binary scores, 261–268
    Bayesian estimates of false
      positive fraction, 261, 263,
      264–268, 282, 287, 290–296
    Bayesian estimates of true
      positive fraction, 261, 263,
      264–268, 282, 287, 290–296
  inverse probability weighting for
      verification bias, 279
    Bayesian analysis without
      missing at random
      assumption, 287
    testing for missing at random,
      287
  ordinal scores, 268–291
    Bayesian estimates of area of
      ROC curve, 270–292
    mammography example, 279
    mammography example with
      several sites, 272
    staging melanoma by as
      surgeon and a
      dermatologist, 273
    two ordinal tests with
      covariates, 273, 274

W
WinBUGS, 451–460
  download, 452
  examples
    estimating ROC area, 458, 459
    estimating true and false
      positive fractions, 453
    exercise stress test, 453
    mammography, 459
  execution of WinBUGS, 454
    sample monitor tool, 453
    burn in(beg box), 455
    specification tool, 454

update tool, 455
    refresh rate, 455
    size of simulation, 455
list statements and data input,
    452
list statements and initial values,
    452, 453
introductory example, 453
main body, 452
monitoring the execution of
    analysis
    auto-correlation of simulated
        values, 455

    MCMC errors, 459
    trace plots, 455
output, 456
    posterior characteristics, 459
        95% credible intervals, 459
        posterior density plots, 459
        posterior mean, 459
        posterior median, 459
        posterior standard
            deviation

X
X-ray, 9–11